D0857507

Veterinary Protozoology

Veterinary Protozoology

NORMAN D. LEVINE, Ph.D.

Iowa State University Press • Ames

NORMAN D. LEVINE is professor emeritus, Department of Pathobiology, College of Veterinary Medicine, University of Illinois.

 Composed and printed by The Iowa State University Press, Ames, Iowa 50010

First edition, 1985

Library of Congress Cataloging in Publication Data

Levine, Norman D.
Veterinary protozoology.

Based on: Protozoan parasites of domestic animals and of man.
Includes index.
1. Veterinary protozoology. 2. Medical protozoology. 3. Protozoa, Pathogenic. I. Levine, Norman D. Protozoan parasites of domestic animals and of man. II. Title.
SF780.6.L49 1985 636.089′6016 84-27867
ISBN 0-8138-1861-3

DEDICATED TO THE MEMORY OF MY PARENTS:

To Max Levine, whose example guided me
to a career in science

To Adele Levine, whose example helped me
to an interest in the humanities

CONTENTS

PREFACE

THIS book is based on my earlier book, *Protozoan Parasites of Domestic Animals and of Man,* the first edition of which was published in 1961 and the second edition in 1973. During the interval since 1973, an increasingly large number of scientific papers has been published on parasitic protozoa and our knowledge of them has grown extensively. Although this book is much shorter, no essential information has been eliminated.

In the interest of brevity and economy, no papers published before 1965 are listed in the Literature Cited sections and no surveys are reported. Earlier references can be found in the *Index-Catalogue of Medical and Veterinary Zoology, Zoological Record, Biological Abstracts, Protozoological Abstracts,* and *Veterinary Bulletin.*

In addition to the acknowledgments in the earlier books, I should like to express my appreciation to Marjorie M. Hildreth, Joyce A. Amacher, Mary J. Schwenk, Shirley Pelmore, Tanya Brown, Tina M. Butler, Wilma K. Ellis, Dianne Langevin, and Sherry Hopper for typing the manuscript; to Drs. B. Becker, K. T. Friedhoff, A. O. Heydorn, A. Liebisch, M. B. Markus, H. Mehlhorn, E. Schein, E. Scholtyseck, and M. Warnecke; and to the *Proceedings of the First Japanese-German Symposium, Protistologica, Tierärztliche Praxis, Tropenmedicin und Parasitologie, Zeitschrift für Parasitenkunde, Zentralblatt für Bakteriologie, I. Originale, A,* Gustav Fischer Verlag, and Georg Thieme Verlag for permission to reproduce some of their illustrations.

Veterinary Protozoology

1
Introduction

PROTOZOA form a subkingdom of the kingdom Protista in the five-kingdom classification of living things (Monera, Protista, Plantae, Fungi, Animalia) (see Whittaker 1977). They are more primitive than animals. No matter how complex their bodies may be, and many of them are very complex, all the different structures are contained in a single cell. However, some protozoa have a syncytial stage in their life cycle in which there are no cell walls between the nuclei, and some species form colonies that swim as a unit and contain somatic and reproductive organisms that look different.

Protozoa are microscopic in size, only a few being visible to the naked eye. A few flagellates contain chlorophyll and are considered green algae by some; many species of colorless protozoa differ from green ones only in their lack of chromatophores, however, and loss of chromatophores can be produced experimentally.

Since their discovery by Antony van Leeuwenhoek in the eighteenth century, some 65,000 species of protozoa have been described. They occur in practically all habitats where life can exist and are among the first links of the food chain on which all higher life depends. Floating in the plankton of tropic seas, they cause the luminous glow of waves and shipwakes. Blooming off our coasts, they cause the red tide that deposits windrows of dead fish on shore and the so-called mussel poisoning that sometimes kills people. They abound in ponds and streams and in the soil. Their role in sewage purification is just beginning to be understood. Their skeletons cover the ocean floor and form the chalk we use in classrooms.

As parasites, protozoa play a double role. Malaria is still the world's most important disease. Trypanosomes have interdicted vast African grazing lands for livestock. Amebae cause dysentery in man, and coccidia cause it in their domestic animals. But other protozoa that pack the termite's hindgut almost solidly can digest the cellulose that it eats, feeding it with their wastes and their dead bodies. Fabulous numbers of protozoa swarm in the rumen and reticulum of cattle and sheep and in the cecum and colon of horses.

In this book we are concerned with the protozoan parasites of farm, pet, and laboratory animls and of man. For further information on the protozoa in general, see the references at the end of this chapter and also the summaries of the International Congresses of Protozoology entitled *Progress in Protozoology.*

STRUCTURES

The structures of protozoa are referred to not as organs but as organelles; organs are composed of cells and organelles are differentiated portions of a cell.

Nuclei

Protozoa are *eukaryotic,* that is, with a nucleus enclosed in a membrane, as opposed to the *prokaryotic* bacteria, in which the nuclear apparatus is not separated from the cytoplasm (Fig. 1.1.) (Gooday 1980).

Protozoa contain one or more nuclei, which may be of several types. In protozoa other than ciliates, the nucleus is *vesicular* and all the nuclei in the same individual look alike. There are two types of vesicular nucleus. In one type an *endosome* is present. The endosome is a more or less central body with a negative Feulgen reaction and therefore without deoxyribonucleic acid. The chromatin, which is Feulgen-positive and forms the chromosomes, lies between the nuclear membrane and the endosome. This type of nucleus is found in the trypanosomes, parasitic amebae, and phytoflagellates.

In the other type of vesicular nucleus, there is no endosome but there may be one or more Feulgen-positive *nucleoli.* In these the chromatin is distributed throughout the nucleus. This type of nucleus is found in the Apicomplexa, hypermastigorid flagellates, opalinids, dinoflagellates, and Radiolaria.

In the ciliates there are two types of nucleus, which are different in appearance; each individual has at least one of each. The *micronucleus* is relatively small; it is diploid, divides by mitosis at fission, and apparently controls the reproductive functions of the organism. The *macronucleus* is relatively large; it divides amitotically at fission, is polyploid, and apparently has to do with the vegetative functions of the organism. Both nuclei appear quite homogenous in composition, in contrast to the vesicular nuclei of other protozoa.

Locomotion

Protozoa move by means of flagella, cilia, pseudopods, or undulating ridges. A *flagellum* is a whiplike organelle composed of a central *axoneme* and an outer sheath. The axoneme arises from a *kinetosome* (*basal granule, blepharoplast*) in the cytoplasm. The axoneme has been shown by electron microscopy to be composed of nine double peripheral and two central microtubules (the so-called 9 + 2 structure). In some species a flagellum may pass backward along the body, attached to the body along its whole length or at several points to form an *undulating membrane.* Flagella are found in the Mastigophorasica and in the flagellate stages of the Sarcodinasica and Apicomplexa.

A *cilium* is an eyelashlike organelle, essentially a small flagellum. It has a sheath, basal granule, and axoneme, the last having, like the flagellum, the 9 + 2 structure of microtubules. Cilia are found in the Ciliophora. The less

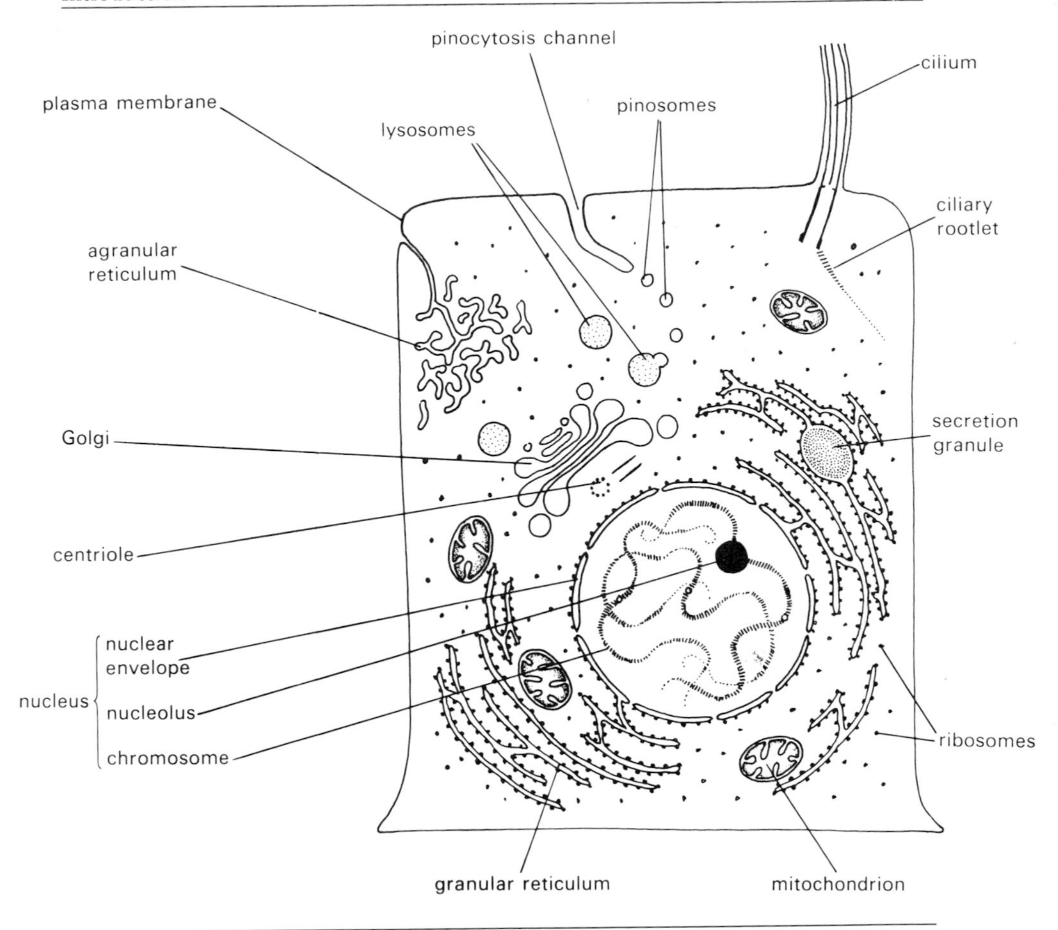

Fig. 1.1. Organelles of a eukaryotic cell. (From Vickerman and Cox 1967)

specialized ciliates have large numbers of cilia, which are arranged in rows and beat synchronously. In the more specialized ciliates, special locomotory organelles have been developed by fusion of cilia. A *cirrus* is a tuft of fused cilia embedded in a matrix. A *membranelle* is a flap formed by the fusion of two or more transverse rows of cilia; membranelles are found especially around the mouth. An *undulating membrane* (not to be confused with the undulating membrane of flagellates) is formed by the fusion of one or more longitudinal rows of cilia; they occur in the oral groove of some ciliates.

A *pseudopod* is a temporary locomotory organelle that can be formed and retracted as needed. There are four types of pseudopods. A *lobopod* is a relatively broad pseudopod with a dense outer layer and a more fluid inner zone; lobopods are found in the amebae and some flagellates. A *filopod* is a slender, hyaline pseudopod that tapers from its base to its pointed tip; filopods tend to anastomose and may fuse locally to produce thin films of cytoplasm; they contain no cytoplasmic granules. A *myxopod* (*rhizopod, reticulopod*) is a filamentous pseudopod with a dense inner zone and a more fluid outer layer in which cytoplasmic granules circulate; myxopods branch and anastomose to form complex networks used for trapping food and also for locomotion; they are found in the Foraminiferorida. An *axopod* is a slender pseudopod that projects from the body without branching or anastomosing; it is composed of a thin outer layer of fluid cytoplasm and an axial filament composed of a fibrillar tube containing a homogenous core; axopods are found in the Actinopodasica (Rinaldi and Grebecki 1979; Stocken 1979).

A number of protozoa effect locomotion by bending, snapping, or twisting the whole body. Subpellicular microtubules may be responsible for this type of movement.

There is still another type of locomotion, *gliding* (exemplified by *Toxoplasma, Sarcocystis,* and coccidian merozoites, gregarines, and *Labyrinthula*), in which the body glides smoothly along without change in shape or other cause visible under the light microscope. In the gregarines, this is due to an undulatory wave passing back in the longitudinal ridges or folds on the outer body surface. Movement of the longitudinal ridges is probably caused by the *subpellicular microtubules* that they contain, which are visible with the electron microscope. Extruded mucus may also play a part. If there are no longitudinal ridges, as in coccidian sporozoites and merozoites, locomotion may also be due to the subpellicular microtubules. However, the mode of locomotion of these organisms is still unclear.

Organelles Associated with Nutrition

Nutrition among the protozoa may be of several types. *Autotrophic nutrition* is a type in which the organism is able to live entirely on inorganic compounds; it synthesizes proteins, carbohydrates, lipids, and other substances from inorganic precursors. Autotrophic nutrition occurs among the phytoflagellates.

In *holophytic nutrition,* which is characteristic of many phytoflagellates, carbohydrates are synthesized by means of chlorophyll carried in *chromatophores,* which vary considerably in size, shape, and number.

In *holozoic nutrition,* particulate food material is ingested through a temporary or permanent mouth. A temporary mouth is formed by amebae when they engulf their food. A permanent mouth is a *cytostome.* It may be simple or it may lead into a *cytopharynx.* In many ciliates the area around the cytostome forms a *peristome,* and a number of other specialized structures may be associated with it. Particulate food passes into a *food vacuole* in the cytoplasm, where it is digested. The indigestible material may be extruded from the body through either a temporary opening or a permanent *cytopyge.*

Many protozoa ingest nutrient material in solution by a process known as *pinocytosis* (the pinching off of vacuoles of fluid through a temporary opening in the body wall).

In some protozoa (e.g., *Eimeria* and *Plasmodium*) a submicroscopic *micropore* (also sometimes called a *micropyle* or *cytostome*) may be visible under the electron microscope. It is apparently used for ingestion of fluids or solids.

In *saprozoic nutrition* no specialized organelles are necessary, nutrients being absorbed through the body wall. This type is found in many protozoa and may be present along with holophytic or holozoic nutrition. How many so-called saprozoic species are actually pinocytotic is still unknown.

Excretory Organelles

Excretion in the Protozoa is either through the body wall or by means of a *contractile vacuole,* which may be simple or may be associated with a system of feeder canals or vacuoles. Contractile vacuoles (sometimes called *water vacuoles*) are probably more important as osmoregulatory organelles than for excretion. They maintain water balance by removing excess water from the cytoplasm and passing it out of the body. They are found in freshwater protozoa but are absent from most marine and parasitic protozoa.

Other Organelles

Protozoa have many other specialized organelles, which are found in different groups. These will be described in the following sections.

REPRODUCTION AND LIFE CYCLES

Reproduction in the Protozoa may be either asexual or sexual. The commonest type of asexual reproduction is *binary fission* (each individual divides into two). The plane of fission is longitudinal in the flagellates and transverse in the ciliates. Cytoplasmic division follows nuclear fission and separation of the daughter nuclei. Vesicular nuclei and micronuclei divide mitotically; macronuclei divide amitotically.

Multiple fission (*schizogony*) is found mostly in the Apicomplexa. In this type of fission, the nucleus divides several times before the cytoplasm divides. The dividing cell is known as a *meront, schizont, agamont,* or *segmenter;* the daughter cells are *merozoites* or *schizozoites.* Nuclear division is mitotic or apparently amitotic.

A third type of asexual division is *budding.* In this process, a small daughter individual separates from the side of the mother and then grows to full size. Division is apparently amitotic. Budding may also be internal, and the number of buds may vary with the species. *Endodyogeny* is a type of internal budding in which two daughter cells form within the mother cell and then break out, destroying it. If more than two daughter cells are formed by internal budding, the process is known as *endopolygeny.* If more than two daughter cells are formed by external budding, the process is known as *ectopolygeny.*

Merozoites are formed from meronts by either schizogony, endodyogeny, endopolygeny, or ectopolygeny. The process, known as *merogony,* occurs in the Apicomplexa.

Of the several types of sexual reproduction that have been described, only two occur in parasitic protozoa. In *conjugation,* which is found among the ciliates, two individuals come together temporarily and fuse along part of their length. Their macronuclei degenerate, their micronuclei divide a number of times, and one of the resultant haploid pronuclei passes from each conjugant into the other. The conjugants then separate, and nuclear reorganization takes place.

In *syngamy* two gametes fuse to form a zygote. If the gametes are similar in appearance, they are called *isogametes;* if they are dissimilar, they are called *anisogametes.* The smaller gamete is ordinarily called a *microgamete,* the larger one a *macrogamete.* The gametes are produced by special cells, the *gamonts,* termed *microgamonts* or *macrogamonts* (or *microgametocytes* or *macrogametocytes*) depending on the type of gamete they produce. The process of gamete formation is known as *gamogony.* It may differ in different groups and will be described in the appropriate places.

Following syngamy, the zygote may or may not divide to form a number of *sporozoites.* The process is known as *sporogony.*

Some protozoa form resistant cysts or spores. A *cyst* results from the formation by the protozoon of a heavy wall around itself. The term *pseudocyst* is used if the wall is formed by the host. A *spore* is produced within the organism by the formation of a heavy wall around one or more individuals. Each spore or cyst may contain one or more *sporozoites.*

The vegetative, motile stage of a protozoon is known as a *trophozoite.*

HISTORY

The first person to see protozoa was the Dutch microscopist Antony van Leeuwenhoek (1632–1723). He ground simple lenses that gave magnifications as high as ×270. His letters to the Royal Society are a classic of biology. Between 1674 and 1716 Leeuwenhoek described many free-living protozoa, among them *Euglena, Volvox,* and *Vorticella* (Dobell 1932). Huygens in 1678 was the first to describe *Paramecium.* Classic work on free-living protozoa was reported by O. F. Müller (1786), Ehrenberg (1830, 1838) and Dujardin (1841).

The first parasitic protozoon to be discovered was *Eimeria stiedai;* Leeuwenhoek found its oocysts in the gallbladder of an old rabbit in 1674. Later, in 1681, Leeuwenhoek found *Giardia lamblia* in his own diarrheic stools, and in 1683 he found *Opalina* and *Nyctotherus* in the intestine of the frog.

The first species of *Trichomonas, T. tenax,* was found by O. F. Müller in 1773 in the human mouth; he named it *Cercaria tenax.* Donné found *T. vaginalis* in the human vagina in 1837, and Davaine found *Trichomonas* and *Chilomastix* in the stools of human cholera patients in 1854.

The first trypanosome was discovered in the blood of the salmon by Valentin in 1841, and the frog trypanosome by Gluge and Gruby in 1842. Lewis found the first mammalian trypanosome, *T. lewisi,* in the rat in 1878. Evans discovered the first pathogenic one, *T. evansi,* in 1881 in India, where it was causing the disease known as *surra* in elephants. Bruce discovered *T. brucei* in Africa in 1895 and described its life cycle and transmission by the tsetse fly in

1897. In 1902 Dutton discovered that African sleeping sickness of man was caused by *T. gambiense*. *Leishmania tropica* was first seen by Cunningham in India in 1885 and was first described and identified as a protozoon by Borovsky in Russia in 1898. Leishman and Donovan independently discovered *L. donovani* in India in 1903.

Histomonas meleagridis, the cause of blackhead of turkeys, was discovered by Theobald Smith in 1895. Its transmission in the eggs of the cecal worm was discovered by Tyzzer and Fabian in 1922 and described in detail by Tyzzer in 1934. The first parasitic ameba, *Entamoeba gingivalis,* was found in the human mouth by Gros in 1849. Lewis found *E. coli* in India in 1870, and Lesh (Lösch) found *E. histolytica* in Russia in 1875. *Balantidium coli* was discovered by Malmsten in 1857.

It was not until 154 years after Leeuwenhoek saw *Eimeria stiedai* that any other Apicomplexa were found. Then in 1828 Dufour described gregarines in the intestines of beetles, and in 1838 Hake rediscovered the oocysts of *E. stiedai.* The most extensive early study of the coccidia was that of Eimer (1870), who described a number of species in various animals. Schaudinn and Siedlecki (1897) described the gamonts and gametes of coccidia and showed that they formed zygotes. Further studies on the life cycle of coccidia were published by Schaudinn in 1898 and 1899. Classic work on the coccidia of gallinaceous birds was reported by Tyzzer (1929) and Tyzzer, Theiler, and Jones (1932).

The human malaria parasite was discovered in 1880 by the French army physician Alphonse Laveran. Golgi (1886, 1889) reported on its schizogony and distinguished the types of fever caused by the different species. MacCallum (1897), working with the closely related *Haemoproteus* of birds, recognized that the exflagellation seen by Laveran was microgamete formation and later observed fertilization and zygote formation in *Plasmodium falciparum.*

Ross worked out the life cycle of the bird malaria parasite, *P. relictum* (*P. praecox*), in India in 1898, showing that it was transmitted by the mosquito *Culex fatigans.* Working independently in Italy, Grassi and his collaborators (1898) almost immediately afterward found that human malaria is transmitted by *Anopheles* mosquitoes.

Babesia bovis was discovered by Babes in 1888. Theobald Smith and F. L. Kilborne described the cause of Texas fever of cattle, *B. bigemina,* in 1893; they proved that it was transmitted by the tick *Boophilus annulatus,* passing through the eggs to the next generation of ticks, which then infected new cattle. This was the first proof of arthropod transmission of a protozoon.

The present century has seen many advances in protozoology, and there are many more ahead. Several times more species of parasitic protozoa have been described since 1900 than were known before. At present about 10,000 species have been named, but these are only a fraction of the total. Exciting new discoveries are being made every year on the physiology and nutritional requirements of protozoa (von Brandt 1973, 1979; Carter 1979; Lloyd and Muller 1979; Levandowsky and Hutner 1979–1980; Walter and Blum 1979). The life cycles, host-parasite relations, and pathogenesis of many species are only now being worked out. Many species of protozoa can be cultivated outside the host in liquid cultures, cell cultures, or chick embryos (Taylor and Baker 1978; Doran 1982). The transmission and scanning electron microscopes and the phase contrast microscope have opened up a whole new field for morphologic study. Chemotherapy is progressing rapidly (Hutner 1977), and new discoveries are being made even in taxonomy, which most people used to consider a dead field.

CLASSIFICATION

The classification of the subkingdom Protozoa used in this book (with some modifications) is that proposed by a committee of the Society of Protozoologists (Levine et al. 1980).

Phylum SARCOMASTIGOPHORA
Single type of nucleus except in Foraminiferorida; sexuality when present essentially syngamy; flagella, pseudopodia or both.

Subphylum MASTIGOPHORA
1 or more flagella typically present in trophozoites; asexual reproduction longitudinal; sexual reproduction known in some groups.

Class PHYTOMASTIGOPHORASIDA
Typically with chloroplasts; if chloroplasts lacking, relationship to pigmented forms clearly evident.

Class ZOOMASTIGOPHORASIDA
Chloroplasts absent.

Order KINETOPLASTORIDA
1 or 2 flagella, typically with paraxial rod in addition to axoneme; single mitochondrion extending length of body, usually containing conspicuous Feulgen-positive (DNA-containing) kinetoplast near flagellar kinetosomes; Golgi apparatus typically in region of flagellar depression.

Suborder BODONORINA
Typically 2 heterodynamic flagella; typical, often large, adbasal kinetoplast or kDNA arranged in several discrete bodies or dispersed throughout mitochondrion.

Family BODONIDAE
Anterior end more or less drawn out; 1 to several contractile vacuoles: *Bodo, Cercomonas, Pleuromonas, Spiromonas.*

Suborder TRYPANOSOMATORINA
1 flagellum, either free or attached to body by means of undulating membrane; kinetoplast relatively small and compact.

Family TRYPANOSOMATIDAE
Body characteristically leaflike but may be rounded: *Blastocrithidia, Crithidia, Endotrypanum, Herpetomonas, Leishmania, Leptomonas, Phytomonas, Rhynchoidomonas, Trypanosoma.*

Order PROTEROMONADORIDA
1 or 2 pairs of flagella without paraxial rods; single mitochondrion,

distant from kinetosomes, curving around nucleus, not extending length of body; without Feulgen-positive kinetoplast; Golgi apparatus encircling band-shaped rhizoplast passing from kinetosomes near surface of nucleus to mitochondrion; cysts present: *Proteromonas.*

Order RETORTAMONADORIDA

2–4 flagella, 1 turned posteriorly and associated with ventrally located cytostomal area bordered by fibril; mitochondria and Golgi apparatus absent; intranuclear division spindle; cysts present: *Chilomastix, Retortamonas.*

Order COCHLOSOMATORIDA

6 flagella; helicoidal anterior ventral sucker; parabasal body and axostyle inconstant (?): *Cochlosoma.*

Order ENTEROMONADORIDA

1 karyomastigont; individual mastigonts with 1–4 flagella; without mitochondria or Golgi apparatus; intranuclear division spindle; cysts present: *Caviomonas, Enteromonas, Trimitus.*

Order DIPLOMONADORIDA

2 karyomastigonts; with twofold rotational symmetry or primarily mirror symmetry; individual mastigonts with 1–4 flagella, typically 1 of them recurrent and associated with cytostome or, in more advanced genera, with organelles forming cell axis; without mitochondria or Golgi apparatus; intranuclear division spindle; cysts present: *Giardia, Hexamita, Octomitus, Spironucleus, Trepomonas.*

Order OXYMONADORIDA

1 or more karyomastigonts, each containing 4 flagella typically arranged in 2 pairs in motile stages; 1 or more flagella may be recurrent, adhering to body surface for at least part of its length; kinetosomes of flagellar pairs connected by paracrystalline structure (preaxostyle) in which are embedded anterior ends of axostylar microtubules; 1 to many axostyles per organism; without mitochondria or Golgi apparatus; division spindle intranuclear; cysts in some species; sexuality in some species: *Monocercomonoides, Oxymonas.*

Order TRICHOMONADORIDA

With 0–6 (typically 4–6) flagella; with Golgi apparatus forming a parabasal body; without mitochondria; division spindle extranuclear; usually without cysts; sexuality unknown.

Family TRIMASTIGIDAE

Body with a ventral depression containing an undulating membrane associated with a recurrent flagellum; 3 free flagella of which 1 is directed anteriorly: *Trimastix.*

Family MONOCERCOMONADIDAE

3–5 anterior flagella; recurrent flagellum free for its entire length or with proximal part adhering for a greater or lesser distance to dorsal body surface; costa absent; pelta present; parabasal apparatus rod-, disk, or V-shaped: *Chilomitus, Dientamoeba, Hexamastix, Histomonas, Hypotrichomonas, Mono-*

cercomonas, Parahistomonas, Protrichomonas, Pseudotrichomonas, Tetratrichomastix, Tricercomitus.

Family TRICHOMONADIDAE

4–6 flagella, of which 1 is recurrent and attached to undulating membrane; costa and axostyle present; parabasal apparatus consists of body associated with 1 or more filaments: *Pentatrichomonas, Pentatrichomonoides, Pseudotrypanosoma, Tetratrichomonas, Trichomitopsis, Trichomitus, Trichomonas, Tritrichomonas.*

INCERTAE SEDIS

Family CALLIMASTIGIDAE

With a compact anterior or lateral group of flagella that beat as a unit: *Callimastix.*

Subphylum OPALINATA

Numerous cilia in oblique rows over entire body surface; cytostome absent; nuclear division acentric; binary fission generally interkinetal; known life cycles involve syngamy with anisogamous flagellated gametes: *Opalina.*

Subphylum SARCODINA

Pseudopodia or locomotive protoplasmic flow without discrete pseudopodia; flagella, when present, usually restricted to developmental or other temporary stages; asexual reproduction by fission; sexuality, if present, associated with flagellate or, more rarely, ameboid gametes.

Superclass RHIZOPODASICA

Locomotion by lobopodia, filopodia, reticulopodia, or protoplasmic flow without production of discrete pseudopodia.

Class LOBOSASIDA

Pseudopodia lobose or more or less filiform but produced from broader hyaline lobe; usually uninucleate; multinucleate forms not flattened or much-branched plasmodia; without sorocarps, sporangia, or similar fruiting bodies.

Subclass GYMNAMOEBIASINA

Without test.

Order AMOEBIDORIDA

Typically uninucleate; mitochondria present; no flagellate stage.

Suborder TUBULINORINA

Body branched or unbranched cylinder; no bidirectional flow of cytoplasm; nuclear division mesomitotic: *Amoeba, Endamoeba, Endolimax, Entamoeba, Iodamoeba, Sappinia.*

Suborder ACANTHOPODORINA

More or less finely tipped, sometimes filiform, often furcate hyaline subpseudopodia produced from a broad hyaline lobe; not regularly diskoid; cysts usually formed; nuclear division mesomitotic or metamitotic: *Acanthamoeba, Hartmannella.*

Order SCHIZOPYRENORIDA

Body with shape of monopodial cylinder, usually moving with more or less eruptive, hyaline, hemispherical bulges; typically uninucleate; nuclear division promitotic; temporary flagellate stages in most species: *Naegleria, Tetramitus, Trimastigamoeba, Vahlkampfia.*

Subclass TESTACEALOBOSIASINA

Body enclosed by test, tectum, or other complex membrane external to plasma membrane and glycocalyx.

Order ARCELLINORIDA

Test, tectum, or other external membrane with single aperture: *Chlamydophrys.*

Phylum APICOMPLEXA

Apical complex (visible with electron microscope) generally consisting of polar ring(s), rhoptries, micronemes, conoid, and subpellicular microtubules present at some stage; nucleus vesicular; cilia absent; sexuality by syngamy; all species parasitic.

Class PERKINSASIDA

With flagellated "zoospores" (sporozoites?); "zoospores" with anterior vacuole; conoid forms incomplete, truncate cone; sexuality absent; homoxenous.

Order PERKINSORIDA

With the characters of the class.

Family PERKINSIDAE

With the characters of the order: *Perkinsus.*

Class SPOROZOASIDA

If present, conoid forms complete truncate cone; reproduction generally both sexual and asexual; oocysts contain infective sporozoites which result from sporogony; locomotion by body flexion, gliding, undulation of longitudinal ridges or flagellar lashing; flagella present only in microgametes of some groups; pseudopods ordinarily absent, if present used for feeding, not locomotion; homoxenous or heteroxenous.

Subclass GREGARINASIDA

Mature gamonts extracellular, large; conoid modified into mucron or epimerite in mature organisms; endodyogeny absent; in invertebrates or lower chordates.

Subclass COCCIDIASINA

Mature gamonts small, typically intracellular; without mucron or epimerite; life cycle usually consisting of merogony, gamogony, and sporogony; most species in vertebrates.

Order EUCOCCIDIORIDA

Merogony present.

Suborder ADELEORINA

Macrogamete and microgamont usually associated in syzygy during development; microgamont produces 1–4 microgametes; sporozoites enclosed in envelope.

Family ADELEIDAE

Zygote inactive; sporocysts formed in oocyst: *Adelina, Klossia.*

Family HAEMOGREGARINIDAE

Zygote active (ookinete); heteroxenous, life cycle involving 1 invertebrate and 1 vertebrate host; merogony in vertebrates; gamonts in vertebrate blood cells; sporogony in invertebrates; gamonts with about 70–80 subpellicular microtubules: *Haemogregarina, Hepatozoon, Karyolysus, Cyrilia.*

Family KLOSSIELLIDAE

Zygote inactive; typical oocyst not formed; a number of sporocysts, each with many sporozoites, develops within a membrane that is perhaps laid down by host cell; microgamont forms 2–4 nonflagellated microgametes; homoxenous, with sporogony and merogony in different locations of the same host: *Klossiella.*

Suborder EIMERIORINA

Macrogamete and microgamont develop independently; syzygy absent; microgamont typically produces many microgametes; zygote not motile; conoid present; sporozoites typically enclosed in sporocyst.

Family CRYPTOSPORIDIIDAE

Development just under surface membrane of host cell or within its brush border and not in the cell proper; meronts with knoblike attachment organelle; microgametes without flagella; homoxenous: *Cryptosporidium.*

Family EIMERIIDAE

Development in host cell proper; oocysts with 0–*n* sporocysts, each with 1 or more sporozoites; homoxenous or at least without asexual multiplication in nondefinitive host; merogony within host, sporogony typically outside; microgametes with 2–3 flagella: *Dorisa, Eimeria, Isospora, Tyzzeria, Wenyonella.*

Family LANKESTERELLIDAE

Development in host cell proper; oocysts with 8 or more sporozoites; microgametes with 2 flagella so far as known; heteroxenous, with merogony, gamogony, and sporogony in the same vertebrate host; sporozoites in blood cells, transferred without developing by an invertebrate (mite, mosquito, or leech); infection by ingestion of invertebrate host: *Lankesterella.*

Family ATOXOPLASMATIDAE

Homoxenous, with merogony and gamogony within the host, sporogony outside; development intracellular (merogony in blood and intestinal cells, gamogony in intestinal cells of the same individual host); transmission by ingestion of sporulated oocysts: *Atoxoplasma.*

Family SARCOCYSTIDAE

Heteroxenous, with a predator-prey life cycle; producing oocysts following syngamy; oocysts with 2 sporocysts, each with 4 sporozoites, in intestine of definitive host; asexual stages in intermediate host; both definitive and intermediate hosts are vertebrates: *Arthrocystis, Besnoitia, Frenkelia, Sarcocystis, Toxoplasma.*

Suborder HAEMOSPORORINA

Macrogamete and microgamont develop independently; without

conoid; syzygy absent; microgamont produces about 8 flagellated microgametes; zygote motile (ookinete); sporozoites naked, with 3-membraned wall; heteroxenous, with merogony in vertebrate host and sporogony in invertebrate; transmitted by blood-sucking insects.

Family PLASMODIIDAE

With the characters of the suborder: *Haemoproteus, Hepatocystis, Leucocytozoon, Plasmodium.*

Suborder PIROPLASMORINA

Piriform, round, rod-shaped, or ameboid; without conoid; without oocysts, spores, or pseudocysts; without flagella; with polar ring(s) and rhoptries; in erythrocytes and sometimes also in other cells; heteroxenous, with merogony in vertebrate and sporogony in invertebrate; vectors are ticks.

Family BABESIIDAE

Piriform, round, or oval; in erythrocytes and sometimes other blood cells; apical complex reduced to polar ring, rhoptries, and subpellicular microtubules; micronemes present in some stages; in mammals: *Babesia.*

Family THEILERIIDAE

Small, round, ovoid, irregular, or bacilliform; in erythrocytes and sometimes other blood cells; merogony in lymphocytes, histiocytes, erythroblasts, or other cells, followed by invasion of erythrocytes; apical complex reduced to rhoptries; in mammals: *Theileria.*

Phylum MICROSPORA

Spores unicellular, with imperforate wall, containing a uninucleate or binucleate sporoplasm and extrusion apparatus with polar tube and polar cap; without mitochondria; obligatory intracellular parasites in nearly all major animal groups.

Class MICROSPORASIDA

Spore with complex extrusion apparatus of Golgi origin; polar tube typically filamentous, extending backward from polar cap and coiling around inside spore wall.

Order MICROSPORORIDA

General tendency toward maximum development and varied specialization of accessory spore organelles (components of extrusion apparatus) and spore wall; mostly in insects: *Encephalitozoon, Nosema, Thelohania.*

Phylum MYXOZOA

Spores of multicellular origin, with 1 or more polar capsules and sporoplasms, with 1–3 (rarely more) valves; all species parasitic, mostly in fishes.

Phylum CILIOPHORA

With cilia or compound ciliary organelles in at least 1 stage of life cycle; with subpellicular infraciliature; with 2 types of nucleus (macronucleus and micronucleus); binary fission transverse; sexuality involving conjugation or autogamy and cytogamy.

Class KINETOFRAGMINOPHORASIDA
Oral infraciliature only slightly distinct from somatic infraciliature.
Subclass GYMOSTOMATASINA
Cytostomal area superficial, apical, or subapical; without vestibulum.
Order PROSTOMATIDORIDA
Cytostome apical or subapical.
Suborder ARCHISTOMATORINA
Cytostome apical.
Family BUETSCHLIIDAE
With anterior concretion vacuole (possibly a statocyst), 1 or more contractile vacuoles, and posterior cytopyge: *Alloiozona, Ampullacula, Blepharoconus, Blepharoprosthium, Blepharosphaera, Blepharozoum, Buetschlia, Bundleia, Didesmis, Holophryoides, Paraisotrichopsis, Polymorphella, Prorodonopsis, Sulcoarcus.*
Subclass VESTIBULIFERASINA
Vestibulum commonly present.
Order TRICHOSTOMATORIDA
Somatic kinetics not reorganized at level of vestibulum.
Suborder TRICHOSTOMATORINA
Somatic ciliature not reduced.
Family ISOTRICHIDAE
Without anterior tuft of longer cilia; in ruminants: *Dasytricha, Isotricha.*
Family PARAISOTRICHIDAE
With anterior tuft of longer cilia; in ruminants: *Paraisotricha.*
Family BALANTIDIIDAE
With cytostome at base of anterior vestibulum: *Balantidium.*
Family PYCNOTRICHIDAE
With long groove leading to cytostome, which may be near the middle or posterior end of body; in herbivores: *Buxtonella, Infundibulorium.*
Suborder BLEPHAROCORYTHORINA
Somatic ciliature reduced.
Family BLEPHAROCORYTHIDAE
With the characters of the suborder: *Blepharocorys, Charonina, Ochoterenaia.*
Order ENTODINIOMORPHIDORIDA
Simple somatic ciliature absent; with membranellar tufts or zones of cilia; pellicle firm, sometimes drawn out into spines; in herbivores.
Family OPHRYOSCOLECIDAE
Ciliary tufts limited principally to oral or adoral area plus 1 internal group; ciliary tufts retractable: *Caloscolex, Campylodinium* (syn., *Amphacanthus*), *Cunhaia, Diplodinium, Diploplastron, Elytroplastron, Enoploplastron, Entodinium, Eodinium, Epidinium, Epiplastron, Eremoplastron, Eudiplodinium, Metadinium, Ophryoscolex, Opisthotrichum, Ostracodinium, Polyplastron.*

Family CYCLOPOSTHIIDAE
Adoral ciliature retractable; dorsal and caudal ciliature nonretractable; often with skeletal plates; in equids: *Cycloposthium, Polydiniella, Prototapirella, Trifascicularia, Tripalmaria, Triplumaria.*
Family SPIRODINIIDAE
Adoral zone ciliature retractable; without skeletal plates; in equids: *Cochliatoxum, Spirodinium.*
Family DITOXIDAE
Adoral zone ciliature nonretractable; with 4 somatic, nonretractable synciliary "ribbons"; without skeletal plates; in equids: *Ditoxum, Tetratoxum, Triadinium.*
Family TROGLODYTELLIDAE
Adoral zone ciliature retractable; with large skeletal plates; in primates: *Troglodytella.*
Subclass SUCTORIASINA
With suctorial tentacles.
Order SUCTORIORIDA
With the characters of the subclass.
Suborder ENDOGENORINA
With endogenous budding.
Family ACINETIDAE
Tentacles in few fascicles or even reduced to a single organelle: *Allantosoma.*
Suborder EVAGINOGENORINA
Budding evaginative, with development of a single larva; tentacles either scattered singly or in fascicles at ends of arms or trunks.
Family CYATHODINIIDAE
Adult stalkless; adult stage fleeting: *Cyathodinium.*
Class OLIGOHYMENOPHORASIDA
Oral ciliature clearly distinct from somatic ciliature; with membranelles, peniculi, or polykineties on left side.

LITERATURE CITED

For citations before 1965, see the *Index-Catalogue of Medical and Veterinary Zoology, Zoological Record, Biological Abstracts, Protozoological Abstracts,* and *Veterinary Bulletin.*

Carter, R. 1979. Proc. 5th Int. Congr. Protozool. 59–65.

Doran, D. J. 1982. In The Biology of the Coccidia, ed. P. L. Long, 229–85. Baltimore: University Park Press.

Gooday, G. W., ed. 1980. The Eukaryotic Microbial Cell. Society for General Microbiology Symposium 30. Cambridge, Engl.: Cambridge University Press.

Hutner, S. H. 1977. J. Protozool. 24:475–78.

Levandowsky, M., and S. H. Hutner. 1979–1980. Biochemistry and Physiology of Protozoa. 2d ed. Vols. 1–3. New York: Academic.

Levine, N. D., J. O. Corliss, F. E. G. Cox, G. Deroux, J. Grain, B. M. Honigberg,

G. F. Leedale, A. R. Loeblich III, J. Lom, D. Lynn, E. G. Merinfeld, F. C. Page, G. Poljansky, V. Sprague, J. Vavra, and F. G. Wallace. 1980. J. Protozool. 27:37–58.
Lloyd, D., and M. Muller. 1979. Proc. 5th Congr. Protozool., 147–59.
Rinaldi, R. A., and A. Grebecki. 1979. Proc. 5th Int. Congr. Protzool., 180–83.
Stocken, W. J. 1979. Proc. 5th Int. Congr. Protozool., 131–39.
Taylor, A. E. R., and J. R. Baker, eds. 1978. Methods of Cultivating Parasites in Vitro. New York: Academic.
von Brandt, T. 1973. Biochemistry of Parasites. 2d ed. New York: Academic.
———, ed. 1979. Biochemistry and Physiology of Endoparasites. New York: Elsevier North-Holland.
Walter, R. D., and J. J. Blum. 1979. Proc. 5th Int. Congr. Protozool., 124–30.
Whittaker, R. H. 1977. In *Parasitic Protozoa,* vol. 1, ed. J. P. Kreier, 1–34. New York: Academic.

GENERAL REFERENCES

Aikawa, M., and C. R. Sterling. Intracellular Parasitic Protozoa. New York: Academic. Paperback.
Buetow, D., ed. 1968. The Biology of Euglena. New York: Academic.
Bulla, L. A., Jr., T. C. Cheng, J. Vavra, and V. Sprague, eds. 1976–1977. Comparative Pathobiology. Vol. 1, Biology of the Microsporidia; vol. 2, Systematics of the Microsporidia. New York: Plenum.
Chen, T.-T., ed. 1967–1972. Research in Protozoology. 4 vols. New York: Pergamon.
Dogiel, V. A., J. I. Poljanskii, and E. M. Chejsin. 1965. General Protozoology. 2d ed. Oxford: Clarendon.
Farmer, J. A. 1980. The Protozoa: Introduction to Protozoology. St. Louis: Mosby. Paperback.
Gutteridge, W. E., and G. H. Coombs. 1977. Biochemistry of Parasitic Protozoa. Baltimore: University Park Press. Paperback.
Jahn, T. L., E. C. Bovee, and F. F. Jahn. 1979. How to Know the Protozoa. 2d ed. Dubuque, Ia.: Brown. Paperback.
Kreier, J. P., ed. 1977–78. Parasitic Protozoa. 4 vols. New York: Academic.
Kudo, R. R. 1966. Protozoology. 5th ed. Springfield, Ill.: Thomas.
Leedale, G. F. 1967. Euglenoid Flagellates. Englewood Cliffs, N.J.: Prentice Hall.
Long, P. L., ed. 1982. The Biology of the Coccidia. Baltimore: University Park Press.
Manwell, R. D. 1968. Introduction to Protozoology. 2d ed. New York: Dover. Paperback.
Scholtyseck, E. 1979. Fine Structure of Parasitic Protozoa. Berlin: Springer-Verlag. Soft cloth.
Vickerman, K., and F. E. G. Cox. 1967. The Protozoa. Boston: Houghton Mifflin. Paperback.
Weinman, D., and M. Ristic, eds. 1968. Infectious Blood Diseases of Man and Animals. Diseases Caused by Protista. Vol. I. New York: Academic.
Wenyon, C. M. 1926. Protozoology. 2 vols. New York: Wood.

2

Flagellates: The Hemoflagellates

THE flagellates belong to the subphylum Mastigophora. They have 1 or more flagella, and a few have pseudopods as well. Their nutrition is holophytic, holozoic, or saprozoic. They multiply by longitudinal binary fission, and some produce cysts.

The subphylum is divided into two classes, Phytomastigophorasida and Zoomastigophorasida. The former contains the phytoflagellates, the great majority of which are free-living and holophytic. Those of parasitic interest will be discussed in Chapter 5.

It is convenient for the purposes of this book to divide the Zoomastigophorasida into two groups: the hemoflagellates, which live in blood, lymph, and tissues, and the other flagellates (see Chaps. 3–5), which live in the intestine and other body cavities.

FAMILY TRYPANOSOMATIDAE

The hemoflagellates all belong to the family Trypanosomatidae. Members of this family have a leaflike or sometimes a rounded body containing a vesicular nucleus (Fig. 2.1). A varying number of subpellicular microtubules lies just beneath the outer membrane; they resemble flagellar microtubules except that they are single, and their number varies with the species as well as at different levels in the same cell. Trypanosomatids have a single flagellum that arises from a kinetosome or basal granule posterior to the end of an elongate blind pouch (flagellar pocket or reservoir). There may be a second (barren) basal body near the first. A contractile vacuole often opens into the reservoir; both structures can be seen only with the phase or electron microscope and not with the ordinary light microscope. The flagellar axoneme is composed of 9 double peripheral and 2 central microtubules. An undulating membrane is present is some genera; the flagellum lies in its outer border. In some if not all bloodstream forms there is a subpellicular organelle composed of 4 microtubules near the point of attachment of the flagellum. Posterior to the kinetosome is a rod-shaped or spherical *kinetoplast* containing DNA. Varying in structure with the species, it consists of a double membrane within which is a group of fibrils

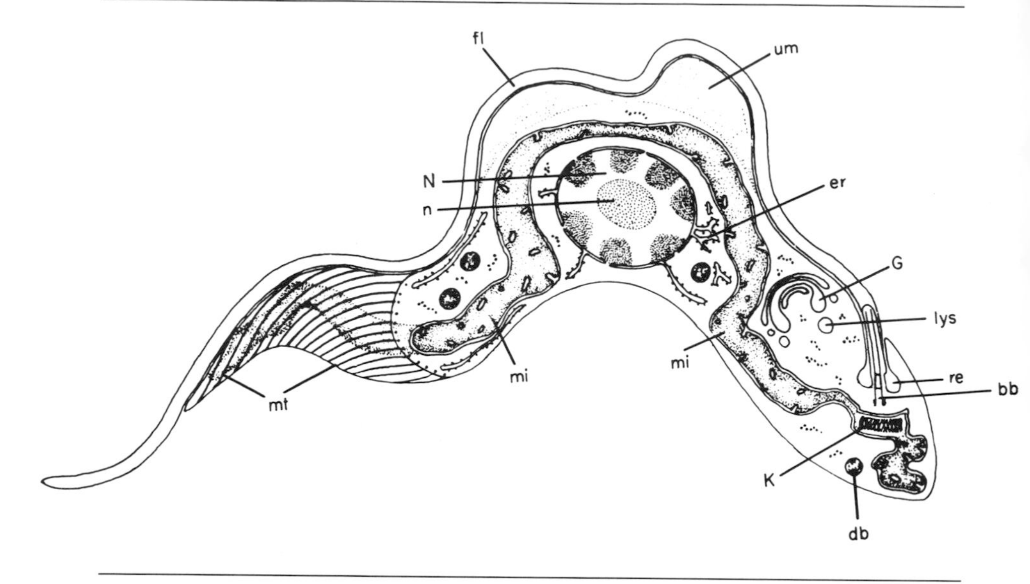

Fig. 2.1. Principal features of a bloodstream trypanosome of the *Trypanosoma brucei* type: *bb* = kinetosome (basal body), *db* = membrane-bound dense body, *er* = endoplasmic reticulum, *fl* = flagellum, *G* = Golgi apparatus, *K* = kinetoplast, *lys* = lysosome, *mi* = mitochondrion, *mt* = subpellicular microtubules (shown only at the anterior end), *N* = nucleus, *n* = nucleolus, *re* = reservoir, *um* = undulating membrane. (From Rudzinska and Vickerman 1968, after Vickerman and Cox 1967)

(which may actually be a single coiled fibril); it is part of a mitochondrion that runs the whole length of the body. It divides by binary fission and presumably has to do with respiration. Dyskinetoplastic or akinetoplastic strains without a stainable kinetoplast can be produced in some trypanosomes by treatment with acriflavine or otherwise; in these the outer membranes remain but inner fibrils have disappeared. Under the ordinary light microscope the kinetoplast and kinetosome are often so close together that they appear fused.

Nuclear division is not closely akin to eukaryote mitosis although it may resemble nuclear division in *Euglena*. The nuclear membrane does not disappear, and chromosomes are not distinguishable. The endosome elongates, becomes encased by a spindle of microtubules, and the spindles seem to push the two halves of the nucleus apart.

A Golgi apparatus, mitochondria, lysosomes, endoplasmic reticulum, ribosomes, and various unidentified granules, vesicles, and vacuoles can be seen with the electron microscope; "volutin" granules can be stained with methylene blue.

Members of this family were originally parasites of the intestinal tract of insects (and possibly leeches), and many are still found only in insects. Others are heteroxenous, spending part of their life cycle in a vertebrate and part in an invertebrate host.

In the course of their life cycles, members of one genus may pass through forms structurally similar to those of other genera. In the *trypomastigote* form, which is perhaps the most advanced, the kinetoplast and kinetosome are near the posterior end and the flagellum forms the border of an undulating membrane that extends along the side of the body to the anterior end. In the *opisthomastigote* form, the kinetoplast and kinetosome are near the posterior end, the flagellum runs through the body to emerge at the anterior end, and there is no undulating membrane. In the *epimastigote* form, the kinetoplast and kinetosome are just posterior to the nucleus and the undulating membrane runs forward from there. In the *promastigote* form, the kinetoplast and kinetosome are still further posterior in the body and there is no undulating membrane. In the *amastigote* form, the body is rounded and the flagellum has degenerated into a tiny fibril that remains inside the body or it may be absent (Fig. 2.2). In the *choanomastigote* form (the "barleycorn" form that occurs in insect parasites of the genus *Crithidia*), the kinetoplast and kinetosome are anterior to the nucleus and the flagellum emerges from the body through a wide, funnel-shaped reservoir. In the *paramastigote* form, the kinetoplast is beside the nucleus (Hoare and Wallace 1966; Janovy et al. 1974; Wallace 1977).

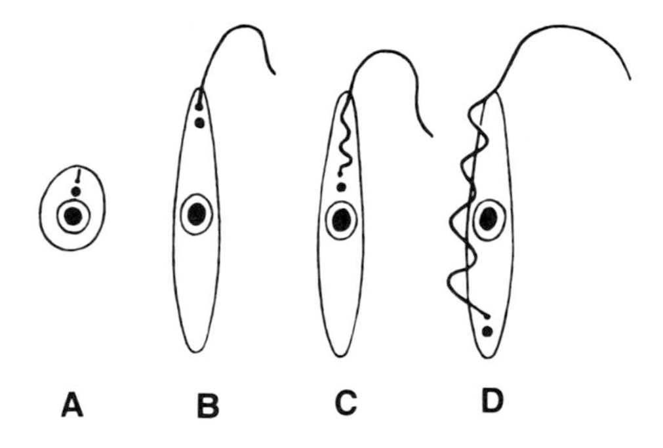

Fig. 2.2. Forms of the Trypanosomatidae: A. amastigote form; B. promastigote form; C. epimastigote form; D. trypomastigote form.

There are 10 genera in the family Trypanosomatidae (McGhee 1968; Hoare 1972). Members of the genus *Trypanosoma* are heteroxenous and pass through amastigote, promastigote, epimastigote, and trypomastigote stages in their life cycles. In some species only trypomastigote forms are found in the vertebrate hosts; in others, presumably more primitive, both amastigote and trypomastigote forms are present.

Members of the genus *Blastocrithidia* are monoxenous in arthropods and other invertebrates. They pass through epimastigote, amastigote, and presumably promastigote stages in their life cycles.

Members of the genus *Crithidia* are monoxenous in arthropods. They have only choanomastigote and amastigote stages; the former has a characteristic barleycorn shape.

Members of the genus *Leptomonas* are monoxenous in invertebrates. They pass through promastigote and amastigote stages in their life cycles.

Members of the genus *Leishmania* are heteroxenous, passing through the amastigote stage in their vertebrate host and the promastigote stage in their invertebrate host or in culture.

Members of the genus *Herpetomonas* are monoxenous in invertebrates. They pass through opisthomastigote, promastigote, amastigote, and presumably epimastigote stages in their life cycles. Their flagellum lies in a long reservoir that runs the whole length of the body and opens at the anterior end, whereas in the trypomastigote form of *Trypanosoma* the reservoir is very short and opens laterally near the posterior end so that the undulating membrane runs along the side of the body.

Members of the genus *Rhynchoidomonas* are monoxenous in insects. Their kinetoplast and kinetosome are a little posterior to the nucleus, and they have no undulating membrane. Little is known about them.

Members of the genus *Phytomonas* are heteroxenous in the latex of plants and hemipterous insects. They pass through promastigote and amastigote stages in their life cycles. They are found in milkweeds and related plants and cause the normally milky sap to become clear.

Members of the genus *Proleptomonas* occur in soil and are coprozoic in feces. Whether they also occur in insects is unknown. Their known stage is promastigote.

Members of the genus *Endotrypanum* occur within the erythrocytes of sloths. They are heteroxenous and have the epimastigote stage in the vertebrate host and the promastigote and amastigote stages in the invertebrate host (a sand fly).

The only genera parasitic in domestic animals and man are *Trypanosoma* and *Leishmania*. Since, however, their stages in the invertebrate vector are structurally similar to those of the genera confined to invertebrates, anyone finding an infected invertebrate cannot be positive whether it is infected with a parasite of vertebrates or with one of its own. It is possible, too, that some of the forms that we now think are confined to invertebrates may actually be normal parasites of some wild vertebrates. See Vickerman (1976), Vickerman and Preston (1976), Newton (1976), Hutner et al (1979), and Wallace (1979).

Genus *Trypanosoma* Gruby, 1843

Members of this genus occur in all classes of vertebrates. They are parasites of the circulatory system and tissue fluids, but some, such as *T. cruzi*, may actually invade cells. Almost all are transmitted by bloodsucking invertebrates. Most species are probably nonpathogenic, but the remainder more than make up for their fellows. (See Fig. 2.3.)

Trypanosomosis is one of the world's most important diseases of livestock and man. Trypanosomes cause African sleeping sickness and Chagas' disease in

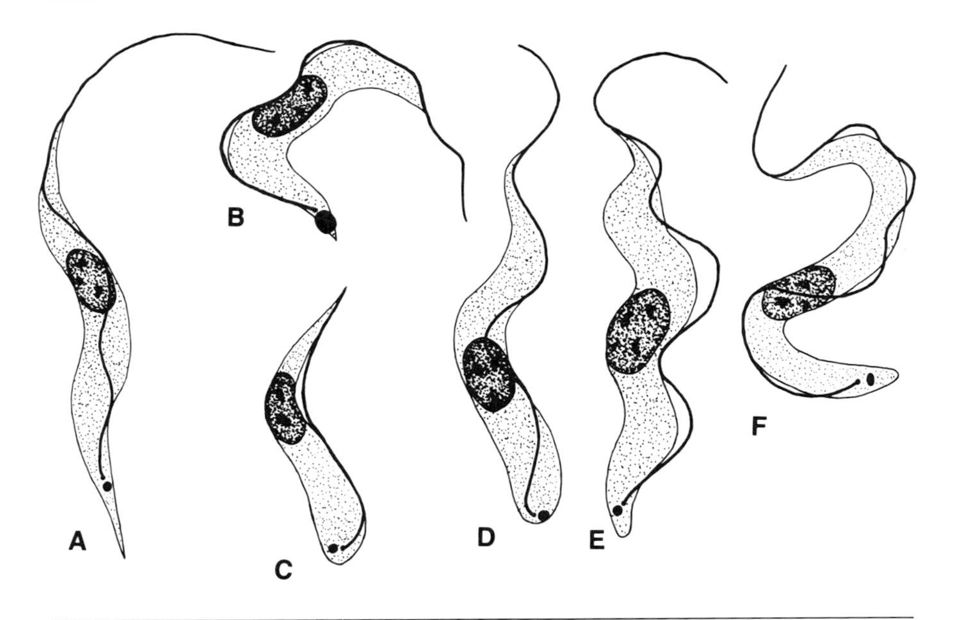

Fig. 2.3 Species of *Trypanosoma:* A. *T. theileri;* B. *T. cruzi;* C. *T. congolense;* D. *T. vivax;* E. *T. brucei equiperdum;* F. *T. b. brucei.* (×2,800)

man and a whole series of similar diseases in domestic animals. They make it practically impossible to raise livestock in many parts of the tropics that would otherwise be ideal. Some, but not all, of the African species are transmitted by tsetse flies. These flies occupy about 4.5 million mi^2 (about 7 million km^2), an area larger than the United States, and this whole region is under the threat of trypanosomosis.

The traditional wheelless agriculture of Central Africa is due to the fact that livestock cannot survive there to pull wagons. Trypanosomosis protected the Central Africans against the horse-riding Arabs from the north and the immigrant Boer farmers from the south, but it also isolated them from industrial knowledge.

Trypanosomosis is much more important today as a disease of domestic animals than of man. However, because it prevents domestic animals from being raised in so much of Africa, it is the essential cause of protein deficiency in this area. Many million African children suffer from kwashiorkor, due to weaning when a younger sibling is born coupled with an inadequate level of protein in the diet.

SEXUALITY. There is no evidence of sex in *Trypanosoma* or other trypanosomatids.

CULTIVATION. *Trypanosoma* is relatively easy to cultivate in artificial media.

IMMUNOLOGY. Trypanosomes are extremely plastic antigenically. Members of the subgenus *Trypanozoon*, which have been studied especially, change in antigenic constitution between the original infection and the first relapse, and between successive relapses. Apparently the external and released antigens vary as the body produces antibodies against them. A succession of variations takes place, and the parasite manages to survive, albeit somewhat altered. However, it reverts to its original antigenic structure upon cyclic transmission through a vector. As many as 20 or more variants have been reported by various workers. A heavy surface coating present on many species in the trypomastigote stage only is responsible for the antigenic changes in successive relapse populations in the blood (Fig. 2.4). When the organisms enter the tsetse fly they lose their antigenic identity, but they regain it, reverting to the original antigenic formula, when they become metacyclic forms in the salivary glands (Fig. 2.5).

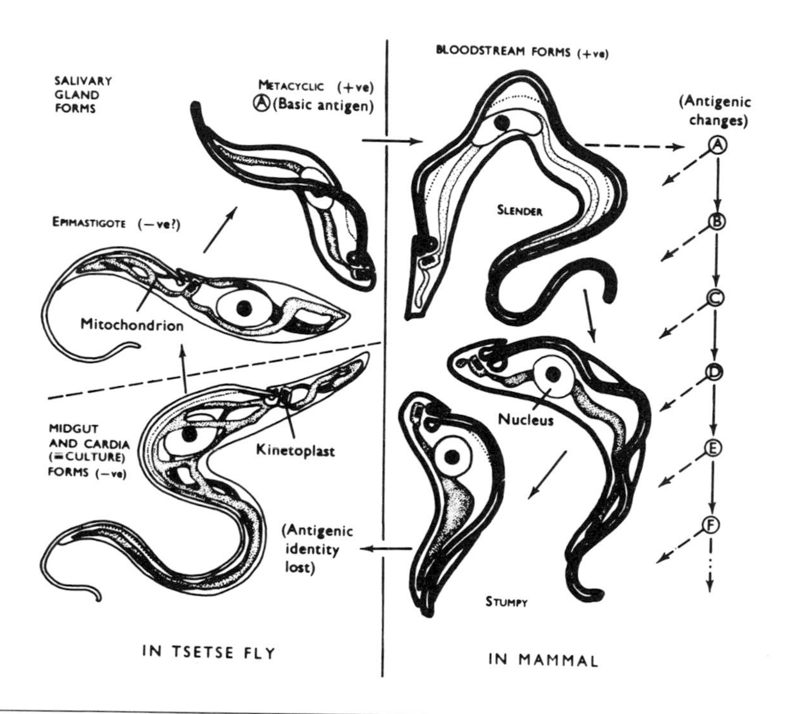

Fig. 2.4. Stages in the life cycle of *Trypanosoma brucei rhodesiense* showing cyclic changes in the surface coat and other features. Forms with a surface coat have a heavy outline. The surface charge is indicated by + or −. *A–F* refer to the sequence of antigenic variants in successive relapse populations in the blood, within each of which is a transformation from slender to stumpy trypomastigotes. When the organisms enter the tsetse fly they lose their antigenic identity; when they become metacyclic forms in the salivary glands they reacquire the surface coat and revert to the basic antigen A. (From Vickerman 1969a)

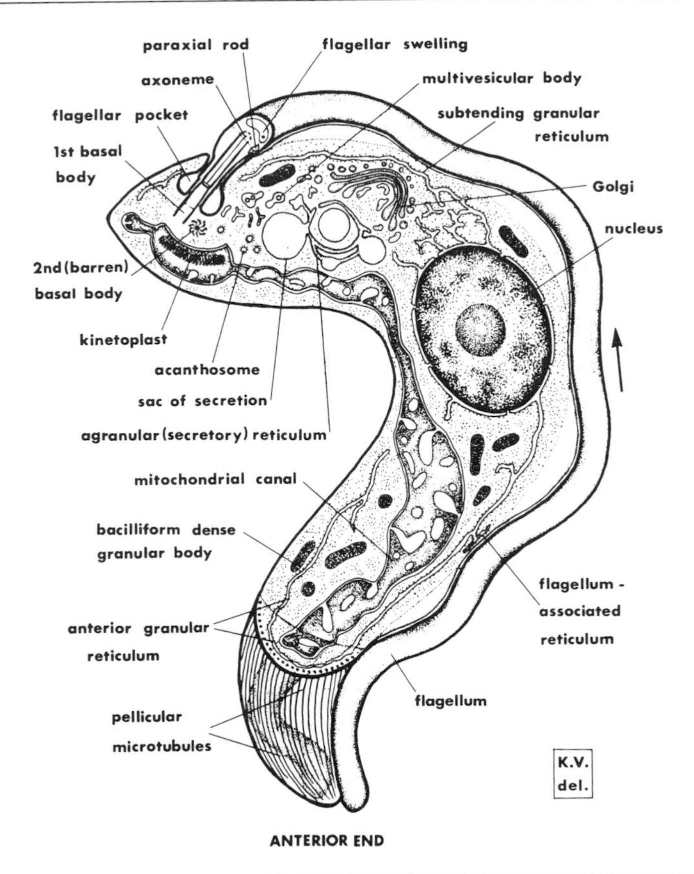

Fig. 2.5. Fine structure of *Trypanosoma congolense*, trypomastigote (bloodstream) phase, sagittal section except for anterior end and most of flagellum shaft. Subpellicular microtubules that lie beneath the whole body surface are shown only at the anterior end. The arrow indicates the usual direction of flagellar wave travel. (From Vickerman 1969b)

The culture forms also lack this coating. In the nonpathogenic *T. lewisi* the coat is filamentous and more diffuse; again, it is present only in the bloodstream and not in the culture forms. See D'Alesandro (1970), Desowitz (1970), Seed (1974), Gray and Luckins (1976), Terry (1976), Afchain et al. (1977), Doyle (1977), Murray and Urquhart (1977), Herbert and Parratt (1979), Parratt and Herbert (1979), Agabian et al. (1980), Borst et al. (1980), Cross (1980), Cross et al. (1980a,b), Holmes (1980), Simpson et al. (1980), Williams et al. (1980).

So far as vaccination is concerned, Holmes (1980) said, "The general prospects for the development of an effective vaccine remain bleak but not entirely hopeless." Cross et al. (1980a,b) echoed this view. Nevertheless, the attempt is being made. Perhaps it will eventually turn out to be successful.

PRESERVATION. Trypanosomes, like many flagellates and certain other protozoa, can be preserved for months or years by slow freezing and storage at −196 C in the presence of about 7.5% glycerol or dimethylsulfoxide. This technique keeps the trypanosomes unchanged, thus stabilizing them.

CLASSIFICATION. It seems impossible to arrive at a classification that is agreeable to everyone. It appears to be a case of "the more we know, the less we know." We have recognized more and more overlaps as our knowledge has increased, and increased details have added to the confusion. Not only have subgenera been introduced, but subspecies as well. The full name of a strain of *T. brucei* derived from some ruminant and not infectious for man might now be *Trypanosoma* (*Trypanozoon*) *brucei brucei* Plimmer and Bradford 1899 SERENGETI/58/EATRO/1716 [Stabilate TREU 386] (in which SERENGETI is the locality of the primary isolation, 58 the year of isolation, EATRO the isolation laboratory code, 1716 the specific number of the isolation, and Stabilate TREU 386 the designation of the particular stabilate). It is permissible to shorten this name and speak familiarly of *T. brucei*, *T. gambiense*, etc.

A large number of species of *Trypanosoma* has been named; Levine (1972) included more than 195 in his tabulation, while Lom (1979) listed 149 from fish alone. At one time it was customary, and still is to some extent, to give different names to trypanosomes from different hosts. Many of these names are still valid; but as we have learned more about the host-parasite relations and epidemiology of the trypanosomes, many others have fallen into synonymy. No attempt is made here to list all the synonyms of each species.

The following outline classification of *Trypanosoma* species of veterinary and medical importance is based on many references. Isozymes have been used most recently to differentiate species and strains, but they are far from satisfactory (Kreutzer and Sousa 1981; Goldberg and Pereira 1983).

Trypanosomes of Mammals

SECTION STERCORARIA

Development in the vector in the posterior station (posterior part of the alimentary tract); transmission by contamination with feces (except for *T. rangeli*, which develops also in the anterior station and in which transmission can be either by fecal contamination or inoculative); free flagellum always present in trypomastigote form; kinetoplast large and not terminal; posterior end of body pointed; multiplication in the mammalian host discontinuous, typically taking place in the epimastigote or amastigote stages; trypanosomes typically not pathogenic.

Subgenus *Megatrypanum* Hoare, 1964

Large mammalian trypanosomes, with kinetoplast typically situated near the nucleus and far from the posterior end of the body; affinities with some corresponding parasites of amphibians and reptiles; known vectors hippoboscid or tabanid flies.

T. (M.) theileri Laveran, 1902, of cattle and antelopes
T. (M.) cervi Kingston and Morton, 1974, of elk, mule deer, and white-tailed deer
T. (M.) melophagium Flu, 1908, of sheep
T. (M.) ingens Bruce et al., 1909, of cattle and wild ruminants
T. (M.) minasense Chagas, 1909, of marmosets and New World monkeys
T. (M.) grayi Novy, 1906, of crocodiles

Subgenus *Herpetosoma* Doflein, 1901

Trypanosomes of medium size, with subterminal kinetoplast, lying at some distance from the pointed posterior end of the body; reproduction (in the mammalian host) in amastigote and/or epimastigote stages; most known vectors fleas.

Metabolism: Glucose consumption rate low; respiratory quotient high; cytochromes present; sensitivity to cyanide high; sensitivity to sulfhydryl reagents moderate; relative activity of glycerophosphate, succinate, and DPNH oxidase systems 1:1:1; acetic, succinic, and lactic acids produced aerobically; acetic, succinic, pyruvic, and lactic acids produced anaerobically; glycerol not produced either aerobically or anaerobically.

T. (H.) lewisi (Kent, 1880) Laveran and Mesnil, 1901, of *Rattus*
T. (H.) musculi Kendall, 1906, of *Musculus*
T. (H.) primatum Reichenow, 1928, of anthropoid apes
T. (H.) nabiasi Railliet, 1895, of *Oryctolagus*
T. (H.) sylvilagi Levine, 1973, of *Sylvilagus*
T. (H.) rangeli Tejera, 1920, of man, dogs, opossums, and monkeys

Subgenus *Schizotrypanum* Chagas, 1909

Relatively small, typically C-shaped trypanosomes; voluminous kinetoplast very near the short, pointed posterior end of the body; multiplication in mammalian host typically intracellular, primarily in amastigote stage, secondarily in epimastigote stage; homogeneous assemblage of structurally indistinguishable species; known vectors reduviid bugs.

T. (S.) cruzi Chagas, 1909, of man, dogs, cats, armadillos, opossums, etc.

SECTION SALIVARIA

Transmission inoculative (except *T. equiperdum*); free flagellum present or absent; kinetoplast terminal or subterminal; posterior end of body usually blunt; reproduction in mammalian host continuous, taking place in trypomastigote stage; trypanosomes pathogenic for certain mammals.

Subgenus *Duttonella* Chalmers, 1918

Monomorphic trypanosomes with a free flagellum; kinetoplast large and usually terminal; development in tsetse fly vector occurs only in proboscis. Vectors tsetse flies.

Metabolism: Glucose consumption rate high to very high; respiratory quotient high; cytochromes absent; insensitive to cyanide; moderate to high sensitivity to sulfhydryl reagents; relative activity of glycerophosphate, succinate, and DPNH oxidase systems 1:1:1; acetic, pyruvic, and lactic acids produced both aerobically and anaerobically; glycerol produced both aerobically and anaerobically.

T. (*D.*) *vivax vivax* Ziemann, 1905, of ruminants and equids
T. (*D.*) *v. viennei* Lavier, 1921, of cattle
T. (*D.*) *v. uniforme* Bruce et al., 1911, of cattle, sheep, goats, and antelopes

Subgenus *Nannomonas* Hoare, 1964

Small forms, usually without free flagellum; kinetoplast medium-sized, typically marginal; development in tsetse fly vector in midgut and proboscis.

Metabolism: Glucose consumption high; respiratory quotient high; cytochromes absent; sensitivity to cyanide and sulfhydryl reagents moderate; relative activity of glycerophosphate, succinate, and DPNH oxidase systems 1:1:1; acetic and succinic acids produced both aerobically and anaerobically; lactic acid not produced; glycerol produced both aerobically and anaerobically.

T. (*N.*) *congolerise congolense* Broden, 1904, of ruminants, equids, and other animals
T. (*N.*) *c. simiae* Bruce et al., 1912, of swine, monkey, and cattle

Subgenus *Trypanozoon* Lühe, 1906

Pleomorphic (slender, intermediate, and stumpy) forms, with or without free flagellum; kinetoplast small, subterminal (invisible with light microscope in *T. equinum*); development in tsetse fly vector in midgut and salivary glands; some forms transmitted mechanically by tabanid vectors or transmitted by contact.

Metabolism: Glucose consumption rate very high; respiratory quotient very low; cytochromes absent; not sensitive to cyanide; very sensitive to sulfhydryl reagents; relative activity of glycerophosphate, succinate, and DPNH oxidase systems 20:1:1; pyruvic acid produced both aerobically and anaerobically; acetic, succinic, and lactic acids not produced either aerobically or anaerobically; glycerol produced both aerobically and anaerobically.

T. (*T.*) *brucei brucei* Plimmer and Bradford, 1899, of all domestic and many wild game mammals
T. (*T.*) *b. gambiense* Dutton, 1902, of man
T. (*T.*) *b. rhodesiense* Stephens and Fantham, 1910, of man and wild ruminants and cattle
T. (*T.*) *b. evansi* (Steel, 1885) Balbiani, 1888, of camels, equids, bovids, dogs, etc.
T. (*T.*) *b. equiperdum* Doflein, 1901, of equids

Subgenus *Pycnomonas* Hoare, 1964

Stout, monomorphic forms with short free flagellum and small, subterminal kinetoplast; development in tsetse fly vector occurs in midgut and salivary glands.

T. (*P.*) *suis* Ochmann, 1905, of pigs

Trypanosomes of Birds

Subgenus *Trypanomorpha* Woodcock, 1906

Very pleomorphic forms, sometimes attaining great size; relatively easy to cultivate; cyclic development in biting Diptera; multiplication in promastigote and epimastigote stages; transmission to birds via the feces. (Whether all avian trypanosomes belong in this subgenus is open to question; they are placed here merely as a matter of convenience).

T. (*T.*) *avium* Danilewsky, 1885, of owls and other birds
T. (*T.*) *paddae* Laveran and Mesnil, 1904, of Java sparrow *Padda oryzivora* and other birds

Trypanosomes of Reptiles

These trypanosomes are so poorly known that it is not advisable to assign most of them to subgenera. *T. grayi* of the crocodile is in the subgenus *Megatrypanum*, as is *T. cecili* Lainson, 1977, of caymans.

Trypanosomes of Amphibia

Subgenus *Trypanosoma* Gruby, 1843

The type species of the genus *Trypanosoma* is *T. rotatorium* (Mayer, 1943) Gruby, 1843. The name of its subgenus must be the same as that of its genus, so that it is *Trypanosoma* (*Trypanosoma*) *rotatorium*. It is not known to what subgenus (or subgenera) the other trypanosomes of amphibia belong.

Trypanosomes of Fish

Subgenus *Haematomonas* Mitrofanov, 1983

Lom (1979) listed 149 named species of *Trypanosoma* in fish and divided them into pleomorphic and monomorphic species. However, he said that many of them were synonyms of others and that it would be premature to introduce further taxa at present (see also Becker 1977).

Specific Trypanosomes

For further information on trypanosomes, see Hoare (1972), Losos and Ikede (1972), Bowman and Flynn (1976), Marinkelle (1976a,b), Lumsden (1977), Mansfield (1977), Soltys and Woo (1977), Boreham (1979), Henson and Noel (1979), Lumsden and Ketteridge (1979), WHO (1979), and Baker (1982). The Centre for Overseas Pest Research (London) issues an abstract journal quarterly, *Tsetse and Trypanosomiasis Information.*

Trypanosoma (Trypanozoon) brucei brucei Plimmer and Bradford, 1899

SYNONYM. *T. pecaudi.*

DISEASE. Trypanosomosis, nagana.

HOSTS. Horse, mule, donkey, ox, zebu, sheep, goat, camel, pig, dog, and many wild game animals. Antelopes, the natural hosts of *T. b. brucei,* are reservoirs of infection for domestic animals.

The chimpanzee can be infected, but attempts to infect man have failed. The small laboratory animals can be infected, as can the turtle *Pseudemys scripta elegans* and the cayman *Caiman sclerops*.

The dog and cat can be infected by eating goats with subclinical infections.

LOCATION. Blood stream, lymph, cerebrospinal fluid.

GEOGRAPHIC DISTRIBUTION. Widely distributed in tropical Africa between 15°N and 25°S latitude, coinciding with the distribution of its vector, the tsetse fly.

PREVALENCE. *T. b. brucei* is one of the commonest and most important parasites of domestic animals in Africa.

STRUCTURE. This species is polymorphic, with slender, intermediate, and stumpy forms. The undulating membrane is conspicuous, the kinetoplast small and subterminal. Slender forms average 29 μm in length but range up to 42 μm; the posterior end is usually drawn out, tapering almost to a point, with the kinetoplast up to 4 μm from the posterior end, with a long, free flagellum. Stumpy forms are stout, 12–26 μm (mean 18) μm long; the posterior end broad, obtusely rounded, with the kinetoplast almost terminal and a free flagellum typically absent. Intermediate forms average 23 μm long and have a medium thick body with a blunt posterior end; a moderately long free flagellum is always present. A fourth form with a posterior nucleus often appears in laboratory animals.

LIFE CYCLE. When it is first introduced into the body, *T. b. brucei* multiplies in the blood and lymph by longitudinal binary fission in the trypomastigote stage, being particularly common in the lymph nodes. Later the trypanosomes pass through the so-called blood-brain barrier into the cerebrospinal fluid and multiply there and between the cells of the brain. Amastigote forms have also been reported in the heart muscle of infected monkeys and in the liver and spleen of infected mice.

The vector is a tsetse fly (genus *Glossina*). *T. b. brucei* is generally transmitted by members of the *morsitans* group of this genus (i.e., *G. morsitans, G. swynnertoni*, and *G. pallidipes*). Both males and females feed on blood and are vectors. Only a small percentage of the tsetse flies that feed on an infected animal become infected, most apparently being resistant.

When ingested by a tsetse fly, *T. b. brucei* localizes in the posterior part of the midgut and multiplies in the trypomastigote form for about 10 days. At first the trypanosomes are relatively broad. On the 10th and 12th days, slender forms appear and migrate slowly toward the proventriculus, where they are found on days 12–20. They then migrate forward into the esophagus and pharynx, thence into the hypopharynx, and finally into the salivary glands. Here they attach themselves to the walls by their flagella or lie free in the lumen, and turn into the epimastigote form. These multiply further and then transform into the metacyclic trypomastigote form, which is small and stumpy and may or may not have a short free flagellum. These are the infective forms. They are injected into the blood with the saliva when the fly bites. The whole life cycle in the tsetse fly takes 15–35 days.

This type of development, in which the trypanosomes are found in the anterior part of the vector and are introduced by its bite, is known as development in the anterior station. It occurs in the Salivaria, in contrast to development in the posterior station or hindgut, which occurs in the Stercoraria and in which infection is by contamination with feces.

PATHOGENESIS. The signs and effects of the trypanosomes in domestic animals are more or less similar. Different hosts are affected to different degrees. Horses, mules, and donkeys are very susceptible. Affected animals have a remittent fever; edematous swelling of the lower abdomen, genitalia, and legs; a watery discharge from the eyes and nose; and anemia. They become emaciated although their appetite is good. Muscular atrophy sets in, and eventually incoordination and lumbar paralysis develop, followed by death. The course of the disease is 15 days to 4 mo, and untreated domestic equids rarely recover.

The disease in sheep, goats, camels, and dogs is also severe. The signs are much the same as in horses. In the dog, fever may appear as soon as 5 days after inoculation (DAI), and the parasites often cause conjunctivitis, keratitis, and blindness.

The disease is usually more chronic in cattle. There is remittent fever with swelling of the brisket, anemia, gradual emaciation, and discharge from the eyes and nose. The animals may survive for several months, and many recover. *T. b. brucei* is not as pathogenic for cattle as *T. congolense*. Swine are more resistant than cattle and usually recover.

The exact way in which trypanosomes act to kill their victims is unknown, although several theories have been advanced. It is known that they have a high glucose metabolism, so one theory was that they rob the body of glucose so that death is due to hypoglycemia. Since the serum potassium level increases in trypanosomosis, another theory was that the effects are due to the high potassium level. However, the latter is a result of the disease and not a cause; it is due to the destruction of red cells with consequent release of potassium into the plasma, and the observed levels are not too harmful.

EPIDEMIOLOGY. The epidemiology of the diseases caused by *T. b. brucei* and other tsetse-borne trypanosomes depends on the bionomics and distribution of their vectors. This is such a vast subject that no attempt will be made to cover it here. In general, tsetse flies occupy about 4.5 million mi^2 of Africa. They occur in woodlands, bush, or forested areas with sufficient rainfall and a mean annual temperature above 20 C. Not all species are good vectors, and trypanosomosis does not occur every place that tsetse flies do. See also Buxton (1948, 1955a,b), Hornby (1949, 1952), Davey (1958), Ashcroft (1959), Nash (1960), Morris (1960), Willet (1963), Glasgow (1963), van Riel (1964), and WHO (1979).

DIAGNOSIS. In the acute or early stage of the disease, trypanosomes can be found in the peripheral blood. Thick blood smears are preferable to thin ones. The protozoa are found even more often in the lymph nodes. They can be detected in fresh or stained smears of fluid obtained by puncture of the nodes. In the later stages of the disease, trypanosomes can be found in the cerebrospinal fluid. Laboratory animals such as the rat can also be inoculated. Microscopic examination by dark-field illumination of the buffy layer of separated blood cells is valuable. Various serologic tests are also used but are not as reliable as finding the trypanosomes themselves.

CULTIVATION. Trypanosomes can be successfully cultivated in a number of media. A common one is NNN medium, which is essentially a 25% blood agar slant. See Cunningham and Honigberg (1977), Hirumi et al. (1977), Evans (1978), Hill et al. (1978), Cunningham and Taylor (1979), Nyindo et al. (1979), and Gardiner et al. (1980).

Trypanosomes can also be cultivated in developing chick embryos or in tissue culture.

TREATMENT. Many different drugs have been used in the treatment of trypanosomosis. Indeed, the first synthetic organic compound of known composition ever used to cure an experimental disease was trypan red, which was developed by Ehrlich and Shiga (1904). Since that time thousands of drugs have been found to have some activity, but the number of satisfactory ones is very small. The number of papers on the subject is great; see Williamson (1962) for a review with 722 references.

To cure *T. b. brucei* infections in horses and dogs, the World Health Organization (WHO 1979) recommended quinapyramine (Antrycide). It is administered subcutaneously as the sulfate in a 10% solution in cold water; the dose is 5 mg per kg body weight. It said that diminazene aceturate (Berenil) and isometamidium chloride (Samorin, Trypamidium) could possibly be used. Diminazene aceturate is administered subcutaneously or intramuscularly in a 7% solution in cold water; the dose is 3.5 mg per kg. Isometamidium chloride is administered deep intramuscularly in a 1–2% solution in cold water; the dose is 0.25–1.0 mg per kg. Quinapyramine sulfate is well tolerated by cattle, sheep, goats, and camels but may possibly cause local reactions in horses and general reactions in dogs. Diminazene aceturate is well tolerated by cattle, sheep, and goats but may possibly cause local reactions in horses and general reactions in horses, camels, and dogs. Isometamidium chloride is well tolerated by cattle, sheep, goats, horses, and dogs but may possibly cause local reactions in cattle.

For prophylactic use the World Health Organization (WHO 1979) recommended quinapyramine prosalt (quinapyramine chloride + sulfate; Antrycide prosalt) for horses; it is administered subcutaneously as a solution of 3.5 g in 15 ml cold water at the rate of 7.4 mg per kg. It is said to be well tolerated by horses, camels, and cattle and possibly to cause local reactions in horses.

CONTROL. Preventive measures against trypanosomosis include measures directed against the intermediate hosts, livestock management, elimination of reservoir hosts, avoidance of accidental mechanical contamination, land use management, and biological control.

Measures directed against parasites include continuous survey and treatment or slaughter of all affected animals and periodic mass prophylactic treatment of all animals. Fly traps and fly repellents have been used without much success in attempting to control tsetse flies.

Elimination of breeding places has been practiced on a wide scale in many areas. Since the tsetse flies breed under brush along streams or in other brushy localities, such measures consist essentially of brush removal. Two methods have been used. (1) *Eradicative clearing* aims at eradication of tsetse flies throughout an area. All the species of trees and shrubs under which the flies survive through the dry season are removed. When this is done thoroughly over a large area, the flies disappear completely. (2) *Protective clearing* is more limited. It is designed to break the contacts between tsetse flies and domestic animals and man at the places where transmission is taking place. Fly-free belts are established. In addition, inspection stations known as deflying houses may be set up on traffic routes to remove any flies carried on vehicles or animals. Clearing can be quite successful. However, it is expensive, requires a large amount of labor, and the initial effort must be followed up faithfully as new growth occurs. The cost of labor is now so high as to be prohibitive for use on a large scale.

Biological control has been suggested. The use of predators on, and parasites of, tsetse flies and of genetic methods has been suggested, but these are still in the laboratory stage. Release of sterile males to control *G. palpalis* is encouraging and may be used in the near future (WHO 1979).

Ground spraying with DDT or dieldrin is still the most favored method of tsetse control in many areas (WHO 1979). (See Folkers 1965.)

Airplane spraying has given good results in some areas. *G. pallidipes* was eradicated from Zululand by spraying with DDT and benzene hexachloride at a total cost of 2.5 million pounds, or slightly less than 2 shillings per acre (DuToit 1959). *G. morsitans* was controlled in Botswana and Zambia by applying aerosols of endosulfan, but various organophosphorus compounds and some pyrethroids were not satisfactory (WHO 1979). Ultralow volume (ULV) spraying with endosulfan is used to control *G. palpalis, G. tachinoides*, and *G. morsitans* in Nigeria (WHO 1979).

Cattle can be sprayed with DDT or another insecticide in order to kill any tsetse flies that light on them. Since tsetse flies bite only in the daytime, African farmers have adopted the practice of grazing cattle at night. The animals are held in a protected corral during the day. The elimination of reservoir hosts, such as wild game in Africa, has been advocated and practiced in the past in some regions despite the protests of many people interested in game preservation. The latter appear to have won, and game eradication is no longer practiced. Game cropping has also been suggested (Lambrecht 1966).

Land use management has raised high hopes. At present, it seems to center primarily around reclamation projects and the need "to accomodate a rapidly expanding human population aspiring to a higher standard of living," i.e., to clear the land to develop housing for the growing population (WHO 1979).

Since trypanosomes can be transmitted mechanically by inoculation of infected blood or lymph, there is danger of transmission by the use of contaminated instruments in bleeding, castration, etc.

A great deal has been written on trypanosomosis control. For further information see Willet (1963, 1972, et seq.), MacLennan (1975), WHO (1979), and the proceedings of the meetings of the International Scientific Committee for Trypanosomiasis.

Trypanosoma (Trypanozoon) brucei gambiense Dutton, 1902, and *Trypanosoma (Trypanozoon) brucei rhodesiense* Stephens and Fantham, 1910

HOSTS. These two subspecies cause African sleeping sickness in man and are also transmissible to the chimpanzee and various monkeys. Reservoirs are cattle, certain wild antelopes, hyenas, lions, and possibly sheep. All the usual small laboratory animals are susceptible to infection with both subspecies. For further discussion of these subspecies, see any textbook of human parasitology. There are now only about 10,000 new cases a year in all of Africa; in earlier years there were hundreds of thousands. See Figure 2.4 for their life cycles.

The problem of distinguishing the strains of *T. brucei* that can infect humans (*T. b. gambiense* and *T. b. rhodesiense*) from those that cannot (*T. b. brucei*) is a difficult one. Rickman and Robson (1970) developed the blood incubation infectivity test (BIIT) for this purpose. It is based on the fact that normal human blood destroys the infectivity for rats (or mice) of *T. b. brucei* but not of *T. b. gambiense* or *T. b. rhodesiense*. A suspect trypanosome strain is incubated with human blood or plasma and then injected into rats (or mice) to determine its subspecies. However, this test is often not successful. There are intermediate strains between the strongly BIIT-negative and BIIT-positive ones.

Trypanosoma (Trypanozoon) brucei evansi (Steel, 1885) Balbiani, 1888

SYNONYMS. *T. aegyptium, T. annamense, T. cameli, T. elephantis, T. equinum, T. hippicum, T. marocanum, T. ninae kohl-yakimov, T. soudanense, T. venezuelense*.

HOSTS. Camel, horse, donkey, ox, zebu, goat, pig, dog, water buffalo, elephant, capybara, tapir, mongoose, ocelot, deer, and other wild animals. Many laboratory and wild animals can be infected experimentally.

LOCATION. Blood, lymph.

DISEASE. Trypanosomosis due to *T. b. evansi* has been given different names in different localities. The most widely used name, surra, is applied to the disease in all hosts. The disease in camels is called el debab in Algeria, mbori in the Sudan, and guifar or dioufar in Chad. In horses it is called murrina in Panama or derrengadera in Venezuela. *T. "equinum,"* now recognized as simply a dyskinetoplastic strain of *T. b. evansi*, causes a disease of horses in South America knonw as mal de Caderas. See Shaw (1977) and Woo (1977) for recent reviews.

GEOGRAPHIC DISTRIBUTION. It is found in northern Africa (north of 15° N latitude in the west and central part of the continent but extending almost to the equator in the east), Asia Minor, the USSR from the Volga River east into Middle Asia, India, Burma, Malaysia, "Indochina," parts of southern China, Indonesia, the Philippines, Central America, and South America. The original distribution of *T. b. evansi* coincided with that of the camel (Hoare 1956). In Africa its southern boundary coincides roughly with the northern boundary of tsetse fly distribution. It now extends far to the east of the camel's range in the Eastern Hemisphere. Often associated with arid deserts and semiarid steppes, it may occur in other types of climate as well. Horseflies are found in the endemic areas and are absent in the disease-free areas.

PREVALENCE. *T. b. evansi* is an important cause of disease over a large part of its range. It is especially common in northern Africa, Asia, and northern South America.

STRUCTURE. The mean length of different host and geographic strains varies considerably. However, the typical forms are 15–34 μm long, with a mean of 24 μm. Most are slender or intermediate in shape, but stumpy forms occur sporadically. All forms are structurally indistinguishable from the corresponding ones of *T. b. brucei*. Strains that lack a kinetoplast visible with the light microscope have occasionally arisen spontaneously or can be produced by treatment with certain dyes, drugs such as diminazene (Berenil), or even by frozen storage. See Vickerman (1977) for a review.

LIFE CYCLE. *T. b. evansi* is transmitted mechanically by biting flies. No cyclic development takes place in the vectors, the trypanosomes remaining in the proboscis. The usual vectors are horseflies of the genus *Tabanus*, but *Stomoxys, Haematopota,* and *Lyperosia* can also transmit it. In Central and South America, the vampire bat is a vector and the disease is known as murrina.

PATHOGENESIS. Surra is nearly always fatal in horses in the absence of treatment; death occurs in 1 wk to 6 mo. The disease is also severe in elephants, but less so in cattle and water buffalo. Surra in camels is similar to the disease in horses but more chronic. In dogs, *T. b. evansi* causes a chronic disease with a high mortality rate; untreated dogs usually die in 1–2 mo.

The signs of surra include intermittent fever, urticaria, anemia, edema of the legs and lower parts of the body, loss of hair, progressive weakness, loss of condition, and inappetence. Conjunctivitis may occur, and abortion is common in camels.

The lesions include splenomegaly, enlargement of the lymph nodes and kidneys, leukocytic infiltration of the liver parenchyma, and petechial hemorrhages and parenchymatous inflammation of the kidneys.

DIAGNOSIS. Same as for *T. b. brucei.*

CULTIVATION. Same as for *T. b. brucei.*

TREATMENT. Treatment of *T. b. evansi* is similar to that of *T. b. brucei.* The World Health Organization/Food and Agriculture Organization (WHO 1979) recommends quinapyramine for curative treatment in both horses and camels. A single subcutaneous dose of 5 mg per kg or even less is effective. A dose of 3 mg per kg has given good results in cattle. A single injection of 2 g is effective in camels. For prophylactic use in both horses and camels, the WHO-FAO recommends quinapyramine prosalt or suramin and quinapyramine. The dose of suramin for horses is 4 g per 1,000 lb body weight intravenously. Camels tolerate this drug well, and a single intravenous injection of 4–5 g is effective in these animals.

Tartar emetic has been superseded in other animals but a single intravenous injection of 200 ml of a 1% solution is still used in camels. This drug is also widely used in cattle in India because it is cheap. I do not recommend it.

CONTROL. The same measures used against *T. b. brucei* can be used against *T. b. evansi.*

Trypanosoma (Trypanozoon) brucei equiperdum Doflein, 1901

This species is structurally indistinguishable from *T. b. brucei*. Dyskinetoplastic strains can be produced in the laboratory.

DISEASE AND HOSTS. *T. b. equiperdum* causes a disease of horses and asses known as dourine; it can also infect laboratory and other animals. Dourine is a venereal disease, transmitted by coitus. It is similar to nagana but runs a more chronic course of 6 mo to 2 yr. The incubation period is 2–12 wk. The first sign of the disease is edema of the genitalia and often of the dependent parts of the body. There is slight fever, inappetence, and a mucous discharge from the urethra and vagina. Circumscribed areas of the mucosa of the vulva or penis may become depigmented.

The second stage of the disease, characterized by urticaria, appears after 4–6 wk. Circular, sharply circumscribed urticarial plaques about 3 cm in diameter arise on the sides of the body, remain 3–4 days, and then disappear. They may reappear later. Muscular paralysis later ensues. The muscles of the nostrils and neck are affected first; the paralysis then spreads to the hind limbs and finally to the rest of the body. Incoordination is seen first, followed by complete paralysis. Dourine is usually fatal unless treated, although mild strains of the parasite may occur in some regions.

GEOGRAPHIC DISTRIBUTION. *T. b. equiperdum* is found in parts of Asia, South Africa, India, and the USSR. It was once common in western Europe and North America but has been eradicated from these regions. The last place where it was known to occur north of Mexico, the Papago Indian Reservation in Arizona, was released from quarantine in 1949. A serologic reactor found in California in 1962, a horse that had originated in Idaho, was killed; no other reacting horses have been found in the United States since.

DIAGNOSIS. Dourine can be diagnosed by finding the parasites in smears of fluid expressed from the urticarial swelling, lymph, and mucous membranes of the genitalia, or the blood. The signs of the typical disease are characteristic enough to permit diagnosis in endemic areas.

The complement fixation test is invaluable in detecting early or latent infections, and it was by its use that dourine was eradicated from North America. All horses imported into the United States must be tested for dourine before they are admitted.

TREATMENT. To treat dourine in horses, a single subcutaneous dose of 5 mg per kg quinapyramine or two intravenous injections of 2 g suramin each, 15 days apart, can be used.

Trypanosoma (Pycnomonas) suis Ochmann, 1905

This species is poorly known. It differs from *T. brucei* in being monomorphic and in having only stout forms 14–19 μm long, with a short free flagellum. The kinetoplast is very small and marginal.

T. (P.) suis occurs in pigs in Africa, causing a chronic infection in adults and a more acute disease with death in less than 2 mo in young pigs. It cannot be transmitted to the goat, sheep, cat, dog, white rat, guinea pig, rabbit, chimpanzee, monkey, ass, or to cattle. Its vector is the tsetse fly *Glossina brevipalpis*, in which it develops first in the intestine and proventriculus and then in the salivary glands. Metacyclic, infective trypanosomes appear in the hypopharynx on the 28th day.

Trypanosoma (Nannomonas) congolense congolense Broden, 1904

SYNONYMS. *T. cellii, T. confusum, T. dimorphon, T. frobeniusi, T. montgomeryi, T. nanum, T. pecorum, T. ruandae, T. somaliense.*

DISEASE. The African disease of cattle known as nagana is ordinarily caused by *T. congolense*, often in mixed infection with *T. brucei* and/or *T. vivax*. Other names given the disease are paranagana, Gambia fever, ghindi, and gobial.

HOSTS. Cattle, equids, sheep, goats, camels, dogs, and swine may be infected. Some breeds of cattle in Africa are trypanotolerant and can be raised in humid tropical areas where nontolerant cattle cannot be raised. Antelopes, giraffes, zebras, elephants, and warthogs are also infected and act as reservoirs. The mouse and other laboratory animals can be infected.

LOCATION. This species develops almost exclusively in the blood. It does not invade the lymph or central nervous system.

GEOGRAPHIC DISTRIBUTION. Widely distributed in tropical Africa between 15° N and 25° S latitude, coinciding with the distribution of the tsetse flies that are its vectors.

PREVALENCE. *T. c. congolense* is the commonest and most important trypanosome of cattle in tropical Africa. The WHO (1979) published tables of the trypanosomes in livestock and the tsetse situation in each of the African countries in 1969–1975.

STRUCTURE. *T. c. congolense* is small, 8–20 μm long; the mean lengths of different populations range from 11–16 μm. It lacks a free flagellum or has a short one and has an inconspicuous undulating membrane and a medium-sized kinetoplast that lies some distance from the posterior end and is typically marginal (Fig. 2.5).

LIFE CYCLE. The vectors of *T. c. congolense* are various species of *Glossina*, including *G. morsitans, G. palpalis, G. longipalpis, G. pallidipes,* and *G. austeni.* After the trypanosomes have been ingested by the tsetse flies, they develop in the midgut as long trypomastigotes without a free flagellum. They then migrate to the proventriculus and thence to the proboscis, where they assume the epismastigote form without a free flagellum. They are attached at first to the wall of the proboscis and multiply there for a time. Later they pass into the hypopharynx, where they turn into metacyclic, infective trypomastigotes similar in appearance to the blood forms. They are injected into the bloodstream when the fly bites. Development to the infective stage in *Glossina* takes from 15 to well over 20 days at 23–34 C.

T. c. congolense can also be transmitted mechanically by other biting flies in tsetse-free areas, but this is apparently rather uncommon.

PATHOGENESIS. There are many strains of *T. congolense*, which differ markedly in virulence and also in antigenic properties. In cattle the parasite may cause an acute, fatal disease resulting in death in about 10 wk; a chronic condition with recovery in about 1 yr; or a mild, almost asymptomatic condition. The disease is similar in sheep, goats, camels, and horses. Swine are more resistant.

The signs of trypanosomosis due to this species are similar to those caused by other trypanosomes, except that the central nervous system is not affected.

The lymph nodes are edematous, the liver is congested, the marrow of the long bones is largely destroyed, and there are hemorrhages in the heart muscle and renal medulla. In cattle treated with quinapyramine or dimidium the lesions are more chronic: the spleen enlarged; the liver swollen and sometimes fibrous; the lymph nodes hypertrophied, edematous, and somewhat fibrotic; the kidneys showing chronic degeneration; and the marrow of the long bones largely destroyed.

EPIDEMIOLOGY. The epidemiology of *T. c. congolense* infections depends on the ecology of its vector tsetse flies.

DIAGNOSIS. This disease can be diagnosed by detection of the parasites in blood smears. Repeated examinations may be necessary in chronic cases. Inoculation of rats or guinea pigs may give positive results when blood examinations are negative.

CULTIVATION. Much the same as for *T. b. brucei.*

TREATMENT. For cure of *T. c. congolense* infections in ruminants the World Health Organization/Food and Agriculture Organization (WHO 1979) recommended diminazene or isometamidium; in horses and dogs it recommended isometamidium. For prophylaxis it recommended isometamidium or pyrithidium in ruminants and isometamidium in horses and dogs. For curative treatment 3.5 mg per kg diminazene aceturate (Berenil) is given as a 7% solution in cold water either subcutaneously or intramuscularly; it is well tolerated by cattle, sheep, and goats. Isometadinium chloride (Samorin, Trypamidium) is given as a 1-2% solution in cold water at the rate of 0.25-1.0 mg per kg deep intramuscularly; it is well tolerated by cattle, sheep, goats, dogs, and horses. For prophylactic treatment 0.5-1.0 mg per kg isometamidium is given as a 1-2% solution deep intramuscularly; 2 mg per kg pyrithidium bromide (Prothidium) is given as a 2% solution deep intramuscularly; it is well tolerated by cattle, sheep, goats, and dogs, although there are possible local reactions in cattle and horses. Protection with both of these drugs lasts 2-4 mo.

Whenever a drug is used continuously for prophylaxis, there is danger that drug-fast strains of parasites may appear because the blood level becomes so low that relatively resistant individuals can survive. This is one reason that quinapyramine is no longer recommended.

CONTROL. Same as for *T. b. brucei.*

Trypanosoma (Nannomonas) congolense simiae Bruce et al., 1912

SYNONYMS. *T. ignotum, T. porci, T. rodhaini.*

HOSTS. *T. c. simiae* was first found in a monkey, but its natural reservoir host is the warthog *Phacophoerus aethiopicus.* It is highly pathogenic for the pig and camel, causing a peracute disease with death usually in a few days. This is the most important trypanosome of domestic swine.

T. c. simiae is only slightly pathogenic for sheep and goats, and apparently nonpathogenic for cattle, horses, or dogs, although it may infect them. The rabbit appears to be the only susceptible laboratory animal. There is a great deal of variation in pathogenicity between strains; indeed, marked changes can occur in the pathogenicity of a single strain.

GEOGRAPHIC DISTRIBUTION. *T. c. simiae* occurs mostly in East Africa and eastern Zaire, but it has also been found in other parts of sub-Saharan Africa.

STRUCTURE. It differs structurally from *T. c. congolense* in being polymorphic instead of monomorphic. It is 12–24 μm long. About 90% of its forms are long and stout with a conspicuous undulating membrane, about 7% are long and slender with an inconspicuous undulating membrane, and about 3% are short with an inconspicuous undulating membrane. A free flagellum is usually absent.

LIFE CYCLE. This species is transmitted in the warthog reservoir by tsetse flies, including *G. palpalis, G. morsitans, G. austeni,* and *G. brevipalpis.* It develops in the midgut and proboscis. Tsetse flies also transmit it to swine, but once it has been introduced into a herd it can apparently be transmitted mechanically by horseflies and other bloodsucking flies.

TREATMENT. *T. c. simiae* is more resistant to drugs than are other African trypanosomes. The World Health Organization/Food and Agriculture Organization (WHO 1979) did not recommend any drug for its cure but suggested that isometamidium in high dose or quinapyramine plus diminazene might be used. For its prophylaxis it recommended 40 mg per kg of quinapyramine in 5% solution in cold water or a suramin-quinapyramine complex subcutaneously; this is well tolerated and protects young pigs for 3 mo and adults for 6 mo.

CONTROL. Control measures are the same as for *T. b. brucei.* In addition, horseflies and other biting flies should be controlled.

Trypanosoma (Duttonella) vivax vivax Ziemann, 1905

SYNONYMS. *T. angolense, T. bovis, T. caprae, T. cazalboui.*

DISEASE. *T. v. vivax* causes souma. It is also found in mixed infections of cattle with *T. c. congolense* and *T. b. brucei.*

HOSTS. Cattle, sheep, goats, camels, and horses are hosts; antelopes and the giraffe are reservoir hosts. The pig, dog, and monkey are refractory to infection. The small laboratory rodents and rabbits are difficult to infect.

LOCATION. It is found in the blood and lymph. Central nervous system infections have been described in sheep.

GEOGRAPHIC DISTRIBUTION. *T. v. vivax* is found throughout the tsetse fly belt in Africa and has, as well, spread to other parts of Africa. See the distribution of *T. c. congolense* for further information.

STRUCTURE. *T. v. vivax* is 20–27 μm long and monomorphic. The posterior end is typically rounded, a free flagellum always present, the kinetoplast large and terminal, and the undulating membrane inconspicuous.

LIFE CYCLE. The most important vectors of *T. v. vivax* are tsetse flies, including *Glossina morsitans, G. tachinoides,* and other species. Development takes place only in the proboscis. The trypanosomes turn into the epimastigote form, multiply in this form, and then turn into metacyclic, infective trypanosomes that pass to the hypopharynx and infect new hosts when the tsetse flies bite. The flies become infective as early as 6 days after they themselves have been infected.

PATHOGENESIS. *T. v. vivax* is most important in cattle, in which the disease is similar to that caused by *T. c. congolense*. The infections in East Africa are usually mild, but in West Africa they are usually fatal in some types of cattle, including zebus.

Camels are less seriously affected than cattle. *T. v. vivax* is apparently more pathogenic for sheep than other trypanosomes and may be found in the central nervous system. It is apparently less pathogenic for goats. It causes a chronic disease, often with spontaneous recovery, in horses. It is not pathogenic for dogs, pigs, and monkeys, and only slightly so for the common laboratory rodents and rabbits.

The signs of disease are similar to those caused by *T. c. congolense*. There is a wide variation in virulence between different strains.

DIAGNOSIS. *T. v. vivax* is detected most readily in lymph node smears. Large numbers are found in the blood only in early infections. Inoculation of laboratory animals is not particularly satisfactory; inoculation of sheep or goats is better, the trypanosomes appearing in 7–10 days.

CULTIVATION. Methods and results are the same as for other trypanosomes.

TREATMENT. The World Health Organization/Food and Agriculture Organization (WHO 1979) recommended the same drugs in the same dosages for *T. v. vivax* as for *T. c. congolense.*

CONTROL. Control measures are the same as those for *T. b. brucei* infections. In areas where tabanids are the vectors, measures directed against these flies should be practiced.

Trypanosoma (Duttonella) vivax viennei Lavier, 1921

SYNONYM. *T. guyanense*.

This subspecies occurs in cattle, horses, etc. in northern South America, Central America, the West Indies, and Mauritius. The deer *Odocoileus gymnotis* is a reservoir in Venezuela. The pig, dog, and monkey are refractory to infection, and the small laboratory rodents are difficult to infect.

This subspecies, undoubtedly derived from *T. v. vivax,* occurs in the New World. Everything that has been said about *T. v. vivax* applies to it except that it is not transmitted by tsetse flies but mechanically by horseflies and other tabanid flies.

Trypanosoma (Duttonella) vivax uniforme Bruce et al., 1911

This species is similar to *T. v. vivax*, but smaller. It is 12–20 μm long, with a mean of about 16 μm. It occurs in cattle, sheep, goats, and antelopes, causing a disease similar to that caused by *T. v. vivax*. Laboratory rodents are refractory to infection. *T. v. uniforme* occurs only in Uganda and Zaire. It is transmitted by tsetse flies in the same way as *T. v. vivax*.

Trypanosoma (Schizotrypanum) cruzi cruzi Chagas, 1909

SYNONYMS. *Schizotrypanum cruzi, T. lesourdi, T. prowazeki, T. rhesii, T. vickersae.*

DISEASE. American human trypanosomosis, Chagas' disease.

HOSTS. Man is the most important host of *T. cruzi*. It is estimated that more than 10 million people are infected, including 4–6 million in Brazil, 2.3 million in Argentina, 2 million in Colombia, and 0.6 million in Venezuela. Many species of wild and domestic animals have been found infected (see Epidemiology, below), and are reservoirs of infection for man. Goble (1970) listed 79 species of natural hosts and 20 of experimental ones occurring in the mammalian orders Marsupialia, Carnivora, Rodentia, Lagomorpha, Edentata, Primata, and Artiodactyla. Miles (1979) said that, apart from the *T. cruzi*-like organisms that are cosmopolitan in bats, well over 100 species of mammals of eight orders are known to harbor *Schizotrypanum* in the Western Hemisphere.

T. c. cruzi was found in nine fatal cases in dogs in Texas (Williams et al. 1977) and has also been found in dogs in Louisiana (Snyder et al. 1980) and even Indiana (Anon. 1979). This last case may have been due to introduction of infected raccoons into the area as part of a wildlife replenishment scheme of local hunting clubs.

LOCATION. The trypanosomes are found in the blood early in an infection. Later they invade the cells of the reticuloendothelial system, heart, striated muscles, and other tissues. In central nervous system infections, they are found in the neuroglial cells. Trypomastigote forms occur in the blood; amastigote, epimastigote, and trypomastigote forms are found within the cells.

GEOGRAPHIC DISTRIBUTION. *T. c. cruzi* occurs in South America from Argentina north, the Antilles, Central America, and the southern United States. See Dias (1953), and Cançado (1968).

T. c. cruzi occurs especially in woodrats in the southwestern states (Texas, Arizona, New Mexico, southern California); it also occurs in raccoons, opossum, skunks, and gray foxes in the southeastern states.

STRUCTURE. The forms in the blood are monomorphic, 16–20 μm long, with a pointed posterior end and a curved, stumpy body. The kinetoplast is subterminal and larger than that of any other trypanosome of domestic animals or

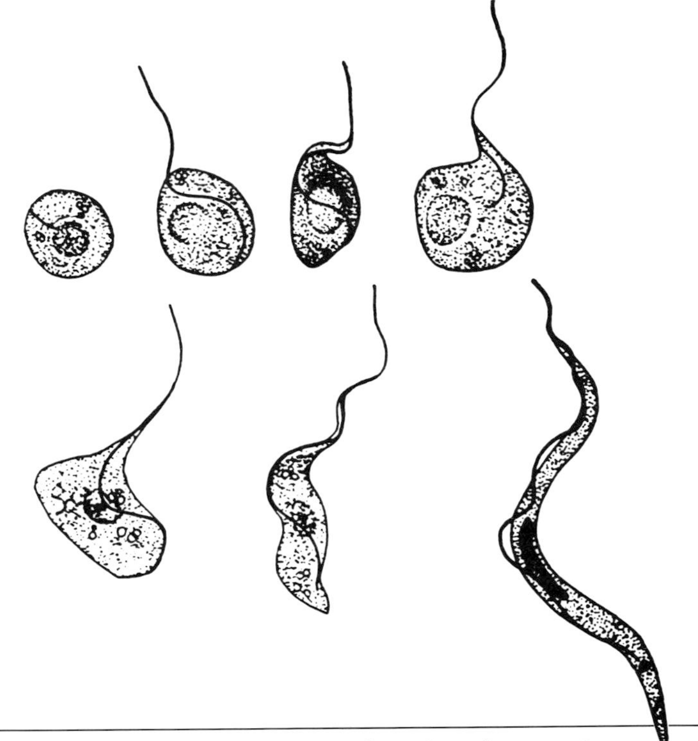

Fig. 2.6. Successive stages in the transformation of an amastigote form of *Trypanosoma cruzi* into a metacyclic trypomastigote form. The metacyclic trypanosome (trypomastigote) (*lower right*) is from a Giemsa-stained smear; the other stages are from living preparations of culture material viewed with the phase microscope. (From Noble 1955)

man, often causing the body to bulge around it. The undulating membrane is narrow with only two or three undulations. There is a moderately long free flagellum. The amastigote forms in the muscle, heart, and other tissue cells are 1.5–4.0 μm in diameter and occur in groups. The late epimastigotes and trypomastigotes in the heart and skeletal muscle have a fuzzy layer of carbohydrates on the outside, while the amastigote and early epimastigote stages do not. (See Fig. 2.6.)

LIFE CYCLE. Although the trypomastigote form of *T. c. cruzi* is common in the blood in the early stages of Chagas' disease, it does not multiply in this form. The trypomastigote forms enter the cells of the reticuloendothelial system, striated muscles, and especially heart muscle, where they round up and turn into amastigote forms. These multiply by binary fission, destroying the host cells and forming nests of parasites. Some apparently turn into epimastigotes. The amastigote forms turn into trypomastigote forms, which reenter the blood. Different strains differ in virulence.

The vectors of *T. c. cruzi* are kissing (conenose) bugs, members of the hemipteran family Reduviidae, subfamily Triatominae. The parasites are also able to develop to the infective stage in bedbugs.

Miles (1979) listed 90 species of Triatominae in 12 genera. The most important vector in South America is probably *Panstrongylus* (syn., *Triatoma*) *megistus* or *Triatoma infestans*. Others are *P. megistus, Rhodnius prolixus, P. herreri, T. phyllosoma, T. barberi, R. pallescens, T. rubrovaria, T. spinolai,* and *T. dimidiata* (Carcavallo 1970; Cedillos 1975). In the United States *T. protracta, T. sanguisuga* (syn., *T. gerstaeckeri*), *T lectularius, T. longipes, T. neotomae,* and *T. rubida* have been found infected.

Both the nymphs and adults of these reduviids can be infected and can transmit the disease. In addition, it is possible to infect sheep keds, *Ornithodoros* ticks, and bedbugs experimentally.

After they have been ingested by the triatomins along with a blood meal, the trypanosomes pass to the midgut. Here they turn into amastigote forms, which multiply by binary fission and turn into either metacyclic trypomastigote or epimastigote forms. The epimastigote forms multiply further by binary fission and extend into the rectum. Here the epimastigotes turn into metacyclic trypomastigotes, which pass out in the feces. The life cycle in the invertebrate host takes 6–15 days or longer, depending on the insect species or stage and on the temperature.

The infective trypomastigotes can actively penetrate the mucous membrane or skin. Triatomins commonly defecate after feeding, and most human infections occur when the feces are rubbed into the eyes or mucous membranes following a bite. Animals can become infected by licking their bites or by eating infected bugs or rodents. Transmission by blood transfusion or via the placenta may occur, as may oral transmission by eating infected mammals or vectors or accidentally ingesting fragments of them, transmission via infected maternal milk, fly-borne contamination, contamination with urine or saliva from heavily infected animals, or laboratory accident.

CULTIVATION. *T. c. cruzi* can be cultivated readily in many different media and in tissue culture.

PATHOGENESIS. *T. cruzi* may cause an acute or chronic disease in laboratory animals, depending on the strain of the parasite and the age and genetic strain of the host. Puppies and kittens are most susceptible, followed in order by mice and guinea pigs. The reservoir hosts are apparently not seriously affected, nor are farm animals. See Gutteridge and Rogerson (1979) and Tafuri (1979).

EPIDEMIOLOGY. Human infections with *T. cruzi* are common in many parts of tropical America, including Brazil, Bolivia, northern Chile, Colombia, northern Argentina, French Guiana, Paraguay, Uruguay, and Venezuela. Chagas' disease is primarily a disease of communities characterized by poverty, ignorance, and poor standards of hygiene (Prata 1974; Mott et al. 1978).

Chagas' disease is a zoonosis, with infections occurring widely in animals and man. The armadillo was thought by Hoare (1949) to have been the original source of the human disease in South America, but the dog, cat, opossum, black rat, and many other wild animals may also be infected (Pifano 1974; Zeledon et al. 1975; Fife 1977; Goldsmith et al. 1978; Schenone et al. 1978).

For further information on the epidemiology of *T. cruzi* infections, see the reviews listed above and also papers by Dias and by S. F. Wood. Miles and Rouse (1970) wrote a 200-page bibliography on the disease.

IMMUNOLOGY. Both cellular and humoral factors are active in immunity. There are antigenic differences between strains of *T. cruzi*

In a memorandum developed by the World Health Organization (WHO 1974), 19 persons reviewed our knowledge of immunity against Chagas' disease and discussed the possibility of developing a vaccine against it. At present, however, it is still at the laboratory animal stage. See also Holmes (1980).

DIAGNOSIS. In the acute stage of the disease, *T. cruzi* can be found in thick blood smears. In chronic or light infections, other methods must be used. *T. cruzi* can be differentiated from the nonpathogenic *T. rangeli* (see below) by its smaller size and large kinetoplast.

One of the most important methods is *xenodiagnosis*, the inoculation of susceptible vector hosts. Laboratory-reared, parasite-free triatomins are allowed to feed on suspected individuals, and their droppings or intestines are examined 30, 60, and 90 days later for developing trypanosomes.

Laboratory animals can be inoculated. In descending order of susceptibility, these are puppies, kittens, mice, and guinea pigs. The trypanosomes can also be cultivated in NNN medium or other media.

The trypanosomes can be found in biopsy examinations of affected lymph nodes or, on necropsy, in sections of heart muscle. Anthony et al. (1979) said that the ELISA test was the test of choice for sero-epidemiologic studies. Kagan (1980b) reviewed immunologic diagnosis, giving specific directions for various tests. He (1980c) tabulated 10 commercial sources for reagents for the tests and recommended the complement fixation, indirect hemagglutination, indirect immunofluorescence, and direct agglutination tests for diagnosis.

TREATMENT. No really satisfactory drug is known for the treatment of *T. cruzi* infections. Perhaps the best is the nitrofurfurylidene derivative nifurtimox (Bayer 2502, Bay 2502, Lampit).

CONTROL. Prevention of human *T. cruzi* infections depends on elimination of poverty and ignorance, improving housing, and eliminating triatomins from houses. The last will also largely prevent infections among domestic dogs and cats. Dusting or spraying houses with residual lindane or dieldrin has given good results.

Trypanosoma (Herpetomonas) rangeli Tejera, 1920

SYNONYMS. *Schizotrypanum rangeli, T. ariarii, T. cebus, T. escomeli, T. guatemalense.*

Hoare (1972) considered *T. rangeli* to be a stercorarian species in process of adaptation to the salivarian pattern of development. It was first found in the triatomin *Rhodnius prolixus* in Venezuela. It was later found in children in Guatemala and still later in Colombia, Chile, and El Salvador. It is quite common in dogs, cats, and man in certain areas of Venezuela, Colombia, Guatemala, and Panama and is sometimes found in mixed infections with *T. cruzi*. It has also been found in the New World monkeys *Cebus fatuellus* and *C. capucinus* and in marmosets and anteaters. Young mice, rats, and rhesus monkeys can be infected experimentally.

The trypanosomes in the blood are considerably larger than *T. cruzi*, being 26–36 μm long. The nucleus is anterior to the middle of the body near its center, the undulating membrane is rippled, and the kinetoplast is small and subterminal.

The most common vector is *Rhodnius prolixus*, but *Triatoma dimidiata* and other triatomins have also been found infected. A piriform stage about 7 μm long has been found in the foregut, and epimastigote and metacyclic trypomastigotes develop in the hindgut. The epimastigote stages may be extremely long, 32–70 or even over 100 μm long. The metacyclic trypomastigote forms have a well-developed undulating membrane and a long free flagellum. They may pass into the hemolymph and thence to the salivary glands. They are usually transmitted by bite but can also be transmitted by fecal contamination.

T. rangeli does not appear to be pathogenic for vertebrates. It can be easily cultivated.

For further information regarding this species, see Pifano (1948, 1954), Groot (1954), Zeledon (1954), Hoare (1972), Grewal (1969), D'Alesandro (1976), and Molyneux (1976).

Trypanosoma (*Megatrypanum*) *theileri* Laveran, 1902

SYNONYMS. *T. americanum, T. falshawi, T. franki, T. himalayanum, T. indicum, T. lingardi, T. muktesari, T. rutherfordi, T. scheini, T. transvaaliense, T. wrublewskii.*

T. theileri occurs in the blood of cattle. It is worldwide in distribution. It is probably quite common but is rarely found in blood smears. Other, similar species have been found in antelopes in Africa and deer in North America. Among these are *T. cervi* Kingston and Morton, 1975, in the mule deer *Odocoileus hemionus*, white-tailed deer *O. virginianus*, and elk *Cervus elaphus* (syn., *C. canadensis*) in the United States.

T. theileri is relatively large, ordinarily 60–70 μm long; forms, however, up to 120 μm long and smaller ones 25 μm long often occur. The posterior end is long and pointed. There is a medium-sized kinetoplast some distance anterior to it. The undulating membrane is prominent, and a free flagellum is present. Both trypomastigote and epimastigote forms may occur in the blood. Multiplication occurs by binary fission in the epimastigote form in the lymph nodes and various tissues.

T. theileri is transmitted by various tabanid flies, including *Tabanus* and *Haematopota*. Cattle are infected by contamination of the mucous membranes by small, metacyclic trypomastigotes that have developed in the hindgut. Intrauterine infection may occur.

T. theileri is ordinarily nonpathogenic, but under conditions of stress it may cause serious signs and even death. It may also be associated with abortion.

Since *T. theileri* is rarely seen in blood smears, discovery ordinarily depends on cultivation. It can be cultivated in NNN and other media at room temperature.

Trypanosoma (Megatrypanum) melophagium (Flu, 1908) Noller, 1917

SYNONYM. *T. woodcocki.*

This parasite is very common in sheep throughout the world. It is nonpathogenic, and infections are so sparse that it can ordinarily be found only by cultivation. The trypanosomes in the blood resemble *T. theileri* and are 50–60 μm long.

T. melophagium is transmitted by the sheep ked *Melophagus ovinus* and can be readily found in its intestine. Epimastigote forms are abundant in the ked midgut, and amastigote forms occur there also. Both multiply by binary fission. The epimastigote forms change into small, metacyclic trypomastigote forms in the hindgut. Sheep are infected when they bite into the keds; the trypanosomes pass through the intact buccal mucosa. Because infections in sheep are so sparse, it has been suggested that no multiplication occurs in this host.

Trypanosoma (Megatrypanum) theodori Hoare, 1931

This nonpathogenic species was found in goats in Israel. It resembles *T. melophagium* and has a similar life cycle, except that its vector is another hippoboscid fly, *Lipoptena caprina.* It is known only from its developmental stages in the fly, and not from the goat.

Trypanosoma (Herpetomonas) lewisi (Kent, 1880) Laveran and Mesnil, 1901

SYNONYMS. *T. eburneense, T. longocaudense, T. rattorum, T. sanguinis.*

This species occurs quite commonly in the black rat, Norway rat, and other members of the genus *Rattus* throughout the world. It is not normally transmissible to mice.

T. lewisi is 26–34 μm long (Fig. 2.7). Its vector is the rat flea *Nosopsyllus fasciatus*, in which it develops in the gut and in which the metacyclic, infective forms occur in the rectum. Rats become infected by eating infected fleas or flea feces. *T. lewisi* is nonpathogenic, and rats infected with it actually grow better than uninfected rats, apparently due to production of thiamine.

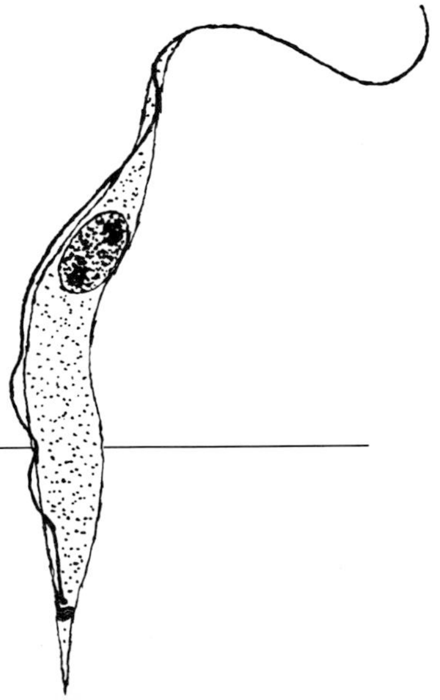

Fig. 2.7. *Trypanosoma lewisi.* (×2,800)

A great deal of research has been done on this species since it is easy to handle and its host is a convenient one.

Trypanosoma (*Herpetomonas*) *musculi* Kendall, 1906

SYNONYM. *T. duttoni.*

This species occurs in the house mouse and other species of *Mus* throughout the world. It is not normally transmissible to rats. It is 28-34 μm long, and structurally indistinguishable from *T. lewisi*. Its natural vector has not been determined, but is probably a mouse flea, *Ctenophthalmus, Nosopsyllus* or *Leptopsylla*. Its life cycle is the same as that of *T. lewisi*. It is nonpathogenic.

Trypanosomes of Birds

Trypanosomes have been reported under a large number of names from many species of birds. They all look more or less alike.

T. avium Danilewsky, 1885, was first described from owls (scientific name not given) and roller-birds *Coracias garrulus* in Europe and has since been reported from a wide variety of birds.

T. calmettei Mathis and Leger, 1909, was described from the chicken in southeast Asia; it is about 36 μm long. *T. gallinarum* Bruce et al., 1911, was described from the chicken in central Africa; it is about 60 μm long. *T. hannai* Pittaluga, 1905, was described from the pigeon, *T. numidae* Wenyon, 1908, from the guinea fowl, and *T. dafilae* from the pintail duck.

Avian trypanosomes are very pleomorphic, sometimes attaining great size. They may be 26–69 μm long or even longer. The kinetoplast is generally a long distance from the posterior end. There is a free flagellum, and the body often appears to be striated. There are many subpellicular microtubules as in other trypanosomes.

Bloodsucking arthropods, including mosquitoes, simuliids, and hippoboscids are the vectors. Bird trypanosomes multiply little in the avian host. They persist over winter, more or less restricted to the bone marrow, and reappear in the peripheral blood in the spring.

Avian trypanosomes are presumably nonpathogenic. Most can be readily cultivated on several media, including NNN medium.

Genus *Leishmania* Ross, 1903

Members of this genus occur primarily in mammals, although 10 species have been named from lizards in the Old World. They cause disease in man, dogs, and various rodents, including gerbils and guinea pigs. *Leishmania* occurs in about 50 countries, and there are about 400,000 new human cases of leishmaniasis each year. According to the United Nations Development Program/World Bank/World Health Organization (1982), the number of people with chronic or long-standing forms of the disease is unknown. In addition, at least one species has been named from bats. *Leishmania* is heteroxenous, transmitted by sandflies. It is found in the amastigote stage in the cells of its vertebrate hosts and in the promastigote stage in the intestine of the sandfly and in culture.

The literature on *Leishmania* is vast. Heyneman et al. (1980) gave a two-volume bibliography with 688 pages of references on it and the diseases it causes. Among reviews since 1970 are those of Bray (1974), Zuckerman (1975), Molyneux (1977), Zuckerman and Lainson (1977), Gardener (1977), Lainson and Shaw (1973, 1979), Sergiev (1979), Killick-Kendrick (1979), Zahar (1979, 1980), Marinkelle (1980), and Marr (1980).

STRUCTURE. All species of *Leishmania* look alike, although there are size differences between different strains or species. The amastigote stage is ovoid or spherical, usually 2.5–5.0 × 1.5–2.0 μm although smaller forms occur. Only the nucleus and kinetoplast are ordinarily visible in stained preparations, but a trace of an internal fibril (the flagellum) can sometimes be seen. This flagellum and the kinetosome (basal granule) from which it arises can also be seen in electron micrographs. The promastigote forms in culture and in the invertebrate host are spindle-shaped, 14–20 × 1.5–3.5 μm.

The kinetoplast becomes irregular and enlarges after a few hours in vitro, developing branches that probably become new mitochondria.

LIFE CYCLE. In the vertebrate host, *Leishmania* is found in the macrophages and other cells of the reticuloendothelial system in the skin, spleen, liver, bone marrow, lymph nodes, mucosa, etc. It may also be found in the leukocytes, especially the large mononuclears, in the bloodstream. It multiplies by binary fission in the amastigote form.

The invertebrate hosts of *Leishmania* are sandflies of the genus *Phlebotomus* in the Old World and *Lutzomyia* in the New World. When the sandflies suck blood, they ingest the amastigote forms. These pass to the midgut, where they assume the promastigote form and multiply by binary fission. They may be either free in the lumen or attached to the walls. They may also be found in the foregut and hindgut.

The location of their further development varies with the particular species of *Phlebotomus* or *Lutzomyia* and with the strain of *Leishmania*; it has been used in their classification.

CULTIVATION. *Leishmania* can be cultivated readily in a number of different media. The NNN medium is commonly used. The organism multiplies in the promastigote form in culture.

IMMUNOLOGY. An individual who has recovered from leishmaniosis is normally immune. This is especially true of *L. tropica*.

Antigens against *Leishmania* are not particularly specific. Stauber (1970) reviewed immunity in the leishmaniases. Kagan (1980b) reviewed the serodiagnosis of the human diseases and tabulated the sources of antigens. He (1980c) preferred the indirect immunofluorescence test for diagnosis.

SPECIATION. The speciation of *Leishmania* has been discussed by many authors, but no final classification has been agreed upon.

The following classification, that of Lainson and Shaw (1979), is merely tentative.

Section Hypopylaria. Primitive species; sandfly infection becomes established in the hindgut (pylorus, ileum, rectum); vertebrate hosts lizards; transmission by eating infected sandflies; found so far only in Old World.

Section Peripylaria. Development in both sandfly hindgut and foregut; vertebrate hosts both lizards and especially mammals; transmission of mammalian forms by bite of sandfly; mode of transmission of lizard forms unknown.

- *L. adleri*
- *L. braziliensis* complex
  - *L. braziliensis braziliensis*
  - *L. braziliensis guyanensis*
  - *L. braziliensis panamensis*

Section Suprapylaria. Development in sandfly only in midgut and foregut; vertebrate hosts only mammals; transmission by bite of sandfly; all in mammals.

- *L. mexicana* complex (in New World)
  - *L. mexicana mexicana*
  - *L. mexicana amazonensis*
  - *L. mexicana pifanoi*
  - *L. mexicana aristidesi*
  - *L. mexicana* subsp. in Trinidad (Tikasingh 1974)
  - *L. enriettii*
- *L. hertigi* complex (in New World)
  - *L. hertigi hertigi*
  - *L. hertigi deanei*
- *L. chagasi* (in New World)
- *L. donovani* complex (in Old World)
  - *L. donovani*
  - *L. infantum*
- *L. tropica* complex (in Old World)
  - *L. tropica*
  - *L. major*
  - *L. aethiopica*

SECTION UNDETERMINED

L. peruviana

Species of *Leishmania*

Leishmania braziliensis complex

Members of this complex are neotropical Peripylaria known to infect man; they have a prolific phase of growth in the pyloric region of the sandfly gut, have nuclear buoyant densities of 1.716 and 1.717 and kinetoplast buoyant densities of 1.691 and 1.694, and have small amastigotes (2.4 × 1.8 μm). All human diseases caused by members of this complex are zoonoses; lower animals are the normal hosts and man is an accidental host.

Leishmania mexicana complex

Members of this species readily infect hamsters, grow luxuriantly in several simple blood-agar media, have nuclear and kinetoplast DNA buoyant densities of 1.697–1.718, and have relatively large amastigotes (3.4 × 2.2 μm). As with the *L. braziliensis* complex, all human diseases caused by members of this complex are zoonoses; lower animals are the normal hosts and man is an accidental host.

Leishmania enriettii Muniz and Medina, 1948

This species, a member of the *L. mexicana* complex, occurs in the domestic guinea pig *Cavia porcellus* in Brazil, where it causes ulcers in the skin, especially of the paws, ears, and nose; it does not invade the viscera. It is not transmissible to the wild guinea pig *Cavia aperea*, to man, or to the laboratory rat or mouse. Golden hamsters and puppies can be infected with difficulty. It is twice as large as the human species. No proven vector is known. Readily cultivated, it is a convenient screening agent for drugs used against strains of *Leishmania* that cause cutaneous lesions in man.

Leishmania hertigi complex

Members of this complex have been found in neotropical porcupines, and no human or canine infections have been reported.

Leishmania chagasi Cunha and Chagas, 1937

This species, which was formerly considered a strain or subspecies of *L. donovani* (see below) causes visceral leishmaniosis or kala-azar in man. It is found from Mexico to northern Argentina but is probably most important in Brazil. It is probably transmitted by *Lutzomyia longipalpis* from a natural host such as the dog and occurs mostly among persons with low economic status, particularly their children. It is found primarily in rural northeastern Brazil. The dog is the principal urban reservoir and the most important source of human infection, while the bush dog or fox *Lycalopex vetulus* is probably the principal rural one.

Leishmania peruviana Velez, 1913

This species occurs in the barren mountain slopes on the western side of the Andes in Peru and also in Bolivia, where it causes a disease in man known as uta, with numerous small skin lesions. The domestic dog is the most important natural host, and the vector possibly *L. noguchii.*

Leishmania donovani complex

Members of this complex cause visceral leishmaniosis (kala-azar) in man. They occur in the Old World. Two species are recognized at present, but there are undoubtedly others, for there are at least five types.

PATHOGENESIS. Kala-azar is an important and highly fatal disease of man, particularly in India. After an incubation period of several months, it starts with an irregular fever lasting weeks to months. The spleen and liver hypertrophy. In untreated cases, the mortality is 75–95%, a little higher in adults than in infants. Death occurs in a few weeks to several years, often resulting from intercurrent disease. In treated cases 85–95% recover. Mediterranean kala-azar in children is similar, but the disease usually runs a shorter course.

Kala-azar is essentially a reticuloendotheliosis. The reticuloendothelial cells are increased in number and invaded by the parasites. The cut surface of the enormously enlarged spleen is congested and purple or brown, with prominent Malpighian corpuscles. The liver is enlarged, and there is fatty infiltration of the Kupffer cells. The macrophages, myelocytes, and neutrophils of the bone marrow are filled with parasites. The lymph nodes are usually enlarged and the intestinal submucosa is infiltrated with macrophages filled with parasites.

In dogs *L. donovani* may cause either visceral or cutaneous lesions; the latter are much more common. The disease is usually chronic, with low mortality, although an acute, highly fatal type is known. There may be emaciation and anemia. There is an abundant scurfy desquamation of the skin, and in some dogs more or less numerous cutaneous ulcers.

DIAGNOSIS. The only sure diagnostic method is finding the parasites themselves, although serologic and other tests have also been used and are of suggestive value. Smears made from biopsy samples of spleen pulp, liver pulp, superficial lymph nodes, bone marrow, or thick blood smears can be stained with Giemsa stain and examined microscopically. In visceral leishmaniosis the spleen is most often positive, but a certain amount of danger is associated with puncturing a soft, engorged, enlarged spleen. Thick blood smears are more often positive in man than in dogs. Examination of bone marrow obtained by sternal puncture is becoming increasingly popular.

In the cutaneous form of the disease, scrapings should be made for staining from the lesions or from the dermis, with as little bleeding as possible. This is probably the method of choice for dogs, since in them the cutaneous disease is more common than the visceral form. *L. donovani* can often be found in apparently normal skin of dogs.

Leishmania can be cultivated in NNN medium or a similar medium. Promastigote forms are present in culture. *Leishmania* can also be found in chicken embryos and in tissue culture.

Various serologic tests have also been used. Kagan (1980a) indicated that the indirect hemagglutination, indirect immunofluorescence, complement fixation, and direct agglutination tests are all satisfactory.

TREATMENT. Leishmanial infections can be treated successfully with various organic antimony compounds. The cheapest is tartar emetic, which must be administered intravenously since intramuscular injection causes tissue destruction. Others are neostibosan, neostam, solustibosan, and urea stibamine. The aromatic diamidines pentamidine and stilbamidine have been used in man, but they are apparently not very effective in dogs.

CONTROL. Prevention of leishmanial infections depends on breaking the life cycle by eliminating sandflies. This can be done by residual spraying of houses, barns, and outside resting places with DDT, other chlorinated hydrocarbons, or organic phosphorus insecticides. Residual spraying for mosquitoes has also markedly decreased the number of sandflies in malarious areas.

In addition to spraying, insect repellents such as dimethylphthalate can be rubbed on the skin, houses can be screened with very fine mesh wire, and decaying vegetation and other breeding places can be cleaned up.

In regions where kala-azar is a zoonosis, treatment of infected dogs and destruction of strays will eliminate the reservoir of infection for man.

Leishmania donovani (Laveran and Mesnil, 1903) Ross, 1903

This species causes kala-azar ("black" fever, dum-dum fever, visceral leishmaniosis) in man in India, Kenya, the Sudan, Ethiopia, and China. For further information about it see the preceding discussions of the *L. donovani* complex and any text on human parasitology.

Leishmania infantum Nicolle, 1909

This species is the cause of Mediterranean or infantile kala-azar of man in countries of the Mediterranean basin plus southern Europe and Middle Asia. Dogs are much more commonly infected than man, and 90% of the affected people are children less than 5 yr old. The prevalence in dogs may reach 20% in some countries, and infection rates as high as 40% have been reported in Greece and Samarkand. Even in such countries, the infection rate in children is only 1–2%. Other hosts are the jackal *Canis aureus*, fox, wolf, porcupine *Hystrix* sp., and fennec fox *Fennecus zerda*. It is transmitted principally by *P. perniciosus*. Other suspected sandflies are *P. longicuspis, P. chinensis, P. mongolensis,* and *P. caucasicus*.

Five cases of visceral leishmaniosis have been reported in dogs in the United States: one in Alabama; one in Washington, D.C.; one in Boston; one in Virginia; and one in Florida. Four of the dogs had been imported into this country from Greece and one from Spain. It has also been found in Canada in dogs from Portugal and Spain.

Leishmania tropica complex

SYNONYMS. *Helcosoma tropicum, Sporozoa furunculosa, Ovoplasma orientale, Plasmosoma jerichaense, L. wrighti, L. cunninghami, L. nilotica, L. recidiva.*

DISEASE. Members of this complex cause cutaneous leishmaniosis (Oriental sore, Aleppo button, Jericho boil, Pendinsk ulcer) in man. The organisms occur in the monocytes and other cells of the reticuloendothelial system, in cutaneous lesions, and in the skin. They may also occur in the lymph nodes and in the mucous membranes. For further information about the human disease, see any textbook of human parasitology.

IMMUNITY. Persons who have recovered spontaneously from classical Oriental sore due to *L. tropica* have a solid immunity. There is no cross-immunity between *L. tropica* and *L. donovani*.

DIAGNOSIS. The same methods used in diagnosing *L. donovani* complex infections are used in diagnosing *L. tropica* complex infections, except for the tissues examined. The parasites are usually abundant in dry Oriental sore (caused by *L. tropica*) but are scanty in wet Oriental sore (caused by *L. major*).

According to Kagan (1980a), the indirect immunofluorescence test is being used routinely for this disease.

TREATMENT. Organic antimony compounds are effective against cutaneous leishmaniosis. The same ones are used as for kala-azar.

CONTROL. The same measures used to prevent kala-azar are effective against cutaneous leishmaniosis. Elimination of gerbil colonies is the most effective measure against *L. major.*

Leishmania tropica (Wright, 1903) Luhe, 1906

This species causes classical, or "dry," Oriental sore in man in regions with a hot, dry climate, from the Mediterranean basin to central and northern India. The incubation period is several months. Ulcers or sores are found on exposed parts of the body in man. After some months or a year, connective tissue is formed, but a permanent scar is left. The disease is very seldom fatal.

The disease is urban in distribution, and dogs are commonly infected. The lesions in the dog are similar to those in man. They are probably confined to the skin.

The disease is transmitted by *P. perfiliewi, P. papatasi, P. sergenti,* and possibly *P. longicuspis.*

Leishmania major (Yakimov and Shokhov, 1915) Bray, Ashford, and Bray, 1973

This organism is the cause of "moist" or "wet" Oriental sore of man in central Asia and southern USSR. There may or may not be cross-immunity between this species and *L. tropica.* The incubation period is 1–6 wk. The lesions are wet and ulcerative but do not extend to the mucous membranes, and relatively few parasites can be found in them. The lymph nodes are often involved.

The disease is rural in distribution. The natural hosts are various desert rodents, the great gerbil *Rhombomys opimus* being the most important. Other natural hosts are the gerbils *Meriones erythrourus, M. lybicus, M. meridianus,* the suslik *Spermophilopsis leptodactylus, Allactaga servtzovi,* and the sand rat *Psammomys obesus.* The hedgehog *Hemiechinus albulus* and the mustelid carnivores *Mustela* spp. and *Vormela peregusna* have also been found infected, but the dog has not. The most important vector, at least for rodents, is the sandfly *P. caucasicus,* which lives in the gerbil burrows. Probably the most important vector for man is *P. papatasi*, and other species of *Phlebotomus* have also been incriminated.

LITERATURE CITED

For citations before 1965, see the *Index-Catalogue of Medical and Veterinary Zoology, Zoological Record, Biological Abstracts, Protozoological Abstracts,* and *Veterinary Bulletin.*

Afchain, D., et al. 1977. Bull. WHO 55:703–13.
Agabian, N., L. Thomashow, M. Milhausen, and K. Stuart. 1980. Am. J. Trop. Med. Hyg. 29:1043–49.
Anonymous. 1979. CDC Vet. Publ. Health Notes 1979, Jan.:5.
Anthony, R. L., C. M. Johnson, and O. E. Sousa. 1979. Am. J. Trop. Med. Hyg. 28:969–73.
Baker, J. R., ed. 1982. Perspectives in Trypanosomiasis Research. Somerset, N.J.: Wiley.
Becker, C. D. 1977. In Parasitic Protozoa, vol. 1, ed. J. P. Kreier, 357–416. New York: Academic.
Boreham, P. F. L. 1979. In Biochemistry and Physiology of Protozoa, 2d ed., vol. 2, ed. M. Levandowsky and S. H. Hutner, 429–57. New York: Academic.
Borst, P., A. C. C. Frasch, A. Bernards, J. H. J. Hoeijmakers, L. H. T. van der Ploeg, and G. A. M. Cross. 1980. Am. J. Trop. Med. Hyg. 29:1033–36.
Bowman, I. B. R., and I. W. Flynn. 1976. In Biology of the Kinetoplastida, vol. 1, ed. W. H. R. Lumsden and D. A. Evans, 435–76. New York: Academic.
Bray, R. S. 1974. Annu. Rev. Microbiol. 28:189–217.
Cançado, J. R., ed. 1968. Doença de Chagas. Hosp. Clinicas, Belo Horizonte, Brazil.
Carcavallo, R. U. 1970. H. D. Srivastava Commem. Vol., 381–90.
Cedillos, R. A. 1975. Bull. Pan Am. Health Org. 9:135–41.
Cross, G. A. M., et al. 1980a. Am J. Trop. Med. Hyg. 29:1023–49.
Cross, G. A. M., A. A. Holder, G. Allen, and J. C. Boothroyd. 1980b. Am. J. Trop. Med. Hyg. 29:1027–32.
Cunningham, I., and B. M. Honigberg. 1977. Science 197:1279–82.
Cunningham, I., and A. M. Taylor. 1979. J. Protozool. 26:428–32.
D'Alesandro, P. A. 1970. In Immunity to Parasitic Animals, vol. 2, ed. G. J. Jackson et al., 691–738. New York: Appleton-Century-Crofts.
D'Alesandro, A. 1976. In Biology of the Kinetoplastida, vol. 1, ed. W. H. R. Lumsden and D. A. Evans, 327–403. New York: Academic.
Desowitz, R. 1970. In Immunity to Parasitic Animals, vol. 2, ed. G. J. Jackson et al., 551–96. New York: Appleton-Century-Crofts.
Doyle, J. J. 1977. In Immunity to Blood Parasites of Animals and Man, ed. L. H. Miller et al., 31–63. New York: Plenum.
Evans, D. A. 1978. In Methods of Cultivating Parasites in Vitro, ed. A. E. R. Taylor and J. R. Baker, 55–88. New York: Academic.
Fife, E. H., Jr. 1977. In Parasitic Protozoa, vol. 1, ed. J. P. Kreier, 135–73. New York: Academic.
Folkers, C. 1965. Tijdschr. Diergeneeskd. 90:1192–1200.
Gardener, P. J. 1977. Trop. Dis. Bull. 74:1069–88.
Gardiner, P. R., L. C. Lamont, T. W. Jones, and I. Cunningham. 1980. J. Protozool. 27:182–85.
Goble, F. C. 1970. In Immunity to Parasitic Animals, vol. 2, ed. G. J. Jackson et al., 597–689. New York: Appleton-Century-Crofts.
Goldberg, S. S., and A. A. Silva Pereira. 1983. J. Parasitol. 69:91–96.
Goldsmith, R. S., R. Zarate, I. Kagan, J. Cedeno-Ferreira, M. Galindo-Vasconcelos, and E. A. Paz. 1978. Salud Publ. Mexico 22:439–52.

Gray, A. R., and G. Luckins. 1976. In Biology of the Kinetoplastida, vol. 1, ed. W. H. R. Lumsden and D. A. Evans, 493–542. New York: Academic.
Grewal, M. S. 1969. Res. Bull. Panjab Univ. 20:449–86.
Gutteridge, W. E., and G. W. Rogerson. 1979. In Biology of the Kinetoplastida, vol. 2, ed. W. H. R. Lumsden and D. A. Evans, 619–52. New York: Academic.
Henson, J. B., and J. C. Noel. 1979. Adv. Vet. Sci. Comp. Med. 23:161–82.
Herbert, W. J., and D. Parratt. 1979. In Biology of the Kinetoplastida, vol. 2, ed. W. H. R. Lumsden and D. A. Evans, 481–521. New York: Academic.
Heyneman, D., H. Hoogstraal, and A. Djigounian. 1980. Bibliography of Leishmania and Leishmanial Diseases, vols. 1 and 2. Cairo, Egypt: Spec. Publ. NAMRU 3.
Hill, G. C., S. P. Shimer, B. Caughey, and L. S. Sauer. 1978. Science 202:763–65.
Hirumi, H., J. J. Doyle, and K. Hirumi. 1977. Science 196:992–94.
Hoare, C. A. 1972. The Trypanosomes of Mammals. Oxford: Blackwell.
Hoare, C. A., and F. G. Wallace. 1966. Nature 212:1385–86.
Holmes, P. H. 1980. Symp. Br. Soc. Parasitol. 18:75–105.
Hutner, S. H., C. J. Bacchi, and H. Baker. 1979. In Biology of the Kinetoplastida, vol. 2, ed. W. H. R. Lumsden and D. A. Evans, 653–91. New York: Academic.
Janovy, J., Jr., K. W. Lee, and J. A. Brumbaugh. 1974. J. Protozool. 21:53–59.
Kagan, I. G. 1980a. In Manual of Clinical Microbiology, 3d ed., ed. E. H. A. Lennette et al., 724–50. Washington, D.C.: American Society for Microbiology.
———. 1980b. In Immunological Investigation of Tropical Parasitic Diseases, ed. V. Houba, 49–64. New York: Churchill.
———. 1980c. In Manual of Clinical Immunology, 2d ed., ed. N. R. Rose and H. Friedman, 573–604. Washington, D.C.: American Society for Microbiology.
Killick-Kendrick, R. 1979. In Biology of the Kinetoplastida, vol. 2, ed. W. H. R. Lumsden and D. A. Evans, 395–460. New York: Academic.
Kreutzer, R. D., and O. E. Sousa. 1981. Am. J. Trop. Med. Hyg. 30:308–17.
Lainson, R., and J. J. Shaw. 1973. Bull. Pan Am. Health Org. 7(4):1–19.
———. 1979. In Biology of the Kinetoplastida, vol. 2, ed. W. H. R. Lumsden and D. A. Evans, 1–116. New York: Academic.
Lambrecht, F. L. 1966. East Afr. Wildl. J. 4:89–98.
Levine, N. D. 1972. In Research in Protozoology, vol. 4, ed. T.-T. Chen, 291–350. Elmsford, N. Y.: Pergamon.
Lom, J. 1979. In Biology of the Kinetoplastida, vol. 2, ed. W. H. R. Lumsden and D. A. Evans, 269–337. New York: Academic.
Losos, G. J., and B. O. Ikede. 1972. Vet Pathol. 9(Suppl.):1–71.
Lumsden, W. H. R. 1977. Protozoology 3:25–32.
Lumsden, W. H. R., and D. S. Ketteridge, eds. 1979. In Biology of the Kinetoplastida, vol. 2, ed. W. H. R. Lumsden and D. A. Evans, 693–721. New York: Academic.
MacLennan, K. J. R. 1975. WHO/TRYP/75.42.
Mansfield, J. M. 1977. In Parasitic Protozoa, vol. 1, ed. J. P. Kreier. New York: Academic.
Marinkelle, C. J. 1976a. In Biology of the Kinetoplastida, vol. 1, ed. W. H. R. Lumsden and D. A. Evans, 175–216. New York: Academic.
———. 1976b. Biology of the Kinetoplastida, vol. 1, ed. W. H. R. Lumsden and D. A. Evans, 217–56. New York: Academic.
———. 1980. Bull. WHO 58:807–18.
Marr, J. J. 1980. In Biochemistry and Physiology of Protozoa, 2d ed., vol. 3, ed. M. Levandowsky and S. H. Hutner, 313–40. New York: Academic.
McGhee, R. B. 1968. In Infectious Blood Diseases of Man and Animals, vol. 1, ed. D. Weinman and M. Ristic, 307–41 New York: Academic.

Miles, M. A. 1979. In Biology of the Kinetoplastida, vol. 2, ed. W. H. R. Lumsden and D. A. Evans, 117–96. New York: Academic.
Miles, M. A., and J. E. Rouse. 1970. Chagas' Disease: A Bibliography. London: Bureau of Hygiene and Tropical Diseases.
Molyneux, D. H. 1976. In Biology of the Kinetoplastida, vol. 1, ed. W. H. R. Lumsden and D. A. Evans, 285–325. New York: Academic.
______. 1977. Adv. Parasitol. 15:1–82.
Mott, K. E., T. M. Muniz, J. S. Lehman, Jr., R. Hoff, R. H. Morrow, Jr., T. S. de Oliveira, I. Sherlock, and C. C. Draper. 1978. Am. J. Trop. Med. Hyg. 27:1116–22.
Murray, M., and G. M. Urquhart. 1977. In Immunity to Blood Parasites of Animals and Man, ed. L. H. Miller et al., 209–41. New York: Plenum.
Newton, B. A. 1976. In Biology of the Kinetoplastida, vol. 1, ed. W. H. R. Lumsden and D. A. Evans, 405–34. New York: Academic.
Nyindo, M., M. Chimtawi, J. Owor, J. S. Kaminjolo, N. Patel, and N. Darji. 1979. J. Parasitol. 64:1039–43.
Parratt, D., and W. J. Herbert. 1979. In Biology of the Kinetoplastida, vol. 2, ed. W. H. R. Lumsden and D. A. Evans, 523–45. New York: Academic.
Pifano, F. C. 1974. Proc. 3d Int. Congr. Parasitol. 1:236–37.
Prata, A. 1974. Proc. 3d Int. Congr. Parasitol. 1:226–27.
Rickman, L. R., and J. Robson. 1970. Bull. WHO 42-911-16.
Rudzinska, M. A., and K. Vickerman. 1968. In Infectious Blood Diseases of Man and Animals, vol. 1, ed. D. Weinman and M. Ristic, 217–306. New York: Academic.
Schenone, H., F. Villarroel, and E. Alfaro. 1978. Bol. Chil. Parasitol. 33:2–7.
Seed, J. R. 1974. J. Protozool. 21:639–46.
Sergiev, V. P. 1979. In Biology of the Kinetoplastida, vol. 2, ed. W. H. R. Lumsden and D. A. Evans, 197–212. New York: Academic.
Shaw, J. J. 1977. Protozoology 3:119–28.
Simpson, L., A. M. Simpson, G. Kidane, L. Livingston, and T. W. Spitchill. 1980. Am. J. Trop. Med. Hyg. 29:1053–81.
Snyder, T. G., III, R. G. Yaeger, and J. Dellucky. 1980. J. Am. Vet. Med. Assoc. 177-247–49.
Soltys, M. A., and P. T. K. Woo. 1977. In Parasitic Protozoa, vol. 1, ed. J. P. Kreier, 239–68. New York: Academic.
Stauber, L. 1970. In Immunity to Parasitic Animals, vol. 2, ed. G. J. Jackson et al., 739–65. New York: Appleton-Century-Crofts.
Tafuri, W. L. 1979. In Biology of the Kinetoplastida, vol. 2, ed. W. H. R. Lumsden and D. A. Evans, 547–618. New York: Academic.
Terry, R. J. 1976. In Biology of the Kinetoplastida, vol. 1, ed. W. H. R. Lumsden and D. A. Evans, 477–92. New York: Academic.
Tikasingh, E. S. 1974. Bull. PAHO 8:232–42.
United Nations Development Program/World Bank/World Health Organization. 1982. Newsletter: Special program for research and training in tropical diseases, 5–8.
Vickerman, K. 1969a. J. Cell Sci. 5:163–93.
______. 1969b. J. Protozool. 16:54–69.
______. 1976. In Biology of the Kinetoplastida, vol. 1, ed. W. H. R. Lumsden and D. A. Evans, 1–34. New York: Academic.
______ 1977. Protozoology 3:57–69.
Vickerman, K., and F. E. G. Cox, 1967. The Protozoa. Boston: Houghton Mifflin.
Vickerman, K., and T. M. Preston. 1976. In Biology of the Kinetoplastida, vol. 1, ed. W. H. R. Lumsden and D. A. Evans, 35–130. New York: Academic.

Wallace, F. G. 1977. Protozoology 3:51-56.

———. 1979. In Biology of the Kinetoplastida, vol. 2, ed. W. H. R. Lumsden and D. A. Evans, 213–40. New York: Academic.

Willett, K. C. 1972. Bull. WHO 47:747-49.

Williams, G. D., L. G. Adams, R. G. Yaeger, R. K. McGrath, W. K. Read, and W. R. Gilderback. 1977. J. Am. Vet. Med. Assoc. 171:171–77.

Williams, R. O., J. R. Young, P. A. O. Majivra, J. J. Doyle, and S. Z. Shapiro. 1980. Am. J. Trop. Med. Hyg. 29:1037–42.

Woo, P. T. K. 1977. In Parasitic Protozoa, vol. 1, ed. J. P. Kreier, 269–96. New York: Academic.

WHO (World Health Organization). 1974. Bull. WHO 50:459–72.

———. 1979. The African Trypanosomiases. WHO Tech. Rep. Ser. 635. Geneva, Switzerland: World Health Organization.

Zahar, A. R. 1979. WHO/VBC/79.749.

———. 1980. WHO/VBC/80.786.

Zeledon, R., G. Solano, L. Burstin, J. C. Swartzwelder. 1975. Am J. Trop. Med. Hyg. 24:214–25.

Zuckerman, A. 1975. Exp. Parasitol. 38:370–400.

Zuckerman, A., and R. Lainson. 1977. In Parasitic Protozoa, vol. 1, ed. J. P. Kreier, 58–133. New York: Academic.

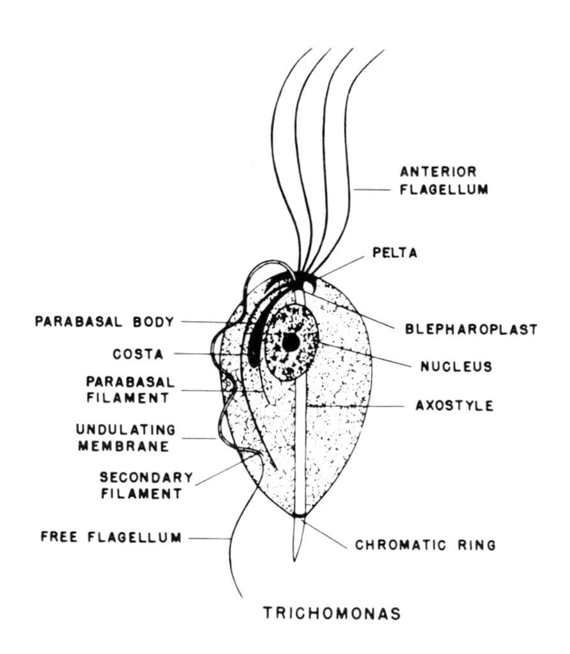

Fig. 3.1. Structures of *Trichomonas*.

3
Flagellates: The Trichomonads

THE trichomonads belong to the family Trichomonadidae within the order Trichomonadorida. The body is usually piriform, with a rounded anterior end and a pointed posterior end (Fig. 3.1). There is a single nucleus in the anterior part of the body. Anterior to the nucleus is a *blepharoplast* composed of several *basal granules,* or *kinetosomes.* Three to five anterior flagella and a posterior (recurrent) flagellum arise from the blepharoplast. The posterior flagellum passes along the border of an *undulating membrane* that extends along the side of the body; a *secondary* or *accessory filament* may be associated with it. The posterior flagellum may or may not extend beyond the undulating membrane as a free flagellum. A filamentous *costa* arises from the blepharoplast and runs along the base of the undulating membrane; its presence differentiates this family from all others in the order. The costa is a rod containing transverse striations. A *parabasal body* (actually the Golgi apparatus) lies posterior to the blepharoplast and appears to arise from it when viewed under the light microscope; there may or may not be one or more cross-striated *parabasal filaments* running through and beyond it. A clear, rodlike *axostyle* also arises from the blepharoplast and passes through the center of the body to emerge from the posterior end. The anterior end of the axostyle is enlarged to form a *capitulum.* There may or may not be a *chromatic ring* at the point of emergence of the axostyle. Just anterior to the blepharoplast and lying along the anterior margin of the body may be a *pelta,* which stains with silver; it apparently arises from (or overlaps) the capitulum of the axostyle. In addition to these structures, there may be various granules within or along the axostyle, along the costa, or in other locations.

There are three subfamilies in the Trichomonadidae. Two of these are of interest to vertebratologists; the third (Pentatrichomonoidinae) occurs in termites. In the subfamily Trichomonadinae there is a typical pelta. This subfamily contains four genera. In the genus *Trichomonas* the posterior flagellum is not free, whereas it is in the other three; *Trichomonas* has four anterior flagella (see Fig. 3.2). In the genus *Trichomitus* there are three anteriorr flagella and the parabasal body is usually V-shaped (occasionally rod-shaped). In the genus

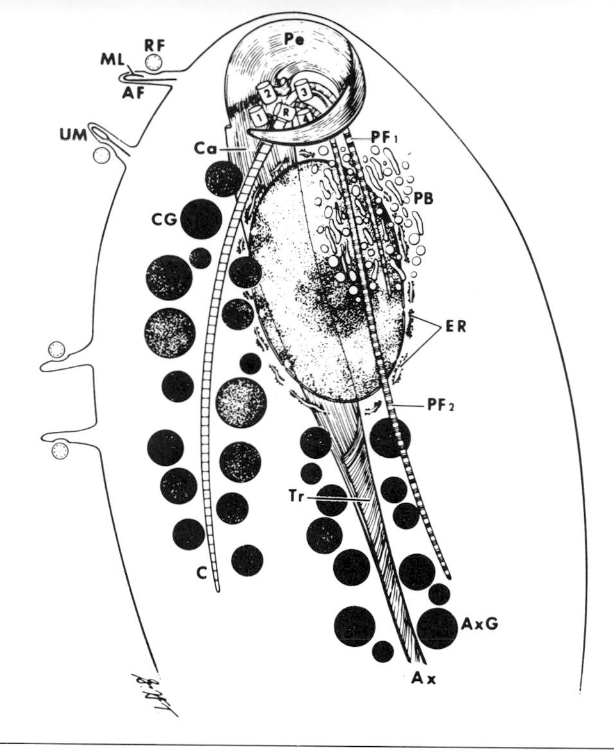

Fig. 3.2. Composite schematic diagram of *Trichomonas gallinae* fine structure; dorsal and slightly right view: *AF* = accessory filament; *Ax* = axostyle; *AxG* = paraxostylar granules; *C* = costa; *Ca* = capitulum of axostyle; *CG* = paracostal granules; *ER* = endoplasmic reticulum; *ML* = marginal lamella; *Pe* = pelta; *PB* = parabasal body; *PF* = parabasal filament; *R* = kinetosome of *RF*, the recurrent flagellum; *Tr* = trunk of axostyle; *UM* = undulating membrane; 1–4 = kinetosomes of the anterior flagella. (From Mattern et al. 1967)

Tetratrichomonas there are four anterior flagella in mature individuals and the parabasal body is usually disk-shaped. In the genus *Pentatrichomonas* there are five anterior flagella, four grouped together at the base and one independent; the parabasal body appears to be composed of small granule(s).

The subfamily Tritrichomonadinae contains forms without a pelta. There are three genera in this subfamily, but two of them occur in termites and are of no concern in this discussion. In the genus *Tritrichomonas* there are three anterior flagella.

There are several species of trichomonads in domestic animals and man, but the nomenclatorial status and host-parasite relations of many of them are not yet clear. They have been found in the cecum and colon of practically every species of mammal or bird examined for them, and they also occur in reptiles, amphibia, fish, and many invertebrates. Peculiarly, none are known from rabbits or other lagomorphs. Many of the cecal trichomonads look alike, and cross-transmission studies have shown that many can be easily transmitted from one host species to another. Some mammalian trichomonads have even been transmitted successfully to day-old chicks, although they will not become established in older birds. Further and extensive studies are needed to establish the correct names and host spectra of many trichomonads.

Most trichomonads are nonpathogenic commensals, but a few are important pathogens. None of the cecal trichomonads has ever been proven to be pathogenic, although some people have thought that they were because they were found in animals that had enteritis or diarrhea. However, the mere presence of an organism in a diseased animal does not mean that the organism caused the disease; the disease may simply have set up conditions favorable to the organism's growth and multiplication. This is especially true of the cecal trichomonads, which flourish in a fluid or semifluid habitat.

The life cycle of trichomonads is simple. They reproduce by longitudinal binary fission. No sexual stages are known. There are no cysts, although degenerating or phagocytized individuals (or entirely different organisms, such as *Blastocystis*) have been mistaken for them. Levine (1973) and Honigberg (1978a) reviewed the trichomonads of veterinary and (Honigberg 1978b) medical importance. See their works for references.

Genus *Tritrichomonas* Kofoid, 1920

Members of this genus have 3 anterior flagella and lack a pelta.

Tritrichomonas foetus (Riedmüller, 1928) Wenrich and Emmerson, 1933

SYNOMYMS. *Trichomonas uterovaginalis, T. vitulae, T. bovis, T. genitalis, T. foetus, T. bovinus, T. mazzanti.*

DISEASE. Bovine trichomonad abortion, bovine genital trichomonosis.

HOSTS. Ox, zebu, possibly pig, horse, roe deer.

LOCATION. Genital tract.

GEOGRAPHIC DISTRIBUTION. Worldwide.

PREVALENCE. Although *T. foetus* is known to be widely distributed, few studies have been made on its incidence. In the United States it is probably third to brucellosis and leptospirosis as a cause of abortion in cattle.

The USDA Agricultural Research Service (1965) estimated that bovine trichomonosis caused an average annual loss from 1951 to 1960 of $8.04 million. Fitzgerald et al. (1958) estimated that losses in production because of reproductive disorders in cows bred by infected western range bulls might amount to about $800 annually per infected bull. Subsequent inflation would make this much higher, and the annual loss due to an infected bull in an artificial insemination ring would be even higher.

STRUCTURE. The body is spindle- to pear-shaped, 10–25 μm long and 3–15 μm wide. It has 3 anterior flagella and a posterior flagellum that trails as a free flagellum about as long as the anterior flagella. The undulating membrane extends almost the full length of the body and has an accessory filament along its margin. The costa is prominent. The axostyle is thick and hyaline, with a capitulum containing endoaxostylar granules and a chromatic ring at its point of emergence from the posterior end of the body. The parabasal body is sausage- or ring-shaped. There is no pelta. (See Fig. 3.3.)

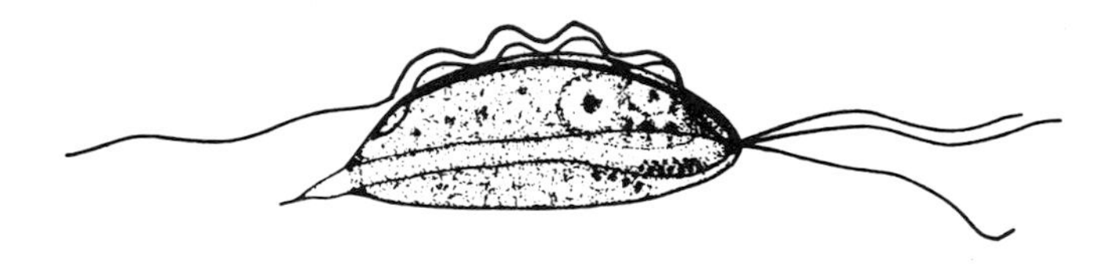

Fig. 3.3. *Tritrichomonas foetus*. (×3,400) (From Wenrich and Emmerson 1933, J. Morphol. 55:193)

PATHOGENESIS. A great deal has been written about bovine trichomonosis. It is a venereal disease transmitted by coitus. It can also be transmitted by artificial insemination. Nonvenereal transmission is rare under natural conditions. After infection of the female, the trichomonads multiply at first in the vagina, causing a vaginitis; they are most numerous there 14–18 days after infection. They invade the uterus through the cervix. They may disappear from the vagina or may remain there, producing low-grade inflammation and catarrh.

Early abortion 1–16 wk after breeding is characteristic of bovine trichomonosis. The foetus is often so small that it is not observed by the owner, who, not realizing that abortion has occurred, may believe that the animal failed to conceive and that its heat periods are irregular.

If the placenta and fetal membranes are completely eliminated following abortion, the cow usually recovers spontaneously. This is the most common course. If, however, part of the placenta or membranes remain, a chronic catarrhal or purulent endometritis results that may cause permanent sterility.

Sometimes the animal does not abort, but the fetus dies and becomes macerated in the uterus. Pyometra results, and the uterus may contain several liters of a thin, grayish white fluid swarming with trichomonads. In the absence of bacteria, this fluid is almost odorless. The cervical seal may remain intact, or it may allow a small amount of fluid to escape when the animal is lying down. Animals with pyometra seldom come in heat, and the owner may believe them to be pregnant. In long-standing cases, the trichomonads may disappear from the uterine fluid. Normal gestation and calving may occur in an infected animal, but this is rare.

In the bull the most common site of infection is the preputial cavity, although the testes, epididymis, and seminal vesicles may sometimes be involved. Spontaneous recovery is rare; bulls remain infected permanently unless treated.

IMMUNOLOGY. Cows or heifers that recover from infection are usually relatively immune, although reinfection can occur. A number of investigators have studied various immunologic responses to trichomonad infection (Honigberg 1978a). There is more than one serologic strain of *T. foetus.*

EPIDEMIOLOGY. Bovine trichomonosis is a venereal disease transmitted at coitus. *T. foetus* is known to occur in cattle, but whether it is also present in other animals and whether it may be transmitted from them to cattle by a nonvenereal route remain to be determined.

T. foetus may or may not survive in frozen bovine semen, depending on the conditions. It survives in some media but not in others. Rapid freezing and high salt concentration are deleterious, and the stage in the population growth curve is important, the protozoa being much more sensitive to injury when frozen during the initial and logarithmic phases than at the peak of the curve and for some time thereafter. Temperature fluctuation during storage is deleterious. Glycerol appears to be toxic at refrigerator temperatures but not at either subfreezing or incubator (37 C) temperatures.

Many different laboratory animals can be infected experimentally in various ways with *T. foetus.* Successful vaginal infections have been established in the rabbit, golden hamster, guinea pig, dog, goat, and pig. The golden hamster is the laboratory animal of choice for experimental vaginal infections. Vaginal inflammation occurs in guinea pigs, and abortion may occur in some pregnant guinea pigs. Laboratory mice and rats are refractory to vaginal infection, but abscesses can be produced by subcutaneous injection of mice.

Trichomonads similar to *T. foetus* have been found in the genital tract and aborted fetuses of swine, roe deer, and horses.

The relation of *T. foetus* to the trichomonads of swine still remains to be elucidated. The pig nasal trichomonad *Tritrichomonas suis* greatly resembles *T. foetus;* vaginal infections have been readily established with it in cattle, and cecal infections with *T. foetus* have been produced in young pigs. The two species appear to be the same. If so, the correct name would be *T. suis,* since it has priority.

DIAGNOSIS. Although the mucus agglutination test, skin tests, and a number of other immunologic procedures have been suggested for diagnosing *T. foetus* infections, the only sure method is to find and identify the protozoa microscopically either directly or in culture.

In heavy infections, particularly of females, the trichomonads can be seen by direct examination of mucus or exudate from the vagina or uterus, amniotic or allantoic fluid, fetal membranes, placenta, fetus stomach contents, oral fluid or other fetal tissues; or, in bulls, of washings from the preputial cavity and rarely seminal fluid or semen. If trichomonads cannot be found by direct microscopic examination, cultures should be made in CPLM (cystein-peptone–liver extract–maltose-serum), BGPS (beef extract–glucose-peptone-serum), or Diamond's media.

Samples can be obtained from the vagina by washing with physiologic

NaCl solution in a bulbed douche syringe. They can be obtained from the preputial cavity with a cotton swab or, better, by washing with physiologic NaCl solution in a bulbed pipette or syringe. The washings should be allowed to settle for 1–3 hr or should be centrifuged before examination.

The external genitalia should be cleaned thoroughly before samples are taken in order to avoid contamination with intestinal or coprophilic protozoa that might be mistaken for *T. foetus,* such as *T. enteris, Monocercomonas ruminantium, Protrichomonas ruminantium, Bodo foetus, B. glissans, Spiromonas angusta, Cercomonas crassicauda, Polytoma uvella, Monas obliqua,* and *Lembus pusillus.* In identifying *T. foetus,* it must be distinguished from these.

Trichomonads are most numerous in the vagina 2–3 wk after infection. Their numbers fluctuate in bulls, the interval between peaks being 5–10 days.

A single examination is not sufficient to warrant a negative diagnosis. A cow can be considered uninfected if, after at least three negative examinations, she has two normal estrous periods and subsequently conceives and bears a normal calf; she should be bred by artificial insemination to avoid infecting the bull. A bull can be considered negative if, after at least six negative examinations at weekly intervals, he is bred to two or more virgin heifers and they remain negative.

CULTIVATION. *T. foetus* can be readily cultivated in a number of media. Among them are CPLM, BGPS (modified Plastridge's) medium, TYM (trypticase–yeast extract–maltose-cysteine-serum) medium, thioglycollate broth plus 1% beef serum, and skim milk containing antibiotics. *T. foetus* can also be cultivated in tissue culture and the chorioallantoic sac of chick embryos.

TREATMENT. Since trichomonosis is ordinarily self-limiting in females, treatment is unnecessary. No satisfactory treatment is known for these infections. Treatment of bulls is expensive, tedious, and time-consuming; unless a bull is exceptionally valuable, it is best to sell it for slaughter. A salve containing trypaflavin and surfen or acriflavin in an ointment base can be rubbed into the penis and prepuce following pudendal nerve block or relaxation of the retractor penis muscles with a tranquilizer. Massage should be continued for 15–20 min, using 120 ml of the ointment. In addition, 30 ml of 1% acriflavin solution should be injected into the urethra. Dimetridazole solution can also be used.

CONTROL. Control of bovine trichomonosis depends on proper herd management. Most infected bulls should be slaughtered. Infected cows should be given breeding rest and then be bred by artificial insemination to avoid infecting clean bulls.

Proper management of bulls used for artificial insemination is especially important, since they may spread the infection widely. They should be examined for *T. foetus* before purchase, and the herds from which they originated should be studied at the same time. In addition, they should be examined repeatedly while in use. Freezing the semen in the presence of glycerol cannot be expected to kill the trichomonads.

Tritrichomonas suis (Gruby and Delafond, 1843)

SYNONYM. *Trichomonas suis.*

COMMON NAME. Large pig trichomonad, pig nasal trichomonad.

HOST. Pig.

LOCATION. Nasal passages, stomach, cecum, colon, occasionally small intestine.

GEOGRAPHIC DISTRIBUTION. Worldwide.

PREVALENCE. *T. suis* is common in swine.

STRUCTURE. *T. suis* is characteristically elongate or spindle-shaped, occasionally piriform or rotund, 9–16 × 2–6 (mean 11 × 3) μm, with 3 anterior flagella about equal in length, 7–17 (mean 13) μm, that end in a round to spatulate knob. The undulating membrane runs the full length of the body and has 4–6 subequal folds. Its marginal filament continues as a posterior free flagellum 5–17 μm long. An accessory filament is present. The costa runs the full length of the body, and fine subcostal granules are present. The axostyle is a hyaline rod 0.6 μm in diameter with a bulbous capitulum 1.7 μm in diameter. It extends 1–17 μm beyond the body as a cone-shaped projection narrowing abruptly to a short tip. There is a chromatic ring around its point of exit. The parabasal body is usually a single, slender, tubelike structure 2–5 μm long. The nucleus is ovoid or elongated, 2–5 × 1–3 μm, with a large, conspicuous endosome surrounded by a relatively clear halo. (See Fig. 3.4.)

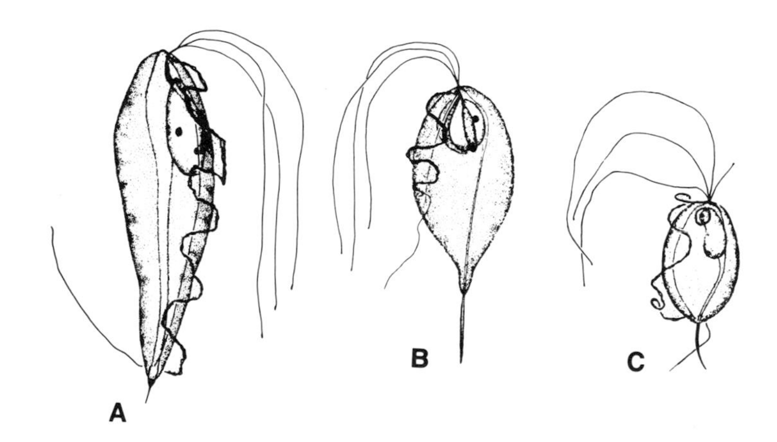

Fig. 3.4. Trichomonads of swine: A. *Tritrichomonas suis* (×3,500); B. *Trichomitus rotunda* (×2,360); C. *Trichomonas buttreyi* (×2,630). (From Hibler et al. 1960)

CULTIVATION. This trichomonad can be readily cultivated in any of the media used for *T. foetus.*

PATHOGENESIS. It was once thought that *T. suis* causes atrophic rhinitis in pigs, but it is now generally agreed that this trichomonad is not pathogenic for pigs.

T. suis may cause abortion in heifers with experimental infections of the reproductive tract. It also causes inflammation in the vagina of guinea pigs.

CULTIVATION. This trichomonad can be readily cultivated in any of the media used for *T. foetus.*

REMARKS. Cattle and swine are often raised together, and the broad host ranges and structural, metabolic, and serologic similarities between *T. suis* and *T. foetus* suggest that they may have had a common origin, if they are not indeed the same. If they are, their correct name would be *T. suis.*

Tritrichomonas enteris (Christl, 1954) Levine, 1961

SYNONYM. *Trichomonas enteris.*

HOSTS. Ox, zebu.

LOCATION. Cecum, colon.

GEOGRAPHIC DISTRIBUTION. Probably worldwide.

PREVALENCE. Common in Bavaria.

STRUCTURE. The body is 6–12 × 5–6 μm, with 3 anterior flagella of equal length arising from a single blepharoplast. The flagellum at the edge of the undulating membrane is single, without an accessory filament. The undulating membrane extends three-fourths of the body length, and a free flagellum extends beyond the undulating membrane. The axostyle is straight, slender, bent like a spoon around the nucleus, and extends at most a quarter of the body length beyond the body. Subcostal granules are present. (See Fig. 3.5.)

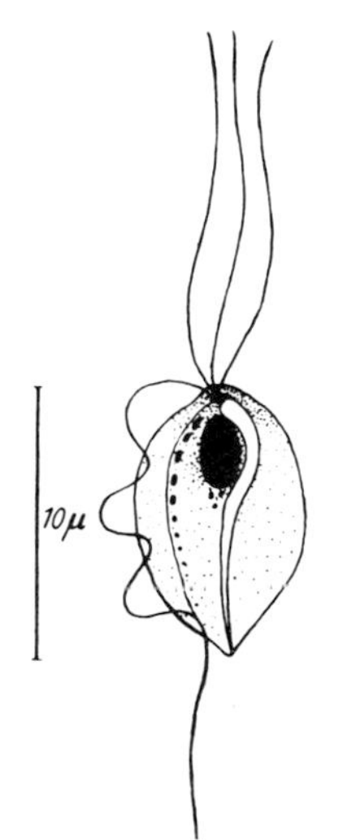

Fig. 3.5. *Tritrichomonas enteris.* (×1,950) (From Christl 1954, in Z. Parasitenkd., published by Springer-Verlag)

REMARKS. This may turn out to be a synonym of *Pentatrichomonas hominis*. However, until a further study is made, it is considered best to retain this name.

Tritrichomonas equi (Fantham, 1921)

SYNONYMS. *Trichomonas equi,* presumably *"Trichomonas" caballi.*

HOST. Horse.

LOCATION. Cecum, colon.

GEOGRAPHIC DISTRIBUTION. Presumably worldwide.

STRUCTURE. *T. equi* is about 11×6 μm and seems to have 3 anterior flagella and an undulating membrane; the axostyle is slender.

PATHOGENESIS. This species has been accused, without any proof, of causing acute equine diarrhea. Its mere presence in the feces, however, does not mean that it caused the diarrhea.

Tritrichomonas sp. (Marganti and Bitti, 1968)

HOST. Dog.

LOCATION. Feces.

GEOGRAPHIC DISTRIBUTION. Europe (Italy).

STRUCTURE. This form is lanceolate piriform, $9-24 \times 3-10$ (mean 17×6) μm, with 3 anterior flagella, an oval nucleus, an axostyle originating in a large bulb and extending slightly beyond the body of the protozoon, a short terminal filament, an undulating membrane of 4–5 undulations with a marginal filament that extends beyond the membrane as a free flagellum, and an accessory filament. It has a costa but not a pelta.

Tritrichomonas eberthi (Martin and Robertson, 1911) Kofoid, 1920

SYNONYM. *Trichomonas eberthi.*

HOSTS. Chicken, turkey, perhaps duck.

LOCATION. Ceca.

GEOGRAPHIC DISTRIBUTION. Worldwide.

PREVALENCE. Common.

STRUCTURE. The body is carrot-shaped, $8-14 \times 4-7$ μm, with vacuolated cytoplasm and 3 anterior flagella. The undulating membrane is prominent, extending the full length of the body. The posterior flagellum extends about half the body length beyond the undulating membrane. An accessory filament is present. The blepharoplast is composed of 4 equidistant granules, but tends to stain as a single body, and 5–12 or more subcostal granules are present. The axostyle is massive, hyaline, with its anterior end broadened to form a capitulum that contains siderophilic, argentophilic granules. Other endoaxostylar granules are also present. A ring of chromatic granules surrounds the axostyle at its point of emergence from the body. The parabasal body is shaped like a flattened rod, sometimes lumpy, of variable length. There are 5 chromosomes. (See Fig. 3.6.)

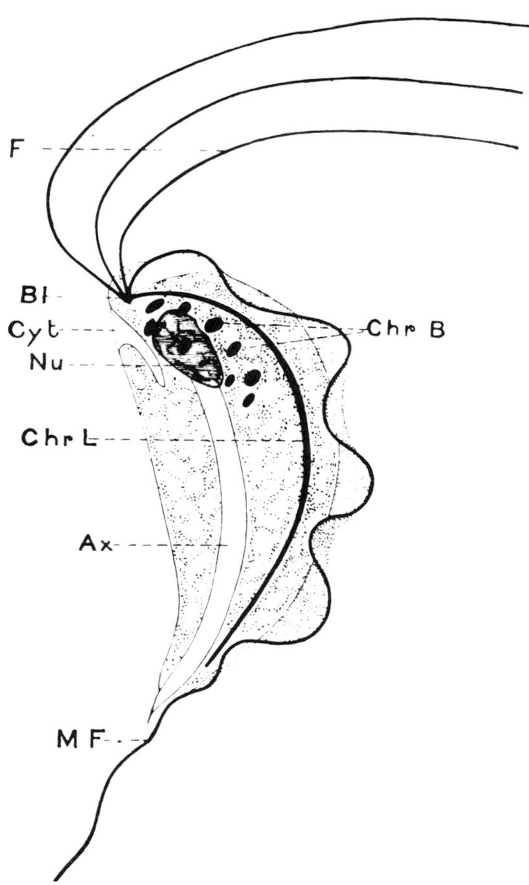

Fig. 3.6. *Tritrichomonas eberthi: F* = free flagellum (three), *Bl* = blepharoplast, *Cyt* = cytostome, *Chr B* = chromatic block (two), *Nu* = nucleus, *Chr L* = chromatic line, *Ax* = axostyle, *MF* = membrane flagellum.(×4,700) (From Martin and Robertson 1911)

PATHOGENESIS. Nonpathogenic.

Other Species of *Tritrichomonas*

Tritrichomonas muris (Grassi, 1879) (syn., *T. criceti*) occurs in the cecum, colon, and sometimes small intestine of the Norway rat, black rat, house mouse, golden hamster, and a large number of wild rodents. It is 16–26 × 10–14 μm.

T. minuta (Wenrich, 1924) occurs in the cecum and colon of the Norway rat, house mouse, golden hamster, and mountain marmot *Marmota flaviventer.* It is 4–9 × 2–5 μm.

T. caviae (Davaine, 1875) occurs in the cecum and colon of the guinea pig. It is 10–22 × 6–11 μm.

Tritrichomonas sp. Nie, 1950, occurs in the cecum of the guinea pig. It is 6–13 × 4.5–6.5 μm.

Genus *Trichomitus* Swezy, 1915

Members of this genus have 3 anterior flagella, a posterior free (trailing) flagellum, and a pelta.

Trichomitus rotunda (Hibler, Hammond, Caskey, Johnson, and Fitzgerald, 1960) Honigberg, 1963

HOST. Pig.

LOCATION. Cecum, colon.

GEOGRAPHIC DISTRIBUTION. Presumably worldwide.

STRUCTURE. *T. rotunda* is typically broadly piriform and only occasionally ovoid or ellipsoidal. It is 7–11 × 5–7 (mean 9 × 6) μm. Cytoplasmic inclusions are frequently present. The 3 anterior flagella are about equal in length, 10–17 (mean 15) μm long, and terminate in a knob or spatulate structure. The blepharoplast appears to consist of a single granule. The undulating membrane is relatively low. It and the costa extend about one-half to two-thirds the length of the body, and its undulation pattern varies from smooth to tightly telescoped or coiled waves. The accessory filament impregnates heavily with silver. The posterior free flagellum is generally shorter than the body. The axostyle, a narrow, straight, nonhyaline rod with a crescent- or sickle-shaped capitulum, extends a relatively long distance beyond the body (0.6–6.3 μm with a mean of 4.3 μm). There is no chromatic ring at its point of exit from the body. The nucleus is practically spherical, 2–3 μm in diameter, with an endosome surrounded by a clear halo. The parabasal body is 2–3 × 0.4–1.3 μm; it is composed of two rami, which form a V; each ramus has a parabasal filament.

PATHOGENESIS. Nonpathogenic.

Trichomitus fecalis (Cleveland, 1928) Honigberg, 1963

SYNONYM. *Tritrichomonas fecalis.*

This species has 3 very long flagella; a heavy undulating membrane; a long, coarse axostyle; and a costa with two rows of granules. It occurs in man.

Other Species of *Trichomitus*

T. wenyoni (Wenrich, 1946) occurs in the cecum and colon of the Norway rat, house mouse, golden hamster, rhesus monkey, and Chacma baboon. It is $4-16 \times 2.5-6$ μm.

Genus *Trichomonas* Donné, 1837

Members of this genus have 4 anterior flagella but no trailing flagellum.

Trichomonas tenax (Müller, 1773) Dobell, 1939

SYNONYMS. *Cercaria tenax, Tetratrichomonas buccalis, Trichomonas buccalis, T. elongata.*

HOSTS. Man, chimpanzee, monkeys (*Macaca mulatta, M. irus, Papio sphinx*).

LOCATION. Mouth, especially between gums and teeth.

GEOGRAPHIC DISTRIBUTION. Worldwide.

PREVALENCE. Common.

STRUCTURE. The body is ellipsoidal, ovoid, or piriform, $4-16 \times 2-15$ μm. Different strains differ in size. The 4 anterior flagella, 7–15 μm long, originate in a basal granule complex anterior to the nucleus and terminate in little knobs or rods. The undulating membrane is usually shorter than the body. An accessory filament is present. There is no free posterior flagellum. The costa is slender and accompanied by a group of large paracostal granules. The parabasal apparatus consists of a typically rod-shaped body and a long filament extending posteriorly from it. The axostyle is slender and extends a considerable distance beyond the body. There is no periaxostylar ring at its point of exit nor is it accompanied by paraxostylar granules. The capitulum of the axostyle is somewhat enlarged and spatulate. The pelta is of medium width. (See Fig. 3.7.)

PATHOGENESIS. None.

REMARKS. Hinshaw (1928) infected a dog that had gingivitis with *T. tenax.*

Trichomonas equibuccalis Simitch, 1939

HOSTS. Horse, donkey.

LOCATION. Mouth, around gums and teeth.

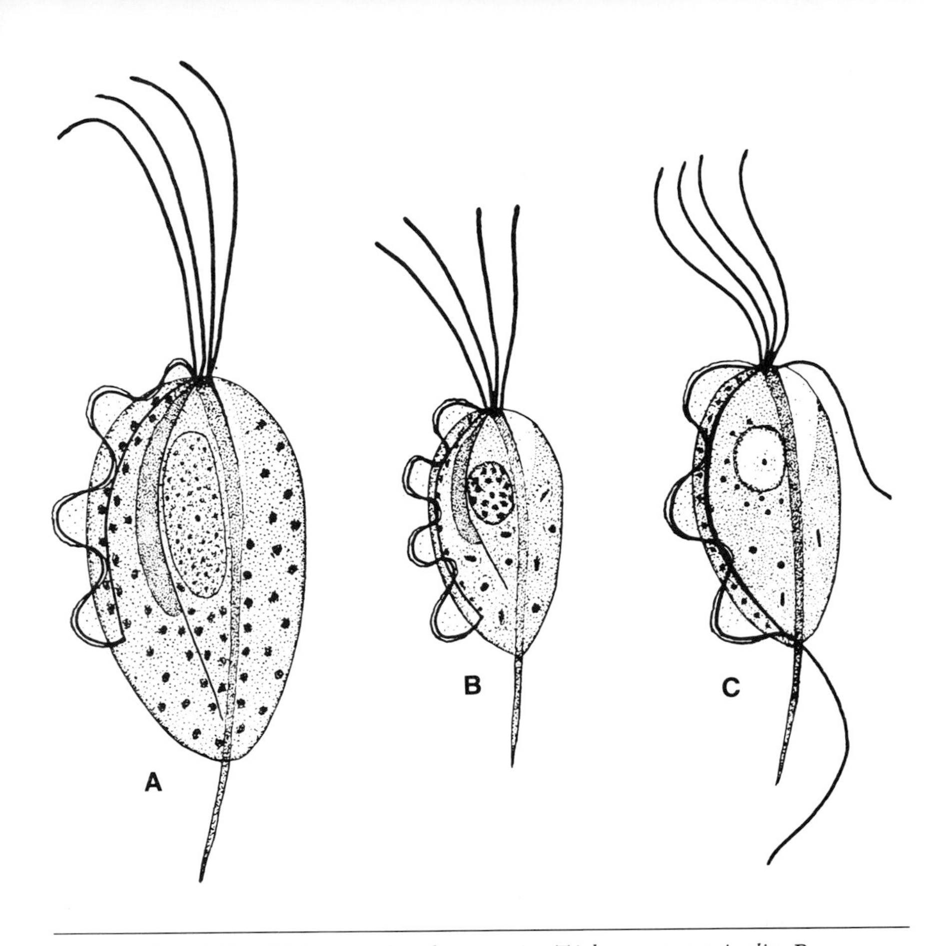

Fig. 3.7. Trichomonads of man. A. *Trichomonas vaginalis*; B. *Trichomonas tenax;* C. *Pentatrichomonas hominis.* (×2,500) (From Wenrich 1947)

STRUCTURE. The body is piriform or ovoid, 7–10 μm long. It has a single blepharoplast and 4 anterior flagella 10–15 μm long. The undulating membrane is relatively short, rarely reaching the posterior end. There is no free flagellum. The costa is slender and not always visible. The axostyle is apparently slender and extends beyond the body.

PATHOGENESIS. Nonpathogenic.

REMARKS. Simitch (1939) transmitted *T. equibuccalis* readily from the horse to the donkey and vice versa but was unable to infect cattle, sheep, and goats with it. This name may perhaps be a synonym of *T. tenax.*

Trichomonas vaginalis Donné, 1836

HOSTS. Man. The golden hamster, guinea pig, and rat can be infected intravaginally. Mice can be infected subcutaneously or intravaginally.

LOCATION. Vagina, prostate gland, urethra.

GEOGRAPHIC DISTRIBUTION. Worldwide.

PREVALENCE. Common.

PATHOGENESIS. *T. vaginalis* causes vaginitis in women. It is transmitted by intercourse. See any textbook of human parasitology for further details.

Trichomonas gallinae (Rivolta, 1878) Stabler, 1938

SYNONYMS. *Cercomonas gallinae, C. hepaticum, T. columbae, T. diversa, T. halli.*

DISEASE. Avian trichomonosis, upper digestive tract trichomonosis.

HOSTS. The domestic pigeon is the primary host, but *T. gallinae* also occurs in a number of other birds, including hawks, falcons, and eagles, which feed on pigeons. A number of other birds have been infected experimentally, including the bobwhite quail, canary, house sparrow, barn swallow, goldfinch, song sparrow, Tovi parakeet, Verraux's dove, and cardinal *Richmondena cardinalis.* Parenteral infections have also been produced experimentally in mice, guinea pigs, rats, and kittens. It causes abscesses following subcutaneous injection into mice and can be found in the macrophages.

PREVALENCE. *T. gallinae* is extremely common in domestic pigeons, in which it often causes serious losses. It is fairly common in the turkey and rare in chickens.

STRUCTURE. The body is roughly elongate ellipsoidal or piriform and 5–19 × 2–9 μm. It is quite plastic. Four anterior flagella 8–13 μm long arise from the blepharoplast. The axostyle is narrow and protrudes 2–8 μm from the body. Its anterior part is flattened into a spatulate capitulum. Anterior to it is a crescent-shaped pelta. There is no chromatic ring around its point of emergence. The parabasal body is sausage- or hook-shaped, about 4 μm long, with a parabasal filament. The costa, a very fine rod, runs two-thirds to three-fourths the length of the body. Two rows of paracostal granules stain with hematoxylin. The undulating membrane does not reach the posterior end of the body. An accessory filament is present, and a free trailing flagellum is absent. There are 6 chromosomes. (See Figs. 3.2 and 3.8.)

PATHOGENESIS. In the pigeon, trichomonosis is essentially a disease of young birds; 80–90% of the adults are infected but show no signs of disease. The disease varies from a mild condition to a rapidly fatal one with death 4–18 days after infection (due in part to differences in virulence of different strains). Severely affected birds lose weight, stand huddled with ruffled feathers, and may fall over when forced to move. A greenish fluid containing large numbers of trichomonads may be found in the mouth.

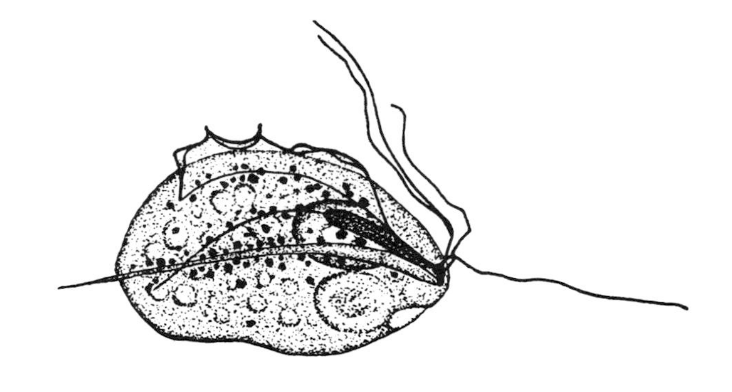

Fig. 3.8. *Trichomonas gallinae*. (×3,400) (From Stabler 1947)

Lesions are found in the mouth, sinuses, orbital region, pharynx, esophagus, crop, and even proventriculus but do not involve the digestive tract beyond the proventriculus. They often occur in the liver and to a lesser extent in other organs, including the lungs, air sacs, heart, pancreas, and more rarely other areas, such as the spleen, kidneys, trachea, bone marrow, and navel region.

The early lesions in the mouth are small, yellowish, circumscribed areas in the mucosa. They increase in number and become progressively larger, finally developing into very large, caseous masses that may invade the roof of the mouth and sinuses and may even extend through the base of the skull to the brain. The early lesions in the pharynx, esophagus, and crop are small, whitish to yellowish caseous nodules that also grow. They may remain circumscribed and separate or form thick, caseous, necrotic masses that may occlude the lumen. The circumscribed, disk-shaped lesions are often described as "yellow buttons." A large amount of fluid may accumulate in the crop. The lesions in the liver, lungs, and other organs are solid, yellowish, caseous nodules ranging up to 1 cm or more in diameter.

IMMUNOLOGY. Previous infection bestows a varying degree of immunity; adult pigeons that have survived infection as squabs are symptomless carriers. Infection with a relatively harmless strain produces immunity against virulent strains, as does injection of plasma from infected pigeons.

EPIDEMIOLOGY. In pigeons and mourning doves, trichomonosis is transmitted from the adults to the squabs in the pigeon milk, which is produced in the crop. The squabs are infected within minutes after hatching. Hawks and other wild raptors become infected by eating infected birds. Turkeys and chickens are infected through contaminated drinking water; feral pigeons and other columborid birds are the original source of infection. The trichomonads pass into the water from the mouths of infected birds, not from their droppings. *T. gallinae* has no cysts and is very sensitive to drying, so direct contamination is necessary.

DIAGNOSIS. Upper digestive tract trichomonosis is diagnosed by observation of the lesions together with finding of the protozoa. It must be differentiated from other conditions that may cause somewhat similar lesions, including fowl pox, vitamin A deficiency, moniliosis (thrush), and capillariosis.

CULTIVATION. *T. gallinae* can be cultivated readily in any of the customary trichomonad media.

TREATMENT. The best treatment for *T. gallinae* is perhaps 2-amino-5-nitrothiazole (enheptin); 6.3 g enheptin soluble per gal drinking water or 0.125% enheptin can be given for 7–14 days to nonbreeding pigeons.

CONTROL. Control of trichomonosis in pigeons depends on elimination of the infection from the adults by drug treatment. Prevention in turkeys and chickens is based on preventing wild pigeons and doves from drinking at their watering places.

Genus *Tetratrichomonas* Parisi, 1910

Members of this genus have 4 anterior flagella, a posterior free (trailing) flagellum, and a pelta.

Tetratrichomonas pavlovi (Levine, 1961) Levine, 1973

SYNONYM. *Trichomonas bovis* Pavlov and Dimitrov, 1957, [*non*] *T. bovis* Riedmüller, 1930.

HOST. Ox.

LOCATION. Large intestine.

STRUCTURE. This species is piriform, usually 11–12 × 6–7 μm. The 4 anterior flagella are about the same length as the body. The undulating membrane is well developed, with 2–4 waves, and extends almost to the posterior end of the body. A posterior free flagellum, an accessory filament, and a costa are present. The nucleus is round-ovoid or ovoid. The axostyle is relatively weak and slender, broadening to form a capitulum at the anterior end, and extending about one-fourth of its length from the posterior end of the body. There are many food vacuoles in the cytoplasm.

REMARKS. Further study is necessary to be sure whether this species is valid. Pending such a study, it is considered best to retain it.

Tetratrichomonas buttreyi (Hibler, Hammond, Caskey, Johnson, and Fitzgerald, 1960) Honigberg, 1963

HOSTS. Pig, ox.

LOCATION. Cecum, colon, rarely small intestine, rumen.

STRUCTURE. *T. buttreyi* is ovoid or ellipsoidal, 4–7 × 2–5 (mean 6 × 3) μm. Cytoplasmic inclusions are frequently present. There are 4 or 3 anterior flagella, which vary in length from a short stub to more than twice the length of the body; each ends in a knob or spatulate structure. The undulating membrane runs the full length of the body and has 3–5 undulations. The accessory filament is prominent, the costa relatively delicate. A posterior free flagellum is present. The axostyle is relatively narrow, with a spatulate capitulum, and protrudes 3–6 μm beyond the body. There is no chromatic ring at its point of exit. A pelta is present. The nucleus is frequently ovoid but varies considerably in shape; it is 2–3 × 1–2 μm and has a small endosome. The parabasal body is a disk 0.3–1.1 μm in diameter.

PATHOGENESIS. Nonpathogenic.

Tetratrichomonas gallinarum (Martin and Robertson, 1911) Honigberg, 1963

SYNONYMS. *Trichomonas gallinarum, T. pullorum.*

HOSTS. Chicken, turkey, guinea fowl, and possibly other gallinaceous birds, such as the quail, pheasant, and chukar partridge.

LOCATION. Ceca, sometimes liver, rarely blood.

GEOGRAPHIC DISTRIBUTION. Worldwide.

PREVALENCE. Common.

STRUCTURE. The body is piriform, 7–15 × 3–9 μm, with 4 anterior flagella and a posterior flagellum that runs along the undulating membrane and extends beyond it. An accessory filament is present. The axostyle is long, pointed, and slender, without a chromatic ring at its point of emergence. Supracostal granules, but no subcostal or endoaxostylar ones, are present. The pelta is elaborate, ending abruptly with a short ventral extension more or less free from the ventral edge of the axostyle. The parabasal body is usually a ring of variously spaced granules plus 1 or 2 fibrils or rami. There are apparently 5 chromosomes. A rather uniform perinuclear cloud of argentophilic granules is usually present.

PATHOGENESIS. *T. gallinarum* probably cannot cause disease by itself, although it has been accused (without proof) of doing so.

EPIDEMIOLOGY. Birds become infected by ingestion of trichomonads in contaminated water or feed.

CULTIVATION. *T. gallinarum* is readily cultivated in the usual trichomonad media.

Tetratrichomonas anatis (Kotlan, 1923)

SYNONYM. *Trichomonas anatis.*

HOST. Domestic duck.

LOCATION. Posterior part of intestinal tract.

GEOGRAPHIC DISTRIBUTION. Europe.

STRUCTURE. The body is broadly beet-shaped, 13–27 × 8–18 μm, with 4 anterior flagella, an undulating membrane extending most of the length of the body, a free trailing flagellum, a costa, and a fibrillar axostyle.

Tetratrichomonas anseris (Hegner, 1929) Levine, 1973

SYNONYM. *Trichomonas anseris.*

HOSTS. Domestic goose, baby chick (experimental).

LOCATION. Ceca.

STRUCTURE. The body is ovoid, 6–9 × 3.6–6.5 (mean 8 × 5) μm. Four anterior flagella appear to arise in pairs from 2 blepharoplasts. The undulating membrane extends almost the full length of the body. A free trailing flagellum and a costa are present. The axostyle is broad and hyaline, extending a considerable distance beyond the body; there is no chromatic ring at its point of emergence. The nucleus is characteristic, completely filled with minute chromatin granules and also with a single large karyosome, usually at one side. Many specimens have large bacteria in their endoplasm.

Tetratrichomonas ovis (Robertson, 1932) Honigberg, 1963

SYNONYMS. *Ditrichomonas ovis, Trichomonas ovis.*

HOST. Domestic sheep.

LOCATION. Cecum, rumen.

STRUCTURE. The 4 anterior flagella are of unequal length, averaging 14, 11, 9, and 7 μm, respectively. The body is piriform, 6–9 × 4–8 (mean 7 × 6) μm, with a slender, hyaline axostyle protruding a mean of 5 μm beyond the body and gradually tapering to a point; without a chromatic ring at its point of exit from the body; and with an anterior nucleus, a prominent pelta at the anterior end, a prominent undulating membrane extending from three-fourths to the full length of the body, a free flagellum extending beyond the undulating membrane, a prominent costa, several irregular rows of paracostal granules, an ovoid or club-shaped parabasal body averaging 2 × 1 μm and containing an intensely chromophilic body, and a parabasal filament.

PATHOGENESIS. None.

Tetratrichomonas felistomae (Hegner and Ratcliffe, 1927) Honigberg, 1978

SYNONYM. *Trichomonas felistomae.*

HOST. Cat.

LOCATION. Mouth.

STRUCTURE. The body is piriform, 6–11 × 3–4 (mean 8 × 3) μm, with 4 anterior flagella longer than the body. The undulating membrane extends most of the body length. There is a free posterior flagellum. The axostyle extends a considerable distance beyond the body.

PATHOGENESIS. Nonpathogenic.

Tetratrichomonas canistomae (Hegner and Ratcliffe, 1927) Honigberg, 1978

SYNONYM. *Trichomonas canistomae.*

HOST. Dog.

LOCATION. Mouth.

STRUCTURE. The body is piriform, 7–12 × 3–4 μm. Four anterior flagella about as long as the body arise in pairs from a large blepharoplast. The undulating membrane extends almost the length of the body, and the free posterior flagellum is about half as long as the body. The costa is apparently slender. The axostyle is threadlike, staining black with hematoxylin, and extends a considerable distance beyond the body. Subcostal granules are absent.

PATHOGENESIS. Nonpathogenic.

REMARKS. This species is apparently different from *T. tenax* of man. *T. canistomae* and *T. felistomae,* however, may well be the same; further study is needed to determine this.

Tetratrichomonas sp. (Morganti and Bitti, 1968)

SYNONYM. *Trichomonas* sp.

HOST. Dog.

LOCATION. Feces.

GEOGRAPHIC DISTRIBUTION. Europe (Italy).

REMARKS. Honigberg (1978a) thought this form might actually be *Pentatrichomonas hominis.*

Other Species of *Tetratrichomonas*

T. macacovaginae (Hegner and Ratcliffe, 1927) Levine, 1973 (syn., *Trichomonas macacovaginae*) occurs in the vagina of the rhesus monkey. It is 8–16 × 3–6 μm and has a free posterior flagellum, a feature that differentiates it from *T. vaginalis.*

T. microti (Wenrich and Saxe, 1950) Honigberg, 1963, occurs in the cecum of the Norway rat, house mouse, golden hamster, vole *Microtus pennsylvanicus,* and other wild rodents. It is 4–9 μm long. It can be transmitted experimentally to other rodents, the dog, and the cat.

Genus *Pentatrichomonas* Mesnil, 1914

Members of this genus have 5 anterior flagella and a pelta.

Pentatrichomonas hominis (Davaine, 1860)

SYNONYMS. *P. felis, P. ardin delteili, P. canis auri, Cercomonas hominis, Monocercomonas hominis, Trichomonas intestinalis, T. confusa, T. felis, T. parva, T. anthropopitheci.*

HOSTS. Man, gibbon, chimpanzee, orangutan, macaques, various other monkeys and baboons, dog, cat, rat, mouse, golden hamster, guinea pig, ground squirrels, and ox.

LOCATION. Cecum, colon, rumen.

GEOGRAPHIC DISTRIBUTION. Worldwide.

PREVALENCE. Common.

STRUCTURE. The body is usually piriform, 8–20 × 3–14 μm. Five anterior flagella are ordinarily present, although some organisms may have 4 and a few 3. Four of the anterior flagella are grouped together, and the 5th is separate and directed posteriorly; a 6th flagellum runs along the undulating membrane and extends beyond it as a free trailing flagellum. The undulating membrane extends the full length of the body. An accessory filament, a costa, and paracostal granules are present. The axostyle is hyaline, thick, with a sharply pointed tip but without a chromatic ring at its point of exit. The parabasal body is small and ellipsoidal. The blepharoplast is composed of 2 granules. The pelta is crescent-shaped, prolonged dorsally in a filament that passes posteriorly in the cytoplasm dorsal to the nucleus. There are 5 or 6 chromosomes.

PATHOGENESIS. Nonpathogenic.

CULTIVATION. *P. hominis* is readily cultivable in the usual trichomonad media.

Pentatrichomonas sp. Allen, 1936

SYNONYM. *P. gallinarum* Auct.

HOSTS. Chicken, turkey, guinea fowl.

LOCATION. Ceca, liver.

GEOGRAPHIC DISTRIBUTION. Probably worldwide.

STRUCTURE. *Pentatrichomonas* sp. resembles *T. gallinarum* structurally except that it has 5 anterior flagella; 4 of these are of equal length and the 5th about half as long as the others. The body is usually spherical, sometimes more or less piriform; fixed specimens are 5–8 × 3–7 (mean 7 × 5) μm. The undulating membrane extends the full length of the body, with a free flegallum at its end. A costa is present. A row of paracostal granules runs between the costa and the undulating membrane. The axostyle is slender, projecting from the posterior end but not discernible in rounded-up specimens. The blepharoplast is composed of a group of small granules.

PATHOGENESIS. Probably nonpathogenic, although it has been accused of causing disease.

LITERATURE CITED

For citations before 1965, see the *Index-Catalogue of Medical and Veterinary Zoology, Zoological Record, Biological Abstracts, Protozoological Abstracts,* and *Veterinary Bulletin.*

Honigberg, B. M. 1978a. In Parasitic Protozoa, vol. 2, ed. J. P. Kreier, 163–73. New York: Academic.
Honigberg, B. M. 1978b. In Parasitic Protozoa, vol. 2, ed. J. P. Kreier, 275–454. New York: Academic.
Levine, N. D. 1973. Protozoan Parasites of Domestic Animals and of Man. 2d ed. Minneapolis: Burgess.
Mattern, C. F. T., B. M. Honigberg, and W. A. Daniel. 1967. J. Protozool. 14:320–39.
USDA. 1965. Agric. handb. 291. Washington, D.C.: USDA.

4

Flagellates: *Histomonas, Dientamoeba,* and Related Forms

ALL genera discussed in this chapter are atypical trichomonads. *Histomonas, Parahistomonas,* and *Protrichomonas* all belong to the subfamily Protrichomonadinae in the family Monocercomonadidae. All three genera occur in domestic animals.

Genus *Histomonas* Tyzzer, 1920

The body is actively ameboid, usually rounded, sometimes elongate, with a single nucleus and 1 flagellum arising from a basal granule close to the nucleus. In recent studies a small pelta, an axostyle, and a V-shaped parabasal body have been found. A single species, *H. meleagridis,* is recognized. McDougald and Reid (1978) reviewed the genus.

Histomonas meleagridis (Smith, 1895) Tyzzer, 1920

DISEASE. Histomonosis, infectious enterohepatitis, blackhead. Lund (1969) reviewed the disease.

HOSTS. Turkey, pheasant, chicken, peafowl, guinea fowl, ruffed grouse, chukar partridge, Japanese quail, and various other gallinaceous birds.

LOCATION. Ceca, liver.

GEOGRAPHIC DISTRIBUTION. Worldwide.

PREVALENCE. Common. This parasite is practically ubiquitous in chickens although not a common cause of disease in them. It is one of the most important causes of disease in turkeys and drove many turkey raisers out of the business before control measures were developed. The USDA (1965) estimated that from 1951 to 1960 histomonosis caused annual losses of $1.9 million in chickens and $1.8 million in turkeys due to mortality, and $2.9 million in chickens and $2.7 million in turkeys due to morbidity.

STRUCTURE. *H. meleagridis* is pleomorphic, its appearance depending on its location and the stage of the disease. The forms in the tissues have no discernible flagella, although there is a basal granule near the nucleus. There are four stages.

(1) The invasive stage is found in early cecal and liver lesions and at the periphery of older lesions. It is extracellular, 8–17 μm long, and actively ameboid, with blunt, rounded pseudopods. Its cytoplasm is basophilic, with an outer zone of clear ectoplasm and finely granular endoplasm. Food vacuoles containing particles of ingested material but no bacteria are present.

(2) The vegetative stage is found near the center of the lesions in slightly older lesions than the invasive stage. It is larger, 12–21 × 12–15 μm. It is less active than the invasive stage and has few if any cytoplasmic inclusions. Its cytoplasm is basophilic, clear, and transparent. The vegetative forms are often packed tightly together and cause disruption of the tissues.

(3) The so-called resistant stage is actually no more resistant than the other stages. There are no cysts. This form is 4–11 μm in diameter, compact, and seems to be enclosed in a dense membrane. The cytoplasm is acidophilic and filled with small granules or globules. These forms may be found singly or may be packed together so that their outlines appear rather angular. They, too, are extracellular but may be taken up by phagocytes or giant cells.

(4) A fourth form is flagellated, occurs in the lumen of the ceca, and is also found in cultures. Its body is ameboid and may be 5–30 μm in diameter. The cytoplasm is usually composed of a clear outer ectosarc and a coarsely granular endosarc. It may contain bacteria, starch grains, and other food particles, including an occasional red blood cell. The nucleus is often vesicular, with a single dense karyosome or as many as 8 scattered chromatin granules (presumably chromosomes). Near the nucleus is a basal granule or kinetosome from which the flagellum arises. This is typically single and short, but as many as 4 have been reported. Movement may be ameboid, and there may be a pulsating, rhythmic, intracytoplasmic movement. The flagella produce a characteristic jerky, oscillating movement resembling that of trichomonads, but *Histomonas* has no undulating membrane. Some individuals have peculiar, cylindrical feeding tubes that sometimes extend out as much as the body diameter and often have internal extensions as long or longer. Electron micrograph studies have also revealed a parabasal body that includes a periodic fiber and Golgi vesicles, a microtubular pelta, and a microtubular axostyle. This form is sometimes present in large numbers in the lumen of the ceca but is ordinarily absent or very difficult to find. (See Fig. 4.1.)

LIFE CYCLE. Reproduction is by binary fission, and there is no evidence of a sexual cycle.

The naked trophozoites are delicate and do not survive more than a few hours when passed in the feces. Turkeys can be infected by ingesting trophozoites, and this mode of infection plays a part in transmitting the parasites

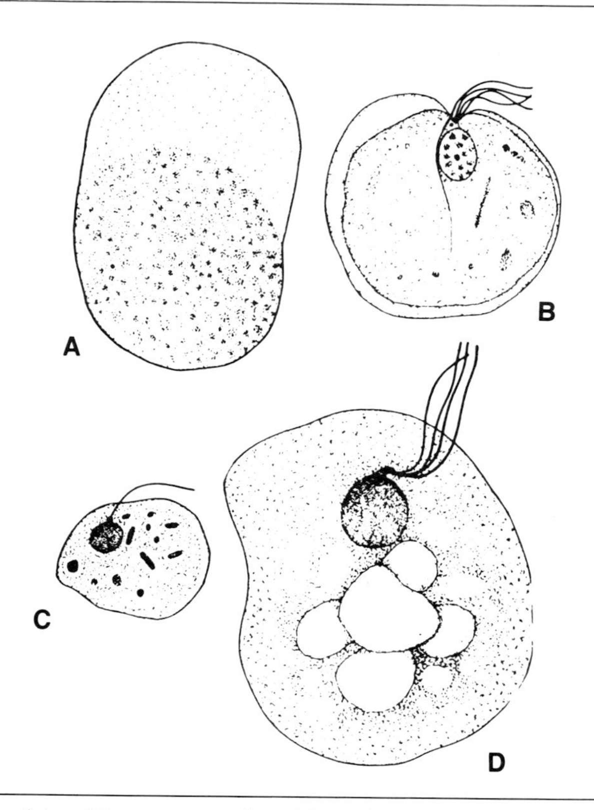

Fig. 4.1. *Histomonas meleagridis* and *Paratrichomonas wenrichi* trophozoites from cecum: A. living trophozoite of *P. wenrichi;* B and D. trophozoites of *P. wenrichi* fixed in Schaudinn's fluid and stained with iron hematoxylin; C. trophozoite of *H. meleagridis* fixed in Schaudinn's fluid and stained with iron alum hematoxylin. (From Wenrich 1943, J. Morphol. 72:279)

once disease has appeared in a flock. However, large numbers of the parasites must be ingested. Protozoa of cecal origin are about 100 times as effective in producing histomonosis in turkeys as those of liver origin (Lund and Chute 1970a), and pretreatment with bile renders the latter noninfective (Lund and Chute 1970b).

By far the most important mode of transmission is in the eggs of the cecal worm *Heterakis gallinarum*. Its discovery by Smith and Graybill (1920) was a milestone in the history of parasitology. This mode of transmission has been amply confirmed by many workers and is the preferred method of producing experimental infections. The protozoa first invade the germinal zone of the nematode ovary, where they multiply extracellularly (Lee 1969) and penetrate the developing oocytes. They feed and divide both here and in the newly formed eggs, becoming gradually smaller.

Infection of *Heterakis* is so common that *Histomonas* infections can be produced with batches of eggs taken from a very high percentage of turkeys or chickens even if the hosts do not appear sick. Not every egg is infected, however. The *Heterakis* eggs must hatch and liberate larvae in order to transmit the protozoa.

Earthworms can transmit both *Heterakis* and *Histomonas* and are probably important for the longtime survival of histomonads in the soil.

Reid (1968) reviewed the etiology and transmission of histomonosis in chickens and turkeys, giving 222 references.

EPIDEMIOLOGY. *Histomonas* is extremely common in *Heterakis*-infected chickens, and these birds are the principal reservoir of infection for turkeys. This accounts for the fact that it is almost impossible to raise turkeys successfully on the same farm with chickens. In addition, wild gallinaceous birds such as the wild turkey, pheasant, bobwhite quail, and ruffed grouse may be infected.

Birds become infected most commonly by ingesting infected *Heterakis* eggs. Infective eggs can survive in the soil for 1–2 yr or even longer. Earthworms can harbor infected *Heterakis* larvae for a long time and be a source of infection for birds.

PATHOGENESIS. Histomonosis can affect turkeys of all ages, the course and mortality of the disease varying with age. Poults less than 3 wk old are refractory, but from this age to about 12 wk the disease is acute and may cause losses averaging 50% of the flock and ranging up to 100%. The birds often die 2–3 days after showing the first signs of disease. In older birds, the disease is more chronic and recovery may take place. Mortality decreases with age. Losses in these birds rarely exceed 25%, but even birds of breeding age may be affected.

Chickens are much less susceptible than turkeys. They ordinarily show no signs of disease, but serious outbreaks may occur in young birds.

Histomonosis occasionally occurs in the peafowl, guinea fowl, bobwhite quail, ruffed grouse, and chukar partridge. Although the parasite occurs in pheasants, it is apparently not very pathogenic for them. The Japanese quail *Coturnix c. japonica* is not a suitable host, but it can be infected.

When the histomonads are released in the cecum they enter the wall and multiply, causing characteristic lesions. Later they pass by way of the bloodstream to the liver.

The incubation period is 15–21 days. The first sign of disease is droopiness; the birds appear weak and drowsy and stand with lowered head, ruffled feathers, and drooping wings and tail. There is a sulfur-colored diarrhea. The head may or may not become darkened. This sign, which is responsible for the name "blackhead," may also occur in other diseases, so the term is misleading.

The principal lesions of histomonosis occur in the cecum and liver. One or both ceca may be affected. Small, raised, pinpoint ulcers containing the parasites are formed first. These enlarge and may involve the whole cecal mucosa. Sometimes the ulcers perforate the cecal wall and cause peritonitis or adhesions. The mucosa becomes thickened and necrotic and may be covered with a characteristic, foul-smelling, yellowish exudate that sometimes consolidates to form a dry, hard, cheesy plug filling the cecum and adhering tightly to its wall.

The ceca are markedly inflamed and often enlarged.

The liver lesions are pathognomonic. They are circular, depressed, yellowish to yellowish green areas of necrosis and tissue degeneration that are not encapsulated but merge with the healthy tissue. Varying in diameter up to 1 cm or more, they extend deeply into the liver. In older birds the lesions are often confluent. Other organs such as the kidney and lung may occasionally be affected.

The parasites can be readily found on histologic examination of the lesions. Hyperemia, hemorrhage, lymphocytic infiltration, and necrosis occur, and macrophages and giant cells are present.

If the birds recover the protozoa disappear from the tissues, and repair takes place. The exudate and necrotic tissue in the ceca are incorporated into the cecal plug, which becomes smaller and is finally passed. If the lesions were not too severe, the ceca may eventually appear entirely normal, but in other cases there may be so much scarring that the lumen is obliterated. In the repair process, the lesions are invaded by blood vessels, lymphoid cells, and connective tissue. The liver lesions may be completely repaired or there may be extensive scar tissue.

Studies with gnotobiotic turkeys have revealed that *H. meleagridis* alone may infect birds but cannot produce disease. Certain bacteria must be present for the protozoon to be pathogenic. Severe disease is caused by a combination of *H. meleagridis* with *Escherichia coli, Clostridium perfringens,* or the normal bacterial flora. No lesions are produced by *H. meleagridis* alone or in combination with various other bacteria.

IMMUNITY. Birds that recover from histomonosis are immune to reinfection. In addition, as mentioned above, susceptibility decreases with age.

DIAGNOSIS. Histomonosis can be diagnosed from its lesions. Typical lesions in the liver are pathognomonic, but diffuse type lesions may be confused with those caused by other conditions. Histologic examination is desirable in case of doubt and in order to differentiate the liver lesions caused by leukosis, tumors, tuberculosis, or mycotic infections. The cecal lesions can be distinguished from those caused by coccidia by microscopic examination of scrapings from the mucosa.

CULTIVATION. *Histomonas* can be cultivated in a number of media, both diphasic and monophasic. Continued cultivation for years reduces its virulence.

PRESERVATION. *H. meleagridis* can be preserved in liquid nitrogen at -196 C in the presence of glycerol or dimethylsulfoxide without loss of virulence.

TREATMENT. Since histomonosis can be prevented by proper management, drug therapy should be regarded as a secondary line of defense against the disease. The chemotherapy of the disease has been reviewed by Reid (1968, 1969), who tabulated the results of 54 reports on chemotherapy.

The following drugs were being marketed in 1975 for the control of histomonosis: Carbarsone, nitrarsone (Histostat-50), furazolidone (NF-180; Furox), dimetridazole (Emtrymix), ipronidazole (Ipropran), nifursol (Salfuride), and ronidazole (Ridazole); the last two were not being marketed in the United States (McDougald and Reid 1978).

CONTROL. Histomonosis can be prevented by good management. Turkeys should be kept separate from chickens, since chickens are carriers. Young turkeys should be kept separate from adults. The same attendants should not care for chickens and turkeys, and persons who go from one flock to another should take care not to carry the infection on contaminated shoes or equipment.

Young birds should be raised on hardware cloth, and their droppings removed regularly. When the poults are old enough to move onto range, they should be placed on clean ground where neither turkeys nor chickens have been kept for 2 yr. The length of time infective cecal worm eggs or earthworms survive in the soil depends on soil type, weather, and the amount of cover provided by vegetation. The eggs will survive only a few weeks on barren soils in warm, dry regions but may remain alive for several years in heavy soils in most climates.

The range should be rotated at regular intervals. Different farmers use different intervals. Many of them move the birds along every week, not returning them to the same place during the season. The frequency of rotation depends on the climate. In cool, damp climates the birds should be moved at least every 10 days, but in hot, dry climates they need to be moved less frequently; it is even possible to raise turkeys successfully without changing the range if the area around the feeders, waterers, roosts, and shelters is kept dry.

Low areas and streams that drain poultry yards should be fenced off. The feeders and waterers should be placed on wire platforms. Most of the droppings are deposited here, and this practice keeps the turkeys from getting at them. Wire should also be used beneath roosts and in shelters to keep the birds from their droppings.

Genus *Parahistomonas* Honigberg, 1969

This genus resembles *Histomonas,* except that it has 4 flagella and a rod-shaped parabasal body; no flagellate tissue-dwelling stage is known. There is one species. McDougald and Reid (1978) reviewed the genus.

Parahistomonas wenrichi (Lund, 1963) Honigberg, 1969

This organism occurs in the ceca of the turkey and probably other birds. It is transmissible to the chicken and ring-necked pheasant. When quiescent it is spherical or ovoid, 9–27 (mean 17) μm in diameter. The ectoplasm is clear and the endoplasm granular, with numerous food vacuoles. The nucleus is vesicular and seldom more than 4 μm in diameter. There are 8 chromosomelike bodies. There are characteristically 4 slender flagella, but locomotion and feeding are by pseudopods. *P. wenrichi* has an axostyle that does not pass out of the body, a parabasal body, and a parabasal filament. (See Fig. 4.2.)

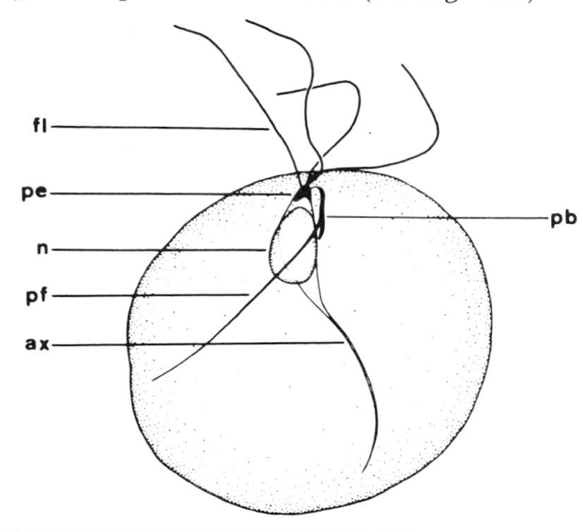

Fig. 4.2. *Parahistomonas wenrichi* trophozoite. Organism from chick liver cell culture; flagella drawn from organism from chick cecum; *ax* = axostyle, *fl* = flagellum; *pb* = parabasal body, *pe* = pelta, *pf* = parabasal filament, *n* = nucleus. Fixed in Hollande's fluid and stained with protargol. (×2,800) (From Honigberg and Kuldova 1969)

P. wenrichi is nonpathogenic. Like *Histomonas,* it can be transmitted in the eggs of *Heterakis gallinarum;* the protozoa are apparently released from the nematode larvae about 6–15 days after the eggs have been ingested by chicks or turkey poults.

Lund (1963) was unable to cultivate this species in the media used for *Histomonas,* trichomonads, parasitic amebae, or other parasitic protozoa.

Because *P. wenrichi* resembles *H. meleagridis* so closely yet is nonpathogenic, it is important to be able to distinguish the two. Perhaps the most important difference (other than number of flagella, which are impossible to see with the light microscope in life and extremely hard to see in stained preparations) is the difference in size; *P. wenrichi* is about 1.5 times as large as *H. meleagridis,* but their ranges overlap.

Genus *Protrichomonas* Alexeieff, 1912

The body is piriform or beet-shaped, with 3 anterior flagella of equal length arising from an anterior blepharoplast. An axostyle also arises from the blepharoplast. The nucleus is anterior and vesicular. Three species have been named from mammals, birds, and a fish.

Protrichomonas ruminantium (Braune, 1913) Levine, 1961, occurs in the rumen of cattle and sheep. It is about 8 μm long, with a nucleus often surrounded by a clear zone, and apparently without a cytostome. It is not pathogenic. This species, which has apparently never been described with the aid of modern techniques, may not be a valid form at all.

Protrichomonas anatis Kotlan, 1923, occurs in the large intestine of the domestic duck and other waterbirds. It is 10–13 × 4–6 μm. Two distinct fibrillae arise from the anterior blepharoplast and pass back through the body, separating to pass around the nucleus and finally leaving the body as a pointed axostyle. The nucleus is often triangular. *P. anatis* is not pathogenic. Like *P. ruminantium,* it has apparently never been studied by modern methods.

Genus *Dientamoeba* Jepps and Dobell, 1918

These organisms were thought to be amebae until their relationship to the trichomonads was discovered. They usually have 2 vesicular nuclei with a delicate membrane and an endosome consisting of several chromatin granules connected to the nuclear membrane by delicate strands. No cysts are known. This genus belongs in the subfamily Dientamoebinae Grassé, 1953, in the trichomonad family Monocercomonadidae Kirby, 1947. There is a single species, *D. fragilis.* McDougald and Reid (1978) reviewed the genus.

Dientamoeba fragilis Jepps and Dobell, 1918

This species occurs in the cecum and colon of man and also of some monkeys and rarely the sheep. It is quite common in man.

Only trophozoites are known. They are 3–22 μm, usually 6–12 μm, in diameter. The ectoplasm is distinct from the endoplasm, which contains food vacuoles filled with bacteria, yeasts, starch granules, and parts of cells. In fresh feces there may be a single, clear, broad pseudopod. About three-fifths of the protozoa contain 2 nuclei, which are connected by a filament or desmose. This appears to be one of the first structures to disappear during degeneration. Each nucleus is vesicular and has an endosome composed of 4–8 granules, from which a few delicate fibers radiate to the nuclear membrane. There is no peripheral chromatin.

Reproduction is by binary fission. There are 4 chromosomes.

At one time *D. fragilis* was thought to be nonpathogenic, and this is true in most cases. However, in some persons it causes a mucous diarrhea and gastrointestinal symptoms. It does not invade the tissues but may cause low-grade irritation of the intestinal mucosa, excess mucus secretion, and hypermotility of the bowel. There may also be mild to moderate abdominal pain and tenderness or discomfort and sometimes an increase in eosinophils.

The mode of transmission of *D. fragilis* is not clear, since there are no cysts and the trophozoites are so delicate. The pinworm *Enterobius vermicularis* might be the vector.

D. fragilis can be readily cultivated in the usual culture media. It is sensitive to most amebicidal drugs, including carbarsone, diodoquin, and erythromycin. It can be preserved by freezing with dimethyl sulfoxide. See any textbook of human parasitology for further information.

LITERATURE CITED

For citations before 1965, see the *Index-Catalogue of Medical and Veterinary Zoology, Zoological Record, Biological Abstracts, Protozoological Abstracts,* and *Veterinary Bulletin.*

Honigberg, B. M., and Kuldova, J. 1969. J. Protozool. 16:526–35.
Lee, D. L. 1969. Parasitology 59:877–84.
Lund, E. E. 1963. J. Protozool. 10:401–4.
______. 1969. Adv. Vet. Sci. 13:355–90.
Lund, E. E., and A. M. Chute. 1970a. J. Protozool. 17:284–87.
______. 1970b. J. Protozool. 17:564–66.
McDougald, L. R., and W. M. Reid. 1978. In Parasitic Protozoa, vol. 2, ed. J. P. Kreier, 139–61. New York: Academic.
Reid, W. M. 1968. Exp. Parasitol. 21:249–75.
______. 1969. Natl. Acad. Sci. pub. 1679, 87–99. Washington, D.C.
USDA. 1965. Agric. handb. 291. Washington, D.C.: USDA.

5

Flagellates: *Spironucleus*, *Giardia*, and Other Flagellates

IN this chapter is discussed a miscellany of flagellates, most of which are found in the digestive tract. Only a few are pathogenic, the great majority being commensal. Some are not parasitic at all but are coprophilic or have been found as contaminants in washings from the sheath of bulls; these are mentioned because they must be differentiated from parasitic forms. A few other species are free-living toxin producers. The classification of the flagellates (to suborders) was revised by Levine et al. (1980), and a revised version is given in Chapter 1.

ORDER PHYTOMONADORIDA

In this order a single, large, green chromatophore is typical. The stored reserves are starch or lipids. No members of this order are parasites of man or domestic animals, but several produce toxins and one other is a pseudoparasite.

Toxic Marine Phytoflagellates

The great majority of phytoflagellates are free-living and holophytic. There are a great many of them. Sournia (1978) gave a manual of phytoplankton. Some of them produce powerful toxins that may kill fish or even man.

Gonyaulax catanella is a marine dinoflagellate found particularly off the coast of California. Its toxin, one of the most powerful known, causes a fatal disease of man known as mussel poisoning. Under conditions still largely unknown, the protozoa multiply tremendously, forming a luminescent bloom in the ocean. Mussels and certain other shellfish feeding on plankton are not harmed by the toxin but accumulate it in their internal organs; people who eat these mussels may then be killed by the toxin.

The blooming of other dinoflagellates, including several species of *Gymnodinium*, cause the "red tide" or "red water" that sometimes kills huge numbers of fish. This condition is particularly common off the coast of Florida, where it is often associated with the discharge of phosphates into the ocean. It also occurs off the Texas coast and elsewhere.

Ciguatera fish poisoning occurs occasionally in people who eat tropical fish. The eating of groupers or red snapper, which are large fish at the top of the food chain, is most commonly associated with it, but other fish may cause it also. The symptoms include nausea, diarrhea, a tingling or burning sensation of the mouth or feet, weakness, numbness, muscle aches, and vomiting (CDC 1980). It is thought to be caused by a toxin produced by the dinoflagellate *Gambierdiscus toxicus*, which occurs on tropical reefs. The fish concentrate the toxin, which is tasteless and not destroyed by cooking, and people become affected by eating the fish.

Another marine phytoflagellate, the chrysomonad *Prymnesium parvum*, has killed fish en masse in brackish fish ponds in Israel and has formed blooms accompanying fish kills in Holland and Denmark.

Other Phytoflagellates

Polytoma uvella Ehrenberg, 1838, occurs in infusions and stagnant water and has been found in bull sheath washings submitted for *Tritrichomonas foetus* diagnosis. It is ovoid to piriform, 15–30 × 9–20 μm, without chromatophores and with numerous starch granules in the posterior part. A red or pink stigma may or may not be present. There are 2 anterior flagella of equal length.

Parasitic Flagellates

The parasitic mode of life has arisen independently a number of times in this group. Free-living species or genera in distantly related families have found suitable living conditions in various hosts. Many of these parasites were previously inhabitants of stagnant water. The fact that so few of them are pathogenic effectively refutes the notion that parasites tend to be pathogenic in their first association with a new host.

Specific Genera

Genus *Retortamonas* Grassi, 1879

The body is usually piriform or fusiform, drawn out posteriorly, and plastic. Near the anterior end is a large cytostome containing in its margin a cytostomal fibril that extends across the anterior end and posteriorly along each side. An anterior flagellum and a posteriorly directed, trailing flagellum emerge from the cytostomal groove (Fig. 5.1). The cysts are piriform or ovoid, have 1 or 2 nuclei, and retain the cytostomal fibril. Species occur in various insects, amphibia, reptiles, and mammals (Kulda and Nohynkova 1978).

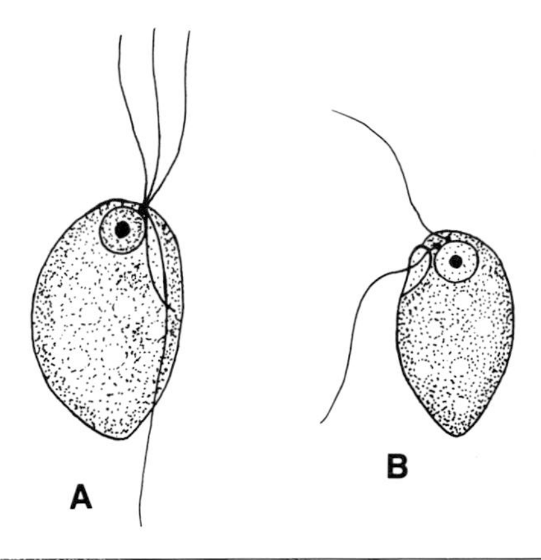

Fig. 5.1. A. *Enteromonas*; B. *Retortamonas* (×2,800)

R. intestinalis (Wenyon and O'Connor, 1917) Wenrich, 1932 (syns., *Embadomonas intestinalis, Waskia intestinalis*) occurs in the cecum of man, the chimpanzee, macaques, and other monkeys.

The trophozoites are elongate piriform, 4–9 × 3–4 μm. The cysts are uninucleate, piriform, 4.5–7 × 3–4.5 μm, and have a rather thick wall. *R. intestinalis* is not common in man and is nonpathogenic. It can be cultivated in the usual culture media for intestinal protozoa.

R. ovis (Hegner and Schumaker, 1928) (syns., *Embadomonas ovis, E. ruminantium*) occurs in the cecum of sheep and cattle. The trophozoites are piriform and average 5.2 × 3.4 μm. *R. ovis* is apparently not pathogenic.

R. cuniculi (Collier and Boeck, 1926) (syn., *Embadomonas cuniculi*) occurs in the cecum of the rabbit. The trophozoites are generally ovoid but occasionally have a taillike process; they are 7–13 × 5–10 μm. The cysts are grape seed–shaped, 5–7 × 3–4 μm. It is apparently nonpathogenic.

R. caviae (Hegner and Schumaker, 1928) occurs in the guinea pig.

Genus *Chilomastix* Alexeieff, 1912

The body is piriform and plastic, with a large cytostomal groove near the anterior end containing in its margin a cytostomal fibril that extends across the anterior end and posteriorly along each side (Fig. 5.2). The nucleus is anterior. There are 3 anteriorly directed flagella and a short 4th flagellum that undulates within the cytostomal cleft. Cysts are formed. *Chilomastix* is found in mammals, birds, reptiles, amphibia, fish, insects, and leeches. All species are apparently nonpathogenic or only slightly pathogenic. Kulda and Nohynkova (1978) reviewed the genus.

Chilomastix mesnili (Wenyon, 1910) Alexeieff, 1912 (syns., *Macrostoma mesnili, Chilomastix suis, C. hominis*), is found in the cecum and colon of man, the orangutan, the chimpanzee, a number of macaques and other monkeys, and the pig. It is quite common.

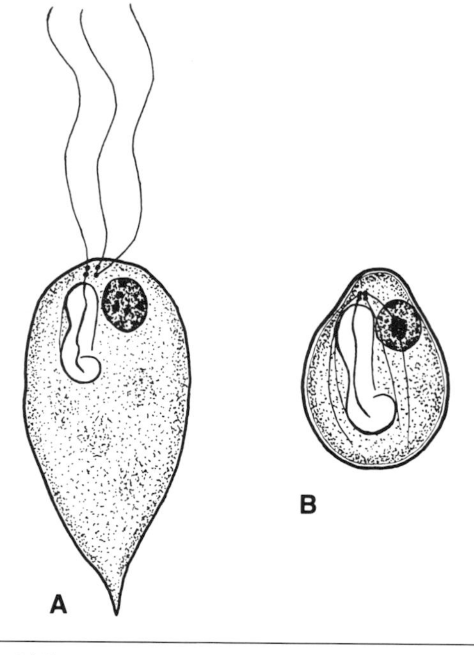

Fig. 5.2. *Chilomastix* sp.: A. *trophozoite;* B. cyst. (×2,800)

Its trophozoites are asymmetrically piriform, with a spiral groove running through the middle half of the body. The posterior end is drawn out when the protozoa are moving. The trophozoites are 6–24 × 3–10 μm. The cytostomal cleft is about 6–8 μm long and 2 μm wide. A complex of 6 minute basal granules lies anterior to the nucleus; from them come the 3 free anterior flagella (2 short and 1 relatively long), the cytostomal flagellum, and the 2 cytostomal fibrils. The cysts are lemon-shaped, 6.5–10 μm long, and contain a single nucleus and the fibrils of the trophozoite. *C. mesnili* is ordinarily considered nonpathogenic.

C. equi Abraham, 1961, was reported from the small (?) intestine of the horse. It is piriform, 16–32 × 6–16 (mean 20 × 12) μm, with a posterior spike 4–17 (mean 6.5) μm long.

C. caprae da Fonseca, 1915, occurs in the rumen of the goat. It is similar to *C. mesnili* and is 8–10 × 4–6 μm.

C. cuniculi da Fonseca, 1915, occurs in the cecum of the domestic rabbit. It is similar to *C. mesnili*, with trophozoites ordinarily 10–15 μm long.

C. gallinarum Martin and Robertson, 1911, occurs commonly in the ceca of the chicken and turkey. The body is pear- or carrot-shaped, 11–20 × 5–12 μm. The nucleus is pressed against the anterior end of the body. The cytostomal pouch is 8-shaped; it spirals toward the left on the ventral side and extends one-half to two-thirds of the body length. Cysts are rare in cecal material but common in culture. They are lemon-shaped, 7–9 × 4–6 μm, with a single nucleus.

Chilomastix sp. was found in the posterior intestine of geese in Chile (Kösters et al. 1976). They transmitted it to chickens and thought that birds had a single species of *Chilomastix.*

C. intestinalis Kuczynski, 1914, *C. wenrichi* Nie, 1948, and *C. megamorpha* Abraham, 1961, occur in the cecum of the guinea pig; *C. bettencourti* da Fonseca, 1915, occurs in the cecum of the laboratory rat, wild Norway rat, domestic mouse, and golden hamster (see also Brugerolle 1973).

Genus *Enteromonas* da Fonseca, 1915

The body is spherical or piriform and is plastic (Fig. 5.1). It has 3 short anterior flagella, one of which may be difficult to see, and a 4th long flagellum that runs along the flattened body surface and extends free for a short distance at the posterior end of the body. A strandlike funis arises from the blepharoplast and extends posteriorly along the body surface; it stains faintly with iron hematoxylin and strongly with protargol. The nucleus is anterior, vesicular, with or without an endosome. There is no cytostome. The cysts are ovoid and are tetranucleate when mature. This genus (syn., *Tricercomonas)* has been reported from a number of mammals. It belongs to the order Enteromonadorida, from which the order Diplomonadorida presumably arose by mirror-image doubling. See Brugerolle (1975) and Kulda and Nohynkova (1978).

Enteromonas hominis da Fonseca, 1915 (syns., *Octomitus hominis, Tricercomonas intestinalis, Diplocercomonas soudanensis, Enteromonas bengalensis*), occurs in the cecum of man, the chimpanzee, macaques, the golden hamster, the Norway rat, the cottontail, and probably other animals throughout the world.

The trophozoite is oval, 4–10 × 3–6 μm, and has many food vacuoles containing bacteria. The cysts are ovoid or ellipsoidal and are usually binucleate but have 4 nuclei when mature. *E. hominis* is readily cultivated in the usual media for enteric protozoa such as LES medium; cysts form in the cultures. It is nonpathogenic.

E. suis (Knowles and Das Gupta, 1929) Dobell, 1935 (syn., *Tricercomonas suis*), occurs in the cecum of pigs. It is shaped like a broad, ovate leaf with a more or less rounded anterior end and a pointed posterior end. It is 9–20 × 6–14 μm and moves sluggishly more or less directly forward, not rotating like *Trichomonas.* The 3 anterior flagella are 8–18 (mean 14) μm long; the posterior flagellum is 9–26 (mean 17) μm long.

E. caviae Lynch, 1922, has been reported from the guinea pig.

Genus *Caviomonas* Nie, 1950

The body is naked, with a vesicular nucleus at the anterior end. One flagellum arises from the nuclear membrane. A bandlike peristyle arises from the nuclear membrane opposite the origin of the flagellum and extends posteriorly along the periphery of the body surface; it stains with hematoxylin and protargol. Cytostome and contractile vacuoles are absent. See Kulda and Nohynkova (1978).

C. mobilis Nie, 1950, occurs in the cecum of the guinea pig. The body is ovoid to elongate carrot-shaped, 2–7 × 2–3 (mean 4 × 3) μm. The posterior end is often pointed.

Genus *Spironucleus* Lavier, 1936

The body is piriform, with 2 nuclei near the anterior end; 6 anterior and 2 posterior flagella; and two independent, hollow pseudoaxostyles. The body is quite symmetrical, with 3 anterior flagella and 1 posterior one arising on each side. There is no cytostome. All members of this genus are parasitic in invertebrates and vertebrates. A partial synonym of this genus, *Hexamita,* is still a valid genus but its members are usually free-living (Brugerolle et al. 1980; Kulda and Nohynkova 1978).

Spironucleus meleagridis (McNeil, Hinshaw, and Kofoid, 1941)

SYNONYM. *Hexamita meleagridis.*

DISEASE. Spironucleosis, hexamitosis, infectious catarrhal enteritis.

HOSTS. Turkey, peafowl, California valley quail, Gambel's quail, chukar partridge, ring-necked pheasant, golden pheasant. *S. meleagridis* has been transmitted from the turkey to the chicken, quail, and domestic duck, and from the ring-necked pheasant, quail, and chukar partridge to the turkey.

LOCATION. Duodenum and small intestine of younger birds; some occur in the cecum and bursa of Fabricius, especially in adults.

GEOGRAPHIC DISTRIBUTION. Worldwide.

PREVALENCE. Spironucleosis occurs in all major turkey-producing areas in the United States and in other countries as well. It appears to be particularly important in California. The USDA (1965) estimated that it caused an annual loss of $1,885 million in 1951–1960; of this, $1,057 million was due to morbidity.

STRUCTURE. The body (Fig 5.3) is 6–12 × 2–5 (mean 9 × 3) μm. The 2 nuclei contain round endosomes two-thirds the diameter of the nucleus. Anterior to each nucleus is a large blepharoplast or group of basal granules, from which 2 anterior and 1 anterolateral flagella arise. Just behind this blepharoplast is a basal granule from which the caudal flagellum arises. The caudal flagella pass posteriorly in a granular line of cytoplasm to their points of emergence near the posterior end of the body. *Spironucleus* moves rapidly without the spiraling characteristic of trichomonads.

LIFE CYCLE. Multiplication is by longitudinal binary fission.

PATHOGENESIS. Spironucleosis is a disease of young birds; adults are symptomless carriers. The mortality in a flock may be 7–80%, but heavy losses seldom occur in poults over 10 wk old. Affected poults appear nervous at first and have a stilted gait, ruffled, unkempt feathers, and a foamy, watery diarrhea. They usually continue to eat but chirp continually. Losing weight rapidly, they be-

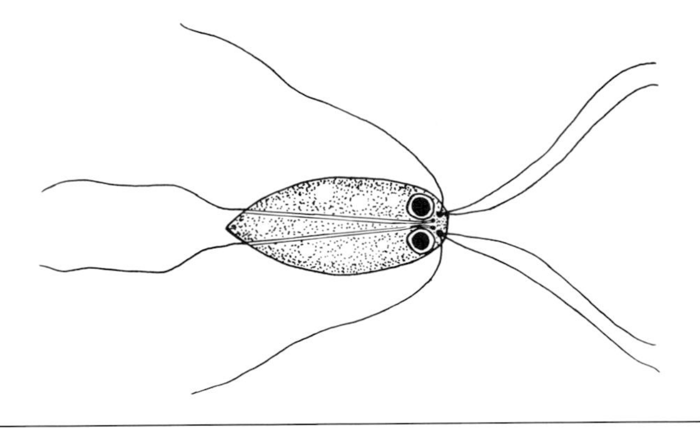

Fig. 5.3. *Spironucleus meleagridis.* (×2,800)

come listless and weak and finally die. Birds often do not appear to be ill until shortly before death, but examination reveals that they are thin and have lowered temperatures. Birds that recover grow poorly; an outbreak may leave many stunted birds in its wake.

The incubation period is 4–7 days. Poults may die within 1 day after signs appear. In acute outbreaks, the mortality reaches a peak in the flock in 7–10 days after the first birds die. In other flocks, deaths may continue for 3 wk.

The principal pathologic changes are found in the small intestine. Catarrhal inflammation with marked lack of tone is present in the duodenum, jejunum, and ileum. The intestinal contents are usually thin, watery, and foamy, with localized bulbous swellings filled with watery fluid. The small intestine, especially the anterior part, is inflamed and edematous. The cecal contents are usually fluid, the cecal tonsils congested.

EPIDEMIOLOGY. *Spironucleus* is transmitted through contaminated feed and water. Carrier adult birds that have survived earlier attacks are the most important source of infection for turkey poults. Sometimes the disease does not appear in the earlier hatches but strikes the later ones after the adults have been sold, possibly because of very light infections in the earlier hatches or perhaps because of a difference in virulence. An outbreak of a strain from carrier turkeys may take several passages in poults to become acute.

Wild quail, pheasants, and chukar partridges sharing the range with turkeys may also be a source of infection. Hot weather and overcrowding may contribute to the severity of an outbreak.

DIAGNOSIS. Spironucleosis can be diagnosed by finding the protozoa in fresh scrapings from the small intestine, particularly from the jejunum and duodenum. *Spironucleus* can be readily differentiated from *Trichomonas, Giardia,* or *Cochlosoma* by its small size, absence of a sucker or undulating membrane, and characteristic motion. Impression smears can also be made of cross sections of fresh small intestine, dried rapidly and stained with Giemsa stain; the protozoa are found in groups in the crypts. *Spironucleus* can also be found in the bursa of Fabricius and cecal tonsils in carrier birds.

TREATMENT. Most drugs are ineffective against *S. meleagridis.*

PREVENTION AND CONTROL. Spironucleosis can be prevented by proper management and sanitation. Poults should be separated from adults, and different caretakers should be used for the two groups. If feasible, breeding birds should be sold 2 wk before any poults are hatched. Separate equipment should be used for different groups of birds, attendants should keep out of the pens, and the feeders and waterers should be placed on wire platforms where they can be reached from the outside. Young birds should be kept on wire. Ranges frequented by pheasants, quail, or partridges should be avoided. General sanitation and fly control should be practiced.

Other Species of *Spironucleus*

S. columbae (Nöller and Buttgereit, 1923) (syns., *Hexamita columbae, Octomitus columbae*) occurs in the duodenum, jejunum, ileum, and large intestine of the pigeon. It is 5–9 × 2.5–7 μm. It is pathogenic, causing a catarrhal enteritis.

Spironucleus sp. was found by Kotlan (1923) in the intestinal mucus of the domestic duck in Hungary. He called it "*Hexamitus intestinalis* (?)." It was usually piriform, 10–13 × 4–5 μm. Kimura (1934) found *Spironucleus* sp. in the large intestine of domestic ducks in California. Both of these may be *S. meleagridis.*

S. muris (Grassi, 1881) (syns., *Hexamita muris, Octomitus muris, Syndyomita muris*) occurs in the posterior small intestine and cecum of the Norway rat, house mouse, golden hamster, and various wild rodents. It is 7–9 × 2–3 μm. It may cause diarrhea and enteritis in laboratory mice.

S. pitheci (da Cunha and Muniz, 1929) (syns., *Hexamita pitheci, Octomitus pitheci*) occurs in the large intestine of rhesus monkeys and probably chimpanzees. It is 2.5–6 × 1.5–4 μm. Wenrich (1933) described *Spironucleus* sp. from the feces of a rhesus monkey in the United States.

Genus *Octomitus* von Prowazek, 1904

The body is piriform, with 2 nuclei near the anterior end, and 6 anterior and 2 posterior flagella. The body is quite symmetrical, with 3 anterior flagella and 1 posterior one arising on each side. There are 2 axostyles that originate at the anterior end and fuse as they pass posteriorly, emerging as a single central rod from the middle of the posterior end. This genus differs from *Spironucleus,* of which it was formerly considered a synonym, in the structure of its axostyles (see also Kulda and Nohynkova 1978).

O. pulcher (Becker, 1926) Gabel, 1954 (syn., *O. intestinalis*), occurs in the cecum of the Norway rat, house mouse, golden hamster, ground squirrels, and other wild rodents. It is 6–10 × 3–7 μm.

Genus *Giardia* Kunstler, 1882

The names given the species of *Giardia* depend on the authorities concerned. Traditionally, it has been believed that *Giardia* is highly host-specific, and different names have been given to almost all the forms in different hosts. According to this view, the species in cattle is *G. bovis*, in goats and sheep *G. caprae*, in the dog *G. canis*, in the cat *G. cati*, in the rabbit *G. duodenalis*, in the guinea pig *G. caviae*, in the Norway rat *G. muris* and *G. simoni*, in the house mouse *G. muris* and *G. microti*, and in man either *G. lamblia* or *G. intestinalis*. However, the view that is gaining adherents is that there are only three species, *G. agilis* of amphibia, *G. muris*, and *G. duodenalis*. A number of outbreaks of giardiosis have occurred among backpackers and hikers in the mountains of the western United States. Their infections were acquired in areas uninhabited by man. Presumably they were due to drinking water contaminated by infected wild animals such as beaver (see Jakubowski and Hoff 1979).

Some cross-transmission experiments have succeeded, while others have failed. Visvesvara et al. (1980a) found that strains of *Giardia* from the cat, guinea pig, and man were antigenically the same. Until further cross-transmission and antigenic studies have been carried out, it is perhaps convenient to use different specific names for most of the forms from different hosts, but it should be remembered that all those of mammals but *G. muris* (and *G. mesocricetus*, which may be a synonym of *G. muris*) look essentially alike.

Associated with this nomenclatorial problem is an important epidemiologic one. If *Giardia* can be transmitted from one host to another, there is obvious danger to man from infections in laboratory and domestic animals, and of infections in one domestic animal from another. Further research is needed on this problem.

The body (Fig. 5.4) is piriform to ellipsoidal, and bilaterally symmetrical. The anterior end is broadly rounded, and the posterior end is drawn out. There is a large sucking (adhesive) disk on the ventral side; the dorsal side is convex. There are 2 anterior nuclei, 2 slender median rods or axostyles, 8 flagella in 4 pairs, and a pair of darkly staining median bodies. It lies flat on the surface of the epithelium. The cysts have 2 or 4 nuclei and a number of fibrillar remnants of the trophozoite organelles. A synonym of this generic name is *Lamblia*. Kulda and Nohynkova (1978) reviewed the genus.

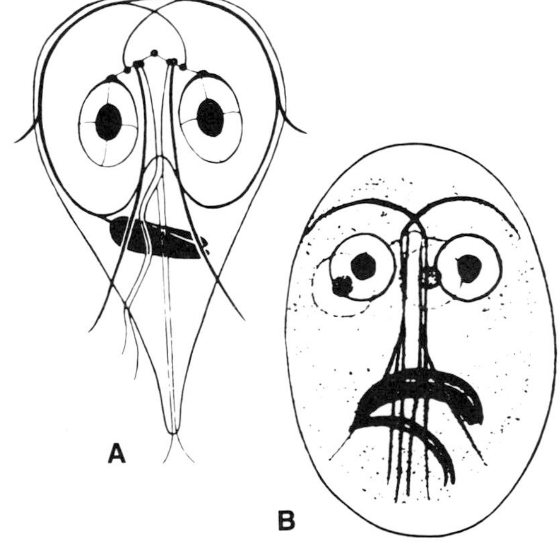

Fig. 5.4. A. *Giardia* sp. trophozoite (×2,800). (After Filice 1952) B. *Giardia bovis* cyst (×2,900). (From Becker and Frye 1929)

Giardia may cause disease on occasion in man and presumably also in other animals. However, the factors that govern its pathogenic activity are still far from clear. *Giardia* can be cultivated axenically in special media (Gillin and Diamond 1979; Meerovitch 1978; Meyer 1979; Visvesvara 1980).

Giardia lamblia Kofoid and Christiansen, 1915

SYNONYMS. *Cercomonas intestinalis, Lamblia intestinalis, G. intestinalis, Megastoma entericum, G. enterica.* European writers still call this species *G. intestinalis*, but there was so much confusion about the availability of the specific names *intestinalis* and *enterica* that Kofoid and Christiansen established the present name in a letter to Stiles. Bray (1964) preferred the name *G. intestinalis.*

DISEASE. Giardiosis.

HOSTS. Man, gorilla, chimpanzee, orangutan, gibbon, pig (?), dog (?), cat (?), guinea pig (?), and possibly many other laboratory, domestic, and wild mammals.

LOCATION. Duodenum, jejunum, upper ileum.

GEOGRAPHIC DISTRIBUTION. Worldwide.

PREVALENCE. *G. lamblia* is common in man. The U.S. Center for Disease Control (CDC 1977) reported it in 3.8% of 414,820 human stool samples examined by state and territorial public health laboratories in 1976, compared with 0.6% for *Entamoeba histolytica*. In 1978 it was found in 3.9% of 332, 312 stools, compared with 0.7% for *E. histolytica*. It is about 3 times as common in children as in adults. See Jakubowski and Hoff (1979) for reports of human outbreaks.

STRUCTURE. The trophozoites are 9–21 (usually 12–15) μm long, 5–15 μm wide, and 2–4 μm thick. The median bodies are curved bars of the *duodenalis* type. The cysts are ovoid, 8–12 × 7–10 μm, and contain 4 nuclei.

PATHOGENESIS. *Giardia* is pathogenic in some persons. Most infections are symptomless, but in a fairly small number there is chronic diarrhea. Waterborne outbreaks of giardiosis are becoming of increasing concern (Jakubowski and Hoff 1979).

DIAGNOSIS. *Giardia* infections can be diagnosed by recognition of the trophozoites or cysts in stained fecal smears. Fixation with Schaudinn's fluid and staining with iron hematoxylin are recommended. Trophozoites alone are generally found in diarrheic stools. Zinc sulfate solution should be used for flotation; sugar and other salt solutions distort the cysts and make them unrecognizable.

The indirect immunofluorescence test might be useful in epidemiologic and immunologic studies (Visvesvara and Healy 1979; Visvesvara et al. 1980b).

TREATMENT. *Giardia* infections in man can be successfully treated with either quinacrine or chloroquine. Three oral doses of 0.1 g each are given daily for 5 days. Reviewing chemotherapy, Wolfe (1979) and Visvesvara (1982) said that quinacrine was the drug of choice.

PREVENTION AND CONTROL. These depend on sanitation. Jarroll et al. (1980) said that halazone, chlorine bleach, globaline, EDWGT, 2% tincture of iodine, and saturated iodine as recommended by the manufacturers were all more than 99.8% effective against *G. lamblia* in drinking water. However, Craun (1979) said that simple water disinfection was ineffective and recommended sedimentation and filtration.

Giardia canis Hegner, 1922

HOST. Dog.

LOCATION. Duodenum, jejunum, upper ileum.

GEOGRAPHIC DISTRIBUTION. Worldwide.

PREVALENCE. Common, especially in young dogs.

STRUCTURE. The trophozoite is 12–17 × 7–10 μm. The median bodies are curved bars of the *duodenalis* type. The cysts are 9–13 × 7–9 μm.

PATHOGENESIS. *G. canis* may or may not cause diarrhea in young dogs.

DIAGNOSIS. Methods are the same as for *G. lamblia.*

TREATMENT. Quinacrine has been found effective against *G. canis.* Dogs may be given 50–200 mg 2 or 3 times a day for 2–7 days, repeating if necessary after 3–4 days. Chloroquine has also been found effective in man; 0.1 g is given 3 times a day for 5 days. Pfeiffer and Supperer (1976) found that 125 mg Flagyl twice daily for 5 days in dogs weighing less than 10 kg was successful.

PREVENTION AND CONTROL. The standard sanitary measures should be used in preventing the transmission of *Giardia.*

Giardia cati Deschiens, 1925

SYNONYM. *Giardia felis.*

HOST. Cat.

LOCATION. Small and large intestines.

GEOGRAPHICAL DISTRIBUTION. Presumably worldwide.

PREVALENCE. Quite common.

STRUCTURE. This name is probably a synonym of *G. lamblia*. The trophozoites are 10–18 × 5–9 (mean 13 × 7) μm. The median bodies are bars of the *duodenalis* type. The cysts are 10.5 × 7 μm.

PATHOGENESIS. Unknown.

Giardia bovis Fantham, 1921

HOST. Ox. Graham (1935) found it alive and active in the digestive tracts of 6 of 21 female *Cooperia oncophora* from a calf from New Jersey.

LOCATION. Duodenum, jejunum, ileum.

GEOGRAPHIC DISTRIBUTION. Worldwide.

PREVALENCE. Unknown.

STRUCTURE. The trophozoites are 11–19 × 7–10 μm. The median bodies are curved bars of the *duodenalis* type. The cysts are 7–16 × 4–10 μm.

PATHOGENESIS. Diarrhea may occur occasionally.

Other Species of *Giardia*

Giardia caprae Nieschulz, 1923 (syns., *G. ovis, G. qadrii*), occurs in goats and sheep, in which it may sometimes cause diarrhea. Dissanaike (1954) found live and active *G. caprae* in the intestines of 50 female and no male *Nematodirus filicollis* from 5 sheep in England.

G. equi Fantham, 1921, occurs in the large intestine of the horse in South Africa. Its trophozoites are 17–21 × 9–12 μm, its cysts 12–16 × 8–9.5 μm.

G. duodenalis (Davaine, 1875) (syns., *Hexamita duodenalis, Lamblia cuniculi*) occurs throughout the world in the anterior small intestine of Old and New World rabbits and cottontails, and also in *Coendu villosus* in Brazil. It occurs sporadically and is presumably not pathogenic. Its trophozoites are 13–19 × 8–11 (mean 16 × 9) μm. The median bodies, curved bars resembling the claws of a claw hammer, lie transversely across the body. The cysts contain 2–4 nuclei. This may be the correct name for all species of *Giardia* with claw-hammer-like median bodies.

G. simoni Lavier, 1924, occurs in the anterior small intestine of the Norway rat, golden hamster, and probably various wild rodents. Its trophozoites are 11–19 × 5–11 μm. Its median bodies are curved bars of the *duodenalis* type.

G. caviae Hegner, 1923, occurs in the anterior small intestine of the guinea pig. Its trophozoites are 8–15 × 6.5–10 μm. Its median bodies are curved bars of the *duodenalis* type.

G. chinchillae Filice, 1952 (syn., *G. duodenalis* race *chinchillae*), occurs frequently in the small intestine of the chinchilla. Its trophozoites are 11–12 × 6–12 μm, its median bodies curved bars of the *duodenalis* type.

G. muris (Grassi, 1879) occurs in the anterior small intestine of the house mouse, Norway rat, black rat, golden hamster, and various wild rodents and is common in laboratory rats and mice. Its trophozoites are 7–13 × 5–10 μm, its median bodies small and rounded. Stevens (1979) found that previous infections with *G. muris* immunized normal (but not athymic) mice against reinfection.

Giardia sp. was found in the feces of a Bactrian camel in the London Zoo (Porter 1953).

Giardia duodenalis race *psittaci* Box, 1981, may cause enteritis in parakeets. The mortality may be high in nestlings. Three oral doses of 1 mg per 20 g body weight or 0.02–0.04% in the drinking water for 5 days of dimetridazole are effective against it.

Genus *Monocercomonoides* Travis, 1932

Members of this genus occur in insects, amphibia, reptiles, and a number of mammals. They have an anterior nucleus, 4 anterior flagella in 2 pairs, a pelta, and an axostyle that is generally filamentous. Nie (1950) described 1–4 strandlike funises that stain with protargol in four species of this genus from the guinea pig. The funis is a costalike structure extending backward just beneath the body surface. Members of this genus are nonpathogenic.

M. caprae (Das Gupta, 1935) Levine, 1961 (syns., *Monocercomonas caprae, Monocercomonoides sayeedi*), was described from the rumen of the goat. Its body is ovoid, 6–12 × 4–8 μm.

M. bovis Jensen and Hammond, 1964, was found in bovine cecal extract cultures. Its body is piriform or ovoid, 4–8 × 2–5 (mean 4 × 3) μm. Two anterior flagella 9 and 10 μm long arise from one blepharoplast, and an anterior flagellum 8.5 μm long and a recurrent flagellum 15 μm long arise from a second blepharoplast. The recurrent flagellum adheres to the body for one-fourth to one-half of its length.

M. caviae (da Cunha and Muniz, 1921) Nie, 1950, *M. quadrifunilis* Nie, 1950, *M. wenrichi* Nie, 1950, and *M. exilis* Nie, 1950, occur in the cecum of the guinea pig.

Monocercomonoides sp. occurs in the laboratory rat and golden hamster.

Genus *Cochlosoma* Kotlan, 1923

The body is ovoid, broadly rounded anteriorly, and narrowly rounded posteriorly. Six flagella of unequal length arise from a blepharoplastic complex at the anterior end; 2 of them are trailing and lie in a longitudinal groove. The nucleus is near the middle of the body. A slender, fibrillar axostyle and a more lateral costa arise from the blepharoplastic complex. On the anteroventral surface is a large sucker that opens on the left side and has a marginal filament. A parabasal body is present.

C. anatis Kotlan, 1923 (syn., *C. rostratum*), occurs in the cloaca, large intestine, and sometimes ceca of the domestic duck, Muscovy duck, wild mallard, and various other wild ducks. Its body is beet-shaped, 6–12 × 4–7 μm. The sucker covers one-third to one-half of the body length. The organism swims forward with an erratic, jerky motion, rotating on its long axis but with very little of the dipping motion of *Giardia*. The parabasal body is sausage-shaped. Reproduction is by longitudinal fission. *C. anatis* has not been cultivated, and its pathogenicity is unknown.

Cochlosoma sp., perhaps a synonym of *C. anatis,* occurs in the intestine of turkeys. Its pathogenicity is unknown.

Genus *Proteromonas* Kunstler, 1883

The body is spindle-shaped. An anterior and a free trailing flagellum arise from two blepharoplasts at the anterior end. The nucleus, anterior to the middle of the body, contains scattered chromatin granules but no endosome. A rhizoplast runs from the blepharoplast to a centrosome on the nuclear membrane. A perirhizoplastic ring surrounds the rhizoplast a short distance behind the blepharoplast; this is considered a parabasal body. A paranuclear body the same size as the nucleus lies beside the nucleus and divides when the nucleus divides; it stains with hematoxylin but not with protargol. All the species are parasitic. They are common in the intestines of reptiles and amphibia. Synonyms of this genus are *Prowazekella, Schizobodo.*

P. brevifilia Alexeieff, 1946, occurs in the cecum of the guinea pig.

Genus *Monocercomonas* Grassi, 1879

The body is piriform, with a rounded anterior end. There are a pelta and a parabasal body. The cytostome and nucleus are anterior. There are 3 anterior flagella and a trailing one. The axostyle usually projects beyond the posterior end of the body. (See Fig. 5.5.) *Trichomastix* and *Eutrichomastix* are synonyms. This genus occurs in mammals, birds, reptiles, amphibia, fish, and insects and other arthropods.

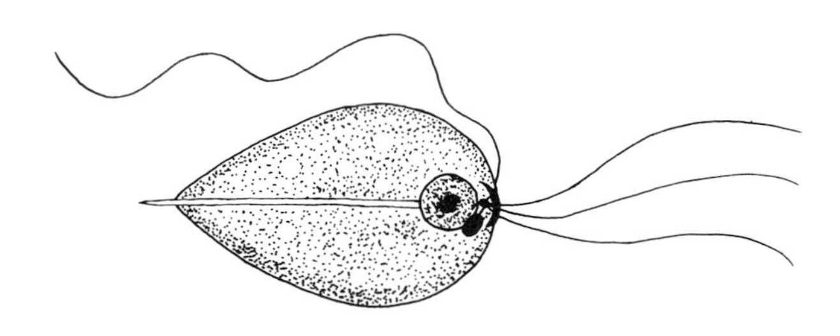

Fig. 5.5. *Monocercomonas*. (×2,800)

M. ruminantium (Braune, 1913) Levine, 1961 (syns., *Trichomonas ruminantium, Tricercomitus ruminantium, Tritrichomonas ruminantium*), occurs in the rumen of cattle and presumably also of sheep. The body is subspherical, 3–8 × 3–7 μm, with 3 anterior flagella 6–8 μm long and a trailing flagellum a little longer. The axostyle is curved and may or may not extend beyond the body, although the posterior end is pointed. A line of granules runs beside the convex side of the axostyle. A pelta is present, as is a parabasal body. This species is nonpathogenic but must be distinguished from *Tritrichomonas foetus*.

M. cuniculi (Tanabe, 1926) (syn., *Trichomastix cuniculi*) occurs in the cecum of the domestic rabbit. It is piriform, 5–14 μm long. Its axostyle is slender, hyaline, and projects from the body.

M. gallinarum (Martin and Robertson, 1911) Morgan and Hawkins, 1948, is said to occur in the ceca of the chicken. Perhaps it is simply a degenerate *Tritrichomonas eberthi*.

Genus *Hexamastix* Alexeieff, 1912

The body is piriform, with a rounded anterior end. The cytostome and nucleus are anterior. There are 6 flagella, one of which trails. A pelta is present, the axostyle is conspicuous, and the parabasal body is prominent. Members of this genus (syn., *Pentatrichomastix*) have been found in mammals, amphibia, and insects.

H. caviae Nie, 1950, and *H. robustus* Nie, 1950, occur in the guinea pig cecum and *H. muris* (Wenrich, 1924) in the cecum of the Norway rat, golden hamster, and other rodents.

Genus *Chilomitus* da Fonseca, 1915

The body is elongate, with a convex aboral surface. The pellicle is well developed. The cuplike cytostome is near the anterior end. Four flagella emerge through it from a bilobed blepharoplast. The nucleus and parabasal body are just below the cytostome. An axostyle is present but may be rudimentary. Cysts may occur. Only a few species have been described, all in mammals. *C. caviae* da Fonseca, 1915, and *C. conexus* Nie, 1950, occur in the guinea pig.

Genus *Tetramitus* Perty, 1852

In this genus the life cycle involves flagellate and ameboid forms; there are also uninucleate cysts. In the flagellate stage the body is ellipsoidal or piriform, with a large, trough-shaped cytostome at the anterior end; a vesicular nucleus with a large endosome; 4 anterior flagella; and a contractile vacuole. Nutrition is holozoic.

T. rostratus Perty, 1852 (syn., *Copromastix prowazeki*), is found in stagnant water and is also coprophilic. It has been found in human and rat feces. The flagellate stage is 14–18 × 7–10 μm. The ameboid stage is 14–48 μm long and usually has a single lobose pseudopod. The cysts are spherical, thin-walled, and 6–18 μm in diameter.

Genus *Trepomonas* Dujardin, 1841

These are free-swimming protozoa with a more or less rounded, bilaterally symmetrical body and with a cytostomal groove on each side of the posterior half. There are 8 flagella, 1 long and 3 short ones on each side. A horseshoe-shaped structure near the anterior margin contains 2 nuclei. Members of this genus are free-living in fresh water and coprophilic or parasitic in amphibia, fish, and turtles.

T. agilis Dujardin, 1841, occurs in stagnant water and the intestine of amphibia and is also coprophilic. It is 7–25 × 2–15 μm, with a flattened body and with the posterior end wider than the rounded anterior end. The flagella emerge near the middle of the body at the anterior end of the cytostome.

Genus *Bodo* Stein, 1875

These are small, more or less ovoid, plastic forms with an anterior cytostome, a central or anterior nucleus, and a kinetoplast (Brugerolle et al. 1979). Cysts are formed.

B. caudatus (Dujardin, 1841) Stein, 1878, is a common coprophilic form and also occurs in stale urine and stagnant water. It is 8–18 × 2.5–6 μm, with a polymorphic, spherical to elongate ovoid body. It has a tiny contractile vacuole, a single vesicular nucleus with a large endosome, and a rounded parabasal body. This species and also *B. foetus* and *B. glissans* have been found in material from bulls submitted for *Tritrichomonas foetus* diagnosis.

Genus *Cercomonas* Dujardin, 1841

These are small forms with a plastic body, one flagellum directed anteriorly and the other running backward over the body to become a trailing flagellum. The nucleus is piriform and is connected with the blepharoplast of the flagella. The cysts are spherical and uninucleate. (See Fig. 5.6.) A number of freshwater and coprophilic species have been described, but it is not clear whether all the species are valid.

C. longicauda Dujardin, 1841, is a common coprophilic flagellate. Its trophozoites are ameboid, 2–15 μm long, have 2 contractile vacuoles, and ingest food by means of pseudopods. Its cysts are 4–7 μm in diameter.

C. heimi (Hollande) is similar to *C. longicauda* but is piriform and has longer flagella.

C. equi (Sabrazes and Muratet, 1908) (syn., *C. asini*) was described from the large intestine of the horse and donkey and also occurs in their feces.

Cercomonas faecicola (Woodcock and Lapage, 1915) (syn., *Helkesimastix faecicola*) was found in the feces of the goat. It is ovoid, with a rigid, pointed anterior end. The anterior flagellum is very short and easily overlooked. The trophozoites are 4–6 μm long, the cysts 3–3.5 μm in diameter.

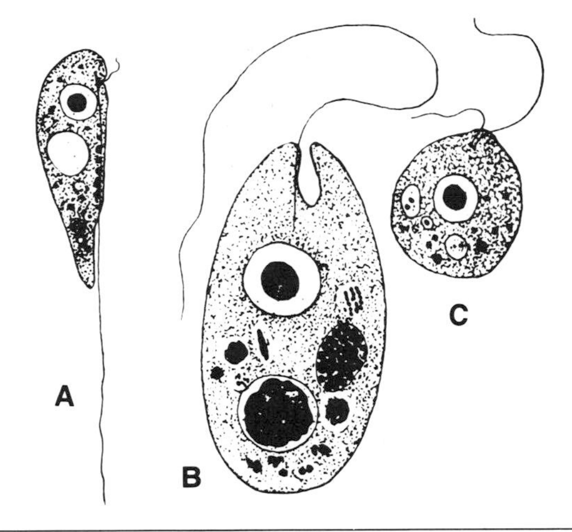

Fig. 5.6. Coprophilic flagellates: A. *Cercomonas* sp. (×4,200); B. *Copromonas subtilis* (×5,100); C. *Monas* sp. (×4,200). (From Noble 1956)

C. crassicauda Dujardin, 1841, occurs in fresh water and is also coprophilic. It has been found in material from bulls submitted for examination for *Tritrichomonas foetus*. Its trophozoites are 10–16 × 7–10 μm.

Cercomonas sp. has been found in fresh bovine and porcine feces and in fecal specimens from Wyoming elk, bison, cattle, horses, and sheep after storage at 4 C for 6–7 days and subsequent cultivation (Noble 1956, 1958).

Genus *Pleuromonas* Perty, 1852

The body is somewhat ameboid. The 2 flagella often appear to emerge separately from the body. The anterior flagellum is very short and often rolled up into a ring. The posterior flagellum is very thick and more than 2–3 times the length of the body. There is a single vesicular nucleus. The cyst is spherical, and 4–8 young individuals apparently emerge from it.

There is a single species in this genus, *P. jaculans* Perty, 1852; it occurs in stagnant water and has been found in the ceca of young chickens. It is 6–10 × about 5 μm.

Genus *Spiromonas* Perty, 1914

Members of this genus (syn., *Alphamonas*) have an elongate, spirally twisted body with 2 anterior flagella. They form spherical cysts in which division into 4 daughter individuals takes place. They live in fresh water.

S. angusta (Dujardin) Alexeieff lives in stagnant water or is coprophilic. It has also been found in bull sheath washings. It is spindle-shaped and about 10 μm long.

Genus *Oikomonas* Kent, 1880

Members of this genus lack chromatophores, lorica, or test. They are solitary. The nucleus is near the center of the body. The single flagellum arises from a basal granule near the body surface. Cysts are formed, at least in the free-living species. This genus is the colorless homolog of *Chromulina*. Its parasitic species are poorly known.

O. communis Liebetanz, 1910, and *O. minima* Liebetanz, 1910, were both described from the rumen of cattle. *O. communis* has also been found in the rumen of goats.

O. equi Hsiung, 1930, occurs in the cecum of horses. It is usually spherical or ovoid and swims in a jerky manner. The nucleus has a large, central endosome, and the cytoplasm is filled with small, dark-staining granules. The body is 3.5–7 × 3–3.5 μm, the flagellum about 20 μm long.

Genus *Sphaeromonas* Liebetanz, 1910

The body is spherical or ellipsoidal, with a more or less central nucleus. A single, long flagellum arises from a basal granule on the nuclear membrane. This genus is poorly known and has apparently not been studied by modern methods. It is closely related to *Oikomonas* and may even be a synonym of that genus. Several species have been named, all parasitic, but most of them are probably the same.

S. communis Liebetanz, 1910 (syns., *Monas communis, S. minima, S. maxima, S. liebetanzi, S. rossica*), occurs in the rumen of the ox and goat and in the cecum and feces of the guinea pig. It may also be coprophilic. The body is spherical or ellipsoidal, 3–14 μm in diameter. The cytoplasm contains many dark-staining granules.

Genus *Monas* Müller, 1773

The body is active and plastic. This genus is the colorless homolog of *Ochromonas*.

Monas sp. has been cultivated from bovine feces and also from fecal samples from elk, bison, and bears. *Monas obliqua* Schewiakoff has been found in material from bulls submitted for *Tritrichomonas foetus* diagnosis.

Genus *Copromonas* Dobell, 1908

The body is elongate ovoid, with an elongate reservoir at the anterior end into which a contractile vacuole discharges. The single flagellum arises from a blepharoplast at the base of the reservoir. The nucleus is vesicular, with a large endosome. Permanent fusion followed by encystment takes place. Nutrition is holozoic on bacteria.

C. subtilis Dobell, 1908 (syn., *Copromonas ruminantium*), was found in the feces of frogs, toads, man, and various domestic and wild animals.

Its trophozoites are 7–20 × 5–6 μm. They are usually ovoid but can change from spindle-shaped to almost round. The flagellum is 1–2 times the length of the body. When the animal swims straight, only the tip of the flagellum moves; sometimes the flagellum is used like a highly flexible probe. The cysts are ovoid or spherical and 7–8 μm long. They have a thin wall and clear contents with a single vesicular nucleus.

Genus *Callimastix* Weissenberg, 1912

The body is ovoid, with a compact central or anterior nucleus. There are 12–15 long flagella that beat in unison near the anterior end. One species occurs in the body cavity of copepods and the others in ruminants and equids. They are nonpathogenic. The form in ruminants may be the swarm spore of a lower fungus.

C. frontalis Braune, 1913, occurs in the rumen of cattle, sheep, and goats. The body is spherical or ovoid, about 12–14 μm in diameter. The nucleus has a large central endosome. The 12 or perhaps more flagella are about 30 μm long; they arise from a row of basal granules on the anterior margin of the body and join to form a single unit distally. It has also been found in material submitted for diagnosis of *Tritrichomonas foetus* infections.

C. equi Hsiung, 1929, occurs in the cecum and colon of the horse. The body is kidney-shaped with the hilus at its anterior third; it is 12–18 × 7–10 (mean 14 × 8) μm. Just behind the hilus is a clear, granule-free area on the margin of which are 12–15 basal granules that give rise to flagella 25–30 μm long; these unite distally and function as a unit. The rest of the cytoplasm is filled with deeply staining granules. The nucleus, 3 μm in diameter, has a large endosome and lies near the center of the body.

LITERATURE CITED

For citations before 1965, see the *Index-Catalogue of Medical and Veterinary Zoology, Zoological Record, Biological Abstracts, Protozoological Abstracts,* and *Veterinary Bulletin*.

Brugerolle, G. 1973. J. Protozool. 20:574–85.

______. 1975. J. Protozool. 22:468–75.

Brugerolle, G., J. Lom, E. Nohynkova, and L. Joyon. 1979. Protistologica 15:197–221.

Brugerolle, G., I. Kunstyr, J. Senaud, and K. T. Friedhoff. 1980. Z. Parasitenkd. 62:47–61.

CDC (Center for Disease Control). 1977. Intestinal parasite surveillance, annual summary 1976. Atlanta: U.S. Department of Health, Education and Welfare.

______. 1980. CDC Mortal. Morb. Wkly. Rep. 29:610–11.

Craun, G. F. 1979. Am. J. Public Health 69:817–19.

Gillin, F. D., and L. S. Diamond. 1979. In Waterborne Transmission of Giardiasis, ed. W. Jakubowski and J. C. Hoff, 270–72. U.S. Environmental Protection Agency, EPA-600/9–79–001.

Hoff, J. C. 1979. In Waterborne Transmission of Giardiasis, ed. W. Jakubowski and J. C. Hoff, 231–39. U.S. Environmental Protection Agency. EPA-600/9-79-001.

Jakubowski, W., and J. C. Hoff, eds. 1979. Waterborne Transmission of Giardiasis. U.S. Environmental Protection Agency. EPA-600/9-79-001.
Jarroll, E. L., Jr., A. K. Bingham, and E. A. Meyer. 1980. Am. J. Trop. Med. Hyg. 29:8–11.
Kösters, J., A. Cubillos, J. Ulloa, and M. I. Montecinos. 1976. Bol. Chil. Parasitol. 31:84–85.
Kulda, J., and E. Nohynkova. 1978. In Parasitic Protozoa, vol. 2, ed. J. P. Kreier, 2–138. New York: Academic.
Levine, N. D., J. O. Corliss, F. E. G. Cox, G. Deroux, J. Grain, B. M. Honigberg, G. F. Leedale, A. R. Loeblich II, J. Lom, D. Lynn, E. G. Merinfeld, F. C. Page, G. Poljansky, V. Sprague, J. Vavra, and F. G. Wallace. 1980. J. Protozool. 27:37–58.
Meerovitch, E. 1978. In Methods of Cultivating Parasites in Vitro, ed. A. E. R. Taylor and J. R. Baker, 19–37. New York: Academic.
Meyer, E. A. 1979. In Waterborne Transmission of Giardiasis, ed. W. Jakubowski and J. C. Hoff, 211–16. U.S. Environmental Protection Agency. EPA-600/9-79-001.
Pfeiffer, H., and R. Supperer. 1976. Wien. Tieraerztl. Monatsschr. 63:1–6.
Sournia, A., ed. 1978. UNESCO Monographs Oceanographic Methods. Vol. 6. Phytoplankton Manual. New York: Unipub.
Stevens, D. P. 1979. In Waterborne Transmission of Giardiasis, ed. W. Jakubowski and J. C. Hoff, 82–90. U.S. Environmental Protection Agency. EPA-600/9-79-001.
USDA. 1965. Agric. handb. 291. Washington, D.C.: USDA.
Visvesvara, G. S. 1980. Trans. R. Soc. Trop. Med. Hyg. 74:213–14.
———. 1982. J. Pediatr. Gastroenterol. Nutr. 1:463–65.
Visvesvara, G. S., and G. R. Healy. 1979. In Waterborne Transmission of Giardiasis, ed. W. Jakubowski and J. C. Hoff, 53–63. U.S. Environmental Protection Agency. EPA-600/9-79-001.
Visvesvara, G. S., G. R. Healy, and E. A. Meyer. 1980a. J. Protozool. 27:38A.
Visvesvara, G. S., P. D. Smith, G. R. Healy, and W. R. Brown. 1980b. Ann. Intern. Med. 93:802–4.
Wolfe, M. S. 1979. In Waterborne Transmission of Giardiasis, ed. W. Jakubowski and J. C. Hoff, 39–52. U.S. Environmental Protection Agency. EPA-600/9-79-001.

6
Amebae

THE amebae belong to the subphylum Sarcodina. Members of this subphylum move by means of pseudopods. They have no cilia and, except in rare instances, no flagella. The group is named for *sarcode,* a term introduced by Dujardin for what was later called protoplasm. Most sarcodines are holozoic, ingesting their prey by means of pseudopods. Their cytoplasm is usually divided into endoplasm, which contains the food vacuoles, nucleus, etc., and relatively clear ectoplasm. The freshwater forms contain one or more contractile vacuoles; these are absent in the saltwater and parasitic species. With a few exceptions reproduction is asexual, by binary fission or rarely by multiple fission, by budding, or by plasmotomy. Most species form cysts.

The Sarcodina originated from the Mastigophora. The group did not arise from a single progenitor but is polyphyletic. One line, for example, passes from *Tetramitus* through *Naegleria* to *Vahlkampfia.* In *Tetramitus,* which is classified among the flagellates, the life cycle includes flagellate and ameboid stages, and the flagellate stage has a permanent cytostome. In *Naegleria,* which is classified among the amebae, the life cycle also includes flagellate and ameboid stages but there is no permanent cytostome. In *Vahlkampfia* there is no flagellate stage, but the amebae are very similar to those of *Naegleria.*

Only a few of the Sarcodina are parasitic. The free-living forms include the most beautiful protozoa of all, the pelagic Radiolaria with their delicate, latticework, siliceous skeletons. One group of Radiolaria has skeletons of strontium sulfate. Another marine group, the Foraminifera, has calcareous shells. Their skeletons form our chalk, and they and the Radiolaria are of great geologic interest; they are used as indicators in oil well drilling. More species of Foraminifera have probably been named than of all the other protozoa put together.

PRIMARY AMEBIC MENINGOENCEPHALITIS

Beginning in 1964, a number of cases of fatal human meningoencephalitis due to small, free-living amebae with nuclei of the *limax* type (discussed below) were reported. Well over a hundred human cases have been reported; most were apparently caused by *Naegleria fowleri,* some by *Acanthamoeba culbertsoni,* and a few perhaps by other species. The disease occurs throughout the world, the great majority of cases associated with swimming in warm water. Among reviews are those by Jadin (1974), Singh (1975), Martinez (1976), Healy and Visvesvara (1977), and Griffin (1978).

Infection is apparently due to water entering the nasal passages. The protozoa pass through the cribriform plate directly to the brain. Symptoms last only a day or so, and death quickly follows. No effective treatment is known.

Genus *Naegleria* Alexeieff, 1912 emend. Calkins, 1913

There is a temporary flagellate stage with 2 flagella arising from separate basal bodies. It can be produced from the ameboid stage by flooding the culture with water and exposing it to air. The ameboid stage has lobopodia and resembles *Vahlkampfia* (discussed later in the chapter). The nucleus is vesicular, with a large endosome. The contractile vacuole is conspicuous. The cysts are uninucleate. *Naegleria* lives on bacteria and is free-living in fresh or stagnant water or coprophilic.

N. gruberi (Schardinger) (syn., *Dimastigamoeba gruberi*) is found in water and is also coprophilic. The active amebae are 8–42 × 8–18 μm, with a single vesicular nucleus 3–4 μm in diameter. The nucleus has a central endosome and sparse granules of peripheral chromatin. The flagellate stage is 5–28 × about 9 μm, and ovoid. The cysts are spherical, 7–21 μm in diameter, translucent, with a single nucleus and several large spherical chromatoid bodies when first formed. The cyst wall is smooth and double, and the outer wall is perforated by 3–8 pores or ostioles. *N. gruberi* is easily cultivated on artificial media.

N. fowleri Carter, 1970 (syns., *N. aerobia, N. invades*), is found in water, feces, and sewage and is pathogenic. It looks like *N. gruberi* and shares some but not all antigens with it, differing in that it grows at a higher temperature and differing also in the polysaccharide structure of the cell membrane and in isoenzymes. It is the principal cause of primary amebic meningoencephalitis of man. Death occurs a day or so after symptoms appear. Following intranasal inoculation, it enters the brain of mice, guinea pigs, sheep, some monkeys, and other animals via the nervous tissues and causes a widespread, acute, necrotizing, hemorrhagic meningoencephalitis and extensive demyelinization of the white matter. There may be large clusters of the amebae in the brain; the effects are due to phagocytosis by the amebae. No effective treatment is known.

N. fowleri has been found in swimming pools, ponds, lakes, thermal effluents from power plants, sewage, soil, and even water supply systems. The heating of water by factories in cooling seems to make the water in their discharges more suitable for *N. fowleri* than before.

Naegleria sp. has been isolated not only from man but also from other animals. Taylor (1977) found it in several species of freshwater fishes in Alabama and Florida. Kingston and Taylor (1976) found it in the snail *Physa gyrina* from a high mountain lake in Wyoming. Some of the forms reported under other names by various investigators may also have been *Naegleria*.

Genus *Acanthamoeba* Volkonsky, 1931

These are relatively small amebae without well-developed ectoplasm. The nucleus is vesicular, with a large endosome. The trophozoite has a broad, anterior lobopod from which are produced singly (or in twos or threes) several (or many) slender, hyaline projections (acanthopods) that taper to a finely rounded end; the acanthopods may move occasionally and are eventually resorbed by the advancing ameba or remain on its surface until resorbed by the posterior end. The cyst wall consists of a more or less polyhedral or stellate endocyst and a wrinkled and more or less closely adherent ectocyst. The endocyst is positive for cellulose. The cyst is uninucleate. There are two essential differences between *Acanthamoeba* and *Hartmannella* (discussed next): (1) the cyst wall of the latter is smooth and composed of a single layer, whereas that of the former is composed of two layers, the inner one of which is apparently composed of cellulose; (2) *Acanthamoeba* forms tapering, hyaline acanthopods whereas *Hartmannella* does not.

A. culbertsoni (Singh and Das, 1970) Sawyer and Griffin, 1975 (syn., *H. culbertsoni*), is facultatively pathogenic. It is structurally similar to *A. castellanii*. It occurs in the same situations as *Naegleria* and has been reported from man a number of times. It causes a fatal meningoencephalomyelitis after intranasal inoculation into mice and other animals, but only a transient infection in monkeys.

A. castellanii (Douglas, 1930) Volkonsky, 1931 (syn., *A. terricola*), has trophozoites that are up to three times as long as wide when moving. They are 21–46 (mean 26) μm long when moving. The cysts are 12–23 (mean about 16) μm long and appear mammillated and either thickly biconvex or polyhedral. The endocyst is polyhedral or stellate. The ectocyst is usually applied loosely to the endocyst. This species occurs throughout the world in fresh water and soil. It has also been found as a contaminant of yeast cultures; it eats the yeast. It is apparently nonpathogenic.

A. polyphaga (Puschkarew, 1913) Page, 1967, has been found in the human eye, human nasal smears, and swimming pools; it is apparently nonpathogenic for man. Kadlec (1978) found it in the bull preputial cavity, cow vagina, rabbit liver, pigeon intestine, and turkey intestine in Czechoslovakia. Taylor (1977) found that a strain was pathogenic in fish.

Acanthamoeba spp. have been reported from fatal human infections and also from lower animals. Ayers et al. (1972) said that an *Acanthamoeba* sp. had killed an army dog in South Vietnam. It caused multifocal necrohemorrhagic foci in the heart, lungs, liver, and pancreas. The infection may perhaps have been the result of invasion of a wound and spread via the blood.

Acanthamoeba has occasionally been encountered as a contaminant of tissue cultures.

Genus *Hartmannella* Alexeieff, 1912

In this genus cysts, if formed, are smooth-walled and rounded. Locomotion is not strongly eruptive.

H. hyalina (Dangeard, 1900) Alexeieff, 1912 (syn., *A. hyalina*), is the type species. A common coprophilic form, it occurs in soil and fresh water and is easily cultivated from old human and animal feces. Its trophozoites are 9–17 μm in diameter when rounded. It has a single contractile vacuole and a single vesicular nucleus with a central endosome and peripheral chromatin. Its cysts are spherical, 10–15 μm in diameter, with a smooth wall.

Singh (1970) mentioned a *Hartmannella* in Indian buffalo. Patton (1969) saw a similar organism in a case of cervical lymphangitis in a Rocky Mountain sheep *Ovis canadensis nelsoni.* Kadlec (1978) found *H. vermiformis* in dog bronchi, turkey trachea, and turkey intestine in Czechoslovakia; none of these strains appeared to be pathogenic. McConnell et al. (1968) described a case of fatal gangrenous pneumonia in a young bull in the Azores due to what they identified histologically as *Hartmannella* (*Acanthamoeba*) sp., but which might have been *Naegleria* or another limax ameba. It is likely that some of the above reports were actually of *Naegleria* or *Acanthamoeba.*

Genus *Vahlkampfia* Chatton and Lalung-Bonnaire, 1912 emend. Calkins, 1913

These are small amebae with a nucleus containing a large endosome and peripheral chromatin, with polar caps during nuclear division. The trophozoites have a single broad pseudopod and move like a slug. The cysts have a perforated wall. The nucleus of this genus closely resembles that of *Naegleria,* but the latter has both flagellate and ameboid stages whereas *Vahlkampfia* has only an ameboid stage. A number of species have been described from fresh water, old feces, vertebrates (mostly lower), and invertebrates, but the taxonomy, nomenclature, and validity of some of them are not certain.

V. lobospinosa (Craig, 1912) Craig, 1913, is coprophilic. Becker and Talbot (1927) found it in the rumen of a cow in Iowa. Its trophozoites are 10–24 μm long. Its cysts have 1 or 2 nuclei and are 7–11 μm in diameter.

V. inornata and *V. avara* were found in the pig nasal cavity and *V. enterica* in the turkey intestine in Czechoslovakia (Kadlec 1978).

Vahlkampfia sp. appeared in fecal samples from Wyoming elk, bison, cattle, horses, sheep, moose, and marmots after storage at 4 C for a few days to a few weeks (Noble 1958). Wilhelm and Anderson (1971) found this form in the feces of cattle, swine, and horses, and also in soil samples in Tennessee. It was common in the soil.

Genus *Sappinia* Dangeard, 1896

The trophozoites have 2 closely associated nuclei with large endosomes. The cysts are binucleate also.

S. diploidea (Hartmann and Nägler, 1908) Alexeieff, 1912, is a common coprophilic ameba in the feces of man and other animals. Its trophozoites are 10–16 μm long, with a thick, smooth, hyaline pellicle; according to Noble (1958) the ectoplasm has fine lines sometimes resembling wrinkles in cellophane. Two nuclei are present, usually pressed tightly together. Each nucleus has a large endosome and frequently a crescentic mass of granules between the endosome and the nuclear membrane. The cytoplasm is usually filled with many food vacuoles. A contractile vacuole is present, formed by the enlargement and coalescence of smaller vacuoles. A single, clear, broad pseudopod is characteristic, although occasionally many pseudopods may be present. The cysts are typically binucleate, 12–18 μm or more in diameter, with thick, uniform walls.

Noble (1958) found that *S. diploidea* appeared in fecal samples from Wyoming elk and bison (but not from cattle, horses, or sheep) after storage at 4 C for a few days to a few weeks. It is readily cultivated.

FAMILY ENDAMOEBIDAE

Members of this family are parasitic in the digestive tract of vertebrates and invertebrates. The genera are differentiated on the basis of nuclear structure. Three genera contain parasites of domestic animals and man, but only 1 of these contains pathogenic species. However, it is important to be able to identify the various species in order to know whether an infection with a pathogenic one is present or not. Albach and Booden (1978) reviewed the group.

Genus *Entamoeba* Casagrandi and Barbagallo, 1895

The nucleus is vesicular, with a comparatively small endosome located at or near its center, with or without periendosomal granules around the endosome, and with a varying number of granules around the periphery of the nucleus. Cysts are usually formed; they contain 1–8 nuclei and may or may not contain chromatoid bodies (rodlike bodies that stain with hematoxylin and that are absorbed and disappear as the cysts mature). The genus occurs in both vertebrates and invertebrates. (See Figs. 6.1 and 6.2.)

Members of the genus found in domestic animals and man can be divided into four groups on the basis of trophozoite and cyst structure. A fifth group includes species about which insufficient structural information is available to determine to which of the other groups they belong. Most of the species within each group are structurally indistinguishable; they are differentiated on the basis of size, hosts, pathogenicity, etc.

1. Histolytica group. The nucleus has a small, central endosome, a ring of small peripheral granules, and a few scattered chromatin granules between them. The cysts have 4 nuclei when mature, and their chromatoid bodies are rods with rounded ends. Glycogen vacuoles, when present in the cyst, are usually diffuse and ill defined.

   E. histolytica of man, other primates, the dog, cat, and rarely the pig
   E. hartmanni of man and the other hosts of *E. histolytica*

E. *equi* of the horse
E. *anatis* of the duck
E. *moshkovskii* of sewage

2. Coli group. The nucleus has a somewhat larger, more eccentric endosome than that of the *histolytica* group and has a ring of coarse peripheral granules and some scattered chromatin granules between them. The cysts have 8 nuclei when mature and splinterlike chromatoid bodies. Glycogen vacuoles, when present in the cysts, may be fairly well defined.

   E. *coli* of man, other primates, the dog, and possibly the pig
   E. *wenyoni* of the goat
   E. *muris* of mice, rats, hamsters, and other rodents
   E. *caviae* of the guinea pig
   E. *cuniculi* of rabbits
   E. *gallinarum* of the chicken, turkey, guinea fowl, duck, and goose

3. Bovis group. The endosome of the nucleus varies in size; it may be as small as that of the *histolytica* group but is generally larger than that of the *coli* group. The ring of peripheral granules in the nucleus may be fine or coarse; the granules are evenly or irregularly distributed. Periendosomal granules may be present. The cysts have 1 nucleus when mature, and their chromatoid bodies are either rods with rounded ends or (less often) splinterlike. Glycogen granules, when present in the cyst, are usually fairly well defined.

   E. *bovis* of cattle
   E. *ovis* of sheep and goats
   E. *dilimani* of goats
   E. *suis* of the pig and perhaps man
   E. *bubalis* of the carabao
   E. *chattoni* of monkeys and man

4. Gingivalis group. The nucleus has a small, central endosome and a ring of small peripheral granules. There are no cysts. Members of this group are found in the mouth.

   E. *gingivalis* of man, other primates, the dog, and the cat
   E. *equibuccalis* of the horse
   E. *suigingivalis* of the pig

5. Insufficiently known species. This group includes E. *gedoelsti* of the horse. Its nucleus resembles that of E. *coli,* but its cysts are unknown.

The only species of *Entamoeba* known to be pathogenic for mammals is E. *histolytica.*

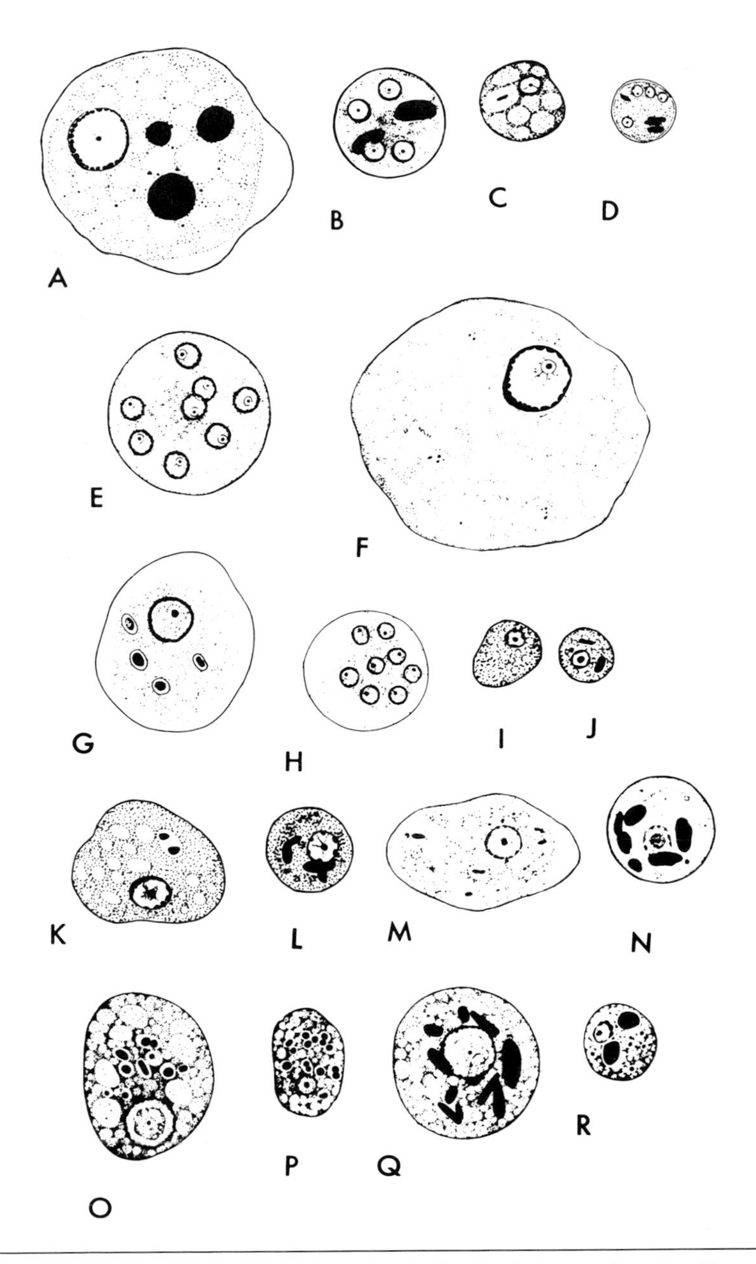

Fig. 6.1. Species of *Entamoeba*: A. *E. histolytica* trophozoite; B. *E. histolytica* cyst; C. *E. hartmanni* trophozoite; D. *E. hartmanni* cyst; E. *E. coli* cyst; F. *E. coli* trophozoite; G. *E. gallinarum* trophozoite; H. *E. gallinarum* cyst; I. *E. bovis* trophozoite; J. *E. bovis* cyst; K. *E. ovis* trophozoite; L. *E. ovis* cyst; M. *E. dilimani* trophozoite; N. *E. dilimani* cyst; O. *E. suis* large trophozoite; P. *E. suis* small trophozoite; Q. *E. suis* large cyst; R. *E. suis* small cyst. (×1,700) (From Hoare 1959, Vet. Rev. Annot.)

Some of the amebae reported as *E. histolytica* from domestic animals may well have been actually *E. hartmanni,* but unless they were specifically described as having small cysts, it is impossible to know which they were.

Before beginning a systematic account of the species of *Entamoeba,* a comment is necessary regarding the *bovis* group. All of these look alike, with minor differences that may not be of taxonomic significance. Different names have been given to the forms in different hosts, but no cross-transmission studies have been attempted; it is quite likely that when they are, some of these names will be found to be synonyms. In this case, *E. bovis* will have precedence over the other names.

The name *E. polecki* has been used for members of the *bovis* group from the pig, goat, and man, but it is a nomen nudum. Von Prowazek's (1912) original description and figures of it are so poor that it is impossible to know whether he was dealing with a member of the genus *Entamoeba* at all.

Entamoeba histolytica Schaudinn, 1903

SYNONYMS. *Amoeba coli, A. dysenteriae, E. caudata, E. dispar, E. dysenteriae, E. nuttalli, E. pitheci, E. tetragena, E. venaticum, Endamoeba histolytica.*

DISEASE. Amebic dysentery.

HOSTS. Man, orangutan, gorilla, chimpanzee, gibbons, langurs, many species of macaques, baboons, guenons, spider and other monkeys, the dog, cat, pig, rat, and possibly cattle; the rat, mouse, guinea pig, and rabbit are often infected experimentally.

LOCATION. Large intestine, sometimes liver, occasionally lungs, and rarely other organs, including brain and spleen.

GEOGRAPHIC DISTRIBUTION. Worldwide.

PREVALENCE. *E. histolytica* is most important as a parasite of man. It has been studied extensively but now takes second place to *Giardia lamblia* in the United States. However, foci of endemic amebiasis continue to exist in the United States. The U.S. Center for Disease Control reported that in 1978 *E. histolytica* had been found by state health department laboratories in 0.7% of 332,312 human stool samples (CDC 1979).

Amebosis has given considerable trouble in laboratory colonies of chimpanzees. The infections appear to be similar to that in man. The infection rate is low in nature but may become high in laboratory colonies. Tissue invasion is relatively rare, but serious or fatal disease may occur.

E. histolytica has been found in naturally infected monkeys of many species throughout the world. The spider monkey is very susceptible, often developing liver infections.

Sporadic cases of amebic dysentery or amebiasis have been reported in dogs; these animals are generally considered to have acquired their infections

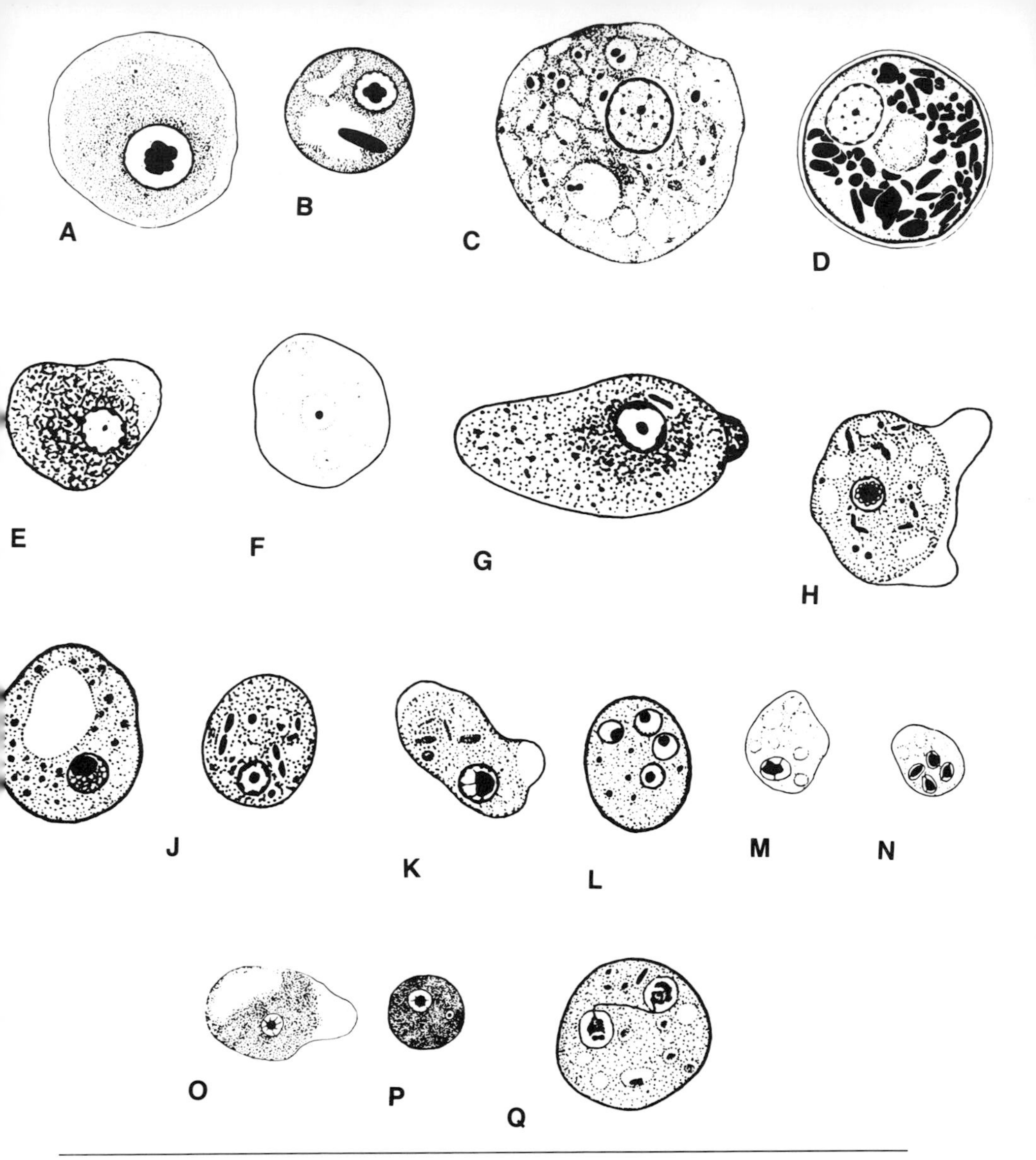

Fig. 6.2. A. *Entamoeba bubalus* trophozoite; B. *E. bubalus* cyst; C. *Entamoeba chattoni* trophozoite; D. *E. chattoni* cyst; E. *Entamoeba gingivalis* trophozoite from dog; F. *Entamoeba gedoelsti* trophozoite; G. *Entamoeba histolytica* ("*E. caudata*") trophozoite; H. *Iodamoeba buetschlii* trophozoite; I. *I. buetschlii* cyst; J. *Entamoeba equibuccalis* trophozoite; K. *Endolimax nana* trophozoite; L. *E. nana* cyst; M. *Endolimax gregariniformis* trophozoite; N. *E. gregariniformis* cyst. (×1,700) (From Hoare 1959, Vet. Rev. Annot.) O. *Vahlkampfia lobospinosa* trophozoite; P. *V. lobospinosa* cyst. (×1,050) (From Becker and Talbott 1927) Q. *Dientamoeba fragilis* trophozoite. (×1,700) (From Wenrich 1944, J. Morphol. 74:467)

from human contacts. Natural *E. histolytica* infections are apparently rare in cats and swine, but natural infections in rats have been reported by a number of workers.

There are two reports of what may have been *E. histolytica* in cattle. Walkiers (1930) reported it in the feces of dysenteric cattle in Zaire. Thiery and Morel (1956) found it in the lungs of a young zebu in Dakar that was slaughtered on account of generalized streptothricosis.

STRUCTURE. The trophozoites of the large, pathogenic race are 20–30 μm in diameter, those of the small, nonpathogenic race 12–15 μm. They have a thick, clear layer of ectoplasm and granular endoplasm. They move rapidly when warm, usually moving forward in a straight line with a single clear pseudopod at the anterior end. When the feces have cooled, the amebae stay in one place and throw out large, clear pseudopods from various parts of their body. The trophozoites often ingest erythrocytes, a feature that differentiates them from those of other amebae. The nucleus is indistinct in living amebae. When stained with hematoxylin, *E. histolytica* has a small, central endosome, a ring of small peripheral granules, and a few scattered chromatin granules in between. The cysts of both the large and small races are 10–20 (mean 12) μm in diameter. They have 4 nuclei when mature and often contain rodlike chromatoid bodies with rounded ends. Diffuse glycogen is present in the young cysts.

LIFE CYCLE. *E. histolytica* multiplies in the trophozoite stage by binary fission. It has 6 chromosomes. Before encysting, the amebae round up, become smaller, and eliminate their food vacuoles. They lay down a cyst wall, and the nucleus divides into 2 and then 4 small nuclei. After the 4-nucleate amebae emerge from the cyst, both the nuclei and cytoplasm divide so that 8 small amebulae result. Each then grows into a normal trophozoite.

PATHOGENESIS. Only the large forms of *E. histolytica* are generally considered pathogenic, although there are reports of mild disease and slight lesions associated with the small form. The organism may cause diarrhea or dysentery, sometimes invading the wall of the cecum and colon to form ragged, undermining, or flask-shaped ulcers that may be confluent. The amebae invade the mucosa at first and multiply to form small colonies that then extend into the submucosa and even into the muscularis. In the absence of bacterial invasion there is little tissue reaction, but in complicated infections there are hyperemia, inflammation, and infiltration with neutrophils.

Some of the amebae may pass into the lymphatics or even the mesenteric venules. Those entering the hepatic portal system pass to the liver, where they may cause abscesses. Those that enter the lymph ducts are generally filtered out by the lymph nodes. Abscesses may be formed in various other organs, including the lungs and brain, depending on the host's resistance.

Different strains of *E. histolytica* differ in pathogenicity, which itself may vary with their concomitant bacteria, the host's condition and nutritional status, and other factors. Actually the relationship between commensalism, frank dysentery, and the whole gamut of effects between is difficult to explain. Kagan (1974, 1975) reviewed the pathogenesis of *E. histolytica;* he thought that it has an immunologic basis.

Amebic dysentery is much more common in the tropics than in the temperate zone. The reason is unknown although several theories have been advanced.

EPIDEMIOLOGY. *E. histolytica* is primarily a parasite of primates; man is the reservoir of infection for domestic animals. This is one of the few zoonoses that man gives to associated animals in return for the many received from them.

Infection is due to ingestion of cysts. Since dysenteric individuals pass trophozoites alone, they are not the important sources of infection that cyst-producing chronic cases and carriers are. The cysts are relatively resistant. They are not affected by water chlorination but can be removed by sand filtration. They are usually transmitted with food or water. Raw vegetables may be a source of infection, and flies may also transmit cysts.

Faulty plumbing and water systems may cause waterborne transmission. The most striking case of this kind took place during the Chicago world's fair in 1933. Food handlers may play an important role in transmission of amebae, even though the cysts rarely survive more than 10 min on the hands, except under the fingernails.

DIAGNOSIS. For accurate identification and differentiation from other species of amebae, fecal smears should be stained with hematoxylin. The smears are generally fixed in Schaudinn's fluid and stained with Heidenhain's iron hematoxylin.

For concentration of cysts, flotation in zinc sulfate solution can be used. The cysts are distorted beyond recognition by the other salt and sugar solutions in common use for flotation of helminth eggs. For concentration by sedimentation, the formalin-triton-ether (FTE) sedimentation technique, merthiolate-iodine-formaldehyde-concentration (MIFC) technique, Scholten (1972) technique, or PVA technique (Markell and Quinn 1977) can be used.

Various serologic and skin tests for amebiasis have been suggested, but none is as satisfactory as finding the amebae themselves. Kagan (1980) tabulated seven commercial sources of diagnostic reagents. He recommended the indirect hemagglutination, double diffusion, indirect immunofluorescence, countercurrent electrophoresis, and immunoelectrophoresis tests for serologic diagnosis.

E. histolytica cannot be differentiated structurally from *E. hartmanni* except on the basis of cyst size, and its differentiation from other intestinal amebae, and especially from *E. coli,* is not an easy task.

A great many conditions have been ascribed to amebae simply because they were there, but the mere fact that amebae are present in a case of diarrhea or dysentery does not necessarily mean that they have caused it.

CULTIVATION. *E. histolytica* can be cultivated in a number of media (Meerovitch 1978). While most strains of *E. histolytica* will not grow below 30

C, the so-called Laredo strain grows as well at 20–25 C as at 37 C. More recently, other strains of human *Entamoeba* have been discovered that will grow at room temperature.

PRESERVATION. *E. histolytica* and other enteric amebae can be preserved by freezing after glycerol or dimethylsulfoxide has been added to the medium.

TREATMENT. Amebiasis can be treated with a number of drugs. The old standard drug, emetine, is not used much because of its toxicity. Other drugs that have been used include (1) the arsanilic acid derivatives carbarsone, glycobiarsol (bismuth glycoarsanilate, Milibis), and thiocarbarsone; (2) the iodoquinoline derivatives diiodohydroxyquin (Diodoquin), chiniofon (Yatren), and iodochlorhydroxyquin (Vioform); (3) the antimalarial drug chloroquine; (4) the antibiotics erythromycin, fumagillin, tetracycline, chlortetracycline, and oxytetracycline (which do not act directly on the amebae but on their concomitant bacteria); (5) entamide furoate (Furamide); and (6) metronidazole (2-methyl-5-nitroimidazole-1-ethanol, Flagyl), which has probably replaced all other drugs. A 750-mg dose is given by mouth 3 times a day for 5–10 days. It is effective against both cysts and trophozoites in the intestine and also in extraintestinal infections. However, as Krogstad et al. (1978) pointed out, "Treatment of the patient with amebiasis remains a quagmire." In addition, metronidazole is suspected of being carcinogenic and mutagenic.

While relatively little work has been reported on the treatment of amebiasis in lower animals, the same drugs are in general effective in them.

PREVENTION AND CONTROL. Infection with amebae can be prevented by sanitation. Water supply systems should be built without cross connections to sewage systems. Water that may be polluted should be boiled or filtered through sand, since ordinary chlorination does not kill the cysts. Food handlers should wash their hands thoroughly after using the toilet. Vegetables grown on polluted ground should be cooked or, if to be eaten raw, should be scalded or soaked in vinegar. Drinks containing ice should be avoided in polluted areas.

REMARKS. *E. histolytica* has been reviewed by Sepulveda and Diamond (1976), Krogstad et al. (1978), and Sepulveda (1980), among others.

Entamoeba hartmanni von Prowazek, 1912

E. hartmanni closely resembles the small race of *E. histolytica.* It can be differentiated by careful examination of hematoxylin-stained preparations. Most trophozoites of *E. hartmanni* are smaller than those of *E. histolytica.* Rounded trophozoites of *E. hartmanni* are 3–10.5 μm in diameter, those of *E. histolytica* 6.5 μm or more in diameter. The trophozoite nucleus of *E. hartmanni* is usually 2.0–2.5 μm in diameter but may be 1.5–3.2 μm, while that of *E. histolytica* is usually 3.0–3.5 μm but may be 2.8–3.8 μm. The peripheral chromatin of *E. hartmanni* is more variable in its arrangement than that of *E. histolytica* and may consist of discrete granules with wide spaces between them, a crescent of granules on one side of the nucleus, or a single large bar of chromatin with several small granules around the membrane. The peripheral chromatin of *E. histolytica,* on the other hand, is generally distributed uniformly along the nuclear membrane.

Most cysts of *E. hartmanni* are smaller than those of *E. histolytica;* they are 3.8–8.0 μm in diameter, while those of small race *E. histolytica* are 5.5 μm or more in diameter. The cyst nuclei of *E. hartmanni* are 1.8–3.0 μm in diameter in uninucleate cysts, 1.3–2.0 μm in binucleate cysts, and 0.7–1.7 μm in tetranucleate cysts; those of small race *E. histolytica* are 2.4–2.8 μm in diameter in uninucleate cysts, 2.0–2.8 μm in binucleate cysts, and 1.4–2.2 μm in tetranucleate cysts. The cysts of *E. hartmanni* seldom contain large glycogen bodies, but nearly all of them have from a few to many small vacuoles; the cysts of *E. histolytica* generally have 1 large glycogen vacuole or no vacuoles. The chromatoid bodies of the two species are similar.

Freedman and Elsdon-Dew (1959) suggested that until an accurate, practical method of separation is devised the mean sizes of 12 μm for trophozoites and 10 μm for cysts be used as the dividing line between *E. histolytica* and *E. hartmanni.*

The incidence of *E. hartmanni* in lower animals and man is unknown because in the past it has ordinarily been lumped with *E. histolytica.*

Entamoeba moshkovskii Chalaya, 1941

This species occurs in sewage and water. It is not a parasite of animals, but of the municipal digestive tract. It has been found in sewage in many parts of the world, including the United States. Although *E. moshkovskii* is not parasitic, the possibility of its accidental presence in fecal samples is of concern in diagnosis.

E. moshkovskii resembles *E. histolytica.* The trophozoites are active, 9–29 (usually 11–13) μm in diameter. The nucleus has a small, central endosome and a peripheral layer of fine granules. The cysts are generally spherical, 7–17 μm in diameter. Each contains a very large glycogen vacuole at first, which is eventually absorbed as the cyst ages. The chromatoid bodies are large, rather elongate, and have rounded ends. Mature cysts have 4 nuclei. The cysts remain viable at 4 C up to 10 mo if not allowed to dry out.

Entamoeba equi Fantham, 1921

Fantham (1921) found this ameba in the feces of two horses in South Africa with signs of intestinal disturbance. It is unusually large, fully extended trophozoites being 40–50 × 23–29 μm and rounded ones 28–35 μm in diameter. The nucleus is of the *histolytica* type but is oval rather than round. Erythrocytes are ingested. The cysts are 15–27 μm in diameter and contain 4 nuclei and chromatoid bars.

Entamoeba anatis Fantham, 1924

Fantham (1924) found this ameba in the feces of a duck that had died of acute enteritis in South Africa. It resembles *E. histolytica,* and its trophozoites ingest erythrocytes. The cysts are spherical or subspherical, thin-walled, 13–14 μm in diameter; they contain 1–4 nuclei and thin, needlelike chromatoid bodies.

Entamoeba coli (Grassi, 1879) Casagrandi and Barbagallo, 1895

SYNONYMS. *Amoeba coli, Endamoeba hominis, Councilmania lafleuri, Entamoeba cynocephalusae.*

This is the commonest species of ameba in man. It also occurs in the gorilla, orangutan, chimpanzee, gibbon and various species of macaques and other monkeys, and has been found in the pig and white-tailed deer.

E. coli occurs in the cecum and colon. It can be cultivated on the usual media. It is nonpathogenic and therefore must be differentiated from *E. histolytica.*

Its trophozoites are 15–50 (usually 20–30) μm in diameter. The cytoplasm is filled with bacteria and debris, and the ectoplasm is thin. The organism moves sluggishly. The nucleus has an eccentric endosome larger than that of *E. histolytica,* and a row of relatively coarse chromatin granules around its periphery. There may also be a few scattered chromatin granules between the endosome and the nuclear membrane. The cysts, which are 10–33 μm in diameter and have 8 nuclei when mature, contain slender, splinterlike chromatoid bodies with sharp, fractured, or square ends, which disappear as the cysts age. The young cysts may each also contain a large, well-defined glycogen globule, which usually disappears before the cysts mature.

Entamoeba wenyoni Galli-Valerio, 1935

This rare species has been reported in goats and camels. The trophozoites are 12 × 9 μm, their protoplasm is fairly granular with no distinction between ectoplasm and endoplasm, and they contain numerous bacteria. The cysts are spherical, 7 μm in diameter, and contain 8 nuclei.

Entamoeba muris (Grassi, 1879)

SYNONYMS. *Amoeba muris, Councilmania muris, C. decumani.*

E. muris occurs commonly in the cecum and colon of rats, mice, the golden hamster, and many wild rodents throughout the world.

E. muris resembles *E. coli.* Its trophozoites are 8–30 μm long. Its cysts are 9–20 μm in diameter and have 8 nuclei when mature. The nucleus is intermediate in structure between those of *E. histolytica* and *E. coli* but more nearly resembles the latter. It varies in diameter from 3–9 μm, with a mean of 4.5 μm. In division, approximately 8 chromosomes are formed. Each binucleate cyst almost always contains a large glycogen vacuole, and the mononucleate cyst very frequently does.

E. muris is nonpathogenic. It is important to the research worker because it must be differentiated from other amebae introduced in experimental infections.

Entamoeba caviae Chatton, 1918

This species is sometimes referred to as *E. cobayae.* Hoare (1959) considered *E. caviae* a synonym of *E. muris.* Perhaps it is. It is common in the ceca of laboratory guinea pigs.

E. caviae resembles *E. coli.* The trophozoites are 10–20 (mean 14.4) μm in diameter, the nucleus 3–5 μm. Its endosome varies in size and shape and may be central or eccentric. In some cases it is composed of several granules. There is a ring of coarse chromatin granules inside the nuclear membrane. The cysts, which are rare, are 11–17 (mean 14) μm in diameter and have 8 nuclei.

E. caviae is nonpathogenic. Because it is so common, it must be differentiated from other amebae in experimentally infected animals.

Entamoeba cuniculi Brug, 1918

This nonpathogenic species occurs quite commonly in the cecum and colon of the domestic rabbit. It resembles *E. coli,* and Hoare (1959) considered it a synonym of *E. muris.*

Entamoeba gallinarum Tyzzer, 1920

This nonpathogenic species occurs in the ceca of the chicken, turkey, and other birds. It closely resembles *E. coli.* The trophozoites are 9–25 μm in diameter, most being 16–18 μm. The endoplasm is highly vacuolated and contains many food vacuoles. The ectoplasm is clear or granular. The nucleus is 3–5 μm in diameter, with an eccentric endosome and a row of granules around the outside. Mature cysts are 12–15 μm in diameter and contain 8 nuclei.

Entamoeba bovis (Liebetanz, 1905)

This nonpathogenic species occurs commonly in the rumen and feces of cattle throughout the world and also in gnus, duikers, and white-tailed deer.

The trophozoites are 5–20 μm in diameter. The smoothly granular cytoplasm is filled with vacuoles of various sizes. The nucleus is large, with a large, central endosome made up of compact granules; there is a conspicuous row of chromatin granules of different sizes around its periphery. The cysts are 4–14 μm in diameter; each contains a single nucleus when mature. Their chromatoid bodies are irregular clumps of varying size, rods, splinters, or granules. A large glycogen vacuole may or may not be present.

Entamoeba ovis Swellengrebel, 1914

SYNONYM. *E. debliecki* in part.

This nonpathogenic species is common in the intestine of sheep, goats, and other ruminants throughout the world.

The trophozoites are 13–14 × 11–12 μm. The nucleus typically contains a large, pale endosome generally composed of several granules, a ring of peripheral chromatin, and numerous small granules between the endosome and the nuclear membrane. In some cases there is very little peripheral chromatin and in others the endosome may be very small. The cysts are 4–13 (mean 7) μm in diameter; each contains a single nucleus when mature. They usually contain numerous chromatoid bodies of varying size, shape, and abundance, and a glycogen vacuole.

It is quite likely that *E. ovis* is a synonym of *E. bovis,* but until cross-infection experiments have been carried out, it is thought best to retain it as a separate species.

Entamoeba dilimani Noble, 1954

This species occurs in the goat. The trophozoites are 12 μm across; they have food vacuoles containing bacteria and broad, rounded pseudopods whose ends have fairly clear ectoplasm. The cysts are 5–16 (mean 10) μm in diameter; each contains a single nucleus. The endosome is usually a small, central dot but may be eccentric. Peripheral chromatin is often absent or may appear as a few large, irregular granules. The entire nucleus is filled with fine granules that may form a ring around the endosome. The cyst contains 1 or more large glycogen vacuoles and from 1 to a large number of chromatoid bodies varying in shape from small, irregular masses to a single, large, sausage-shaped body. This species differs from *E. ovis* of American goats in that the peripheral chromatin rarely forms a heavy ring; the endosome is usually a single, small dot; and a periendosomal ring of chromatin is usually present.

Entamoeba suis Hartmann, 1913

SYNOMYM. *E. debliecki* in part. A number of authors have used the name *E. polecki* for this species, but this name is a nomen nudum. It occurs in the cecum and colon of swine.

The trophozoites are 5–25 μm long. The nucleus varies in appearance. The endosome is central and is usually quite large. It sometimes almost fills the nucleus but may be small and similar to that of *E. histolytica.* There is a rather homogeneous ring of peripheral chromatin within the nuclear membrane. There are ordinarily no chromatin granules between the endosome and the peripheral ring. The cytoplasm is granular, vacuolated, and contains bacteria in its food vacuoles. The cysts are 4–17 μm in diameter; each has a single nucleus when mature. The chromatoid bodies in the cysts vary markedly in shape, from stout rods with rounded ends similar to those of *E. histolytica* to irregular granules of varying size. There may or may not be a glycogen vacuole. Cysts without chromatoid bodies or glycogen vacuoles are also common. *E. suis* is probably nonpathogenic. It can be cultivated in the usual media.

Entamoeba bubalis Noble, 1955

The trophozoites average 12 μm in diameter. The cysts are 5–9 (mean 8) μm in diameter. They may contain 1 or more vacuoles but usually a single large one that crowds the cyst contents to its periphery. The chromatoid bodies are usually small and irregular in shape but may occasionally be large, with rounded ends similar to those of *E. histolytica.* The cysts each contain a single nucleus 2.6 μm in diameter with a large endosome 1.4 μm in diameter that often appears to be a cluster of 4 granules. There is usually a distinct peripheral ring of chromatin, but the amount of peripheral chromatin may vary from practically none to a few isolated clumps, to a ring of dots. There is no periendosomal chromatin.

Entamoeba chattoni Swellengrebel, 1914

SYNONYM. *E. polecki* in part.

HOSTS. This species, generally considered nonpathogenic, occurs in the large intestine of macaques and a number of other monkeys. About 20 human cases have been reported. *E. chattoni* is probably much more common in monkeys than *E. histolytica,* from which it must be distinguished.

The trophozoites are 9–25 μm long. The cysts are 6–18 μm in diameter. The nucleus varies a great deal in structure. It may be indistinguishable from that of *E. histolytica,* with a small, central endosome and a row of fine, peripheral granules. The endosome, however, may be large or small, central or eccentric, compact or diffuse, and composed of 1 or many granules. The peripheral chromatin may be fine or coarse and uniform, irregular, or diffuse; there may or may not be chromatin granules between the endosome and the peripheral chromatin. The cysts are almost always uninucleate when mature; less than 1% are binucleate, and they are never tetranucleate. The chromatoid bodies are usually irregular and small, but may also be rodlike with round or pointed ends, oval, or round. A glycogen vacuole may or may not be present in each.

Entamoeba gingivalis (Gros, 1849) Brumpt, 1914

SYNONYMS. *Amoeba gingivalis, A. buccalis, A. dentalis, A. kartulisi, E. buccalis, E. maxillaris, E. canibuccalis.*

This species occurs commonly in the human mouth, where it lives between the teeth, under the edge of the gums, and in the tartar. It has occasionally been found in infected tonsils. It was once thought to be the cause of pyorrhea, but is now known to be a harmless commensal that finds an ideal home in diseased gums.

E. gingivalis can be transmitted to dogs with gingivitis but not to healthy ones. It is occasionally found in dogs and cats and also occurs in chimpanzees, macaques, and baboons.

E. gingivalis has no cysts. The trophozoites are 5–35 μm (usually 10–20 μm) long. The cytoplasm consists of a zone of clear ectoplasm and granular endoplasm containing food vacuoles. The amebae usually feed on leukocytes and epithelial cells, sometimes on bacteria, and rarely on red blood cells. There are usually a number of pseudopods. The nucleus is 2–4 μm in diameter, with a moderately small endosome, a peripheral layer of chromatin granules, and some delicate achromatic strands extending from the endosome to the nuclear membrane. Reproduction is by binary fission. There are 5 or 6 chromosomes.

Entamoeba equibuccalis Simitch, 1938

SYNONYM. *E. gingivalis* var. *equi.*

This ameba occurs in the mouths of horses. It is structurally identical with *E. gingivalis,* except that its trophozoites are somewhat smaller, 7–14 μm in diameter. It has no cysts.

Entamoeba suigingivalis Tumka, 1959

This ameba occurs on the coating of the teeth of domestic pigs. It resembles a small *E. gingivalis,* being 7–12 (mean 9) μm long when fixed and stained. It is questionable whether this is a separate species.

Entamoeba gedoelsti (Hsiung, 1930)

SYNONYM. *Endamoeba gedoelsti.*

This ameba occurs in the cecum or colon of horses. No cysts have been seen. The trophozoites, 7–13 μm long, contain bacteria in their food vacuoles. The nucleus is similar to that of *E. coli,* with an eccentric endosome surrounded by a halo and a row of peripheral chromatin granules.

Entamoeba caprae Fantham, 1923

This species occurs in the intestine and reticulum of the goat. It is very large, up to 34 × 24 μm. The pseudopods are short and lobose, and red cells may be ingested. The nucleus is oval, 9–10 μm in diameter, with an eccentric endosome. No cysts have been seen. The relationship of this form to other goat amebae remains to be determined.

Entamoeba sp.

Brenon (1953) tabulated three deaths from amebic dysentery among the causes of death he observed in 1,005 chinchillas in California. Since the amebae of chinchillas have apparently not been described, they cannot be assigned to any species.

Genus *Iodamoeba* Dobell, 1919

In this genus the nucleus is vesicular, with a large endosome rich in chromatin, a layer of lightly staining globules surrounding the endosome, and some achromatic strands between the endosome and the nuclear membrane. The cysts each contain a large glycogen body that stains darkly with iodine. They are ordinarily uninucleate. This genus occurs in vertebrates. A single species is recognized.

Iodamoeba buetschlii (von Prowazek, 1912) Dobell, 1919

SYNONYMS. *Entamoeba williamsi* in part, *Endolimax williamsi, E. piloenucleatus, E. kueneni, I. wenyoni, I. suis.*

HOSTS. Pig, man, chimpanzee, gorilla, guenons, mangabeys, macaques, other monkeys and baboons, and the giant forest hog *Hylochoerus meintritzhageni.*

LOCATION. Cecum and colon.

GEOGRAPHIC DISTRIBUTION. Worldwide.

PREVALENCE. *I. buetschlii* is the commonest ameba of swine, and the pig was probably its original host. It is also common in monkeys and man.

STRUCTURE. The trophozoite is usually 9–14 μm long but may be 4–20 μm. It has clear, blunt pseudopods that form slowly, and it moves rather slowly. The ectoplasm is clear but not well separated from the granular endoplasm. Food vacuoles containing bacteria and yeasts are present in the cytoplasm. The nucleus is relatively large, and ordinarily contains a large, smoothly rounded, central endosome surrounded by a vesicular space containing a single layer of periendosomal granules about midway between the endosome and the nuclear membrane. Fibrils extend to the nuclear membrane, but there are no peripheral granules inside the membrane. Some trophozoites have tubelike processes that may be used for feeding.

The cysts are often irregular in form. They are usually 8–10 μm long but may range from 5 to 14 μm. They contain a single nucleus in which the periendosomal granules have usually aggregated into a crescent-shaped group at the side of the endosome, pushing it to one side. They contain a large, compact mass of glycogen that stains deeply with iodine. The glycogen disappears after 8–10 days in feces held at room temperature, and at the same time the cysts die and disintegrate. There are no chromatoid bodies in the cysts, but they may contain small, deeply staining granules something like volutin granules.

LIFE CYCLE. *I. buetschlii* reproduces by binary fission. The haploid number of chromosomes is usually more than 10, possibly 12.

PATHOGENESIS. *I. buetschlii* is nonpathogenic except under unusual circumstances. These have never been noted in the pig but have occasionally occurred in man.

EPIDEMIOLOGY. *I. buetschlii,* like other intestinal amebae, is transmitted by cysts.

CULTIVATION. This species can be cultivated in the usual media.

PREVENTION AND CONTROL. The same preventive measures recommended for *E. histolytica* will also prevent *I. buetschlii* infections.

Genus *Endolimax* Kuenen and Swellengrebel, 1917

These are small amebae. The nucleus is vesicular, with a comparatively large, irregularly shaped endosome composed of chromatin granules embedded in an achromatic ground substance, and with several achromatic threads connecting the endosome with the nuclear membrane. Cysts are present. This genus occurs in both vertebrates and invertebrates.

Endolimax nana (Wenyon and O'Connor, 1917) Brug, 1918

SYNONYMS. *Amoeba limax, Entamoeba nana, Endolimax intestinalis, E. cynomolgi, E. suis, Councilmania tenuis.*

HOSTS. Man, pig, gorilla, chimpanzee, gibbons, guenons, macaques, other monkeys and baboons, capybara *Hydrochoerus capybara,* tree porcupine *Coen-*

dou prehensilis. It is quite likely that *E. ratti* (see below) is a synonym of *E. nana,* so that the latter's host range may be even broader than this.

LOCATION. Cecum and colon.

GEOGRAPHIC DISTRIBUTION. Worldwide.

PREVALENCE. *E. nana* is common in man and other primates and also in pigs.

STRUCTURE. The trophozoites are 6–15 (mean 10) μm in diameter. They move sluggishly by means of a few blunt, thick pseudopods. The endoplasm is granular, vacuolated, and contains bacteria and crystals. The nucleus contains a large, irregular endosome composed of a number of chromatin granules. Several achromatic fibrils run from the endosome to the nuclear membrane. There are ordinarily no peripheral chromatin granules. The cysts are oval, often irregular, and thin-walled; they are usually 8–10 μm long but may be 5–14 μm. The mature cysts contain 4 nuclei and may contain ill-defined glycogen bodies. They have no chromatoid bodies but sometimes have small granules resembling volutin and occasionally a few filaments of uncertain nature.

LIFE CYCLE. Reproduction is by binary fission in the trophozoite stage. The ameba that leaves the cyst is multinucleate, but it divides into uninucleate amebulae that grow into ordinary trophozoites.

PATHOGENESIS. *E. nana* is nonpathogenic.

EPIDEMIOLOGY. *E. nana* is transmitted in the same way as other enteric amebae.

CULTIVATION. This species can be cultivated in the usual media.

PREVENTION AND CONTROL. The same preventive measures recommended for *E. histolytica* will also prevent *E. nana* infections.

Endolimax ratti Chiang, 1925

This species, which may be a synonym of *E. nana,* occurs in the cecum and colon of laboratory and wild rats.

Endolimax caviae Hegner, 1926

This species occurs commonly in the cecum of the guinea pig. It is somewhat smaller than *E. nana,* with trophozoites 5–11 × 5–8 μm, but otherwise resembles it. The cysts are apparently unknown.

Endolimax gregariniformis (Tyzzer, 1920) Hegner, 1929

SYNONYMS. *Pygolimax gregariniformis, E. janisae, E. numidae.*

This nonpathogenic species is found throughout the world in the ceca of the chicken, turkey, guinea fowl, pheasant, goose, duck, and various wild birds, including herons and owls.

Its trophozoites are usually 4–13 μm long with a mean of 9 × 5 μm. They are oval, often with a posterior protuberance, and move sluggishly. The ectoplasm is not clearly separated from the endoplasm. The food vacuoles contain bacteria. The nucleus is similar to that of *E. nana* but tends to have a larger endosome and a more apparent nuclear membrane, often with chromatin granules at the juncture of the achromatic threads with the membrane. The cysts have 4 nuclei when mature; they are 8–11 × 7–8 (mean 10 × 7) μm. They tend to be somewhat lima-bean–shaped instead of truly ovoid and are often highly vacuolated.

LITERATURE CITED

For citations before 1965, see the *Index-Catalogue of Medical and Veterinary Zoology, Zoological Record, Biological Abstracts, Protozoological Abstracts,* and *Veterinary Bulletin.*

Albach, R. A., and T. Booden. 1978. In Parasitic Protozoa, vol. 2, ed. J. P. Kreier, 455–506. New York: Academic.
Ayers, K. M., L. H. Billups, and F. M. Garner. 1972. Vet Pathol. 9:221–26.
CDC (Center for Disease Control). 1979. Intestinal parasite surveillance, annual summary 1978. Atlanta: U.S. Department of Health, Education and Welfare.
Griffin, J. L. 1978. In Parasitic Protozoa, vol. 2, ed. J. P. Kreier, 507–49. New York: Academic.
Healy, G. H., and G. S. Visvesvara. 1977. Proc. Conf. Risk Assess. Health Eff. Land Appl. Munic. Waste Sludges, 196–202.
Jadin, J. B. 1974. Actual. Protozool. 1:179–89.
Kadlec, V. 1975. Folia Parasitol. 22:317–21.
______. 1978. J. Protozool. 25:235–37.
Kagan, I. G. 1974. Arch. Invest. Med. 5(Suppl. 2):457–64.
______. 1975. Z. Klinik Therap. Infekt. 3:96–98.
______. 1980. In Manual of Clinical Immunology, ed. N. R. Rose and H. Friedman, 573–604. Washington, D.C.: American Society for Microbiology.
Kingston, N., and P. C. Taylor. 1976. Proc. Helminthol. Soc. Wash. 43:227–29.
Krogstad, D. J., H. C. Spencer, Jr., and G. R. Healy. 1978. N. Engl. J. Med. 298:4 pp.
McConnell, E. E., F. M. Garner, and J. H. Kirk. 1968. Pathol. Vet. 5:1–6.
Markell, E. K., and P. M. Quinn. 1977. Am. J. Trop. Med. Hyg. 26:1139–42.
Martinez, A. J. 1976. In Amibiasis, ed. B. Sepulveda and L. S. Diamond, 64–81. Mexico City: Instituto Mexicano Seguro Social.
Meerovitch, E. 1978. In Methods of Cultivating Parasites in Vitro, ed. A. E. R. Taylor and J. R. Baker, 19–37. New York: Academic.
Patton, R. A. 1969. Personal communication.
Scholten, T. 1972. J. Parasitol. 58:633–34.
Sepulveda, B., ed. 1980. Octavo Seminario Sobre Amibiasis. Arch. Invest. Med. 11(Suppl. 1):1–328.
Sepulveda, B., and L. S. Diamond, eds. 1976. Amibiasis. Mexico City: Instituto Mexicano Seguro Social.
Singh, B. N. 1970. H. D. Srivastava Commem. Vol., 55–64.
______. 1975. Pathogenic and Non-Pathogenic Amoebae. New York: Halsted.
Taylor, P. W. 1977. J. Parasitol. 63:232–37.
Wilhelm, W. E., and J. H. Anderson. 1971. J. Parasitol. 57:1378–79.

7
Apicomplexa: The Coccidia Proper

ALL members of the phylum Apicomplexa are parasitic. There are about 4,500 named species. They include the gregarines (parasites of invertebrates), the hemogregarines (mostly blood parasites of lower vertebrates), the coccidia (mostly intestinal parasites of vertebrates but also of a few invertebrates), the malaria parasites and their relatives (blood parasites of vertebrates), and the piroplasms (blood parasites of mammals and a few birds).

Electron microscopy has revealed common structures in the group, enabling taxonomists to arrange them better than before. A polar ring, conoid, micronemes, rhoptries, and subpellicular microtubules are usually present in some stage; these make up the apical complex. In addition, one or more micropores may also be present. There is a single type of nucleus. Cilia and flagella are absent except that the microgametes of some groups are flagellated. The phylum probably arose from the Mastigophora.

The coccidia and their relatives belong to the order Eucoccidiorida. In this order, merogony is present and the life cycle involves both sexual and asexual phases.

Becker (1924) wrote a classic review of the coccidia. Pellérdy (1974) reviewed them all, as did Long (1982). The coccidia of rodents, ruminants, carnivores, and artiodactyls were reviewed by Levine and Ivens (1965a, 1970, 1981, 1985). Those of wild birds were reviewed by Todd and Hammond (1971). Scholtyseck (1979) described their fine structure, as did Chobotar and Scholtyseck (1982). Reviews of their taxonomy and life cycles were given by Levine (1982a), of their genetics by Jeffers and Shirley (1982), of their host and site specificity by Joyner (1982), of their biochemistry and physiology by Wang (1982), of their behavior in vitro by Doran (1982), of their pathogenesis by Fernando (1982), of their hosts' immune response by Rose (1982), of their chemotherapy by McDougald (1982) and Ryley (1982), of their drug resistance by Chapman (1982), and of their control by Fayer and Reid (1982). Levine (1982b) compiled a glossary of terms used in discussing them.

The great majority of coccidia proper belong to the suborder Eimeriorina, in which the microgamont and macrogamete develop independently, there is no syzygy, the microgamont typically produces a large number of microga-

NUMBER SPOROCYSTS PER OOCYST	NUMBER SPOROZOITES PER SPOROCYST					
	1	2	4	8	16	VARIED OR MANY
0			CRYPTOSPORIDIUM	SCHELLACKIA TYZZERIA MOST GREGARINES		NO OOCYSTS KLOSSIELLA LANKESTERELLA
1			MANTONELLA	CARYOSPORA		
2		CYCLOSPORA	ATOXOPLASMA BESNOITIA FRENKELIA HAMMONDIA ISOSPORA SARCOCYSTIS TOXOPLASMA	DORISA	SIVATOSHELLA	
4		CRYSTALLOSPORA EIMERIA GOUSSIA	WENYONELLA			
8		OCTOSPORELLA				GOUSSEFFIA
16	SKRJABINELLA	HOARELLA	PYTHONELLA			
VARIED OR MANY		ADELINA	KLOSSIA			

Fig. 7.1. Numbers of sporocysts per oocyst and sporozoites per sporocyst in the genera of the order Eucoccidiorida. In the genera without sporocysts, the numbers of sporozoites per oocyst are given.

metes, the zygote is not motile, and the sporozoites are typically enclosed in sporocysts within the oocyst (Fig. 7.1). Within this suborder most of the coccidia proper belong to the family Eimeriidae, the overwhelming majority homoxenous; they are the subject of this chapter.

FAMILY EIMERIIDAE

Members of this family are homoxenous, with a single host. Merogony and gamogony take place within the host cells, and sporogony ordinarily occurs outside the host's body. The oocysts and meronts lack an attachment organelle. The oocysts contain 0, 1, 2, 4, or many sporocysts, each containing one or more sporozoites. The microgametes have 2 or 3 flagella. The genera are differentiated by the number of sporocysts in their oocysts and the number of sporozoites in each sporocyst (Fig. 7.1).

STRUCTURE. The structure of a typical oocyst, that of *Eimeria,* is shown in Figure 7.2. The oocyst wall is composed of 1 or 2 layers and may be lined by a membrane. It may have a micropyle, which may be covered by a micropylar cap. In this genus there are 4 sporocysts, each containing 2 sporozoites, within the oocyst. There may be an oocyst residuum or a sporocyst residuum in the oocyst and the sporocyst, respectively; these are composed of material left over after the formation of the sporocysts or sporozoites. The sporocyst may have a knob, the Stieda body, at one end, and there may be a substiedal body beneath it. The sporozoites are usually elongate, with one end rounded and the other (the anterior end) tapered, or they may be sausage-shaped. They may contain 1 or more proteinaceous clear globules (refractile bodies, eosinophilic globules) of unknown function.

Merozoites are formed by merogony (ecto- or endopolygeny) within the

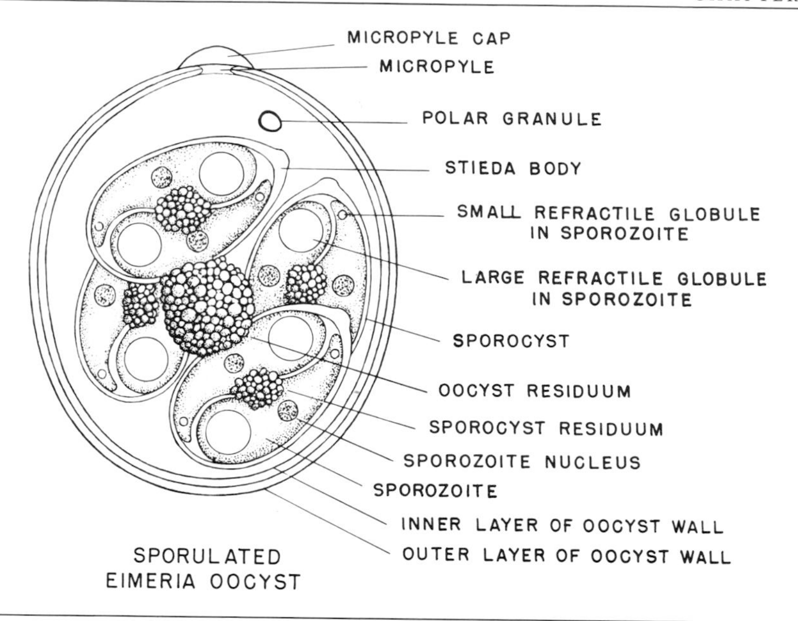

Fig. 7.2. Structures of sporulated *Eimeria* oocyst.

host cell (see the life cycle section). Both they and the sporozoites have an apical complex. Sporozoites, merozoites, and sporocyst residua all contain stored carbohydrate as small bodies of amylopectin with a chain length of about 20 glucose residues; there are also lipid droplets in the sporocyst residua. The sporozoites and merozoites are each covered with a pellicle consisting of a continuous outer limiting membrane and an inner membrane that terminates at the polar ring; they each have 22–26 subpellicular microtubules, a protrusible conoid composed of spirally arranged microtubules, 1 or 2 rings anterior to the conoid, a polar ring, a nucleus with or without a nucleolus, rhoptries, micronemes, clear globules, endoplasmic reticulum, Golgi apparatus, mitochondria with tubular cristae, micropores, lipoidlike bodies, oval polysaccharide (amylopectin) bodies, and ribosomes.

LOCATION. Most coccidia are intracellular parasites of the intestinal tract, but some occur in other organs such as the liver and kidney. Each species is usually found in a specific location within the intestinal tract (e.g., in the cecum, in the duodenum, in the ileum, etc.). They may invade different cells in these locations. Some species are found in the epithelial cells, others in the submucosa. Their location within the host cell also varies. Some species are found above the host cell nucleus, while others are found beneath it and a few within it. Some species enlarge the host cell only slightly; others cause it to become enormous. The host cell nucleus is also often greatly enlarged even though it may not be invaded.

LIFE CYCLE. The life cycles of the Eimeriidae are similar and can be illustrated by that of *Eimeria tenella,* which is found in the ceca of the chicken (Fig. 7.3).

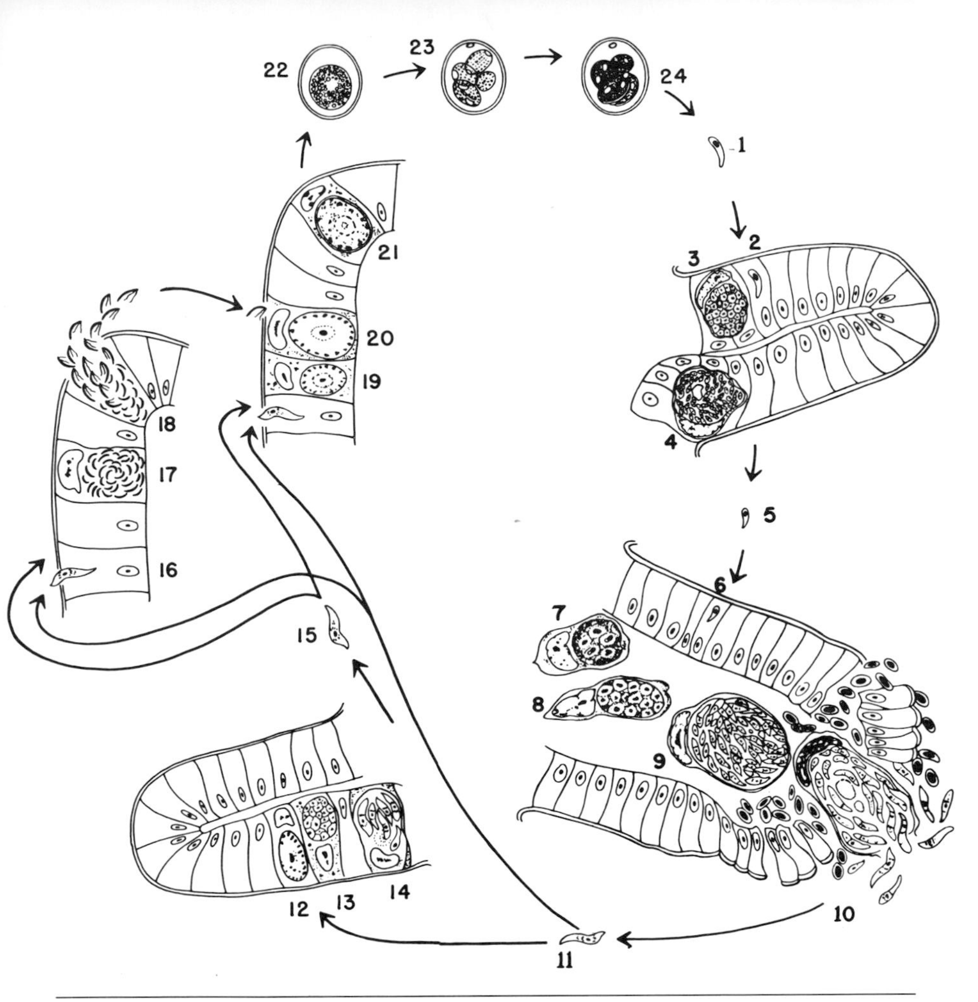

Fig. 7.3. Life cycle of the chicken coccidium *Eimeria tenella*. A sporozoite (*1*) enters an intestinal epithelial cell (*2*), rounds up, grows, and becomes a first generation meront (schizont) (*3*). This produces a large number of first generation merozoites (*4*), which break out of the host cell (*5*), enter new intestinal epithelial cells (*6*), round up, grow, and become second generation meronts (*7, 8*). These produce a large number of second generation merozoites (*9, 10*), which break out of the host cell (*11*). Some enter new host intestinal epithelial cells and round up to become third generation meronts (*12, 13*), which produce third generation merozoites (*14*). The third generation merozoites (*15*) and the great majority of second generation merozoites (*11*) enter new host intestinal epithelial cells. Some become microgamonts (*16, 17*), which produce a large number of microgametes (*18*). Others turn into macrogametes (*19, 20*). The macrogametes are fertilized by the microgametes and become zygotes (*21*), which lay down a heavy wall around themselves and turn into young oocysts. These break out of the host cell and pass out in the feces (*22*). The oocysts then begin to sporulate. The sporont throws off a polar body and forms 4 sporoblasts (*23*), each of which forms a sporocyst containing 2 sporozoites (*24*). When the sporulated oocyst (*24*) is ingested by a chicken, the sporozoites are released (*1*).

The oocysts are passed in the feces; at this time they contain a single cell, the sporont. They must have oxygen in order to develop to the infective stage, a process known as sporulation or sporogony. The sporont, which is diploid, undergoes reduction division and throws off a refractile polar body. The sporont divides directly into 4, forming 4 sporoblasts, each of which then develops into a sporocyst. Two sporozoites develop within each sporocyst. Sporulation takes 2 days at ordinary temperatures. The oocysts are then infective and ready to continue the life cycle.

When they are eaten by a chicken, the oocyst wall breaks in the gizzard, releasing the sporocysts. The sporozoites within them are activated by bile or trypsin after the sporocysts reach the small intestine; they then escape from the sporocysts.

The sporozoites enter the epithelial cells, either directly or via white blood cells. Once there, each sporozoite rounds up and becomes a first generation meront. By a process of asexual multiple fission (endopolygeny, ectopolygeny, merogony, schizogony) each meront forms about 900 first generation merozoites, each about 2–4 μm long. These get their name from the Greek word *meros,* a part. They break out into the lumen of the cecum about 2.5–3 days after infection. Each first generation merozoite enters a new host cell and rounds up to form a second generation meront, which lies above the host cell nucleus. By multiple fission it forms about 200–350 second generation merozoites about 16 μm long (Figs. 7.4 and 7.5), which are found 5 days after inoculation (DAI). Some of them enter new intestinal cells and round up to form third generation meronts, which lie beneath host cell nuclei and produce 4–30 third generation merozoites about 7 μm long. Others are engulfed and digested by macrophages.

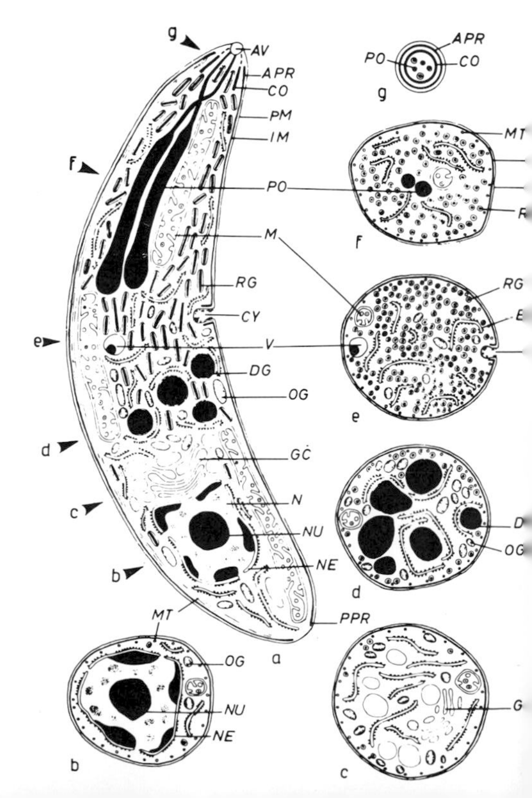

Fig. 7.4. Second generation merozoite of *Eimeria miyairii:* a. longitudinal section; b. cross section at level of nucleus; c. cross section at level of Golgi complex; d. cross section at level of large dense granules; e. cross section at level of rhoptries; f. cross section through rhoptry; g. cross section at level of conoid; *APR* = anterior polar ring, *AV* = anterior vesicle, *CO* = conoid, *CY* = micropore (cytostome), *DG* = dense granule, *ER* = endoplasmic reticulum, *GC* = Golgi complex, *IM* = inner membrane, *M* = mitochondrion, *MT* = subpellicular microtubule, *N* = nucleus, *NE* = nuclear envelope, *NU* = nucleolus, *OG* = ovoid granule, *PM* = plasma membrane, *PO* = rhoptry (paired organelle), *PPR* = posterior polar ring, *RG* = microneme (rod-shaped granule), *V* = vacuole. (From Andreassen and Behnke 1968)

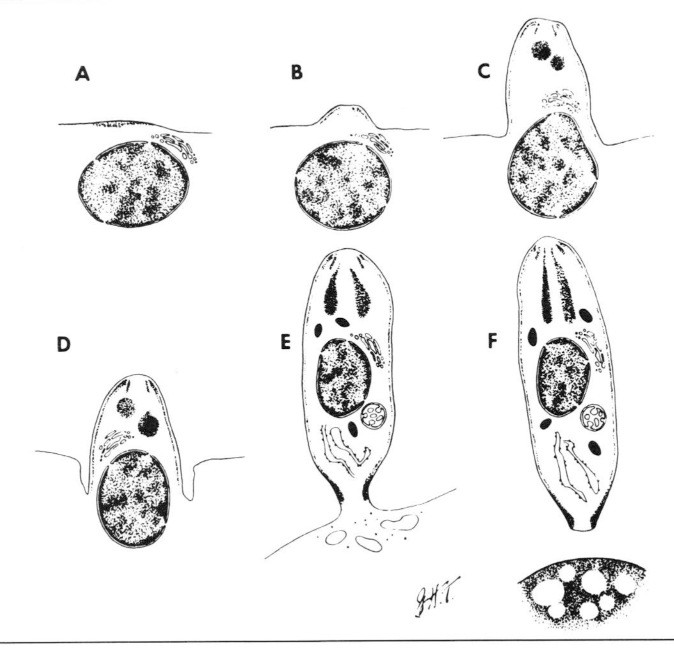

Fig. 7.5. Development by ectopolygeny of first generation merozoite of *Eimeria bovis*: A. condensation of material along blastophore membrane opposite nucleus; B. conical elevation of blastophore surface enclosing inner membrane and conoids; C. elongated anterior end of merozoite containing dense bodies; D. invagination of blastophore membrane around nucleus; E. nearly developed merozoite still attached to blastophore at posterior end; F. free merozoite after separation from blastophore, which remains as a residual body. (From Sheffield and Hammond 1967)

Many of the second generation merozoites enter new host cells and begin the sexual phase of the life cycle, known as gamogony or gametogony. Most of these merozoites turn into macrogametes (female gametes, often called macrogamonts or macrogametocytes). These simply grow until they reach full size (Fig. 7.6). Some of the merozoites turn into microgamonts (male gamonts or microgametocytes) (Fig. 7.7). Both the macrogametes and microgamonts lie below the host cell nuclei. Within each microgamont a large number of tiny biflagellate microgametes is formed (Fig. 7.8). They bud off of the microgamonts (leaving a residuum of cytoplasm), break out, and fertilize the macrogametes.

The resultant zygotes lay down walls around themselves in the following way. The macrogametes contain one or two sets of eosinophilic plastic granules (wall-forming bodies) composed of mucoprotein in their cytoplasm. These granules pass to the periphery, flatten out, and coalesce to form the oocyst wall after fertilization. One set forms the outer layer and the other the inner layer. (See Fig. 7.9.)

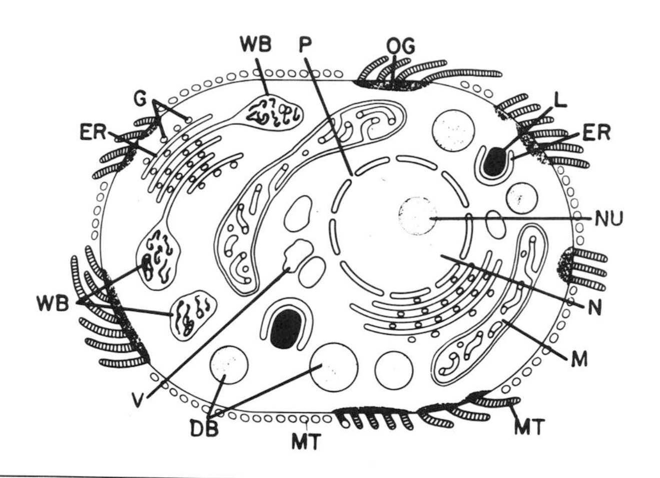

Fig. 7.6. Young macrogamete (macrogamont) of *Eimeria perforans: DB* = dark body, *ER* = endoplasmic reticulum, *G* = Golgi apparatus, *L* = lipoid droplet, *M* = mitochondrion, *MT* = microtubule, *N* = nucleus, *NU* = nucleolus, *OG* = osmiophilic granules, *P* = pore in nuclear membrane, *V* = vacuole, *WB* = wall-forming body. (From Scholtyseck 1964)

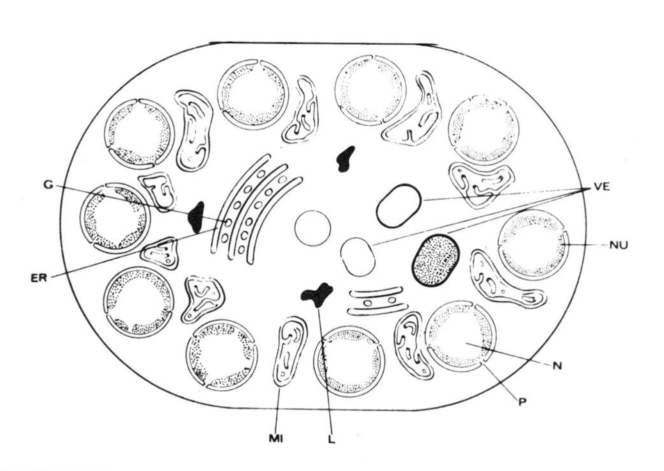

Fig. 7.7. Young microgamont of *Eimeria*: *ER* = endoplasmic reticulum, *G* = glycogen granules, *L* = lipid inclusion, *MI* = mitochondrion, *N* = nucleus, *NU* = nuclear substance, *P* = pore in nuclear membrane, *VE* = different kinds of vesicles. (From Hammond et al. 1967)

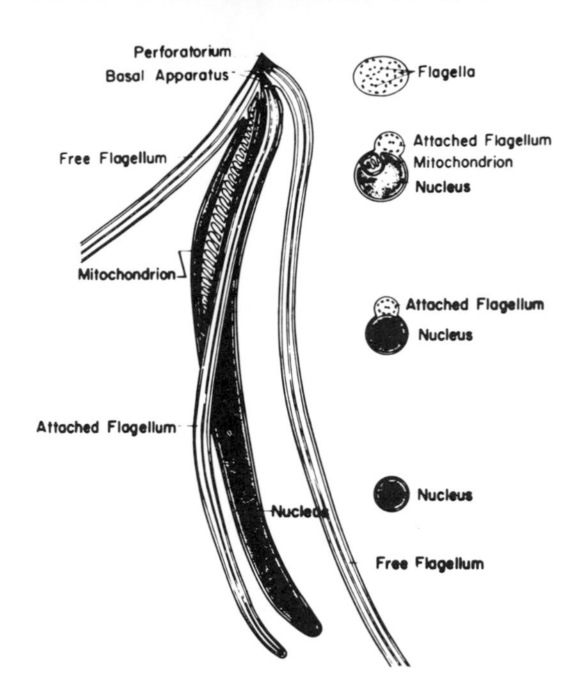

Fig. 7.8. Microgamete of *Eimeria perforans*. Side view (*left*) and cross sections at various levels (*right*). (From Scholtyseck 1965)

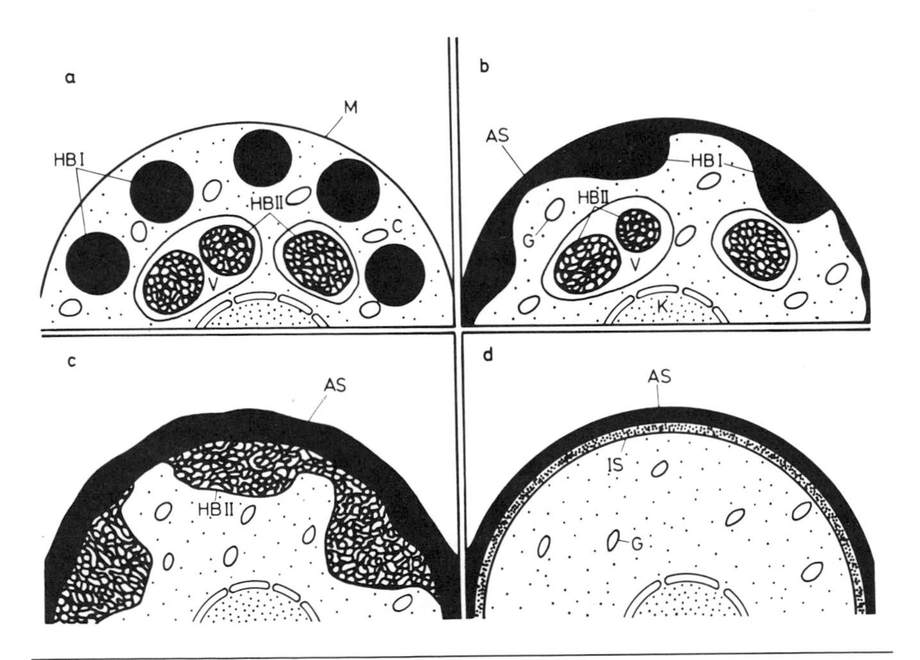

Fig. 7.9. Formation of oocyst of *Eimeria stiedai*: *AS* = outer layer of oocyst wall, *C* = cytoplasm, *G* = glycogen, *HB I* and *HB II* = wall-forming bodies, *IS* = inner layer of oocyst wall, *K* = nucleus, *M* = macrogamete membrane, *V* = vacuole surrounding wall-forming body. (From Scholtyseck et al. 1969)

The oocysts then break out of their host cells, enter the intestinal lumen, and pass out in the feces. The prepatent period, from the time of inoculation to the appearance of the first oocysts in the feces, is 7 days. Oocysts continue to be discharged for a number of days thereafter, because the sporozoites do not all enter the host cells immediately but may remain in the lumen for some time and also because many of the oocysts are retained in a plug of material in the ceca for some days before they are eliminated.

In the absence of reinfection, coccidial infections are self-limiting. Asexual reproduction does not continue indefinitely, as it does, for example, in *Plasmodium.* In *E. tenella,* three generations of merozoites are produced; in other species there may be one, two, or more. After this the life cycle enters its sexual phase, oocysts are formed and eliminated from the body, and the infection is over. Reinfection may take place, but the host develops more or less immunity following the primary infection.

The number of oocysts produced in an animal per oocyst fed depends in part on the number of merozoite generations and the number of merozoites per generation. A single oocyst of *E. tenella* containing 8 sporozoites is theoretically capable of producing 2,520,000 second generation merozoites (8 × 900 × 350), each of which can develop into a macrogamete or microgamont.

In *E. bovis* of cattle, there are two asexual generations. The first is a giant meront containing about 120,000 merozoites and the second produces 30–36 merozoites, so that a single oocyst could produce about 3.5 million second generation merozoites. In the rat, *E. nieschulzi* is theoretically capable of producing 1,500,000 oocysts per oocyst fed, *E. miyairii* 38,016, and *E. separata* only 1,536. In *E. nieschulzi* there are four generations of merozoites, while in the latter two species there are only three. In the rabbit, *E. magna* can produce 800,000 oocysts per oocyst fed, *E. media* 150,000, and *E. coecicola* 100,000.

The actual numbers of oocysts produced per oocyst fed are usually considerably lower than the theoretical ones. If the host is resistant or immune, it destroys many sporozoites and merozoites; many others pass out in the feces before they have time to enter host cells. The infecting dose is also an important factor in determining the number of oocysts produced. The greater the infecting dose, the smaller the number of oocysts usually produced per oocyst fed. For example, Hall (1934) obtained a yield of 1,455,000 oocysts of *E. nieschulzi* per oocyst fed when the infecting dose was 6 oocysts, 1,029,666 when it was 150 oocysts, and 144,150 when it was 2,000. If the infecting dose is too small, however, relatively small numbers of oocysts may be produced. Hall (1934) found that when only a single oocyst was fed, the yield was 62,000.

Host age has a marked effect on the number of oocysts produced. Rose (1967) found that in chicks 0 wk old an oral inoculum of 500 oocysts produced 36,400 oocysts per oocyst fed, one of 5,000 oocysts produced 20,300, and one of 50,000 oocysts produced 103. In chicks 14 days old, the numbers of oocysts per oocyst fed produced by these inocula were 41,600, 15,700, and 1,500, respectively; the numbers produced in chicks 35 days old were 151,200, 25,500 and 8, respectively. Data were given for other ages as well.

All the factors responsible for these findings are not known. More effective mobilization of the host's defenses and immunologic competence are probably important, but lack of enough epithelial cells to parasitize, sloughing of patches of epithelium, increased intestinal motility with resultant diarrhea and elimination of merozoites before they can reach a host cell, and entrapment of merozoites in tissue debris and cecal cores may also play a part.

PATHOGENESIS. While many species of coccidia are pathogenic, many others are not. Pathogenicity depends on a number of factors, some of which are probably still unknown. Among those that might be mentioned are the number of host cells destroyed per infecting oocyst, which depends on the number of merozoite generations, the number of merozoites per generation, and the location of the parasite in the host tissues and within the host cell. The size of the infecting dose or doses, the degree and time of reinfection, and the degree of acquired or natural immunity of the host are also important.

The great majority of coccidian species develop in the epithelial cells of the intestine. In considering coccidian pathogenicity, it is well to remember that these host cells are continually passing out of the intestine and being replaced. In the chick the cells constituting a whole crypt of Lieberkuehn are replaced in 2 to 5 days, although the turnover is slower in large animals. If a host cell is discarded naturally, the parasite within it has little chance to harm the host. Hence the type of host cell invaded and the length of time it would normally remain functional in the host must be considered in assessing the pathogenicity of a coccidian species.

Coccidiosis is generally not a problem in the wild or on sparse pastures. Under these conditions, the usual result is coccidiasis (i.e., a light, nonpathogenic infection). When animals are brought together under conditions in which the infecting dose can be high, however, disease may result. Coccidiosis, in other words, is a result of monoculture and of the raising of animals in confinement. It is a disease of the feedlot and not of the range.

Even with a pathogenic species, the final effect on the host depends on interplay between many factors. It may range from rapid death in susceptible animals to an imperceptible reaction in immune ones.

If disease is present, the signs are those of a diarrheal enteritis. There may or may not be blood in the feces, depending on the parasite species and severity of infection. Affected animals gain weight poorly, become weak and emaciated, or may even die, depending again on the parasite species and the size of the infecting dose. Young animals are much more commonly affected than older ones. Those animals that recover develop an immunity to the species that infected them. However, this immunity is seldom absolute and recovered adult animals are often continuously reinfected so that they carry light infections that do not harm them but make them a source of infection for the young. In addition, under conditions of stress their immunity may be broken down and they may suffer the disease again.

DIFFERENTIATION OF SPECIES. Both structural and biological characters are used to separate the species of coccidia. Both the endogenous and exogenous stages of the life cycle may differ structurally. However, since the endogenous stages of many species are unknown, the structure of the oocyst is most commonly used.

A second group of criteria is the location of the endogenous stages in the host. This has been discussed in the life cycle section. Host specificity, a third criterion, varies with the protozoan genus and to some extent with the species. The host range of most *Eimeria* species is relatively narrow. A single species rarely infects more than one host genus unless the latter genera are closely related.

Cross-immunity studies are also used in differentiating the coccidia of a particular host from each other. Infection of an animal with one species of coccidium produces immunity against that species but not against other species that occur in the same host.

DIAGNOSIS. Coccidiosis can be diagnosed by finding the coccidia on microscopic examination. There are several pitfalls in diagnosis. Each species of domestic and laboratory animal has several species of coccidia, some of which are pathogenic and some of which are not. An expert is often needed to differentiate between some of the species; the mere presence of oocysts in the feces, even in the presence of disease signs, is not necessarily proof that the signs are due to coccidia and not to some other agent.

Following recovery from a coccidial infection, an animal is relatively immune to reinfection with the same species. This immunity is not so solid that the animal cannot be reinfected at all, but it does mean that the resultant infection will be low-grade (except possibly under conditions of stress) and will not harm the host. Such low-grade infections are extremely common (i.e., the animals have coccidiasis rather than coccidiosis). Hence, the presence in the feces of oocysts of even highly pathogenic species of coccidia does not necessarily mean that the animal has clinical coccidiosis.

On the other hand, coccidia may cause severe symptoms and even death early in their life cycle before any oocysts have been produced. This occurs commonly, for example, with *E. tenella* of the chicken and *E. zuernii* of the ox. Consequently, failure to find oocysts in the feces in a diarrheal disease does not necessarily mean that the disease is not coccidiosis.

The only sure way to diagnose coccidiosis, then, is by finding lesions containing coccidia at necropsy. Scrapings of the lesions should be mixed on a slide with a little physiologic NaCl solution and examined microscopically. It is not enough to look for oocysts; meronts, merozoites, gametes, and gamonts inside the host cells must be sought for and recognized as well.

Some species of coccidia can be identified from their unsporulated oocysts, but study of the sporulated oocysts is often desirable. Oocysts can be sporulated by mixing the feces with several volumes of 2.5% potassium bichromate solution, placing the mixture in a thin layer in a Petri dish, and allowing it to stand for 1 day to 2 wk or more, depending on the species. The potassium bichromate prevents bacterial growth that might kill the protozoa, and the thin layer is necessary so that oxygen can reach the oocysts.

TREATMENT. Many different drugs have been used for therapy, particularly against *E. tenella* of the chicken. These include not only sulfonamides but also derivatives of phenylarsonic acid, diphenylmethane, diphenyldisulfide, diphenylsulfide, nitrofuran, triazine, carbanilide, imidazole, and benzamide.

Several thousand papers have probably been published on coccidiostatic drugs, and their use in poultry production is so common in the United States that it is difficult to obtain a commercial poultry feed that does not contain one or another of them. They are used to a considerably lesser extent for other classes of livestock.

None of these drugs will cure a case of coccidiosis once signs of the disease have appeared. They are all prophylactic and must be administered at the time of exposure or soon thereafter in order to be effective. They act against the meronts and merozoites and occasionally against the sporozoites, preventing the life cycle from being completed, but are usually not effective against the gametes. Since exposure in nature is continuous, these drugs must be used continuously in the nonimmune host. This is usually done by mixing them with the feed or water.

Nowhere is knowledge of the normal course of the disease more important than in interpreting the results of treatment of coccidiosis, and nowhere is the controlled experiment more important than in research in this field. Coccidiosis is self-limiting not only in the individual patient but also in a flock or herd. In a typical outbreak, signs of disease appear in only a few animals at first, the number of affected animals builds up rapidly to a peak, and then the disease subsides spontaneously. In the early stages, most farmers do little, thinking that the condition is unimportant and will soon be over. When more animals become affected and losses increase, it takes a little time to establish a diagnosis, so treatment is often not started until the outbreak has reached its peak. Under these circumstances, it matters little what treatment is used—the disease will subside. This is the reason so many quack remedies have in the past received glowing testimonials from satisfied users.

A similar course of events is encountered by the small animal practitioner. The patient with coccidiosis is not brought in until it is already sick. By that time it is too late for any anticoccidial drug to be of value, although supportive treatment and control of secondary infections may be helpful. If the patient recovers, however, whatever drug happened to be used is often given undeserved credit.

After an animal has been receiving a coccidiostatic drug for some time during exposure to infection, it ordinarily develops an immunity to the coccidia. This occurs because the sporozoites are not affected by the drug but invade the cells and stimulate the host's defenses.

After coccidiostats had been mixed in poultry feeds for a number of years, it was inevitable that drug resistant strains of coccidia would appear. Drug resistance is now a well-known complicating factor in the use of chemotherapeutic agents. A race has developed between the coccidia and the pharmaceutical houses, and some day we may be reduced to sanitation to control coccidia.

MIXED INFECTIONS. All domestic animals have more than one species of coccidia. Some are highly pathogenic, others less so, and still others practically nonpathogenic. Pure infections with a single species are rare in nature, so the observed effect is the result of the combined actions of the particular mixture of coccidia and other parasites present, together with the modifying effects of the nutritional condition of the host and environmental factors such as weather and management practices.

In the remainder of this chapter, each species of coccidium in a particular host animal will be taken up first, and then a general discussion of coccidiosis in each host will follow.

Genus *Eimeria* Schneider, 1875

In this genus the oocyst contains 4 sporocysts, each of which contains 2 sporozoites.

Eimeria alabamensis Christensen, 1941

This species occurs primarily in the small intestine but also in the cecum and upper colon in heavy infections in the ox, zebu, and water buffalo. It is fairly common throughout the world.

OOCYSTS. The oocysts are usually ovoid; 13–25 × 11–17 μm; with a smooth, pale yellow, 1- or 2-layered wall about 0.6–0.7 μm thick; without a micropyle, polar granule, or residuum. The sporocysts are ellipsoidal, 10–16 × 4–6 μm, with a tiny Stieda body and a residuum. The sporozoites lie lengthwise head to tail in the sporocysts and have 1–3 clear globules. Sporulation takes 4–8 days. (See Fig. 7.10.)

LIFE CYCLE. The endogenous stages are in the nuclei of the villar epithelial cells. There is probably more than one generation of meronts. The mature meronts are 8–18 × 5–14 μm 8 DAI and contain 16–32 merozoites. Gamonts can be recognized 6 DAI. The microgamonts are 8–25 × 7–21 (mean 16 × 11.5) μm, the macrogametes 7–20 × 7–12 (mean 12 × 9) μm.

PATHOGENESIS. Under field conditions, *E. alabamensis* is considered essentially nonpathogenic.

Eimeria auburnensis Christensen and Porter, 1939

SYNONYMS. *E. ildefonsoi, E. khurodensis,* probably *E. bombayansis.*

This species occurs commonly in the middle and lower third of the small intestine of the ox, zebu, and water buffalo throughout the world.

OOCYSTS. The oocysts are ovoid, somewhat flattened at the small end; 32–46 × 19–30 μm; with a smooth, heavily mammillated or rarely rough, 2-layered wall 1–1.8 μm thick, the outer layer colorless to yellowish orange, the inner layer greenish; with a micropyle and polar granule; without a residuum. The sporocysts are elongate ovoid, almost ellipsoidal, 15–23 × 6–11 μm, with a wall about 0.2 μm thick, with a Stieda body and residuum. The sporozoites are elongate, rather comma-shaped, 15–18 × 3–5 μm, lie lengthwise head to tail in the sporocysts, and have a large clear globule at the large end and sometimes 1–2 small ones elsewhere. The sporulation time is 2–3 days. (See Fig. 7.10.)

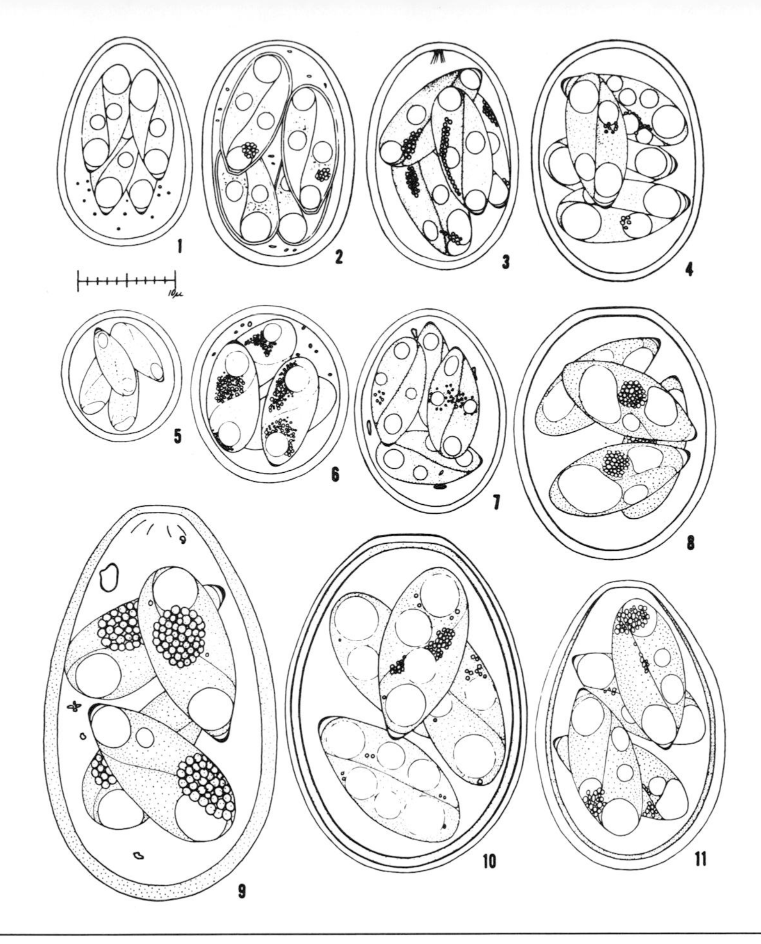

Fig. 7.10. Sporulated oocysts of coccidia from the ox: 1. *Eimeria alabamensis;* 2. *E. cylindrica;* 3 and 4. *E. ellipsoidalis;* 5. *E. subspherica;* 6 and 7. *E. zuernii;* 8. *E. illinoisensis;* 9. *E. auburnensis;* 10. *E. canadensis;* 11. *E. bovis.* (From Levine and Ivens 1967)

LIFE CYCLE. There are two asexual generations. The first generation meronts occur throughout the small intestine, usually in cells of the reticular connective tissue deep in the lamina propria near the muscularis mucosae. When mature 10 days after inoculation they are 134–338 × 95–172 (mean 178 × 114) μm and contain thousands of merozoites about 11 μm long. Second generation meronts occur in the subepithelium in cells of mesodermal origin in the distal part of the villi 2–8 m anterior to the ileocecal valve. When mature 12–15 days after inoculation, they are 6–12 × 5–9 (mean 8.5 × 6) μm and contain 4–11 spindle-shaped merozoites 7–9 × 1–2 μm.

The gamonts occur in the subepithelium in mesenchymal-mesodermal cells of the small intestine. When mature at 18–19 days the microgamonts are 61–151 × 42–109 (mean 103 × 70) μm and contain thousands of microgametes 4–8 × 0.5–0.75 μm with 2 and sometimes 3 flagella. The macrogametes are about 18 μm in diameter at 18 days. The prepatent period is 16–24 days, the patent period usually 2–8 days.

PATHOGENESIS. This species is apparently moderately pathogenic.

Eimeria bovis (Züblin, 1908) Fiebiger, 1912

SYNONYMS. *Coccidium bovis, E. smithi, E. (Globidium) bovis, E. aareyi, Globidium fusiformis* (?).

This species occurs very commonly in the small and large intestines of the ox, zebu, and water buffalo throughout the world.

OOCYSTS. The oocysts are generally ovoid; 23–24 × 17–23 μm; with a smooth, 2-layered wall, the outer layer colorless and about 1.3 μm thick, the inner layer brownish yellow and about 0.4 μm thick; with an inconspicuous micropyle; without a polar granule or a residuum. The sporocysts are elongate ovoid, 13–18 × 5–8 μm, with an inconspicuous Stieda body and a residuum. The sporozoites are elongate, lie lengthwise head to tail in the sporocysts, and usually have a clear globule at each end. The sporulation time is 2–3 days. (See Fig. 7.10.)

LIFE CYCLE. There are two asexual generations. The first generation meronts are in the endothelial cells of the lacteals in the villi of the posterior half of the small intestine. They grow to giant size and become mature 14–18 DAI, at which time they are 207–435 × 134–367 (mean 303 × 281) μm and contain 55,000–170,000 (mean 120,000) merozoites. The meronts are easily visible to the naked eye as whitish balls. The second generation meronts occur in the epithelial cells of the cecum and colon but may extend into the terminal meter of the small intestine in heavy infections. They average 10 × 9 μm when mature and contain 30–36 merozoites averaging 6–7 (mean 6.2) μm in length.

The sexual stages generally occur only in the cecum and colon but may be in the terminal meter of the small intestine in heavy infections. They appear 17 DAI. The microgamonts are 12–18 μm in diameter and produce many biflagellate microgametes. The prepatent period is 16–21 days, the patent period usually 5–15 days.

PATHOGENESIS. *E. bovis* is one of the two most pathogenic of the bovine coccidia. Diarrhea and/or dysentery, tenesmus, and elevated body temperature appear about 18 DAI and are often followed by death. The most severe pathologic changes occur in the cecum, colon, and terminal 0.3 m of the ileum, and are due to the gamonts. At first the mucosa is congested, edematous, and thickened, with petechiae or diffuse hemorrhages. Its lumen may contain a large amount of blood. Later the mucosa is destroyed and sloughed, and a patchy or continuous membrane forms over its surface. The submucosa may also be destroyed. If the animal survives, both mucosa and submucosa are later replaced.

Eimeria brasiliensis Torres and Ramos, 1939

SYNONYMS. *E. boehmi, E. orlovi, E. helenae.*

This species occurs in the ox, zebu, and water buffalo. Its oocysts have been found in the feces; its location in the host is uncertain. It occurs throughout the world but is uncommon.

OOCYSTS. The oocysts are ellipsoidal but sometimes slightly ovoid; 31–49 × 21–33 μm; with a generally smooth, brownish yellow, 1- or 2-layered wall that may be more or less covered by round, partially coalescent, yellowish plaques; with a micropyle covered by a cap; with or without polar granules; without a residuum. The sporocysts are ellipsoidal; 16–22 × 7–10 μm; with or without a small, dark Stieda body; with a residuum. The sporozoites are elongate, lie lengthwise head to tail in the sporocysts, and have a large posterior and a small anterior clear globule. The sporulation time is 6–14 days. (See Fig. 7.11.)

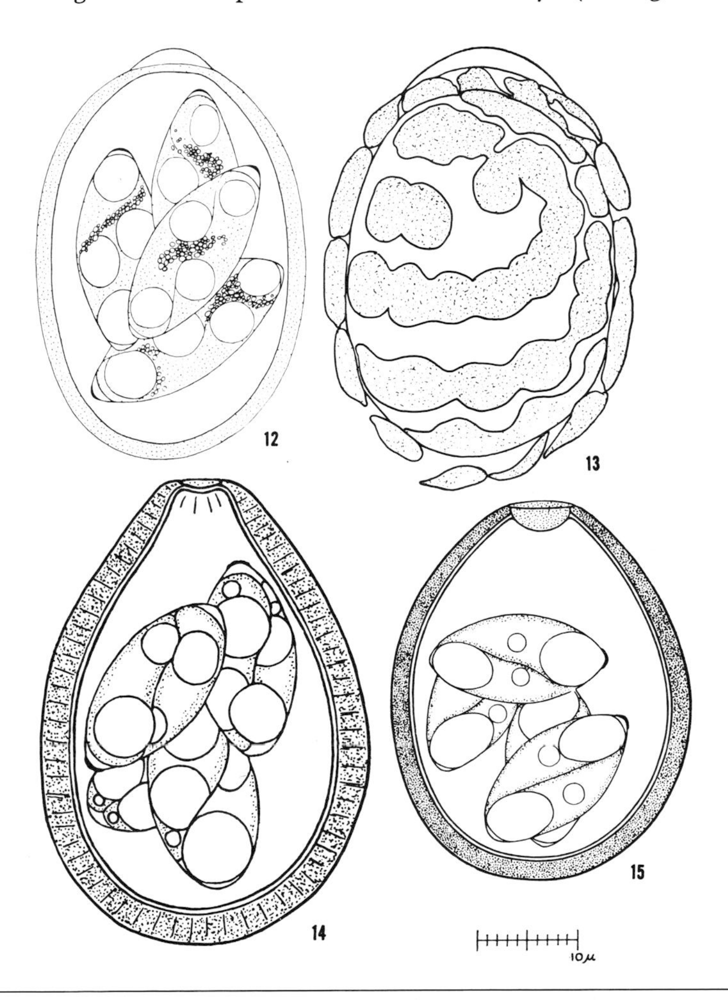

Fig. 7.11. Sporulated oocysts of coccidia from the ox: 12 and 13. *Eimeria brasiliensis;* 14. *E. bukidnonensis;* 15. *E. wyomingensis.* (From Levine and Ivens 1967)

Eimeria bukidnonensis Tubangui, 1931

This species is fairly common in the feces of the ox, zebu, and water buffalo throughout the world. Its location in the host is uncertain.

OOCYSTS. The oocysts are piriform; 34–64 × 26–41 μm; with a yellowish brown, punctate, radially striated, 1- or 2-layered wall about 3–4 μm thick; with a micropyle about 3–5 μm in diameter; with or without a polar granule; without a residuum. The sporocysts are elongate, somewhat pointed at both ends, 12–21 × 9–12 μm, with or without a Stieda body, without a residuum. The sporozoites are elongate, lie lengthwise head to tail in the sporocysts, and have a clear globule at each end. The sporulation time is 4–55 days. (See Fig. 7.11.)

PATHOGENESIS. This species is probably nonpathogenic.

Eimeria canadensis Bruce, 1921

SYNONYM. *E. zurnabadensis.*

This species occurs commonly in the feces of the ox, zebu, and water buffalo throughout the world. Its location in the host is unknown.

OOCYSTS. The oocysts are ovoid or ellipsoidal; 28–39 × 20–29 μm; usually with a smooth, 2-layered wall, the outer layer colorless to yellowish, 0.5 μm thick, the inner layer clear, yellow, 1.3 μm thick; with an inconspicuous micropyle and 1 or more polar granules; generally without a residuum. The sporocysts are elongate ovoid, 15–22 × 6–10 μm, with an inconspicuous Stieda body and a residuum. The sporozoites are elongate, lie lengthwise head to tail in the sporocysts, and have 2–3 clear globules each. (See Fig. 7.10.)

PATHOGENESIS. Apparently slight.

Eimeria cylindrica Wilson, 1931

This species occurs quite commonly in the feces of the ox, zebu, and water buffalo throughout the world.

OOCYSTS. The oocysts are ellipsoidal; 16–34 × 12–19 μm; with a colorless, smooth, 1-layered wall about 1.2 μm thick at the sides and bottom and about 0.7 μm thick at one end; apparently without a micropyle; with a polar granule shattered into many fragments; without a residuum. The sporocysts are elongate ellipsoidal, 12–16 × 4–6 μm, without a Stieda body or with an inconspicuous one, with a thin wall and a residuum. The sporozoites are elongate, lie lengthwise head to tail in the sporocysts, and have 1 or more rather indistinct clear globules. The sporulation time is 2–3 days. The prepatent and patent periods are both 10 days. (See Fig. 7.10.)

PATHOGENESIS. Slightly pathogenic.

Eimeria ellipsoidalis Becker and Frye, 1929

This species occurs commonly in the epithelial cells of the small intestine of the ox, zebu, and water buffalo throughout the world.

OOCYSTS. The oocysts are ellipsoidal to slightly ovoid; 12–32 × 10–29 μm; with a smooth, colorless, 1- or 2-layered wall; without a discernible micropyle; ordinarily without a polar granule; without a residuum. The sporocysts are ovoid, 11–17 × 5–7 μm, with or without an inconspicuous Stieda body, with a residuum. The sporozoites are elongate, 11–14 × 2–3 μm, lie head to tail in the sporocysts, and have 2 clear globules. The sporulation time is 3 days. (See Fig. 7.10.)

LIFE CYCLE. The mature meronts are 9–16 × 7.5–15 (mean 11 × 9) μm and contain 24–36 merozoites 8–11 × 1–2 μm. Mature gamonts can be found 10 DAI distal to the host cell nucleus in epithelial cells near the bottoms of the crypts. Mature microgamonts are 12–17 × 11–17 (mean 15 × 13) μm and contain many microgametes 2–3 μm long. The prepatent period is 8–13 days.

PATHOGENESIS. Mildly to moderately pathogenic.

Eimeria illinoisensis Levine and Ivens, 1967

This species occurs in the feces of the ox, presumably throughout the world. It is not common.

OOCYSTS. The oocysts are ellipsoidal or slightly ovoid; 24–30 × 19–23 μm; with a smooth, colorless, 1-layered wall about 1.3 μm thick with a pale tan inner surface that looks like a membrane in intact oocysts; without a definite micropyle, polar granule, or residuum. The sporocysts are elongate ovoid, 13–17 × 6–8 μm, with a small Stieda body and a residuum. The sporozoites are elongate, lie lengthwise head to tail in the sporocysts, and have 2 or more clear globules. (See Fig. 7.10.)

Eimeria pellita Supperer, 1952

This species occurs in the feces of the ox, presumably throughout the world. It is not common.

OOCYSTS. The oocysts are ovoid; 32–42 × 22–27 (mean 38 × 24) μm; with a 2-layered wall, the outer layer yellowish brown and about 1 μm thick, heavily pitted giving the surface a velvety appearance, the inner layer smooth, light yellow, about 0.75 μm thick, lined by a thin membrane; with a micropyle and a polar granule consisting of several rodlike bodies; without a residuum. The sporocysts are ellipsoidal, 17–20 × 7–9 (mean 18.5 × 8) μm, with a small Stieda body and a residuum. The sporozoites are elongate, with 2 clear globules. The sporulation time is 10–12 days.

Eimeria subspherica Christensen, 1941

This species occurs fairly commonly in the feces of the ox, zebu, and water buffalo throughout the world. Its location in the host is unknown.

OOCYSTS. The oocysts are spherical to subspherical; 9–14 × 8–13 μm; with a 1- or 2-layered, smooth, pale yellowish wall 0.5–1 μm thick; without a micropyle, polar granule, or residuum. The sporocysts are elongate ovoid, 6–10 × 2–5 μm, with a small Stieda body, usually without a residuum. The sporozoites are elongate, lie lengthwise head to tail in the sporocysts, and have a clear globule at the large end. The sporulation time is 4–5 days. (See Fig. 7.10.)

LIFE CYCLE. The prepatent period is 7–18 days, the patent period 4–15 days.

Eimeria wyomingensis Huizinga and Winger, 1942

This species occurs rather uncommonly in the feces of the ox, zebu, and water buffalo, presumably throughout the world. Its location in the host is unknown.

OOCYSTS. The oocysts are ovoid; 36–46 × 26–32 μm; with a yellowish brown, speckled, somewhat rough, 1-layered wall about 2–3.5 μm thick, lined by a membrane; with a micropyle about 5 μm in inside diameter; without a polar granule or residuum. The sporocysts are ellipsoidal with somewhat narrow ends, about 18 × 9 μm, with a tiny Stieda body, generally without a residuum. The sporozoites are elongate, about 7–8 × 5 μm, lie lengthwise head to tail in the sporocysts, and have a large clear globule. The sporulation time is 3–7 days. The prepatent period is 13–15 days, the patent period 1–7 days. (See Fig. 7.11.)

Eimeria zuernii (Rivolta, 1878) Martin, 1909

SYNONYMS. *Cytospermium zurnii, E. zuerni.*

This species occurs commonly in the small and large intestines of the ox, zebu, and water buffalo throughout the world.

OOCYSTS. The oocysts are subspherical, subovoid, ovoid, or sometimes ellipsoidal; 12–29 × 10–21 μm; with a smooth, colorless, 1-layered wall about 0.7 μm thick; without a micropyle or a residuum; without or with 1 or more polar granules (if present, somewhat shattered). The sporocysts are ovoid, 7–14 × 4–8 (mean 11 × 6) μm, with a tiny Stieda body, with or without a residuum. The sporozoites are elongate, lie head to tail in the sporocysts, and have a clear globule at the large end. Free sporozoites are 8–10 × 2–3 μm. The sporulation time is 2-10 days. (See Fig. 7.10.)

LIFE CYCLE. First generation meronts are in the lamina propria of the lower ileum. They are giant, 60–234 μm in diameter when mature at 14–16 days, and contain thousands of merozoites. Second generation meronts are in the epithelial cells of the cecum and proximal colon beginning about day 16. They are 15–22 × 13–21 μm and contain a small number of merozoites 13–17 × 1–3 μm.

Macrogametes and microgamonts can be found in the epithelial cells of the small and large intestines beginning on day 16. The mature macrogametes are 18–24 × 16–24 μm, the mature microgametes 22–26 × 16–19 μm. The latter produce many microgametes. The prepatent period is about 15–17 days, the patent period about 11 days.

PATHOGENESIS. This is the most pathogenic coccidium of cattle. In acute infections it causes a bloody diarrhea of calves. At first the feces are streaked with blood. The diarrhea becomes more severe; bloody fluid, clots of blood, and liquid feces are passed, and straining and coughing may cause this mixture to spurt out as much as 2–3 m. The animal's rear quarters may look as though they had been smeared with red paint. Anemia, weakness, and emaciation accompany the dysentery; secondary infections, especially pneumonia, are common. This acute phase may continue for 3–4 days. If the calf does not die in 7–10 days, it will probably recover.

E. zuernii may also cause a more chronic type of disease. Diarrhea is present, but there may be little or no blood in the feces. The animals are emaciated, dehydrated, weak, and listless, with rough hair coats, drooping ears, and sunken eyes.

A generalized catarrhal enteritis involving both the small and large intestines is present. The lower small intestine, cecum, and colon may be filled with semifluid, bloody material. Large or small areas of intestinal mucosa may be eroded and destroyed, and the mucous membrane may be thickened, with irregular whitish ridges in the large intestine or smooth, dull gray areas in the small intestine or cecum. Diffuse hemorrhages are present in the intestines in acute cases, petechial hemorrhages in mild ones.

One of the most puzzling types of bovine coccidiosis is winter coccidiosis, which is most commonly caused by *E. zuernii*. (See below.)

REMARKS. According to Fitzgerald (1975), bovine coccidiosis is the fifth most important bovine disease in the United States. It occurs most frequently east of the Mississippi River and in beef cattle. About 80,000 cattle or more die of the disease each year. *E. zuernii* is by far the most important species involved.

Eimeria ankarensis Sayin, 1969

This species was found in the feces of the water buffalo in Turkey. Its location in the host is unknown.

OOCYSTS. The oocysts are ovoid; 32–43 × 25–29 μm; with a yellowish brown, 2-layered wall 3.0–3.5 μm thick, the outer layer thick and rough, the inner layer very thick and dark brown; with a micropyle 6 μm in diameter, not covered by the wall's outer layer; without a polar granule or residuum. The sporocysts are elongate, almost ellipsoidal, 18–23 × 8–10 μm, with a Stieda body and residuum. The sporozoites are elongate, rather comma-shaped, lie lengthwise head to tail in the sporocysts, and have 2 clear globules. The sporulation time is 3–4 days at room temperature.

Eimeria bareillyi Gill, Chhabra, and Lall, 1963

This species occurs in the epithelial cells of the jejunum villi of the water buffalo. It is fairly common in Asia and Europe.

OOCYSTS. The oocysts are piriform with a bluntly truncate small end; 24–35 × 15–25 μm; with a smooth, yellowish to darkish brown wall about 1 μm thick; with a micropyle 5–6 μm in diameter and a residuum; without a polar granule. The sporocysts are lemon-shaped, 15–18 × 6–9 μm, with a Stieda body and residuum. The sporozoites are banana-shaped, about 10 × 4 μm, with a clear globule at the large end and sometimes 1–2 smaller ones.

LIFE CYCLE. The macrogametes are 25–35 × 14–17 μm. The microgamonts are 20–37 × 18–25 μm and contain many slightly curved, rodlike microgametes 3 μm long and some small masses of residual bodies.

PATHOGENESIS. Apparently slight, if at all.

Eimeria gokaki Rao and Bhatavdekar, 1959

This species was found in the feces of the water buffalo in India.

OOCYSTS. The oocysts are ovoid; 22–31 × 18–25 μm; with a thin, yellowish brown wall; with a micropyle and micropylar cap. The sporocysts are elongate and probably have a residuum. The sporulation time is less than 7 days.

Eimeria ovoidalis Ray and Mandal, 1961

This species was found in the feces of the water buffalo in India.

OOCYSTS. The oocysts are ovoid, 32–40 × 20–28 μm, with a pinkish orange wall, with a micropyle, without a micropylar cap or residuum. The sporocysts are ovoid, 14–16 × 8–9 μm, with a Stieda body and a residuum. The sporozoites are 8–9 × 4–5 μm. The sporulation time is 90–120 hr.

Eimeria thianethi Gwéllésiany, 1935

This species has been found in the feces of the water buffalo, ox, and zebu in the USSR (Georgia) and India.

OOCYSTS. The oocysts are ovoid, grayish yellow; 34–49 × 26–34 μm; with a 2-layered wall 2 μm thick, the outer layer thin and homogeneous, the inner layer thick, with transverse striations; with a distinct micropyle in some oocysts; apparently without a polar granule or residuum. The sporocysts are lemon-shaped, with pointed ends, 22 × 9.5 μm, with a residuum.

Eimeria ahsata Honess, 1942

This species occurs commonly in the small intestine of the Rocky Mountain bighorn sheep, domestic sheep, and mouflon throughout the world.

OOCYSTS. The oocysts are ellipsoidal to somewhat ovoid, 23–48 × 17–30 μm, with a micropyle and a micropylar cap, ordinarily with 1 or occasionally more polar granules, without a residuum. The sporocysts are 12–22 × 6–10 μm, without a Stieda body, with a residuum. The sporozoites are elongate, lie lengthwise head to tail in the sporocysts, and have 1–3 clear globules each. The sporulation time is 36–72 hr. (See Fig. 7.12.)

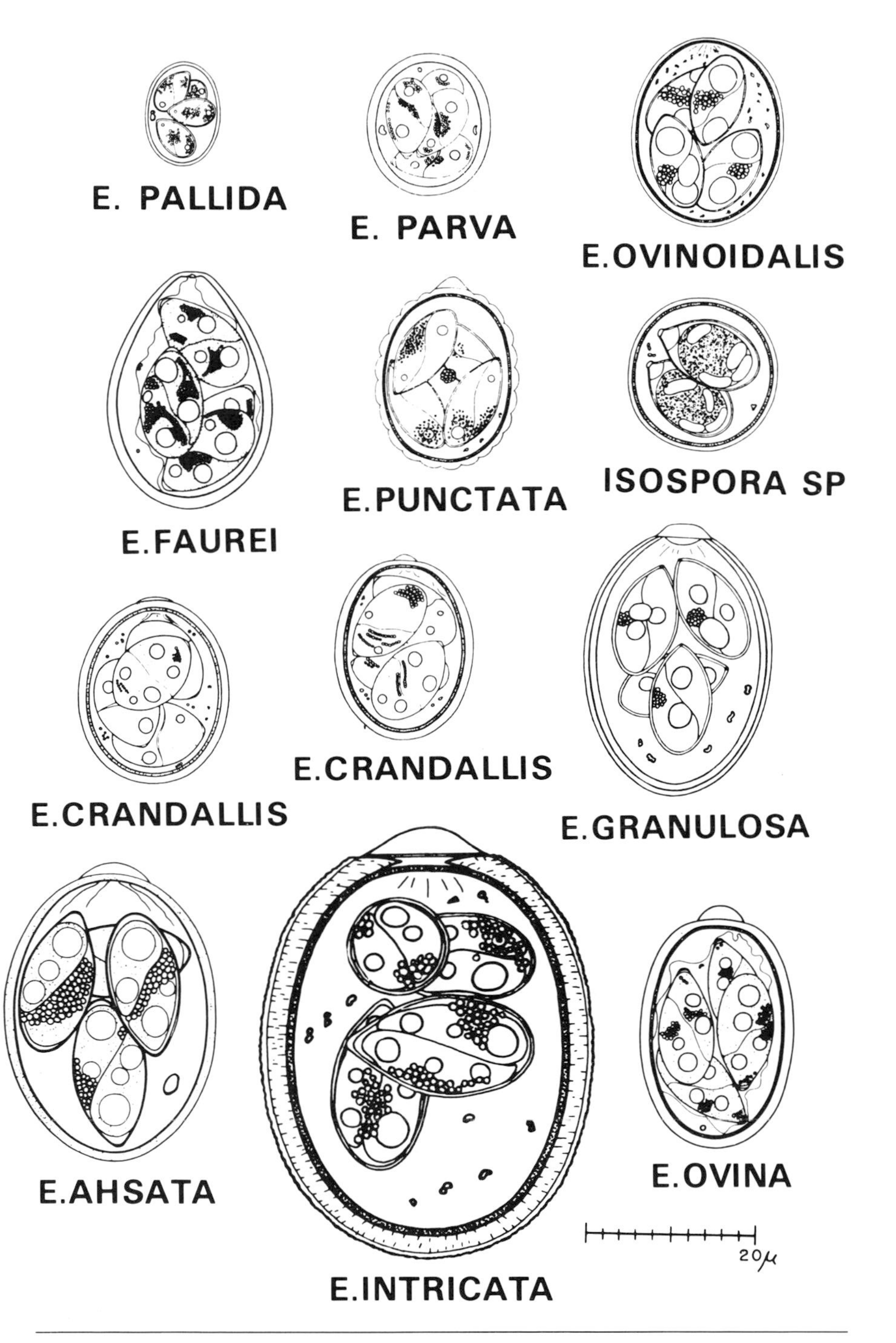

Fig. 7.12. Sporulated oocysts of coccidia from the sheep. (*E. crandallis* and *E. ovina* from Kamalapur 1961, M.S. thesis, Univ. of Ill.; *E. granulosa, E. intricata, E. ovinoidalis, E. pallida, E. punctata,* and *Isospora* sp. from Shah 1963, J. Parasitol. 49:799–807; *E. ahsata* from Levine et al. 1962, Proc. Helminthol. Soc. Wash. 29:87–90)

LIFE CYCLE. The meronts average 184 × 165 μm at 15 days. Gamonts appear at 11 days. The macrogametes are 35–45 μm in diameter and the microgamonts about 36.5 × 26 μm. The prepatent period is 18–21 days, the patent period 10–12 days.

PATHOGENESIS. This is one of the most pathogenic of all sheep coccidia. The intestines of infected lambs have thickened, somewhat edematous areas in the upper part. The Peyer's patches and the last 20–25 cm of the small intestine above the ileocecal valve are inflated.

Eimeria crandallis Honess, 1942

This species occurs commonly in the small intestine anterior to the ileocecal valve of the Rocky Mountain bighorn sheep, domestic sheep, mouflon, and argali throughout the world.

OOCYSTS. The oocysts are subspherical to broadly ellipsoidal; 18–28 × 15–20 μm; with a smooth, 2-layered wall, the outer layer colorless and 1.1 μm thick, the inner layer a darkish yellow membrane; with a distinct to indistinct micropyle and a micropylar cap; usually with 1 or more often shattered polar granules; without a residuum. The sporocysts are broadly ovoid, 8–13 × 6–9 μm, without a Stieda body, with or without a residuum. The sporozoites lie transversely at the ends of the sporocysts, and have 1–2 clear globules. The sporulation time is 1–3 days, the prepatent period 13–20 days. (See Fig. 7.12.)

PATHOGENESIS. Slightly pathogenic.

Eimeria faurei (Moussu and Marotel, 1902) Martin, 1909

SYNONYMS. *Coccidium faurei, C. ovis, E. aemula.*

This species occurs in the small (large?) intestine of the domestic sheep, Rocky Mountain bighorn sheep, argali, and mouflon throughout the world.

OOCYSTS. The oocyst is ovoid; 25–37 × 18–28 μm; with a smooth, greenish, yellowish, or pinkish, 1-layered wall 1.3–1.8 μm thick, lined by a membrane; with a conspicuous micropyle 2–3 μm in diameter; without a micropylar cap, or a residuum; with a polar granule. The sporocysts are ovoid or piriform, 11–17 × 7–9 μm, without a Stieda body or with an inconspicuous one, with a residuum. The sporozoites are elongate, lie lengthwise head to tail in the sporocysts, and have 1–2 large clear globules each. The sporulation time is 1–4 days. (See Fig. 7.12.)

PATHOGENESIS. Mildly pathogenic.

Eimeria gonzalezi Bazalar and Guerrero, 1970

This species occurs quite commonly in the feces of domestic sheep in Peru and Poland.

OOCYSTS. The oocysts are ellipsoidal or ovoid; 26–38 × 18–26 μm; with a smooth, 2-layered wall 1–2 μm thick, the outer layer transparent, the inner layer yellowish brown; with a micropyle, micropylar cap, and polar granule; without a residuum. The sporocysts are ovoid, 12–15 × 6–9 μm, with a slightly perceptible Stieda body and a residuum. The sporozoites have 2–3 clear globules each. The sporulation time is 4–6 days.

Eimeria granulosa Christensen, 1938

This species occurs quite commonly in the feces of the domestic sheep, Rocky Mountain bighorn sheep, and mouflon, presumably throughout the world.

OOCYSTS. The oocysts are piriform, ellipsoidal, or shaped like a stout, broad-shouldered urn; 22–37 × 17–26 μm; with a smooth, 2-layered wall, the outer layer pale and 0.4–0.6 μm thick, the inner layer darker and 0.8 μm thick, lined by a membrane; with a micropyle and micropylar cap at the *broad* end (which differentiates it from all other sheep coccidia); with or without polar granules; without a residuum. The sporocysts are ovoid or elongate ovoid, 13–16 × 8–9 μm, with a faintly perceptible Stieda body and a residuum. The sporozoites are elongate, lie lengthwise head to tail in the sporocysts, and have 1–3 clear globules. The sporulation time is 3–4 days. (See Fig. 7.12.)

Eimeria intricata Spiegl, 1925

This species occurs quite commonly in the small and large intestines of the domestic sheep, Rocky Mountain bighorn sheep, and mouflon throughout the world.

OOCYSTS. The oocysts are ellipsoidal or slightly ovoid; 39–59 × 27–47 μm; with a 2-layered wall, the outer layer irregular, granular, transversely striated, brownish yellow to dark brown, 2–3 μm thick, the inner layer dark brown, 0.4–0.8 μm thick, lined by a membrane; with a micropyle in the outer layer only; generally with a micropylar cap and 1 or more polar granules; without a residuum. The sporocysts are elongate ovoid, 17–22 × 9–14 μm, without a Stieda body or with an extremely tiny one, with a residuum. The sporozoites are elongate, lie lengthwise head to tail in the sporocysts, and have 2–3 clear globules. The sporulation time is 3–12 days. (See Fig. 7.12.)

LIFE CYCLE. The meronts are mostly in the cells lining the lower small intestine crypts. They are up to 65 × 45 μm and contain large merozoites up to 19.5 × 4 μm.

The gamonts, gametes, and oocysts are in the epithelial cells of the crypts of the large and small intestines. The mature macrogametes are 32–54 × 25–36 μm. The mature microgamonts are 61–250 × 36–71 μm and contain many slender, flagellated microgametes 5–6 μm long. The prepatent period is 20–27 days, the patent period 5–11 days.

PATHOGENESIS. Mildly pathogenic.

Eimeria marsica Restani, 1971

This species has been found in the feces of the domestic sheep in Europe. It is probably rare.

OOCYSTS. The oocysts are ellipsoidal; 15–22 × 11–15 μm; with a smooth, colorless to slightly grayish or pale yellow, 2-layered wall 0.5–1 μm thick; with a micropyle 1.5–2 μm in diameter; with or without an inconspicuous micropylar cap; with or without a polar granule (usually shattered); without a residuum. The sporocysts are elongate ovoid, 7–11 × 4–7 μm, with a small or no Stieda body, with a residuum. The sporozoites are elongate, lie lengthwise head to tail in the sporocysts, and have a single, small clear globule. The sporulation time is 3 days.

LIFE CYCLE. The prepatent period is 14–15 days.

PATHOGENESIS. Probably slight.

Eimeria ovina Levine and Ivens, 1970

SYNONYMS. *E. arloingi* in sheep, *E. hawkinsi, E. bakuensis.*

This species occurs very commonly in the small intestine of the domestic sheep, Rocky Mountain bighorn sheep, mouflon, and argali throughout the world.

OOCYSTS. The oocysts are ellipsoidal to ovoid, generally with rather straight sides; 23–36 × 15–24 μm; with a 2-layered wall, the outer layer smooth, greenish or yellow-brown to orange, about 1–1.3 μm thick, the inner layer brownish yellow, dark brown to black, up to 0.4–0.5 μm thick; with a micropyle and micropylar cap; with 1 or more polar granules; without a residuum. The sporocysts are elongate ovoid, 11–17 × 6–9 μm, with or without an inconspicuous Stieda body, with a residuum. The sporozoites are elongate, lie lengthwise head to tail in the sporocysts, and have a large clear globule at the broad end and a smaller one at the narrow end. The sporulation time is 2–4 days. (See Fig. 7.12.)

LIFE CYCLE. The meronts are in the endothelial cells lining the central lacteals of the small intestine villi. They mature 13–21 DAI, at which time they are about 122–146 μm in diameter and contain a large number (perhaps hundreds of thousands) of merozoites about 9 × 2 μm. There is apparently only one generation of meronts and merozoites.

The sexual stages are in the epithelial cells of the small intestine villi. The microgamonts form many microgametes and a large residual mass. The prepatent period is 19–29 days, the patent period about 10 days.

PATHOGENESIS. This species is much less pathogenic than *E. ahsata* or *E. ovinoidalis.* At necropsy, only a few small, slightly hemorrhagic areas scattered throughout the lining of the small intestine occur up to day 13. From then until day 19 the small intestine is more or less thickened and edematous, and thick, white opaque patches made up of groups of heavily parasitized villi are present. The sexual stages do not appear to do as much damage as the asexual stages, since the condition of affected animals appears to improve before oocysts are shed.

Eimeria ovinoidalis McDougald, 1979

SYNONYMS. *E. ninakohlyakimovae* in sheep, *E. galouzoi* in part, *E. ovis* of Musaev, 1970.

This species occurs commonly in the small and large intestines of the domestic sheep, Rocky Mountain bighorn sheep, and mouflon throughout the world.

OOCYSTS. The oocysts are ellipsoidal or subspherical to somewhat ovoid; 16–30 × 13–22 μm; with a 2-layered wall, the outer layer smooth, colorless to slightly greenish, 1 μm thick, the inner layer yellowish brown, 0.4 μm thick, lined by a membrane; with a micropyle, without a micropylar cap; ordinarily with 2 or more polar granules; without a residuum. The sporocysts are elongate ovoid, 10–14 × 4–8 μm, with a Stieda body and a residuum. The sporozoites are elongate, 11–14 × 2–4 μm, lie lengthwise head to tail in the sporocysts, and have 1 large and 1 small clear globule each. The sporulation time is 1–4 days. (Fig. 7.12.)

LIFE CYCLE. There are two generations of meronts. The first generation meronts become mature in the small intestine lamina propria 9 DAI, at which time they average 290 μm in diameter and contain many thousands of merozoites averaging 12 × 2 μm. Second generation meronts are in epithelial cells lining the crypts of the large intestine. They become mature 10–11 DAI, at which time they have a mean diameter of about 12 μm and contain an average of 24 merozoites each, averaging 5.5 × 1.4 μm.

The sexual stages are in the epithelial cells lining the crypts of the ileum, cecum, and upper part of the large intestine and appear 11–15 DAI. The mature macrogametes average 16 × 12 μm; the mature microgamonts average 15 × 12 μm and contain many microgametes arranged peripherally around a central residuum. The prepatent period is 9–15 days, the patent period 7–28 days.

PATHOGENESIS. This is one of the most pathogenic of sheep coccidia. There are differences in pathogenicity between strains, however. Petechial hemorrhages appear in the small intestine 3–7 DAI. The small intestine then becomes thickened and inflamed. There is extensive hemorrhage in the posterior small intestine of severely affected lambs by day 15. The cecum and upper part of the large intestine become thickened and edematous; they are hemorrhagic by day 19. Vast areas of the posterior small intestine of heavily infected lambs are denuded.

Eimeria pallida Christensen, 1938

This species is fairly common in the feces of domestic sheep, presumably throughout the world. It has also been reported from the domestic goat and ibex, but its occurrence in them is not certain.

OOCYSTS. The oocysts are ellipsoidal; 12–20 × 8–15 μm; with a smooth, colorless to very pale yellow or yellowish green, 2-layered wall 0.5 μm in total thickness, the outer layer accounting for almost the whole wall thickness and the inner layer appearing simply as a dark line on the inner surface of the wall; without a perceptible micropyle or micropylar cap; with or without a polar granule; without a residuum. The sporocysts are elongate ovoid, 6–9 × 4–6 μm, without a Stieda body, with a residuum. The sporozoites are elongate, usually lying lengthwise head to tail in the sporocysts but often with a tendency to lie crosswise in them, with a single clear globule. The sporulation time is 1–3 days. (See Fig. 7.12.)

PATHOGENESIS. Apparently nonpathogenic.

Eimeria parva Kotlan, Mocsy, and Vajda, 1929

This species is common worldwide in domestic sheep, mouflon, argali, and Rocky Mountain bighorn sheep. The meronts are in the small intestine and the sexual stages in the cecum and colon and to a lesser extent in the small intestine.

OOCYSTS. The oocysts are subspherical, ovoid, ellipsoidal or spherical; 12–23 × 10–19 μm; with a smooth, pale yellow to yellowish green, brownish yellow, or faint pinkish mauve, 2-layered wall with an outer layer 0.8–1.2 μm thick (thinner at the micropylar end) and an inner layer that is a dark, thin membrane; with an inconspicuous micropyle; without a micropylar cap or a residuum; generally with a polar granule. The sporocysts are ovoid, 6–13 × 5–8 μm, without a Stieda body or with a small one, with a residuum composed of a few fine granules. The sporozoites have 1 clear globule. The sporulation time is 1–8 days. (See Fig. 7.12.)

PATHOGENESIS. Apparently not very pathogenic.

Eimeria punctata Landers, 1955

SYNONYM. *E. honessi* Landers.

This species is uncommon, presumably worldwide, in the feces of the domestic sheep. It has also been reported from the domestic goat, but its occurrence in it is not certain.

OOCYSTS. The oocysts are ellipsoidal or subspherical to ovoid; 18–28 × 16–21 μm; with a 2-layered wall with conspicuous, uniform, cone-shaped pits about 0.4–0.5 μm in diameter, the outer layer 1.4 μm thick and colorless to yellowish, the inner layer 0.4 μm thick and greenish to brownish yellow; with a micropyle and an imperceptible to prominent micropylar cap; ordinarily with a polar granule; usually with a residuum. The sporocysts are elongate ovoid or piriform, with a faintly perceptible Stieda body and a residuum. The sporozoites are elongate, lie lengthwise head to tail in the sporocysts, and have a single clear globule at one end. The sporulation time is 1–2 days. (See Fig. 7.12.)

Eimeria weybridgensis Norton, Joyner, and Catchpole, 1974

SYNONYM. *E. arloingi* "B" of Pout, Norton, and Catchpole, 1973.

This species is common, presumably worldwide, in the small intestine (chiefly jejunum) of the domestic sheep.

OOCYSTS. The oocysts are ellipsoidal to subspherical; 17–31 × 14–19 μm; with a 2-layered wall, the outer layer usually smooth, about 1 μm thick and colorless or pale yellow, the inner layer a thick, dark membrane; with a micropyle, micropylar cap, and polar granule; without a residuum. The sporocysts are elongate ovoid, 13–15 × 6–8 μm, without a Stieda body, with a residuum. The sporozoites are elongate, lie lengthwise head to tail in the sporocysts, and have a clear globule at each end. The sporulation time is 45 hr.

LIFE CYCLE. The prepatent period is 23–33 days, the patent period 9–12 days.

PATHOGENESIS. Slightly if at all pathogenic.

Eimeria gilruthi (Chatton, 1910) Reichenow and Carini, 1937

SYNONYMS. *Gastrocystis gilruthi, Globidium gilruthi.*

Meronts of this species have been found in the abomasum wall of the domestic sheep, domestic goat, and Siberian wild goat. It is common in some regions of the world and rare in others.

OOCYSTS. Unknown.

LIFE CYCLE. The meronts in the connective tissue and mucous membranes of the abomasum are easily visible to the naked eye as whitish nodules 300–900 μm in diameter; they have a wall up to 40 μm thick. The mature meronts contain thousands of crescentic to sickle-shaped merozoites that have been variously described as 4–16 μm long.

REMARKS. Whether this form is truly *Eimeria* remains to be seen. Various investigators have said that it is not *Sarcocystis.*

Eimeria alijevi Musaev, 1970

SYNONYMS. *E. galouzoi* in part, *E. kandilovi, E. parva* in the goat.

This species is common worldwide in the small and large intestines of the domestic goat and ibex.

OOCYSTS. The oocysts are ellipsoidal, subspherical, spherical, or ovoid; 15–23 × 12–22 μm; with a smooth, 2-layered wall, the outer layer 0.6–0.8 μm thick and pale yellowish to colorless, the inner layer 0.3–0.5 μm thick, dark brown to yellowish brown, or a dark, thin membrane; with a usually inconspicuous micropyle; without a micropylar cap or residuum; ordinarily with 1 polar granule (sometimes shattered). The sporocysts are elongate to ovoid, 7–13 × 4–9 μm, with or without a Stieda body, with a residuum. The sporozoites are elongate, lie at an angle or lengthwise head to tail in the sporocysts, and usually have 1–2 clear globules. The sporulation time is 1–5 days.

LIFE CYCLE. Meronts are in the epithelial cells of the villi of the middle part of the small intestine. They are up to 260 × 180 μm and can be easily seen with the naked eye as whitish bodies. Much smaller meronts 15–18 × 9–12 μm occur in the epithelial cells of the crypts of Lieberkuehn of the small intestine. Gamonts and oocysts are in the colon, cecum, and posterior small intestine epithelial cells. The macrogametes are 14–18 × 9–14 μm, the microgamonts 22–25 × 15–20 μm. The prepatent period is 7–12 days, the patent period 6–18 days.

PATHOGENESIS. This species may be markedly pathogenic.

Eimeria apsheronica Musaev, 1970

SYNONYMS. *Coccidium caprae* (?), *E. aemula* in the goat, *E. faurei* in the goat.

This species is common worldwide in the small intestine of the domestic goat, ibex, and probably Siberian mountain goat.

OOCYSTS. The oocysts are ovoid; 24–37 × 18–26 μm; with a 1- or 2-layered, smooth, faint green to greenish yellow-brown wall about 1–2 μm thick; with a micropyle; without a micropylar cap or a residuum; with a polar granule (usually shattered). The sporocysts are piriform or ellipsoidal, 11–17 × 7–11 μm, without a Stieda body or with a vestigial one, with a residuum. The sporozoites are elongate, lie lengthwise head to tail in the sporocysts, and usually have 1–2 large clear globules. The sporulation time is 1–2 days. The prepatent period is 14–17 days, the patent period 4–9 days.

Eimeria arloingi (Marotel, 1905) Martin, 1909

SYNONYMS. *Coccidium arloingi, E. hawkinsi* in part, *C. caprae.*

This species is common worldwide in the small intestine of the domestic goat.

OOCYSTS. The oocysts are ellipsoidal or slightly ovoid; 22–36 × 16–26 μm; with a 2-layered wall, the outer layer smooth, colorless, 1 μm thick, the inner layer brownish yellow, 0.4–0.5 μm thick, often forming a membrane lining the wall; with a micropyle; ordinarily with a prominent micropylar cap and 1 or more polar granules (sometimes shattered); without a residuum. The sporocysts are ovoid, 10–17 × 5–10 μm, without a Stieda body or with a vestigial one, with a residuum. The sporozoites are elongate, lie lengthwise head to tail in the sporocysts, and usually have a large clear globule at the large end and a small one at the small end. The sporulation time is 1–4 days.

LIFE CYCLE. There are two generations of meront. The mature first generation meronts are in the endothelial cells of the lacteals of the villi; in Peyer's patches in the duodenum, jejunum, and ileum; and also in the sinuses of the mesenteric lymph nodes draining these regions. They are 130–359 × 65–243 μm, contain many thousands of merozoites 9–12 × 1–2 μm, and are mature 9–20 DAI. Second generation meronts are in the epithelial cells of the villi and crypts of the small intestine. They are mature at 12 days, at which time they are 11–44 × 9–20 μm and contain 8–24 merozoites 4–10 μm long and sometimes a residuum.

Microgamonts 19–34 × 13–29 μm are found 11–26 DAI in the epithelial cells lining the crypts and the villi of the jejunum and ileum. They produce several hundred microgametes and a large residuum. The macrogametes are in the same locations and are 19–28 × 14–20 μm. The prepatent period is 14–17 days, the patent period 14–15 days.

PATHOGENESIS. *E. arloingi* is markedly pathogenic. Inflammation, edema, epithelial necrosis, leukocyte infiltration, and hyperplasia of varying severity are present in the small intestine.

Eimeria caprina Lima, 1979

This species is found in the domestic goat.

LOCATION. Unknown.

GEOGRAPHIC DISTRIBUTION. Common; in North America so far as is known.

OOCYSTS. The oocysts are ellipsoidal or slightly ovoid; 27–40 × 19–26 μm; with a smooth, 2-layered wall, the outer layer about 1.1 μm thick and dark brown to brownish yellow, the inner layer about 0.6 μm thick and colorless; with a micropyle 5–7 μm in diameter; without a micropylar cap or a residuum; ordinarily with 1 or more polar granules (sometimes shattered). The sporocysts are elongate ovoid, 13–17 × 7–10 μm, with a small Stieda body, with a residuum. The sporozoites are elongate, lie lengthwise head to tail in the sporocysts, and usually have a large clear globule at the large end and a smaller one at the small end. The sporulation time is 2–3 days. The prepatent period is 17–20 days, the patent period 3–6 days.

Eimeria caprovina Lima, 1980

This species is quite common in the feces of the domestic goat in North America. The domestic sheep can be infected experimentally.

OOCYSTS. The oocysts are ellipsoidal, subspherical or slightly ovoid; 26–36 × 21–28 μm; with a smooth, 2-layered wall, the outer layer about 1 μm thick and colorless, the inner layer about 0.6 μm thick and dark brown to brownish yellow; with a micropyle; without a micropylar cap or a residuum; with 1–4 or many polar granules. The sporocysts are elongate ovoid, 13–17 × 8–9 μm, with a Stieda body and a residuum. The sporozoites are elongate, lie lengthwise head to tail in the sporocysts, and usually have a large clear globule at each end. The sporulation time is 2–3 days. The prepatent period is 14–20 days, the patent period 4–9 days.

Eimeria christenseni Levine, Ivens, and Fritz, 1962

SYNONYMS. *E. tirupatiensis, E. ahsata* in goats.

This species is quite common worldwide in the small intestine of the domestic goat.

OOCYSTS. The oocysts are ovoid, sometimes ellipsoidal; 27–44 × 17–31 μm; with a 2-layered wall, the outer layer smooth, colorless to very pale yellowish, 0.8–1.2 μm thick; the inner layer brownish yellow, 0.4–0.7 μm thick, the inner layer forming a membrane; with a micropyle and micropylar cap, with 1 or more polar granules (sometimes partly shattered); without a residuum. The sporocysts are broadly ovoid, 12–18 × 8–11 μm, without a Stieda body or with a vestigial one, with a residuum. The sporozoites are elongate, lie lengthwise head to tail in the sporocysts, and have 1 or more clear globules. The sporulation time is 3-6 days.

LIFE CYCLE. First generation meronts are in the endothelial cells of the lacteals of the jejunum and ileum and in the lamina propria and lymph vessels of the submucosa and mesenteric lymph nodes. When mature at 14 days they are ellipsoidal, 100–277 × 81–130 μm, and contain thousands of straight merozoites about 6–8 × 1–2 μm. Second generation meronts can be seen 16 DAI, mostly in epithelial cells of the crypts and less often in those of the villi in the small intestine and also in the sinuses of the mesenteric lymph nodes. They are 9–20 × 8–12 μm and contain 8–24 merozoites and sometimes a residuum.

Gamonts are present in the epithelial cells of the villi and crypts of the small intestine 16 DAI. The mature macrogametes are 19–35 × 13–25 μm. The mature microgamonts are 19–50 × 12–40 μm and contain hundreds of crescent- or comma-shaped microgametes about 3 × 0.5 μm and a residuum. The prepatent period is 14–23 days, the patent period 3–30+ days.

PATHOGENESIS. This species is one of the most pathogenic coccidia of goats. Lesions begin in the small intestine of kids 4 DAI. Focal infiltration with lymphocytes and plasma cells, epithelial necrosis, and submucosal edema are usually associated with focal aggregates of coccidia, especially gamonts and oocysts, in the jejunum and ileum. Superficial desquamation of the mucosa and superficial necrosis are also present. The capillaries are congested, and there are petechial hemorrhages. The cellular reaction in the lamina propria and submucosa consists of lymphocytes, macrophages, plasma cells, polymorphonuclear leukocytes, and eosinophils. Edema and pericapsular infiltration by lymphocytes are present in the lymph nodes. There are white foci in the intestine consisting essentially of masses of macrogametes, microgamonts, and oocysts in the epithelial cells of the tips and sides of the villi and in the crypts.

Eimeria gilruthi (Chatton, 1910) Reichenow and Carini, 1937

See the discussion of this species above (*E. gilruthi*). Its meronts have been found in both sheep and goats, but no oocysts that can be assigned to it have been seen.

Eimeria hirci Chevalier, 1966

SYNONYM. *E. crandallis* in the goat.

This species is common, presumably worldwide, in the feces of the domestic goat, ibex, and wild goats.

OOCYSTS. The oocysts are ellipsoidal to subspherical or spherical; 17–29 × 14–22 μm; with a smooth, 1- or 2-layered wall, the outer layer colorless and 0.7–1 μm thick, the inner layer light brown to brownish yellow and 0.4–0.6 μm thick, lined by a membrane; with a micropyle and often a micropylar cap; with 1 or more polar granules, without a residuum. The sporocysts are ovoid, about 8–14 × 4–9 μm, with a tiny Stieda body, ordinarily with a residuum. The sporozoites lie lengthwise, or at an angle, or even at the ends of the sporocysts and have 1–2 clear globules. The sporulation time is 1–3 days. The prepatent period is 13–16 days, the patent period 5–14 days.

Eimeria jolchijevi Musaev, 1970

SYNONYM. *E. granulosa* from goats.

This species is fairly common, presumably worldwide, in the feces of the domestic goat.

OOCYSTS. The oocysts are ellipsoidal or ovoid; 26–37 × 18–26 μm; with a smooth, 2-layered wall, the outer layer pale yellow to yellowish brown or colorless and 0.9–1.2 μm thick, the inner layer dark brown to brownish yellow and 0.5–0.8 μm thick; with a micropyle at the broad end; with a prominent micropylar cap; with 1 or more polar granules (sometimes shattered); without a residuum. The sporocysts are ovoid, 12–18 × 6–10 μm, with a small Stieda body, with a residuum. The sporozoites are elongate, lie lengthwise head to tail in the sporocysts, and have 1–2 or more large clear globules. The sporulation time is 2–4 days. The prepatent period is 14–17 days, the patent period 3–10 days.

Eimeria kocharii Musaev, 1970

SYNONYMS. *E. intricata* in goats, *E. nazijrovi*.

This species occurs rarely in the USSR and India in the feces of the domestic goat, ibex, and Siberian mountain goat.

OOCYSTS. The oocysts are ellipsoidal or slightly ovoid; 39–59 × 27–47 μm; with a 2-layered, transversely striated, brownish wall 0.4–0.8 μm thick, lined by a membrane; with a micropyle; with a prominent micropylar cap and 1 or more polar granules; without a residuum. The sporocysts are elongate ovoid with one end bluntly pointed, 17–22 × 9–14 μm, without a Stieda body or with an extremely tiny one, with a residuum. The sporozoites are elongate, lie lengthwise head to tail in the sporocysts, and have 2–3 clear globules. The sporulation time is 4–12 days.

Eimeria ninakohlyakimovae Yakimoff and Rastegaieff, 1930

SYNONYMS. *E. nina-kohl-yakimovi, E. galouzoi* in part.

This species is common worldwide in the small intestine, especially the posterior part, and also the cecum and colon of the domestic goat, wild goat, and ibex.

OOCYSTS. The oocysts are ellipsoidal or subspherical to slightly ovoid; 19–28 × 14–23 μm; with a smooth, 2-layered wall, the outer layer 0.8–1 μm thick and slightly yellowish or colorless, the inner layer 0.4–0.6 μm thick and dark brown to yellowish brown; usually with a micropyle 3-6 μm in diameter; without a micropylar cap or a residuum; with 1 or more polar granules (sometimes shattered). The sporocysts are ovoid, 9–15 × 4–10 μm, with a Stieda body, with a residuum of many scattered granules. The sporozoites are elongate, lie lengthwise head to tail in the sporocysts, and have 2 clear globules. The sporulation time is 1–4 days.

LIFE CYCLE. Meronts are in the epithelial cells of the ileum, cecum, and upper large intestine and are 31–43 × 22–31 μm. The gamonts, gametes, and oocysts are in the same locations. The mature macrogametes are 9–18 × 7–13 μm. The mature microgamonts are 20–25 × 15–18 μm and have whorls of microgametes on their surface and some residual material in their center. The prepatent period is 10–13 days.

PATHOGENESIS. This species causes slight to moderate enteritis, with round, smooth white plaques about 0.2–0.3 mm in diameter in the mucosa of the small intestine, cecum, and colon. The cellular reaction consisted of lymphocytes and polymorphonuclear leukocytes.

Eimeria pallida Christensen, 1938

This species is fairly common, presumably worldwide, in the feces of the domestic goat and ibex. The domestic sheep is the type host. The following discussion pertains only to the goat.

OOCYSTS. The oocysts are ellipsoidal or slightly ovoid; 13–18 × 10–14 μm; with a smooth, colorless to very pale yellow, 2-layered wall, the outer layer 0.6 μm thick, the inner layer merely a dark line on the inner surface of the wall; with an imperceptible micropyle and a polar granule; without a micropylar cap or a residuum. The sporocysts are elongate ovoid, 6–9 × 4–5 μm, without a Stieda body, with a residuum. The sporozoites are elongate, lie lengthwise head to tail in the sporocysts, and have a single clear globule.

REMARKS. It is likely that this name actually refers to different species in the sheep and goat.

Eimeria punctata Landers, 1955

This species is uncommon in the feces of the domestic goat. It has been found in Germany, Turkey, and Somalia. The type host is the domestic sheep; the form in the goat may be a different species.

OOCYSTS. The oocysts are ovoid to ellipsoidal; 21–31 × 15–23 μm; with a several-layered, greenish yellow to brownish green wall pitted like a thimble; with a broad micropyle and a micropylar cap. The sporocysts are slender.

Eimeria bactriani Levine and Ivens, 1970

SYNONYMS. See Levine and Ivens (1970).

This species has been found in the small intestine of the Bactrian camel and dromedary in the USSR.

OOCYSTS. The oocysts are spherical to ellipsoidal, about 21–34 × 20–28 μm, with a light yellowish to yellowish brown wall, with a micropyle, without a micropylar cap or a residuum. The sporocysts are spherical or elongate, 8–9 × 6–9 μm, with a residuum. The sporulation time is 6 days to 3 wk.

LIFE CYCLE. The endogenous stages are in the epithelial cells of the villi of the small intestine. The meronts are 16 × 10 μm and each contains 20–24 merozoites 9 × 2 μm. The mature macrogametes are 25 × 20 μm. The mature microgamonts are 14 μm in diameter or 19 × 12 μm and contain microgametes 4 μm long.

Eimeria cameli (Henry and Masson, 1932) Reichenow, 1952

SYNONYMS. See Levine and Ivens (1970).

This species is presumably common worldwide in the small intestine and to a lesser extent cecum of the dromedary and Bactrian camel.

OOCYSTS. The oocysts are piriform; 80–100 × 55–94 μm; with a rough, brown, opaque, 2-layered wall 12–15 μm thick, lined by a thin membrane; with a micropyle about 10–27 μm wide; with or without a micropylar cap and polar granule; without a residuum. The sporocysts are elongate, pointed at both ends, or ellipsoidal; 30–50 × 14–20 μm; without a Stieda body; with a residuum. The sporozoites are comma-shaped, lie lengthwise head to tail in the sporocysts, and have a clear globule at the large end. The sporulation time is 9–15 days.

LIFE CYCLE. The meronts, giant in size, are in the small intestine. They contain many merozoites about 2 μm long. The macrogametes and microgamonts are in the ileum and also to some extent in the large intestine. The microgamonts are up to 350 μm in diameter.

PATHOGENESIS. Probably mildly pathogenic.

Eimeria dromedarii Yakimoff and Matschoulsky, 1939

SYNONYMS. See Levine and Ivens (1970).

This species is apparently quite common in the feces of the dromedary and Bactrian camel in India, Iraq, and Pakistan.

OOCYSTS. The oocysts are ovoid; 23–33 × 20–25 μm; with a brown, 2-layered wall 0.1–3 μm thick, thickened to form a kind of "cap"; without a polar granule or a residuum. The sporocysts are ovoid or spherical, 8–11 × 6–9 μm, without a Stieda body or a residuum. The sporozoites are comma-shaped, with 1–2 clear globules. The sporulation time is 15–17 days.

Eimeria pellerdyi Prasad, 1960

SYNONYM. *E. pellerdei.*

This species occurs in the feces of the Bactrian camel. Its prevalence and geographic distribution are unknown.

OOCYSTS. The oocysts are ovoid or ellipsoidal; 22–24 × 12–14 μm; with a smooth, colorless, 2-layered wall; without a micropyle, polar granule, or residuum. The sporocysts are ovoid, 9–11 × 4–6 μm, with a small Stieda body and a residuum. The sporozoites are club-shaped, 8–10 × 1–3 μm, with a clear globule at the large end.

Eimeria rajasthani Dubey and Pande, 1963

This species is common in the feces of the dromedary in India.

OOCYSTS. The oocysts are ellipsoidal, 34–39 × 25–27 μm, with a 2-layered wall 2–3 μm thick, the outer layer relatively thick and light yellowish green, the inner layer darker; apparently with a micropyle that is not visible; with a micropylar cap; without a polar granule or a residuum. The sporocysts are almost ovoid, 14–15 × 8–11 μm, with a Stieda body and residuum. The sporozoites are elongate, 10–14 × 3–4 μm, lie lengthwise head to tail in the sporocysts, and have 2 or sometimes more clear globules.

Eimeria alpacae Guerrero, 1967

This species was found in the feces of the alpaca in Peru.

OOCYSTS. The oocysts are ellipsoidal or rarely ovoid; 22–26 × 18–21 μm; with a smooth, 2-layered wall, the outer layer 1.1 μm thick and very pale greenish to bluish; the inner layer 0.4 μm thick, appearing as a dark yellow line; with a micropyle and micropylar cap; with or without 1–3 polar granules; without a residuum. The sporocysts are ovoid, 10–13 × 7–8 μm, with a wall about 0.2 μm thick, with a very faintly perceptible Stieda body and a residuum. The sporozoites are elongate, lie lengthwise head to tail in the sporocysts, and have 1–3 clear globules.

Eimeria lamae Guerrero, 1967

This species was found in the feces of the alpaca in Peru.

OOCYSTS. The oocysts are ellipsoidal or occasionally ovoid; 30–40 × 21–30 μm; with a smooth, 2-layered wall, the outer layer 1.3 μm thick and bluish to greenish yellow, the inner layer 0.5 μm thick and brownish yellow; with a micropyle and micropylar cap; with or without a polar granule; without a residuum. The sporocysts are elongate ovoid, 13–16 × 8–10 μm, with a wall about 0.25 μm thick, with a Stieda body and a residuum. The sporozoites are elongate, lie lengthwise head to tail in the sporocysts, and have 1–3 clear globules.

Eimeria macusaniensis Guerrero, Hernandez, Bazalar, and Alva, 1971

This species was found in the feces of the alpaca in Peru.

OOCYSTS. The oocysts are ovoid or sometimes piriform; 81–107 × 61–80 μm; with a 3-layered wall, the outer layer smooth, colorless, 1 μm thick, the middle layer granular, dark brown, 7.5 μm thick, and the inner layer colorless, 1 μm thick; with a micropyle and micropylar cap; without a polar granule or a residuum. The sporocysts are elongate ovoid, 33–40 × 16–20 μm, with a faintly perceptible Stieda body and a residuum. The sporozoites are elongate, lie lengthwise head to tail in the sporocysts, and have a large clear globule at the large end and usually a small one at the small end.

Eimeria punoensis Guerrero, 1967

This species was found in the feces of the alpaca in Peru.

OOCYSTS. The oocysts are ellipsoidal or occasionally ovoid; 17–22 × 14–18 μm; with a smooth, 2-layered wall, the outer layer blue to purplish and 0.7 μm thick, the inner layer 0.3 μm thick, appearing as a dark line; with a micropyle and micropylar cap; with 1–2 or without polar granules; without a residuum. The sporocysts are somewhat elongate ovoid, 8–11 × 5–7 μm, with a faintly perceptible Stieda body and a residuum. The sporozoites are elongate, lie lengthwise head to tail in the sporocysts, and have 1–3 clear globules.

Eimeria betica Martinez and Hernandez, 1973

This species was found in the feces of the domestic pig in Spain.

OOCYSTS. The oocysts are usually cylindrical; salmon-colored; 20–28 × 10–14 (mean 23 × 13) μm; with a smooth wall 0.9–1.1 μm thick; with a large micropyle without a collar; without a polar granule; with a residuum. The sporocysts are ovoid, 8–14 × 4–6 (mean 11 × 5) μm, with a Stieda body and a residuum. The sporozoites are vermiform, 8–9 μm long. The sporulation time is 4–5 days.

Eimeria debliecki Douwes, 1921

SYNONYMS. *Coccidium suis, E. brumpti* in part, *E. jalina, E. scrofae.*

This species is common worldwide in the epithelial cells of the anterior small intestine of the domestic pig and wild boar.

OOCYSTS. The oocysts are usually ellipsoidal, sometimes cylindrical, occasionally ovoid; 20–30 × 14–19 μm; with a smooth, colorless, 2-layered wall; without a micropyle or a residuum; with a polar granule. The sporocysts are elongate ovoid, 13–20 × 5–7 μm, with a large Stieda body and a large residuum. The sporozoites are vermiform, 14–20 × 3–4 μm when excysted, lie lengthwise head to tail in the sporocysts, and have 2 large clear globules. The sporulation time is 10 days. (See Fig. 7.13.)

LIFE CYCLE. The endogenous stages are in the distal part of the columnar epithelial cells of the tips of the villi in the small intestine posterior to the bile

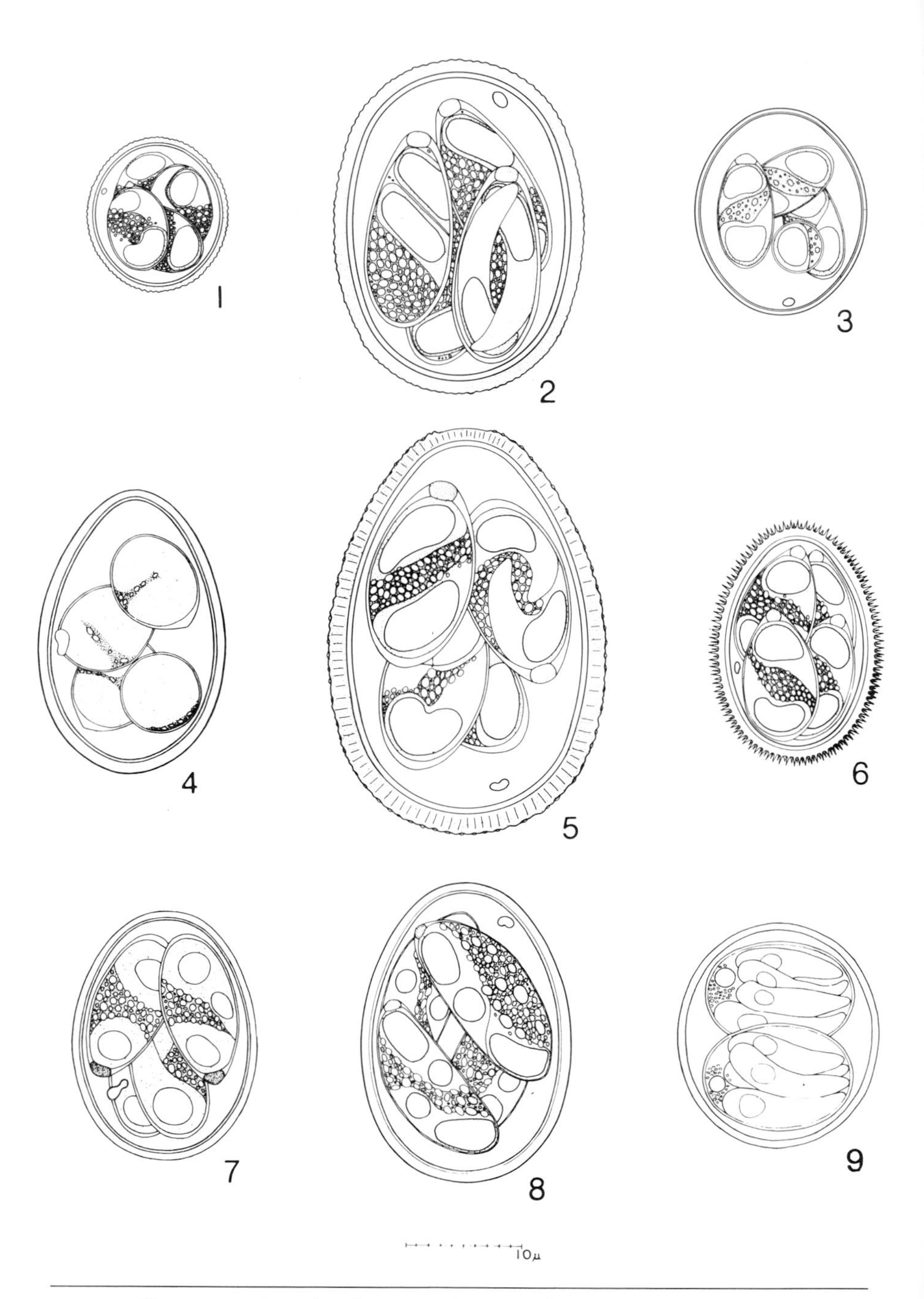

Fig. 7.13. Sporulated oocysts of coccidia from the pig: 1. *Eimeria perminuta;* 2. *E. polita;* 3. *E. suis;* 4. *E. porci;* 5. *E. scabra;* 6. *E. spinosa;* 7. *E. neodebliecki;* 8. *E. debliecki;* 9. *Isospora suis*. (From Vetterling 1965)

duct. The first generation meronts mature at 2 days; they are 8–12 μm in diameter and produce 16 vermiform merozoites 12–15 × 1.8 μm and a large polar residuum. The second generation meronts mature at 4 days, are 13–16 × 10–15 μm, and produce 32 merozoites 6–8 × 1.8 μm and sometimes a small amount of residual material. The macrogametes mature more than 5 DAI and are 12–16 × 9–13 μm. The microgamonts are mature 5 DAI, are 9–14 × 7–9 μm, and produce many biflagellate microgametes 5–6 × 0.6 μm. The prepatent period is 156 hr, the patent period 118 hr. (See Fig. 7.14.)

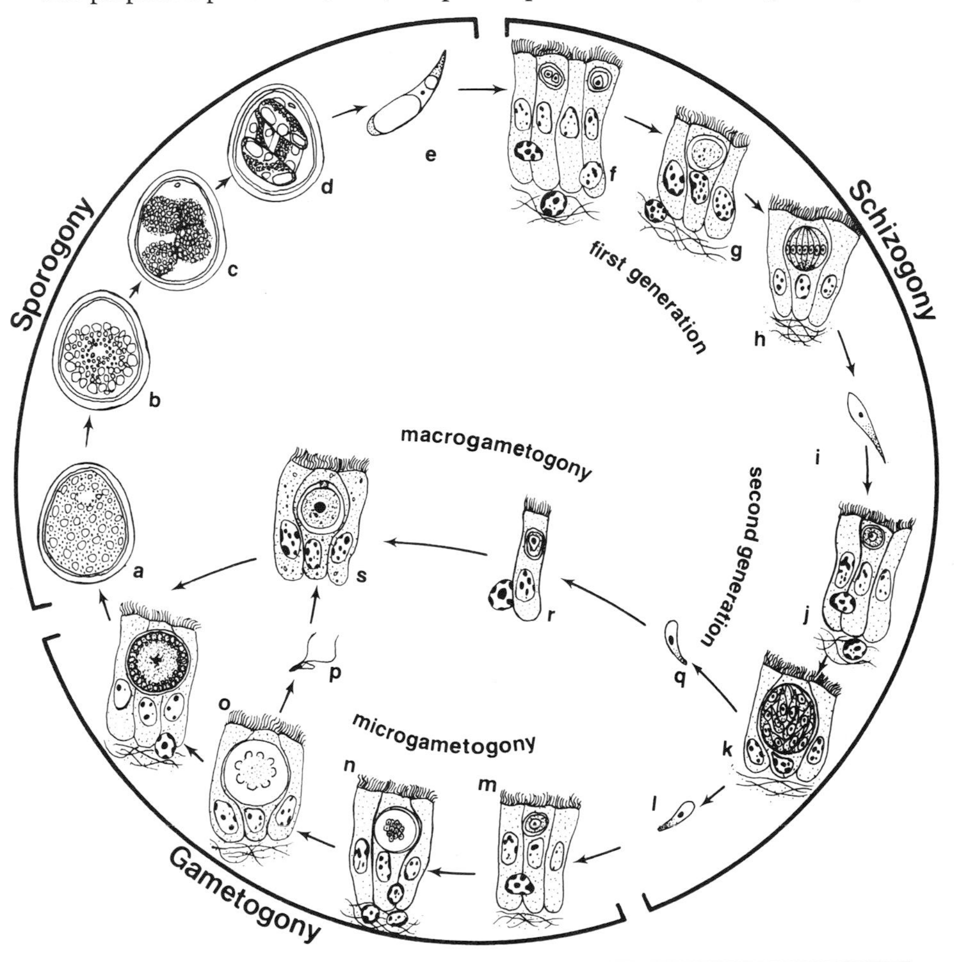

Fig. 7.14. Life cycle of *Eimeria debliecki* from the pig: a–d. oocyst sporulation; e. isolated sporozoite; f. formation and development of first generation meront; g. first generation meront; h. first generation meront containing merozoites; i. first generation merozoite; j and k. formation and development of second generation meront; l and q. second generation merozoite; m–o. formation of microgametes; p. microgamete; r and s. formation of macrogamete; t. fertilized macrogamete (zygote). All stages except the oocyst are found in epithelial cells of the small intestine. (Adapted from Vetterling 1966)

PATHOGENESIS. Slightly pathogenic. Large numbers of oocysts cause catarrhal exudation from the jejunum of pigs 2–7 wk old; small numbers (20,000 or more) have no pathogenic effects, nor do large numbers of oocysts in older pigs.

Eimeria guevarai Romero and Lizcano, 1971

This species was found in the feces of the domestic pig in Spain.

OOCYSTS. The oocysts are piriform; 26–32 × 15–19 μm; with a smooth, slightly tinted, 2-layered wall; without a micropyle or residuum; with a polar granule. The sporocysts are elongate ovoid, 9–11 × 6–8 μm, with a Stieda body and a residuum. The sporozoites are elongate, lie lengthwise head to tail in the sporocysts, and have a clear globule at one end. The sporulation time is 10 days at 20 C.

Eimeria neodebliecki Vetterling, 1965

SYNONYMS. *E. brumpti* in part, *E. debliecki* in part.

This species is common, presumably worldwide, in the feces of the domestic pig.

OOCYSTS. The oocysts are ellipsoidal or ovoid; 17–26 × 13–20 μm; with a smooth, colorless, 2-layered wall 0.7–1.4 μm thick; without a micropyle or a residuum; with a polar granule. The sporocysts are elongate or broadly ovoid, 9–14 × 5–8 μm, with a Stieda body and a residuum. The sporozoites are vermiform, lie lengthwise head to tail in the sporocysts, and have 2 clear globules. The sporulation time is 312 hr at 25 C. The prepatent period is 240 hr, the patent period 144 hr. (See Fig. 7.13.)

Eimeria perminuta Henry, 1931

SYNONYM. *E. perminuta* var. *mathurai.*

This species is more or less common worldwide in the feces of the domestic pig and wild boar.

OOCYSTS. The oocysts are spherical to subspherical; 12–20 × 9–17 μm; with a rough, yellow, 2-layered wall 1 μm thick; without a micropyle or a residuum; with a polar granule. The sporocysts are ellipsoidal to broadly ovoid, 6–8 × 4–6 μm, with a Stieda body and a residuum. The sporozoites are elongate, lie lengthwise head to tail in the sporocysts, and have 2 clear globules. The sporulation time is 7–9 days. (See Fig. 7.13.)

Eimeria polita Pellérdy, 1949

SYNONYMS. *E. cerdonis, E. debliecki* in part.

This species is fairly common worldwide in the epithelium of the villi in the jejunum and ileum of the domestic pig and wild boar.

OOCYSTS. The oocysts are ellipsoidal or rarely ovoid; 22–39 × 17–26 μm, with a smooth or slightly rough, yellowish to colorless, 2-layered wall 1–1.5

μm thick; with an imperceptible micropyle; with or without a polar granule; without a residuum. The sporocysts are ellipsoidal to ovoid, 13–19 × 5–9 μm, with a Stieda body and a residuum. The sporozoites are elongate, lie lengthwise head to tail in the sporocysts, and have 1–2 clear globules. (See Fig. 7.13.)

LIFE CYCLE. Merogony occurs in the partially invaginated epithelial lining of the tips of the villi of the small intestine. The mature meronts are about 14–24 × 11–23 μm and contain 15–30 merozoites about 9–22 × 1.5–3 μm; there are probably two generations. The gamonts and gametes are in the same part of the intestine as the meronts. They mature on days 8 or 9. The macrogametes are 16–29 × 15–25 μm. The microgamonts are 16–29 × 13–29 μm, with a residuum, and produce 60–150 biflagellate microgametes 2.2 × 0.6 μm. The prepatent period is 7–8 days, the patent period 6–8 days.

PATHOGENESIS. Slightly pathogenic.

Eimeria porci Vetterling, 1965

SYNONYM. *E. debliecki* in part.

This species is quite common, presumably worldwide, in the lower jejunum and ileum of the domestic pig.

OOCYSTS. The oocysts are ovoid; 18–27 × 13–18 μm; with a smooth, colorless to yellowish, 2-layered wall 0.9–1.2 μm thick; with a micropyle 2–3 μm wide (which is often indistinct or apparently absent); with a polar granule; without a residuum. The sporocysts are ovoid, 8–12 × 6–8 μm, with a Stieda body and a residuum. The sporozoites are elongate or corpulent, lie either at the ends or lengthwise head to tail in the sporocysts, and have an indistinct clear globule. The sporulation time is 216 hr. (See Fig. 7.13.)

LIFE CYCLE. The asexual cycle occurs beneath the host cell nucleus of the epithelial cells of the lower jejunum and ileum. There are two asexual generations, the first one 1–3 DAI and the second one 3–6 DAI. The sexual stages occur in the same location. Young gamonts can be recognized 5 DAI. The prepatent period is 5–7 days, the patent period 6 days.

Eimeria residualis Martinez and Hernandez, 1973

This species was found in the feces of the domestic pig in Spain.

OOCYSTS. The oocysts are ordinarily ellipsoidal, ochraceous; 28–36 × 14–24 (mean 32 × 20) μm, with a smooth wall 1.3–1.5 μm thick; with a large micropyle surrounded by a collar; apparently without a polar granule; with a residuum. The sporocysts are ellipsoidal, 11–17 × 6–8 (mean 14 × 8) μm, with a Stieda body and a residuum. The sporozoites are 13 μm long. The sporulation time is 8–9 days.

Eimeria scabra Henry, 1931

SYNONYMS. *E. debliecki* in part, *E. romaniae*, *E. scarba*.

This species is quite common worldwide in the epithelial cells of the villi and necks of the crypts of the posterior small intestine and possibly also cecum and colon of the domestic pig and wild boar.

OOCYSTS. The oocysts are ovoid to ellipsoidal; 25–45 × 18–28 μm; with a rough, striated, yellow to brown, 2-layered wall 1.5–3 μm thick; with a micropyle and polar granule; without a residuum. The sporocysts are ovoid, 14–18 × 7–9 μm, with a prominent Stieda body and a residuum. The sporozoites are elongate, lie lengthwise head to tail in the sporocysts, and have 2 clear globules. The sporulation time is 9–12 days. (See Fig. 7.13.)

LIFE CYCLE. There are three generations of meront. The first generation meronts mature 3 DAI, at which time they are 16 × 13 μm and contain 16–24 merozoites and usually no residuum. The second generation meronts mature 5 DAI, at which time they are 16 × 12 μm and contain 14–22 merozoites; about 40% have a residuum. The third generation meronts mature 7 DAI, at which time they are 21 × 16 μm and contain 14–28 merozoites; about 90% have a residuum.

The macrogametes are 18 × 12 μm; the microgamonts are 17 × 13 μm, both in sections. The latter contain many biflagellate microgametes 2.5–4.4 μm long. The prepatent period is 7–11 days, the patent period 4–8 days.

PATHOGENESIS. This species causes temporary constipation or diarrhea when the infecting dose is not less than 200 oocysts.

Eimeria spinosa Henry, 1931

This species is quite common, apparently worldwide, in the ileum and jejunum of the domestic pig.

OOCYSTS. The oocysts are ovoid, occasionally ellipsoidal; 14–26 × 12–21 μm; with a rough, spined, brown, 2-layered wall about 1 μm thick; without a micropyle or a residuum; with a polar granule. The sporocysts are elongate ovoid, 10–14 × 5–7 μm, with a prominent Stieda body and a residuum. The sporozoites are elongate, lie lengthwise head to tail in the sporocysts, and have a clear globule at the large end. The sporulation time is 9–22 days. (See Fig. 7.13.)

LIFE CYCLE. All stages are in the villar epithelial cells, a few are in the crypts, and sometimes they are in the lamina propria of the jejunum and ileum. Mature meronts are 8–10 μm in diameter and contain more than 20 merozoites 4–6 × 1–1.5 μm. The prepatent period is 7 days.

PATHOGENESIS. Somewhat pathogenic.

Eimeria suis Nöller, 1921

SYNONYMS. *E. jalina, E. brumpti* in part, *E. debliecki* in part, *E. tuis.*

This species is quite common worldwide in the feces of the domestic pig.

OOCYSTS. The oocysts are ellipsoidal to subspherical; 13–22 × 11–16 μm; with a smooth, colorless, 2-layered wall 0.5 μm thick; without a micropyle or a residuum; with a polar granule. The sporocysts are elongate ovoid, 8–12 × 4–6 μm, with a prominent Stieda body and a residuum. The sporozoites are elongate, lie lengthwise head to tail in the sporocysts, and have a clear globule at the broad end. The sporulation time is 12 days. The prepatent period is 10 days, the patent period 6 days. (Fig. 7.13.)

Eimeria sp. Desser, 1978

This species was found in the bile duct epithelium of the liver of the domestic pig in New Zealand.

OOCYSTS. The oocysts are smooth, 28–32 × 13–17 μm, without a micropyle. They were apparently described from fixed sections.

REMARKS. Desser (1978) found giant meronts, gamonts, and oocysts in sections of the liver of a pig in New Zealand. He said that they belonged to the "*Eimeria debliecki* group."

Eimeria leuckarti (Flesch, 1881) Reichenow, 1940

SYNONYM. *Globidium leuckarti.*

This species is apparently uncommon worldwide in the small intestine of the horse and ass.

OOCYSTS. The oocysts are ovoid or piriform, somewhat flattened at the small end; 70–90 × 49–69 μm; with a rough, 2-layered wall, the outer layer 5–9 μm thick, dark brown, opaque, and granular, the inner layer about 1 μm thick and colorless; with a micropyle at the small end; without a polar granule or a residuum. The sporocysts are elongate, 30–43 × 12–15 μm, with a Stieda body and a residuum. The sporozoites are elongate, up to 35 μm long, lie lengthwise head to tail in the sporocysts, and have a clear globule at the broad end. Sporulation takes 15–41 days.

LIFE CYCLE. The meronts and merozoites have not been described. Early gamonts are in the cells of the intestinal lamina propria at 14 days. When mature, macrogametes are 41–58 μm in diameter and microgamonts are up to 300 × 170 μm, visible to the naked eye, and contain microgametes 3–5 μm long with 2 unequal anterior flagella about 8–12 μm long. The host cells and host cell nuclei are hypertrophied. The prepatent period in Shetland ponies is 15–33 days, the patent period 12–32 days.

PATHOGENESIS. Diarrhea, loss of weight, and even death have been reported in heavily infected animals, but it is not clear that these were due to the protozoa.

Eimeria coecicola Kheisin, 1947

SYNONYMS. *E. oryctolagi, E. neoleporis* of Pellérdy (1954).

This species is rather rare worldwide in the intestine of the domestic rabbit. The meronts are in the epithelial cells of the ileum, the gamonts in those of the cecum and its vermiform process.

OOCYSTS. The oocysts are ellipsoidal, "cylindrical," or sometimes ovoid; light yellow to light brown; 23–40 × 15–21 μm; with a smooth wall; with a distinct micropyle; without a polar granule; usually with a residuum. The sporocysts are elongate ovoid, about 16–17 × 3–9 μm, with a Stieda body and a residuum. The sporozoites are elongate, lie lengthwise head to tail in the sporocysts, and have a clear globule at the broad end. Sporulation takes 3 days.

LIFE CYCLE. The meronts are 12–15 μm in diameter and contain 8–12 merozoites 10 × 0.8 μm. The number of generations is unknown. The gamonts are usually beneath the host cell nucleus. The microgamonts are 20–22 × 12–17 μm and contain biflagellate microgametes 2.5 × 1.5 μm. The oocysts are in the tunica propria. The prepatent period is 8–9 days, the patent period 7–9 days.

PATHOGENESIS. In heavy infections the crypts become considerably enlarged. There are many white spots at the surface of the cecum. This species is apparently highly pathogenic.

Eimeria elongata Marotel and Guilhon, 1941

SYNONYM. *E. neoleporis* (?).

This species occurs in the feces of the domestic rabbit in Europe and perhaps the cottontail *Sylvilagus floridanus* in North America.

OOCYSTS. The oocysts are elongate ellipsoidal with almost straight sides; slightly grayish; 35–40 × 17–20 μm; with a thin wall; with a broad, easily visible micropyle; without a polar granule or a residuum. The sporocysts are elongate, with a residuum that is almost as long as the sporocysts. The sporulation time is 4 days.

Eimeria exigua Yakimoff, 1934

This species is presumably worldwide in the feces of the domestic rabbit.

OOCYSTS. The oocysts are subspherical; 12–21 × 10–18 (mean 14 × 13) μm; with a smooth, colorless wall; without a micropyle or a residuum; presumably without a polar granule. The sporocysts are 9 × 5 μm, with a residuum but apparently without a Stieda body. They sporulate in 36–48 hr.

Eimeria flavescens Marotel and Guilhon, 1941

SYNONYMS. *E. pellerdyi* Coudert, 1977, *E. hakei, E. irresidua* of Francalani and Manfredini, 1967.

This species is quite common, presumably worldwide, in the lower small intestine, cecum, and colon of the domestic rabbit.

OOCYSTS. The oocysts are ovoid; 25–37 × 14–24 (mean 32 × 21) μm; with a two-layered wall, the outer layer smooth, yellow, about 1.4 μm thick, the inner one a dark line about 0.4 μm thick; with a prominent micropyle at the *broad* end; without a polar granule or a residuum. The sporocysts are elongate ovoid, 13–17 × 7–10 (mean 15 × 9) μm, with a small Stieda body and a residuum. The sporozoites are elongate, 17 × 5 μm after excystation, lie lengthwise head to tail in the sporocysts, and have 1 clear globule. Sporulation takes 38 hr or less.

LIFE CYCLE. The first generation meronts are in the glands of the lower small intestine; the second to fifth generation meronts and gamonts are in the cecum and colon; second, third, and fourth generation meronts are in the superficial epithelium, and fifth generation meronts and gamonts are in the glands. The first generation meronts are 15 × 9 μm at 72 hr and contain 16 merozoites 9 × 2.5 μm. The second generation meronts are 12.5–18 × 8–16 (mean 16 × 13) μm at 88–115 hr and contain 12–20 (mean 16) merozoites 10 × 2 μm. The third generation meronts are 12.5–19 × 9–16 (mean 16 × 12) μm at 120 hr and contain 12–20 (mean 15.5) merozoites 11 × 2 μm. The fourth generation meronts are 10–19 × 7–13 (mean 14.5 × 10) μm at 120–156 hr and contain 12–24 (mean 16) merozoites 17 × 2.5 μm. There are few fifth generation meronts; they are 14–18 × 11–16 (mean 16 × 13) μm at 144–156 hr and contain about 50 merozoites 5 × 2 μm.

Gamonts and gametes appear at 163 hr. The microgamonts produce many flagellated microgametes and a residuum. The macrogametes have many wall-forming bodies. The prepatent period is 8–11 (mean 9) days.

PATHOGENESIS. This species is very pathogenic for young Dutch rabbits, low doses producing severe enteritis with high morbidity and mortality.

Eimeria intestinalis Kheisin, 1948

SYNONYMS. *E. agnosta, E. piriformis* in part.

This species is relatively uncommon, presumably worldwide, in the domestic rabbit. Meronts are in the small intestine and gamonts in both small and large intestines.

OOCYSTS. The oocysts are piriform or ovoid; 21–36 × 15–21 (mean 27–32 × 17–20) μm; with a smooth, light yellow or brownish wall; with a micropyle at the narrow end and a residuum; apparently without a polar granule. The sporocysts are ovoid, 10 × 5 μm, with a small Stieda body and a residuum. The sporozoites are elongate, lie lengthwise head to tail in the sporocysts, and have a clear globule at the broad end. Sporulation takes 1–6 days.

LIFE CYCLE. There are three generations of meront. The first generation meronts are at the base of the villi 10–50 cm anterior to the ileocecal valve; they are 15–25 μm in diameter, can be seen 4–6 DAI, and contain 20–60 first generation merozoites with a pointed anterior end that is 7–8 μm long. There are two types of second generation meront, both in the distal part of the villi. Type I (Type A) can be seen on days 5–9. They are 13–30 μm in diameter and contain 70–120 slender merozoites 6–12 × 0.5–1.0 μm that have a nucleus near the larger end. Type II (Type B) second generation meronts can be seen on days 7 and 8. They are 10–16 μm in diameter and contain 5–20 short, broad merozoites 10–12 μm long. Third generation meronts can be seen on days 7–10, in the distal part of the villi. They are 13–15 μm in diameter and contain 15–35 merozoites 10–12 × 1 μm. All meronts have a residuum.

The gamonts develop beginning on day 8; they are above the host cell nucleus in the epithelial cells of the villi. The microgamonts are ovoid, 20–27 × 18–20 μm when mature, and form about 250–500 microgametes about 3–4 × 0.5 μm with 2 flagella 5–7 μm long. The macrogametes are piriform when mature. The prepatent period is 9–10 days, the patent period 6–10 days.

PATHOGENESIS. This species is nonpathogenic for adult rabbits, even in large doses. It is pathogenic, however, for month-old rabbits, causing more or less severe intestinal catarrh and diarrhea; it might kill young rabbits. At necropsy, edema and grayish white foci that may coalesce to form a homogeneous, sticky, purulent layer may be found in the intestine.

Eimeria irresidua Kessel and Jankiewicz, 1931

This species is quite common worldwide throughout the small intestine of the domestic rabbit. The cottontail *Sylvilagus floridanus* can be infected experimentally.

OOCYSTS. The oocysts are ellipsoidal; 35–42 × 19–28 (mean 38 × 23) μm; with a 2-layered wall, the outer layer smooth, yellow, 1.4 μm thick, the inner layer a dark line 0.4 μm thick; with a micropyle; without a polar granule; with or without a residuum. The sporocysts are ovoid, 15–22 × 7–11 (mean 19 × 9) μm, with a small Stieda body and a large residuum. The sporozoites are elongate, 22 × 4 μm after excystation in bile and trypsin, with 1–3 clear globules; they lie lengthwise head to tail in the sporocysts. Sporulation takes 46 hr or less.

LIFE CYCLE. First generation meronts are in the glands, second generation meronts in the lamina propria, and third and fourth generation meronts and gamonts in the villous epithelium. First generation meronts can be seen at 96 hr. Second generation meronts at 120 hr are 29–39 × 26–35 (mean 36 × 31) μm and contain about 50 merozoites 16 × 2 μm. Third generation meronts at 144 hr and fourth generation ones at 168 hr are 20–29 × 16–25 (mean 23 × 21) μm, with 40–50 merozoites 16 × 2.5 μm and a small residuum. Young gamonts appear at 192 hr and mature during the next day. The microgamonts produce many flagellated microgametes and a central residuum. The prepatent period in the rabbit is 8–10 (mean 9) days.

PATHOGENESIS. This species is said by some to be one of the more pathogenic of the intestinal coccidia of rabbits, causing the usual signs of intestinal coccidiosis, hyperemia and sometimes extravasation of blood, and epithelial sloughing. However, others consider it not or only slightly pathogenic.

Eimeria magna Pérard, 1925

SYNONYM. *E. perforans* var. *magna.*

This species is not particularly common; it occurs worldwide in the middle and lower small intestine of the domestic rabbit; the cottontail *Sylvilagus floridanus* can be infected experimentally.

OOCYSTS. The oocysts are ovoid, frequently yellowish orange or brownish; 27–41 × 17–29 (mean about 35 × 24) μm; with a 2-layered wall of which the outer layer is easily detached; with a micropyle at the small end with a collarlike protrusion around it; apparently without a polar granule; with a large residuum. The sporocysts are ovoid, 11–16 × 6–9 μm, without a Stieda body, with a residuum. The sporozoites are elongate, lie head to tail in the sporocysts, and have a central nucleus and a large, clear globule at the broad end. The sporulation time is 2–5 days.

LIFE CYCLE. The meronts develop in the villar epithelial cells, either above or below the host cell nucleus, from the middle jejunum to the posterior end of the ileum. There are either two or three generations. The microgamonts are about 25–40 × 20–30 μm when mature and contain about 750–1,000 biflagellate microgametes about 3–4 × 0.5 μm and a residuum. The mature macrogametes are 30–35 × 15–20 μm. The prepatent period is about 7–9 days, the patent period 12–21 days.

PATHOGENESIS. Some strains of this species are very pathogenic; others are not. The principal signs are loss of weight, inappetence, and diarrhea. A good deal of mucus may be passed. The infected animals lose their appetites and grow thin. The intestinal mucosa is hyperemic and inflamed, and epithelial sloughing may occur.

Eimeria matsubayashii Tsunoda, 1952

This species occurs primarily in the ileum of the domestic rabbit in Japan and India.

OOCYSTS. The oocysts are ovoid, 22–29 × 16–22 (mean 25 × 18) μm, with a micropyle and a residuum, without a polar granule. The sporocysts are ovoid, about 7 × 6 μm, with a Stieda body and a residuum. The sporozoites are elongate, broader at one end than the other, lie lengthwise head to tail in the sporocysts, and have a clear globule at the large end. The sporulation time is 30–40 hr.

PATHOGENESIS. Heavy infections cause inflammation of the ileum characterized by diphtheritic enteritis.

Eimeria media Kessel, 1929

This species is quite common worldwide, primarily in the small intestine (but also in the large intestine in heavy infections) of the domestic rabbit. The cottontail *Sylvilagus floridanus* has been infected experimentally.

OOCYSTS. The oocysts are ovoid or ellipsoidal; 19–37 × 13–22 μm; with a smooth, pinkish, apparently 1-layered wall; with a micropyle and a residuum; without a polar granule. The sporocysts are ovoid, 17.5 × 7 μm, without a Stieda body, with a residuum. The sporozoites are elongate, lie lengthwise head to tail in the sporocysts, and have a clear globule at the large end. The sporulation time is 2–3 days.

LIFE CYCLE. The endogenous stages are found above or below the host cell nuclei of the epithelial cells of the villi and also occur in the submucosa. There are two asexual generations of merozoites followed by micro- and macrogamete production. Completion of the endogenous cycle takes 6 days. The prepatent period is 5–6 days, the patent period 15–18 days.

PATHOGENESIS. This species is slightly to moderately pathogenic. It may cause the usual signs of intestinal coccidiosis. The affected parts of the intestine may be edematous, with grayish foci.

Eimeria nagpurensis Gill and Ray, 1961

This species has been found in the feces of the domestic rabbit in India and Iran.

OOCYSTS. The oocysts are barrel-shaped, with the long sides parallel at least in their middle third; 20–27 × 10–15 μm; colorless or slightly yellow; with a thin wall; without a micropyle or residuum; presumably without a polar granule. The sporocysts are oat-shaped, with the anterior end sharply pointed; 15 × 5 μm; apparently with a Stieda body; with a residuum. The sporozoites are elongate, 12.5 × 2 μm, with one end wider than the other, lie lengthwise head to tail in the sporocysts, and have a clear globule at the broad end.

Eimeria neoleporis Carvalho, 1942

This species is common in the cottontail *Sylvilagus floridanus,* but the domestic rabbit can be infected both naturally and experimentally. It is presumably worldwide in the large intestine and posterior small intestine.

OOCYSTS. The oocysts are subcylindrical or elongate ellipsoidal, usually tapering somewhat toward the micropyle; pinkish yellow; 33–44 × 16–23 (mean 39 × 20) μm; with a smooth, 2-layered wall; with a micropyle; usually without a residuum; apparently without a polar granule. The sporocysts are ellipsoidal, 17 × 8–9 μm, with a Stieda body and a residuum. The sporozoites are banana-shaped, with one end wider than the other, lie lengthwise head to tail in the sporocysts, and have a clear globule at the large end. The sporulation time is 50–75 hr.

LIFE CYCLE. There are apparently four merogonic generations. Gamont formation begins on day 10 and is completed by day 12. The microgamonts produce a large number of biflagellate microgametes. In the domestic rabbit the prepatent period is 11–14 days, the patent period 8–16 days.

PATHOGENESIS. This species is slightly to markedly pathogenic, depending on the extent of the infection. The affected intestinal mucosa is inflamed and hyperemic, and caseous necrosis may be present.

Eimeria perforans (Leuckart, 1879) Sluiter and Swellengrebel, 1912

SYNONYMS. *Coccidium perforans, E. nana, E. lugdunumensis.*

This species is common worldwide in the epithelial cells of the villi and crypts of the small intestine, especially the middle part, of the domestic rabbit and can also be transmitted experimentally to the cottontail *Sylvilagus floridanus*.

OOCYSTS. The oocysts are ellipsoidal; colorless to pinkish; 15–30 × 11–20 μm; with a smooth, 2-layered wall; with a micropyle that is rarely visible on routine examination; without a polar granule; with a residuum. The sporocysts are ovoid, 5–9 × 3–5 (mean 8 × 4) μm, with a Stieda body and a residuum. The sporozoites are elongate, with one end broader than the other, lie head to tail in the sporocysts, and have a clear globule at the large end. The sporulation time is 1–2 days.

LIFE CYCLE. There are two asexual generations, both with two types of meront, one with 16–36 small, mononucleated merozoites produced by ectopolygeny and the other with 2–12 large, multinucleated merozoites produced by endopolygeny. These are followed by the gamonts and gametes. The prepatent period is 4–6 days, the patent period 12–32 days.

PATHOGENESIS. This is one of the less pathogenic intestinal coccidia of rabbits, but it may nevertheless cause mild to moderate signs if the infection is heavy enough. The duodenum may be enlarged and edematous, sometimes chalky white; the jejunum and ileum may contain white spots and streaks, and there may be petechiae in the cecum.

Eimeria piriformis Kotlan and Pospesch, 1934

This species is fairly common, presumably worldwide, in the large intestine of the domestic rabbit.

OOCYSTS. The oocysts are piriform, yellowish brown to dark brownish; 26–33 × 15–21 (mean 29–32 × 18–19) μm; with a 2-layered wall apparently lined by a membrane; with a conspicuous micropyle; without a polar granule or a residuum. The sporocysts are ovoid, 10–13 μm long, apparently with a Stieda body and a residuum. The sporozoites are elongate, with one end wider than the other, lie lengthwise head to tail in the sporocysts, and have a clear globule at the large end. The sporulation time is 2–6 days.

LIFE CYCLE. There are three generations of meront. Merogony is completed in 11 days; gametogony starts 8 DAI and is over by day 15. The macrogametes are 20 μm in diameter, and the mature microgamonts are 15–18 μm in diameter. The prepatent period is 9 days, the patent period 5–10 days.

PATHOGENESIS. Quite pathogenic.

Eimeria stiedai (Lindemann, 1864) Kisskalt and Hartmann, 1907

SYNONYMS. *Monocystis stiedae, Psorospermium cuniculi, Coccidium cuniculi, C. oviforme, E. stiedae.*

This species is common worldwide in the wall of the bile ducts in the liver of the domestic rabbit, also occurs naturally in the cottontails *Sylvilagus floridanus* and *S. nuttalli,* and has been transmitted experimentally to *S. audubonii.*

OOCYSTS. The oocysts are ovoid or ellipsoidal; 31–42 × 17–25 (mean 37 × 20) μm; with a smooth, colorless, pink, salmon-colored, or reddish orange, 1- or 2-layered wall; with a micropyle 6–10 μm in diameter at the small end; without a polar granule or a residuum. The sporocysts are ovoid, 17–18 × 8–10 μm, with a Stieda body and a residuum. The sporulation time is about 3 days.

LIFE CYCLE. The sporozoites emerge from the sporocysts in the small intestine and migrate to the liver via the lymph vessels. Merogony occurs in the epithelial cells of the bile ducts above the host cell nucleus. The number of asexual generations is uncertain; there are at least six and undoubtedly more. In due time some merozoites become macrogametes while others become microgamonts that produce large numbers of comma-shaped, biflagellate microgametes. These fertilize the macrogametes, which lay down an oocyst wall, break out of their host cell, pass into the intestine with the bile and hence out of the body. The prepatent period is 14–16 days, the patent period presumably 21–30 days.

PATHOGENESIS. In mild cases there may be no signs, but in more severe ones the animals lose their appetites and grow thin. There may be diarrhea, and the mucous membranes may be icteric. The disease, which is more severe in young animals than in old ones, may be chronic, or death may occur in 21–30 days.

Some of the signs are due to interference with liver function. The liver may become markedly enlarged, and white circular nodules or elongated cords may appear in it. At first they are sharply circumscribed but later they tend to coalesce. They are enormously enlarged bile ducts filled with the developing parasites. There is tremendous hyperplasia of the bile duct epithelial cells. Instead of forming a simple, narrow tube, the epithelium is thrown into great, arborescent folds, and each cell contains one or more parasites.

Eimeria acervulina Tyzzer, 1929

SYNONYM. *E. diminuta.*

This species is common worldwide in the anterior small intestine of the chicken and also in the Sri Lanka jungle fowl *Gallus lafayettei.*

OOCYSTS. The oocysts are ovoid, smooth; 12–23 × 9–17 (mean 16–18 × 13–15) μm; with a 2-layered wall; without a micropyle; with a polar granule; without a residuum. The sporocysts are ovoid, with a Stieda body, without a residuum. The sporulation time is 1 day. (See Fig. 7.15.)

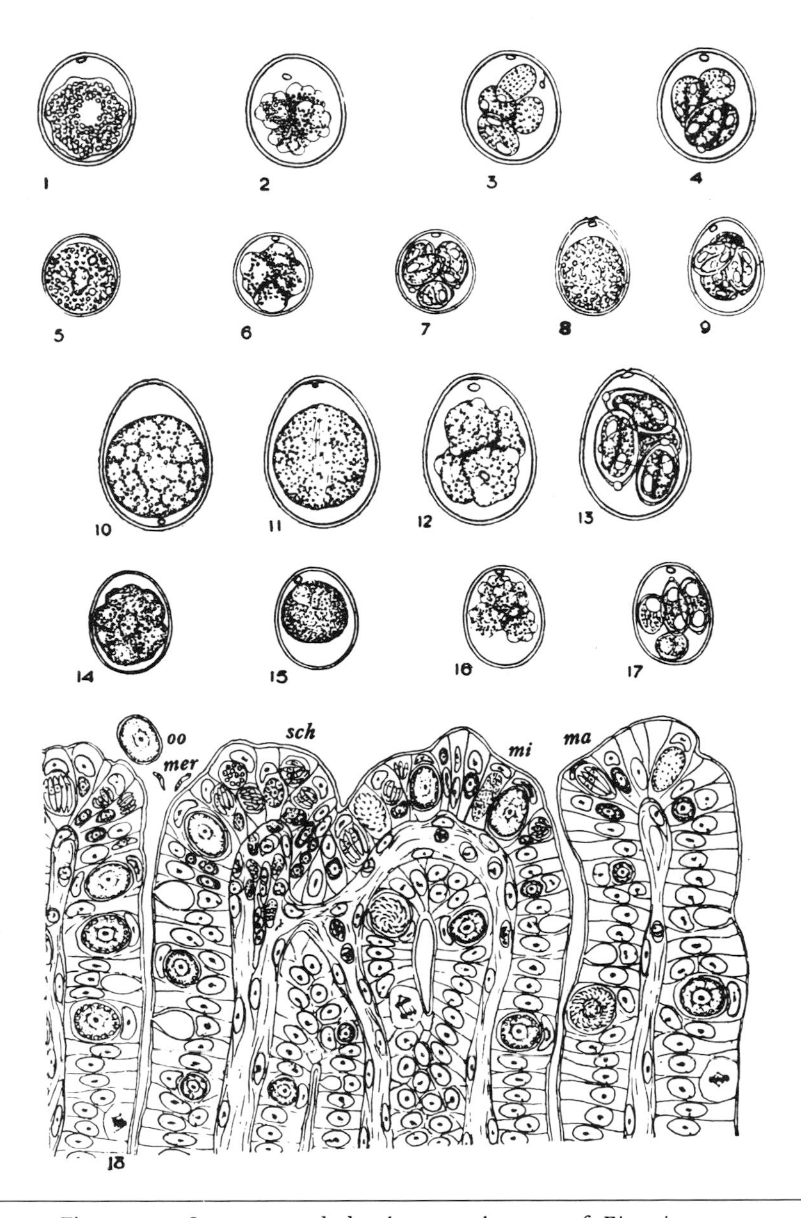

Fig. 7.15. Structure and developmental stages of *Eimeria* spp. from the chicken: 1–4. oocysts of *E. tenella*; 5–7. oocysts of *E. mitis;* 8 and 9. oocysts of *E. acervulina;* 10–13. oocysts of *E. maxima;* 14–17. oocysts of *E. necatrix;* 18. developmental stages of *E. tenella* in cecal epithelium 7–9 days after infection; *oo* = oocyst, *mer* = third generation merozoite, *sch* = third generation meront (schizont), *mi* = microgamont, *ma* = macrogamete. (From Tyzzer 1929, in *American Journal of Hygiene*, published by the Johns Hopkins Press)

LIFE CYCLE. The sporocysts emerge from the oocysts in the gizzard, and the sporozoites are activated and emerge in the small intestine. Most enter the duodenum. The meronts are found in the epithelial cells of the villi of the anterior small intestine, where they lie above the host cell nucleus.

There are four asexual generations. First generation meronts are at the base of the glands of Lieberkuehn, are 9–11 μm long, mature in 36–48 hr, and produce 8–16 merozoites 4–4.5 μm long and a small residuum. Second generation meronts are in the neck of the glands, 5–5.5 μm long, mature 41–56 hr after inoculation, and produce 16 merozoites 4–4.5 μm long and no residuum. Third generation meronts are at the base of the villi, 4–5.5 μm long, mature 56–72 hr after inoculation, and produce 8 merozoites 5–6 μm long and a residuum. Fourth generation meronts are on the sides and tips of the villi, 7–9 μm long, mature 80–96 hr after inoculation, and produce 32 merozoites 9–10 μm long and a large residuum.

The sexual stages lie above the host cell nuclei in the epithelial cells of the villi and to a lesser extent in the gland cells. They appear 4 DAI and require 40 hr to mature. The macrogametes are 14.5–19 μm in diameter, the microgamonts 7–8 μm. The latter produce many triflagellate microgametes 2–3 μm long. The prepatent period is 4 days, the patent period several days or longer.

PATHOGENESIS. *E. acervulina* is generally considered only slightly pathogenic, but very large inocula (many millions of oocysts) may cause severe signs and even death. The intestine may be thickened and a catarrhal exudate may be present, but hemorrhage is rare. The maturing oocysts lie massed in limited areas, and form whitish or gray spots or streaks running transversely in the intestinal mucosa.

Eimeria brunetti P. P. Levine, 1942

This species is quite common, presumably worldwide, in the chicken. Early generation meronts occur throughout the small intestine, and late generation meronts, gamonts, and gametes in the posterior small intestine, rectum, ceca, and cloaca.

OOCYSTS. The oocysts are ovoid, 14–34 × 12–26 (mean 23–25 × 19–20) μm, with a smooth wall, without a micropyle or a residuum, with a polar granule. The sporocysts are elongate ovoid, about 13 × 7.5 μm, with a Stieda body and a residuum. The sporulation time is 18 hr to 2 days.

LIFE CYCLE. There are at least three asexual generations before gamogony. The first generation meronts average 28 × 21 μm in sections and contain an average of 318 merozoites each. They are in the epithelial cells at the base of the villi in the midintestine. The second generation meronts are found at the tips of the villi 3 DAI. They are smaller than the first generation meronts but vary greatly in size and contain 15–120 merozoites similar in size to the first generation merozoites. The third generation meronts appear at 84 hr and are mature by 4 DAI; they are similar to the second generation meronts. Gamonts appear on day 5 and lie at the tips and sides of the villi, either above the host

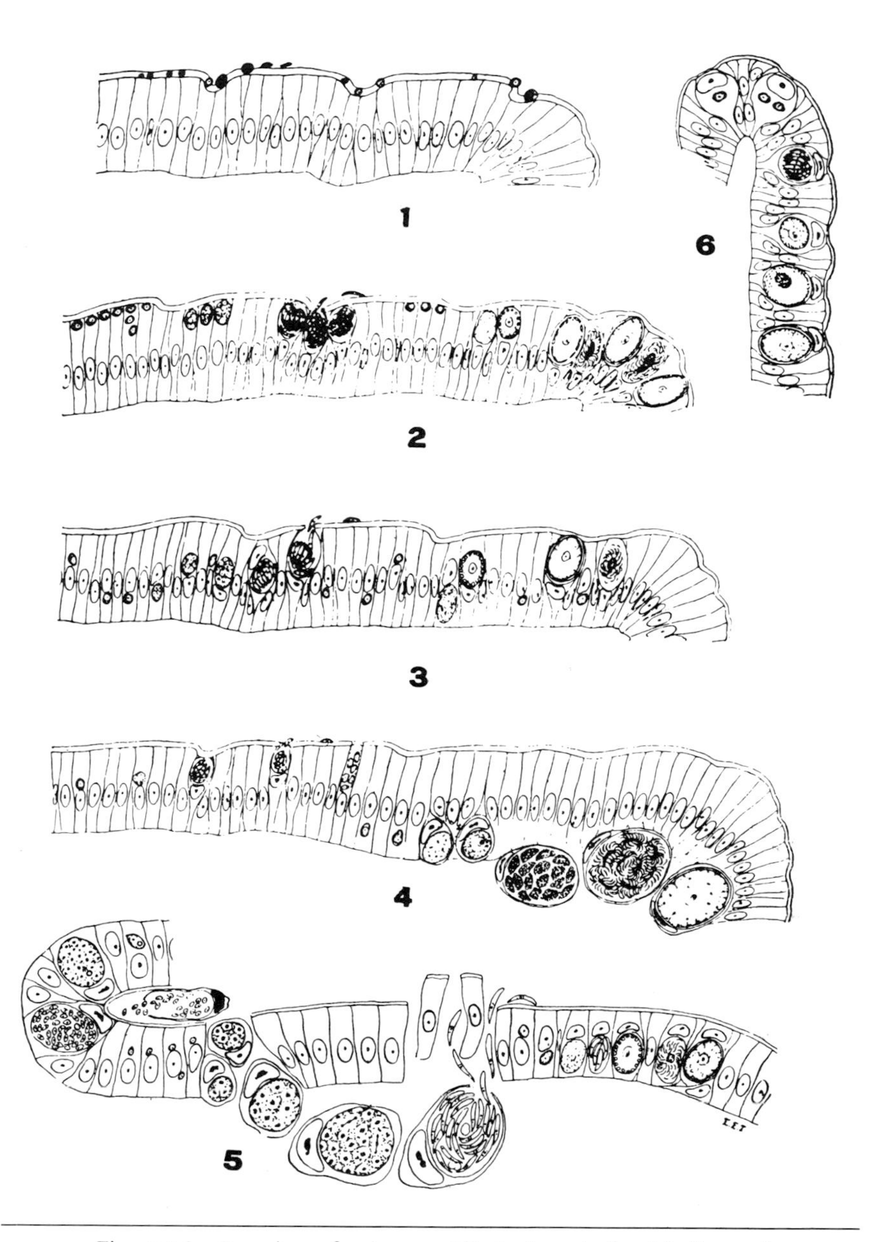

Fig. 7.16. Location of avian coccidia in intestinal epithelium of chicken: 1. *Cryptosporidium meleagridis;* 2. *Eimeria acervulina;* 3. *E. mitis;* 4. *E. maxima;* 5. *E. tenella.* (From Tyzzer 1929, in *American Journal of Hygiene*, published by the Johns Hopkins Press)

cell nuclei or on the basement membrane. They are in the lower small intestine and large intestine, as are the later meronts. The microgamonts contain several centers of microgamete development (blastophores) and are larger than the macrogametes, which are about 25 × 22 μm. The prepatent period is 5 days.

PATHOGENESIS. *E. brunetti* is markedly pathogenic, but its effects depend on the degree of infection. In light infections, no gross lesions may be seen. In heavier infections, the gut wall becomes thickened and a pink or blood-tinged catarrhal exudate appears 4–5 days after experimental inoculation; the droppings are quite fluid and contain blood-tinged mucus and many mucous casts. The birds become somewhat depressed. These signs continue for 5 days and then subside if the birds recover.

In early or light infections, hemorrhagic, ladderlike streaks are present on the mucosa of the lower intestine and rectum. In heavy infections a characteris-

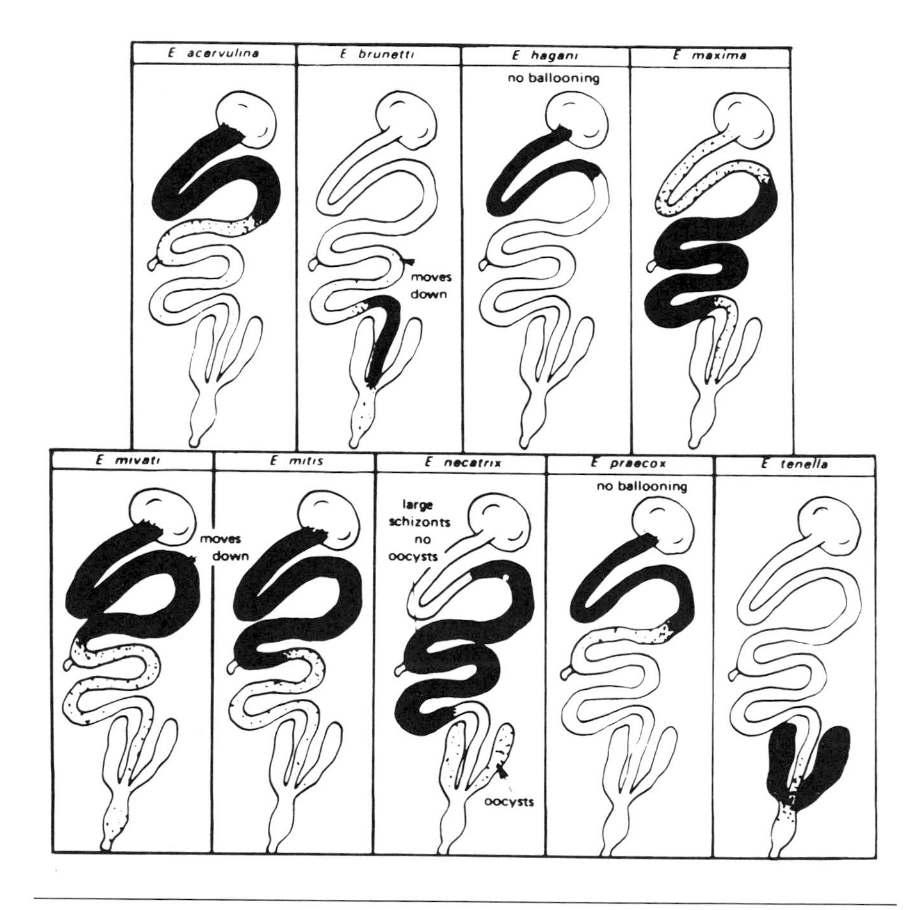

Fig. 7.17. Location of lesions caused by chicken coccidia. (From Reid 1964, Ga. Agric. Exp. Stn. Tech. Bull. 39)

tic necrotic enteritis appears that may involve the entire intestinal tract but is more often found in the lower small intestine, colon, and tubular part of the ceca. A patchy or continuous dry, caseous, necrotic membrane may line the intestine, and the intestine may be filled with sloughed, necrotic material. Circumscribed white patches may be visible through the serosa; there may even be intestinal perforation with resultant peritonitis.

Eimeria hagani P. P. Levine, 1938

This species is apparently rare; it occurs in the anterior half of the small intestine of the chicken and has been found in North America and India.

OOCYSTS. The oocysts are ovoid, 16–21 × 14–19 (mean 19 × 18) μm, with a smooth wall, without a micropyle, with a polar granule. No other structural information appears to be available. The sporulation time is 18 hr to 2 days. The prepatent period is 6–7 days.

PATHOGENESIS. This species is only slightly if at all pathogenic.

Eimeria maxima Tyzzer, 1929

SYNONYMS. *E. indentata, E. tyzzeri.*

This species is common worldwide in the small intestine of the chicken and also occurs in the Malaysian jungle fowl *Gallus lafayettei.*

OOCYSTS. The oocysts are ovoid; 21–42 × 16–30 (mean 29–31 × 21–23) μm; with a smooth or somewhat roughened, yellowish wall; without a micropyle or a residuum; with a polar granule. The sporocysts are ovoid, 15–19 × 8–9 μm, with a Stieda body, without a residuum. The sporozoites are 19 × 4 μm, with a conspicuous clear globule. The sporulation time is 30 hr to 2 days.

LIFE CYCLE. The meronts are above the host cell nuclei or occasionally beside them in the epithelial cells of the tips of the villi of the duodenum and upper ileum. There are three asexual generations. The first generation meronts are deep in the epithelial cells of the deep glands of the duodenum, where they occur about 48 hr after inoculation. They produce 25–50 loosely packed merozoites. The second generation meronts are in the epithelial cells of the small intestine villi also, but near the openings of the crypts of Lieberkuehn. They occur during the third day and produce about 12 merozoites. The third generation meronts are in the epithelial cells of the small intestine villi also, but along the sides of the superficial villi and sometimes in their tips. They occur during the fourth day and produce about 12 merozoites.

The sexual stages are found beneath the host cell nuclei. As they become larger, the host cells are displaced toward the center of the villi and come to lie in their interior. The mature microgamonts are 30–48 × 22–33 μm and form a large number of microgametes. They have 3 flagella and 2–10 (normally 4) additional microtubules running back from the perforatorium and above the single mitochondrion. After fertilization an oocyst wall is laid down, and the oocysts break out of the villi and are passed in the feces. The prepatent period is 5–6 days, the patent period only a few days. The haploid number of chromosomes is 5.

PATHOGENESIS. *E. maxima* is slightly to moderately pathogenic. The asexual stages cause relatively little damage, the most serious effects being due to the sexual stages. Different strains apparently differ in pathogenicity.

The principal lesions are hemorrhages in the small intestine. The intestinal muscles lose their tone, and the intestine becomes flaccid and dilated, with a somewhat thickened wall. Short, fine, hairlike hemorrhages are sometimes present in the intestinal wall. There is a catarrhal enteritis; the intestinal contents are viscid and mucoid, grayish, brownish, orange or pinkish, occasionally but not usually with flecks of blood. Birds that recover soon return to normal.

Eimeria mitis Tyzzer, 1929

SYNONYM. *E. beachi.*

Common worldwide, throughout the small intestine but primarily in its posterior part and even in the tubular part of the ceca of the chicken.

OOCYSTS. The oocysts are subspherical; 10–21 × 9–18 (mean 16 × 13–16) μm; with a smooth, colorless wall; without a micropyle or a residuum; with a polar granule. The sporocysts are ovoid, 10 × 6 μm, with a Stieda body, without a residuum. The sporulation time is 18 hr to 2 days.

LIFE CYCLE. The endogenous stages are in the epithelial cells of the villi and occasionally in the glands of Lieberkuehn. The meronts produce 6–24 (rarely up to 30) merozoites, but the number of meront generations is not known. The merozoites are crescent-shaped, with blunt ends, and about 5 × 1.5 μm.

The microgamonts are about 9–14 μm long; the macrogametes are somewhat larger. In contrast to most other coccidian species, both asexual and sexual stages occur together. The prepatent period is 4–5 days, the patent period 10 days.

PATHOGENESIS. *E. mitis* is slightly pathogenic but is unlikely to be of pathologic significance under normal field conditions.

Eimeria mivati Edgar and Seibold, 1964

This species is fairly common, presumably worldwide, primarily in the upper small intestine but also in the remainder of the small intestine and large intestine of the chicken. Recently, some doubt has been expressed as to the validity of this species.

OOCYSTS. The oocysts are ellipsoidal to ovoid; 11–20 × 12–17 (mean 16 × 13) μm; with a smooth, colorless, 1-layered wall 0.4 μm thick; with a micropyle and polar granule; without a residuum. The sporocysts are elongate ovoid, 7–12 × 5–6 μm, with a Stieda body and a residuum. The sporozoites are crescent- or banana-shaped, with a clear globule at each end.

LIFE CYCLE. The sporocysts break out of the oocysts in the crop, proventriculus, and gizzard. Many sporozoites are free in the duodenum within 20 min after feeding. Entering epithelial cells deep in the glands or crypts of the anterior third of the small intestine, they lie above the host cell nucleus, where they round up and turn into first generation meronts. These mature 36 hr after inoculation, when they are 9–14 μm in diameter and contain 10–30 first generation merozoites 9–11 × 2 μm. These enter new epithelial cells in the crypts and bases of the villi, where they lie above the host cell nuclei and mature 55–67 hr after inoculation. The second generation meronts are 5–11 × 4–8 μm and contain 16–20 second generation merozoites 6–9 × 1 μm and a residual body. These enter new epithelial cells, mostly along the sides and tips of the villi, in the anterior small intestine, and throughout the ileum, even into the ceca and rectum. They lie above and beneath the host cell nuclei. They form third generation meronts that mature 80 hr after inoculation. They are 6–8 × 4–7 μm and contain 10–14 merozoites 6–7 × 1 μm and a residual body. Some of the third generation merozoites enter new host epithelial cells in the same location as the third generation meronts. They form fourth generation meronts that mature about 96–168 hr after inoculation, when they are 10–17 × 10–13 μm and contain 16–20 fourth generation merozoites 7–10 × 1–2 μm and a residual body.

Some of the third generation merozoites and all of the fourth generation ones enter epithelial cells from the duodenum to the rectum and form gamonts. Those anterior to the yolk stalk generally lie above the host cell nucleus, while those posterior to it generally lie below the host cell nucleus. Mature microgamonts are present 93–120 hr after inoculation. They are 10–16 × 9–15 μm and contain 50–100 microgametes 2–4 μm long, each with 2 flagella 5–10 μm long. Mature oocysts appear 93–96 hr after inoculation. The prepatent period is about 4–5 days, the patent period 6–12 days.

PATHOGENESIS. *E. mivati* is more pathogenic than *E. acervulina,* but light infections are not pathogenic and heavy ones produce no more than 10% mortality.

Eimeria necatrix Johnson, 1930

This species is common worldwide in the chicken. The first and second generation meronts are in the small intestine, and the third generation meronts, gamonts, and gametes are in the ceca.

OOCYSTS. The oocysts are ovoid; 12–29 × 11–24 (mean 20 × 17) μm; with a smooth, colorless, 2-layered wall; without a micropyle or a residuum; with a polar granule. The sporocysts are ovoid, with a Stieda body, without a residuum. The sporulation time is 18 hr to 2 days.

LIFE CYCLE. Chickens become infected by ingesting sporulated oocysts. The sporozoites enter the epithelial cells of the small intestine, pass through the epithelium into the lamina propria or core of the villi, and migrate toward the muscularis mucosae. Most of them are engulfed by macrophages en route and are transported by them to the epithelial cells of the fundus. The macrophages invade these cells and appear to disintegrate during or after the invasion process, leaving the sporozoites unharmed. These then round up to form first generation meronts.

Both the first and second generation meronts are found above the host cell nuclei in the epithelial cells of the gland fundi. The first generation merozoites are liberated 2.5–3 DAI and enter adjacent epithelial cells. The second generation meronts are 39–66 × 33–54 (mean 52 × 38) μm. Most of their merozoites are liberated 5–8 DAI, but a few may still be liberated as long as 23 DAI. They are 8–11 × 1.5–2 (mean 9 × 2) μm. They pass to the cecum, where they penetrate the epithelial cells, coming to lie below the host cell nuclei, and turn into third generation meronts. Most of these are found in the surface epithelium, but some enter the glandular epithelium. These third generation meronts are relatively small and contain only 6–16 third generation merozoites.

The third generation and some of the second generation merozoites enter other cecal epithelial cells and become macrogametes or microgamonts. These also lie below the host cell nuclei. Microgametes develop from the microgamonts, fertilization takes place, and oocysts form and are released. The prepatent period is 6–7 days, the patent period about 12 days.

PATHOGENESIS. Next to *E. tenella,* this is the most pathogenic and important species of chicken coccidia. Indeed, with the decrease in importance of *E. tenella* due to the use of coccidiostatic drugs, *E. necatrix* has come to the fore in many areas as the cause of more losses than *E. tenella.*

E. necatrix is often said to cause a more chronic type of coccidiosis than *E. tenella.* This is not because it runs a longer course but because it produces so much scar tissue in the small intestine that its effects are more lasting.

The principal lesions are in the small intestine, the middle third of which is the most seriously affected. Small, white, opaque foci are found here by the 4th DAI. Composed of second generation meronts developing deep in the mucosa, they are so deep that they can be seen from the serosal but not the mucosal surface of the intestine. They are seldom more than 1 mm in diameter but may coalesce and thus appear larger. Severe hemorrhage may appear on the 5th or 6th day. The small intestine may be markedly swollen and filled with clotted or unclotted blood. Its wall is greatly thickened, dull red, and many petechial hemorrhages appear in the white, opaque foci, which by now contain second generation merozoites. The gut wall may lose contractility, become friable, and even appear gangrenous. The epithelium may slough, and by the end of the 6th day a network of fibrin-containing mononuclear cells appears in the destroyed areas. This is later replaced by connective tissue, and permanent scarring results, which interferes with intestinal absorption.

The ceca are not seriously affected. They may be contracted and their contents may be dehydrated.

Death usually occurs 5–7 days after exposure. Many of the birds that recover remain unthrifty and emaciated. The aftereffects of this type of coccidiosis are often so long-lasting that it is not worthwhile to keep birds that have recovered from severe attacks.

Eimeria praecox Johnson, 1930

This species is common, probably worldwide, in the upper third of the small intestine of the chicken. It also occurs in the Sri Lanka jungle fowl *Gallus lafayettei.*

OOCYSTS. The oocysts are ovoid; 20–25 × 16–20 (mean 21 × 17) μm; with a smooth, colorless wall; without a micropyle or a residuum; with a polar granule. The sporocysts have no residuum. The sporulation time is 2 days.

LIFE CYCLE. The endogenous stages occur in the epithelial cells of the villi, usually along the sides of the villi and below the host cell nuclei. There are at least three and probably four generations of meront, the second of which appears as early as 32 hr after inoculation. Later development is irregular, both sexual and asexual stages being seen together. The prepatent period is 2.5–4 days, the patent period 4 days or a little more.

PATHOGENESIS. Essentially nonpathogenic.

Eimeria tenella (Railliet and Lucet, 1891) Fantham, 1909

SYNONYMS. *E. avium, E. bracheti, Coccidium tenellum, C. globosum.*

This species is very common worldwide in the ceca of the chicken.

OOCYSTS. The oocysts are ovoid; 14–31 × 9–25 (mean 25 × 19) μm; with a smooth, two-layered wall; without a micropyle or a residuum; with a polar granule. The sporocysts are ovoid, without a residuum. The sporulation time is 18 hr to 2 days.

LIFE CYCLE. The life cycle of *E. tenella* has already been described as an example of coccidian life cycles. The prepatent period is 6 days.

PATHOGENESIS. This is the most pathogenic of the chicken coccidia and is responsible for heavy losses. Together with other species, it was estimated by the USDA (1965) that coccidiosis caused an annual loss of $34,854,000 in the United States due to death and disease alone. To this should be added the cost of medicating the feeds that are generally fed to poultry, and various labor and other costs entailed by disease outbreaks.

Cecal coccidiosis is found most frequently in young birds. Chicks are most susceptible at 4 wk of age; chicks 1–2 wk old are more resistant, although day-old chicks can be infected. Older birds develop immunity as the result of exposure.

Coccidiosis due to *E. tenella* may vary in severity from an inapparent infection to an acute, highly fatal disease, depending on the infective dose of oocysts, the pathogenicity of the coccidian strain, the breed and age of the chickens, their state of nutrition, and the other disease agents and stresses to which they are concomitantly subjected.

Cecal coccidiosis is an acute disease characterized by diarrhea and massive cecal hemorrhage. The first signs appear when the second generation meronts begin to enlarge and produce leakage of blood into the ceca. Blood appears in the droppings 4 DAI. At this time the birds appear listless. They may become droopy and inactive and eat little, although they still drink. The greatest amount of hemorrhage occurs 5–6 DAI. It then declines, and oocysts appear in the feces 7 DAI if the birds live that long. The oocysts increase to a peak on the 8th or 9th day and then drop off rapidly. Very few are still being shed by the 11th day. A few oocysts may be found for several months.

Coccidiosis is self-limiting. If the birds survive to the 8th or 9th DAI, they generally recover.

The lesions of cecal coccidiosis depend on the stage of the disease. About the 4th DAI the lamina propria becomes infiltrated with eosinophils, there is marked congestion, and the cecal wall is thickened. The epithelium may be torn, and coccidia, blood, and tissue cells may be released into the lumen. Hemorrhage is present throughout the cecal mucosa.

On the 5th day there is extensive epithelial sloughing, and the cecum is filled with large amounts of unclotted or partly clotted blood. This increases on the 6th day, and the cecal contents consolidate to form a cecal core, which begins to form on the 7th day. It adheres tightly to the mucosa at first but soon comes loose and lies free in the lumen.

About 7 DAI, the wall of the cecum changes color from red to mottled reddish or milky white due to the formation of oocysts. It is greatly thickened. The cecal core, which was at first reddish, becomes yellowish or whitish. If it is small enough, it may be passed intact in the feces, but it is usually broken up into small pieces. In a few days the cecum becomes normal in appearance or at most slightly enlarged and thickened. Occasionally the cecum may rupture or adhesions may form.

Epithelial regeneration is complete in light infections, but in severe ones it may not be. There is a marked inflammatory reaction, with extensive lymphoid and plasma cell infiltration, and there may be some giant cells. Connective tissue is increased. The epithelium may not be replaced between the glands, and cysts formed by constriction of the glands during the inflammatory stage may persist.

The loss of blood into the ceca causes anemia.

Birds that recover from coccidiosis may suffer ill effects for some time or even permanently.

Eimeria adenoeides Moore and Brown, 1951

This species is quite common, presumably worldwide, in the lower small intestine, ceca, and rectum of the turkey.

OOCYSTS. The oocysts are ellipsoidal or sometimes ovoid, 19–31 × 13–21 (mean 26 × 17) μm, with a smooth wall, sometimes with a micropyle, with 1–3 polar granules, without a residuum. The sporocysts are elongate ovoid, apparently with a Stieda body and a residuum. The sporozoites contain a clear globule at the large end. The sporulation time is 18 hr to 4 days.

LIFE CYCLE. First generation meronts can be found in the epithelial cells as early as 6 hr after inoculation. They are 30 × 18 μm when mature (at 60 hr) and produce about 700 merozoites 4–7 × 1.5 μm with a central nucleus.

Second generation meronts become mature 96–108 hr after inoculation. They are 10 × 10 μm and produce 12–24 merozoites about 10 × 3 μm with a nucleus a little nearer the rounded than the pointed end.

Sexual stages can be recognized as early as 120 hr after inoculation. The mature macrogametes and microgamonts are about 20 × 18 μm. The prepatent period is 104–132 hr, the patent period 7–20 days.

PATHOGENESIS. Very pathogenic.

Poults develop signs of anorexia, droopiness, and ruffled feathers during the 4th day after experimental inoculation. If death occurs, it is usually on the 5th or 6th day but may be a little later.

The intestines appear quite normal until the 4th day. The walls of the lower third of the small intestine, ceca, and rectum become swollen and edematous, petechial hemorrhages visible from the mucosal but not the serosal surface appear, and the lower intestine becomes filled with mucus. Then edematous changes are seen in the intestine, and infected epithelial cells begin to break off, leaving the villi denuded. The blood vessels become engorged; cellular infiltration of the submucosa and epithelial denudation increase progressively until the 6th day.

During the 5th day most of the terminal intestine is congested and contains large numbers of merozoites and long streaks of blood. By the end of the day the intestine contains caseous material composed of cellular debris, gametes, and a few immature oocysts. A little later the caseous exudate is composed largely of oocysts. The feces in severe cases are relatively fluid and may even be blood-tinged and contain mucous casts 2.5–5 cm long. Caseous plugs are sometimes present in the ceca.

On the 6th to 8th days the terminal intestine may contain white, creamy mucus, and petechiae may be present in its mucosa. By the 9th day the intestinal contents appear normal, although they still contain large numbers of oocysts.

In birds that recover from the disease or that have received relatively few oocysts, resolution is rapid. Vascularity is greatly reduced, the deep glands are almost free of parasites by the 7th day, and the intestine is almost normal by the 9th or 10th day.

Eimeria dispersa Tyzzer, 1929

This species is presumably relatively uncommon, primarily in the duodenum but also in the jejunum and ileum of the turkey, bobwhite quail, ring-necked pheasant, chukar partridge, and (presumably) ruffed grouse and sharp-tailed grouse. It has been found in North America and Britain. The bobwhite quail is its type host.

OOCYSTS. The oocysts are ovoid; 22–31 × 18–24 (mean 26 × 21) μm; with a smooth, 1-layered wall; without a micropyle, polar granule, or residuum. The sporocysts are ovoid, with a Stieda body. The sporulation time is 2 days.

LIFE CYCLE. First generation meronts are present 30 hr after inoculation; they average 14 × 13 μm and contain an average of 19 merozoites about 4.5 × 1 μm. Second generation meronts are present 48 hr after inoculation; they average 8 × 7 μm and contain an average of 13.5 merozoites 4.5 × 1 μm. Third generation meronts are present 72 hr after inoculation; they average 9 × 9 μm and contain an average of 15 merozoites 6 × 2 μm. Fourth generation meronts are present 96 hr after inoculation; they average 12 × 10.5 μm and contain an average of 7 merozoites 8 × 2 μm. Sporozoites and first generation meronts lie close above the epithelial cell nuclei. Second, third, and fourth generation meronts are also above the host cell nucleus but lie near the brush border of the host cell. All are in the villar epithelial cells of the small intestine.

The mature macrogametes, 18–20 μm in diameter, can be found in the small intestine villar epithelial cells 96 hr after inoculation. The microgamonts are slightly smaller but are in the same location; they produce biflagellate microgametes. The prepatent period is 5–6 days.

PATHOGENESIS. *E. dispersa* is usually only slightly pathogenic. With pathogenic strains, the most severe lesions occur 5–6 days after experimental inoculation. The entire small intestine is markedly dilated, and the duodenum and anterior jejunum are creamy white when seen through the serosal surface. The anterior half of the small intestine is filled with creamy, yellowish, sticky, mucoid material. The wall of the anterior intestine is edematous, but there is little epithelial sloughing. The intestinal tract is virtually normal 8 DAI.

Eimeria gallopavonis Hawkins, 1952

This species is uncommon, presumably worldwide, in the ileum, rectum, and to a lesser extent ceca of the turkey.

OOCYSTS. The oocysts cannot be differentiated with certainty from those of *E. meleagridis*. They are ellipsoidal, 22–33 × 15–19 (mean 27 × 17) μm, with a smooth wall, without a micropyle or a residuum, with a polar granule. The sporocysts are ovoid, with a Stieda body and a residuum. The sporulation time is 1 day.

LIFE CYCLE. The endogenous stages are in the epithelial cells at the tips of the villi, where most lie above the host cell nucleus. The first generation meronts occur in the ileum and rectum. They produce approximately 8 merozoites and a residuum 3 DAI. There are apparently two sizes of second generation meronts. The smaller ones occur in the rectum, ileum, and rarely in the ceca. They produce 10–12 merozoites and a residuum 4–5 DAI. The large ones, which occur only in the rectum, are 20 μm in diameter and produce a large, undetermined number of merozoites 4 DAI. There are a few third generation meronts in the rectum that produce about 10–12 merozoites. These and most of the second generation merozoites develop into sexual stages, which are found primarily in the rectum and only occasionally in the ileum and ceca. The macrogametes and microgamonts are similar to those of other turkey coccidia. The prepatent period is 6 days.

PATHOGENESIS. Moderately pathogenic (?).

Eimeria innocua Moore and Brown, 1952

This species is apparently uncommon. It occurs throughout the small intestine of the turkey and has been reported in North America and Bulgaria.

OOCYSTS. The oocysts are subspherical, 19–26 × 17–25 (mean 22 × 21) μm, with a smooth wall, without a micropyle or polar granule. The sporulation time is 2 days.

LIFE CYCLE. The endogenous stages occur in the epithelial cells of the villi. The tips of the villi are most heavily parasitized, while the crypts and deep glands are never affected. The prepatent period is 5 days, the patent period up to 9 days.

PATHOGENESIS. Nonpathogenic.

Eimeria meleagridis Tyzzer, 1927

This species is common worldwide in the turkey. First generation meronts are in the middle small intestine, second generation meronts in the ceca, and sexual stages in the ceca, rectum, and to a slight extent the ileum.

OOCYSTS. The oocysts are ellipsoidal, 19–31 × 14–23 (mean 24 × 17–18) μm, with a smooth wall, without a micropyle or a residuum, with 1–2 polar granules. The sporocysts are ovoid, with a Stieda body and a residuum. The sporulation time is 15 hr to 3 days.

LIFE CYCLE. First generation meronts are present 2–5 DAI. They are 20 × 15 μm and contain 50–100 merozoites 7 × 1.5 μm. Second generation meronts appear 60 hr after inoculation and are mature after 70 hr. They are about 9 μm in diameter and contain 8–16 merozoites 10 × 2 μm. There may be a third asexual generation, but most of the second generation merozoites develop into sexual stages.

Macrogametes and microgamonts appear at 91 hr and are mature 9 DAI. They are about 18 × 13 μm. The microgametes are biflagellate. The prepatent period is 3–5 days.

PATHOGENESIS. Practically nonpathogenic.

Eimeria meleagrimitis Tyzzer, 1929

This species is quite common worldwide in the turkey. It is usually in the small intestine above the yolk stalk but sometimes may be throughout the whole intestine.

OOCYSTS. The oocysts are subspherical, 16–27 × 13–22 (mean 19 × 16) μm, with a smooth wall, without a micropyle or a residuum, with 1–3 polar granules. The sporocysts are ovoid, with a Stieda body and a residuum. The sporozoites have a clear globule at the large end. The sporulation time is 1–3 days.

LIFE CYCLE. The sporocysts emerge from the oocysts in the gizzard, and the sporozoites are activated and emerge from the sporocysts in the small intestine. They invade the tips of the villi and migrate down the villi in the lamina propria until they reach the glands. Young first generation meronts can be found in the gland epithelial cells as early as 12 hr after inoculation, and many are mature by 48 hr. They are usually 17 × 13 μm and enlarge the host cell, pushing its nucleus into the gland lumen. They contain 80–100 merozoites, which are about 4.5 × 1.5 μm and have the nucleus at the larger end.

The first generation merozoites invade the adjacent epithelial cells, forming colonies of second generation meronts. Most of these are mature 66 hr after inoculation. They are 8 × 7 μm and contain 8–16 merozoites 7 × 1.5 μm with a nucleus near the center. Third generation meronts may be recognized as early as 72 hr after inoculation and reach maturity at about 96 hr. They are about 8 × 7 μm and differ from the second generation meronts in having a residuum. They produce 8–16 merozoites about 7 × 1.5 μm with a nucleus much nearer the large end than in the second generation merozoites.

Macrogametes and microgamonts appear 114 hr after inoculation. They are about 15 × 11 μm; the microgamonts contain a residuum, and the microgametes have 2 long flagella. The prepatent period is 3–6 days.

PATHOGENESIS. This species is moderately to markedly pathogenic, causing catarrhal enteritis. The death rate is high in young poults up to 6 wk of age, but older birds are more resistant.

Lesions appear at the end of the 4th DAI. The jejunum is slightly thickened, dilated, and contains an excessive amount of clear, colorless fluid or mucus containing merozoites and small amounts of blood and other cells. By 5–6 DAI the duodenum is enlarged, its blood vessels are engorged, and it contains a reddish brown, necrotic core that adheres firmly to the mucosa and extends a little way into the upper small intestine. The remainder of the intestine is congested, and petechial hemorrhages may be present in the mucosa of most of the small intestine.

Regeneration of the mucosa begins on the 6th or 7th day. A few petechiae are present in the duodenum and jejunum, and there are a few minute streaks of hemorrhage and spotty congestion in the ileum. The posterior part of the jejunum and ileum may contain greenish, mucoid casts 5–10 cm long and 3–6 mm in diameter; necrotic material may be found in the ileum or feces.

Feed consumption begins to drop 2–3 days after exposure; at 4 days the birds huddle together with closed eyes, drooping wings, and ruffled feathers. Their droppings at this time are scanty and slightly fluid. At the peak of the disease 5–6 days after exposure, some of the feces form cylinders 1–2 cm long and 3–6 mm in diameter. The droppings are not bloody, although a few flecks of blood may occasionally be seen. Death usually occurs 5–7 DAI.

Eimeria subrotunda Moore, Brown, and Carter, 1954

This species, apparently uncommon, is found in the turkey and has been reported only from North America. It occurs in the upper small intestine 5 cm or more anterior to the yolk stalk rudiment.

OOCYSTS. This species closely resembles *E. innocua.* The oocysts are subspherical, 16–26 × 14–24 (mean 22 × 20) μm, with a smooth wall, without a micropyle or a polar granule. No other structural information was given. The sporulation time is 2 days.

LIFE CYCLE. The endogenous stages occur in the epithelial cells of the tips of the villi, extend along the sides of the villi to some extent, but never invade the crypts and deep glands. The prepatent period is 4 days, the patent period 12–13 days.

PATHOGENESIS. Apparently nonpathogenic.

Eimeria mandali Banik and Ray, 1964

SYNONYMS. *E. pavonis* Patnaik, 1965, not *E. pavonis* Mandal, 1965; *E. patnaiki.*

This species was found in the feces of the peafowl in India.

OOCYSTS. The oocysts are spherical to subspherical, 14–20 × 14–18 (mean 18 × 17) μm, with a distinct micropyle 1 μm wide and a residuum. The sporocysts are 6–12 × 4–8 μm, with a Stieda body and a residuum (Pellérdy 1974).

Eimeria mayurai Bhatia and Pande, 1966

This species was found in the small intestine, especially duodenum, of the peafowl in India.

OOCYSTS. The oocysts are ellipsoidal, ovoid, or subspherical; 18–27 × 13–20 (mean 20–22 × 14–16) μm; with a smooth, 2-layered wall 1–2 μm thick, the outer layer light blue, the inner layer brownish; without a micropyle (or perhaps with an indistinct one); with a polar granule; without a residuum. The sporocysts are ovoid, 10–13 × 5–7 (mean 12 × 6) μm, with a Stieda body and a residuum. The sporozoites are elongate, 9–11 × 3–4 (mean 10 × 4) μm, lie lengthwise head to tail in the sporocysts, and have a large clear globule near the broad end and sometimes a small one at the other end. The sporulation time is 2 days.

LIFE CYCLE. The endogenous stages lie beneath the host cell nucleus in the epithelial cells, mainly of the villi but also deep in the lamina propria in the gland cells of the crypts of Lieberkuehn. The mature second generation meronts are nearly spherical or ovoid, 9–15 × 6–11 (mean 11 × 9) μm, and contain 16–70 merozoites 5–8 × 1 (mean 6 × 1) μm. The prepatent period is 7 days.

PATHOGENESIS. Moderately pathogenic.

Eimeria pavonina Banik and Ray, 1961

SYNONYM. *E. cristata.*

This species was found in the feces of the peafowl in India.

OOCYSTS. The oocysts are ovoid; with a 2-layered wall, the outer layer brilliant blue and thicker than the inner layer; without a micropyle; with a residuum. The sporocysts are boat-shaped.

Eimeria pavonis Mandal, 1965

SYNONYM. *E. indica.*

This species was found in the feces of the peafowl in India.

OOCYSTS. The oocysts are ovoid, 20–25 × 18 μm, with a 2-layered wall, with a micropyle at the narrow end, with a polar granule, without a residuum. The sporocysts are elongate; more or less ellipsoidal; 12–16 × 7 μm; with a thick wall, a Stieda body, and some scattered granules forming a residuum. The sporozoites are elongate with one end pointed, lie lengthwise head to tail in the sporocysts, and have a clear globule at the large end. The sporulation time is 65–70 hr.

Eimeria gorakhpuri Bhatia and Pande, 1967

This species was found in the tubular part of the ceca and to a lesser extent of the general intestinal mucosa in the guinea fowl in India.

OOCYSTS. The oocysts are ellipsoidal or ovoid; 16–24 × 13–17 (mean 20 × 14) μm; with a smooth, "double-contoured," 2-layered wall 1.5 μm thick, the outer layer yellowish, the inner layer light violet; without a micropyle or a residuum; with a group of polar granules. The sporocysts are ovoid, 8–12 × 5–7 (mean 10 × 6) μm, with a faint Stieda body and a residuum. The sporozoites are nearly comma-shaped, 7 × 3 μm, with a large clear globule at the large end and a smaller one at the other end. The sporulation time is 2 days.

LIFE CYCLE. The meronts are 4.5 × 3–4 μm and have 2–6 merozoites 3–3.5 × 1 μm. Immature macrogametes are 11–15 × 8–10 (mean 12 × 9) μm. The mature microgamonts are 7–9 × 5–9 (mean 8 × 7) μm and contain numerous slightly curved microgametes 13 μm long.

Eimeria grenieri Yvoré and Aycardi, 1967

This species has been found in the intestine of the guinea fowl in Africa and Europe. The meronts are in the small intestine and the sexual stages in the large intestine.

OOCYSTS. The oocysts are ellipsoidal; 15–27 × 12–18 (mean 21 × 15) μm; with a smooth, 2-layered wall; with a micropyle and polar granules; without a residuum. The sporocysts are ovoid, with a Stieda body, apparently with a residuum. The sporozoites are 8 × 2 μm. The sporulation time is 18 hr.

LIFE CYCLE. There are three generations of meront. The first generation meronts are in the epithelial cells of the crypts of Lieberkuehn below the host cell nucleus near the muscularis mucosae of the duodenum; they are 14.5 ×

14 μm and have up to 100 merozoites 2.5–3 μm long. The second generation meronts are in the crypts and lower part of the villi of the upper and middle small intestine. They lie above the host cell nucleus, are 24 × 22 μm, and contain 20–30 merozoites 5–6 μm long. The third generation meronts are in the middle to the tips of the villi of the lower and midintestines. They lie below the host cell nucleus, are 6 × 6 μm, and contain 10–12 small merozoites.

The gamonts and gametes are in the cecal epithelium. The prepatent period is 112 hr, the patent period 3 days.

PATHOGENESIS. Somewhat pathogenic.

Eimeria numidae Pellérdy, 1962

This species was found commonly in Hungary in the small and large intestine mucosa of the guinea fowl.

OOCYSTS. The oocysts are ellipsoidal; 15–21 × 12–17 μm; with a thin, smooth wall; with a button-shaped elevation in the position of a micropyle; with a polar granule; without a residuum. The sporocysts are elongate, pointed at one end, without a Stieda body. The sporulation time is 1–2 days.

LIFE CYCLE. The first generation meronts are in the epithelial cells of the duodenum 2–3 DAI. They are 5–8 μm in diameter and produce 2–10 merozoites 4–5 × 1.5 μm and a residuum; both ends of these merozoites are pointed and their nucleus is central and elongate. The second generation meronts occur in the epithelial cells of the villi of the jejunum and ileum and also in the large intestine as far as the rectum. They are ovoid, 12–14 μm in maximum diameter, and contain 6–24 merozoites 8–9 × 2 μm that have one end bluntly rounded with the nucleus near it and 2–3 clear vacuoles. Some of the second generation merozoites form third generation meronts that resemble the first generation ones and contain 5–8 merozoites. Other second generation merozoites and all the third generation ones form sexual stages, which appear 5 DAI. They are above and below the host cell nuclei in the epithelial cells of the ileum, rectum, and ceca. The mature macrogametes are 18 μm in diameter. The prepatent period is 5 days.

PATHOGENESIS. Not markedly pathogenic, although very heavy infections may kill guinea fowl. The clinical signs are apathy and mucous diarrhea. The small intestine mucosa is inflamed and thickened; its contents are more fluid than normal but contain no blood. Most of the inflammation is associated with gamogony.

Eimeria anseris Kotlan, 1932

This species has been found in the domestic goose in Europe and the USSR and the blue goose *Anser caerulescens* and Richardson's Canada goose *Branta canadensis hutchinsi* in Canada. In light to moderate infections it occurs in the small intestine, mainly the posterior part; in heavy infections it occurs in the ceca and rectum, as well.

OOCYSTS. The oocysts are shaped like a sphere surmounted by a truncate cone; 16–24 × 13–19 (mean 21 × 17) μm; with a smooth, colorless, 1-layered wall about 1 μm thick slightly thickened around the micropyle but incised sharply to form a plate or shelf across the micropyle itself; without a polar granule; with a residuum that is a mass of amorphous material just beneath the micropyle and forms a seal beneath it. The sporocysts are ovoid and almost completely fill the oocyst, 8–12 × 7–9 μm, with a wall that is slightly thickened at the small end and a residuum. The sporozoites often lie more or less transversely at the anterior and posterior ends of the sporocysts. The sporulation time is 1–2 days.

LIFE CYCLE. The endogenous stages occur in compact clumps under the intestinal epithelium near the muscularis mucosae and also in the epithelial cells of the villi. The meronts are spherical; 12–20 μm in diameter; and contain 15–25 slightly curved, crescent-shaped merozoites. There is probably only a single asexual generation. The sexual stages are mostly in the subepithelial tissues of the villi but invade the epithelium in heavy infections. The macrogametes are 12–16 × 10–15 μm. The microgamonts are spherical, 12–26 × 8–18 μm. The prepatent period is 6–7 days, the patent period 2–8 days.

PATHOGENESIS. More or less pathogenic, apparently depending on the strain.

Eimeria kotlani Gräfner and Graubmann, 1964

This species has been found in the large intestine of the domestic goose in East Germany and the USSR.

OOCYSTS. The oocysts are ovoid; 29–33 × 23–25 μm; with a smooth, colorless, 2-layered wall 2 μm thick; with a broad micropyle surrounded by a dependent ring ("lips") formed at the small end by the inner wall; without polar granules or residuum. The sporocysts are elongate ovoid, apparently without a Stieda body, with a residuum. The sporozoites are elongate, lie head to tail in the sporocysts, and apparently have a clear globule at the large end. The sporulation time is 4 days.

LIFE CYCLE. The mature meronts are 15–21 (mean 17) μm in diameter. The microgamonts are 17–31 (mean 21) μm in diameter. The macrogametes, 12–21 (mean 17) μm in diameter, are in the epithelium and subepithelium.

PATHOGENESIS. This species sometimes causes diarrhea and dysentery in geese and may kill them; sometimes it is essentially nonpathogenic. The characteristic changes occur in the cloaca, colon, and tubular part of the ceca; they may extend to the middle of the small intestine in severe cases. The affected intestine is edematous and inflamed, and its mucosa is grayish red and hemorrhagic, with a diphtheroid coating. The intestinal contents, which contain numerous oocysts and meronts, are grayish red to red. Histologically the normal structure of the mucosa has disappeared and there is a great deal of infiltration; lymphocytes, pseudoeosinophils, and neutrophils, as well as erythrocytes, are mixed with epithelial and gland cells. The meronts are the most pathogenic stages. They occur in the mucosa from the epithelium to the muscularis mucosae.

Eimeria nocens Kotlan, 1933

This species occurs in the posterior part of the small intestine (and also in the duodenum and even the ceca and large intestine) of the domestic goose in Europe and the USSR and the blue goose in Canada.

OOCYSTS. The oocysts are ovoid but flattened at the micropylar end; 25–33 × 7–24 (mean 31 × 22) μm; with a smooth, 2-layered wall, the outer layer pale yellow and 1.3 μm thick, the inner layer almost colorless and 0.9 μm thick; with a prominent micropyle apparently only in the inner layer and covered by the outer layer; without a polar granule or residuum (but part of the oocyst wall often forms one or more roundish protuberances just below the micropyle). The sporocysts are broadly ellipsoidal, 9–14 × 8–10 (mean 12 × 9) μm, with a thick wall, sometimes with a very small Stieda body and a large residuum. The sporozoites usually lie head to tail in the sporocysts and contain 2 or more large, clear globules that almost obscure their outline. The sporulation time is 2.5 days or more.

LIFE CYCLE. The endogenous stages are primarily in the epithelial cells at the tips of the villi but may also occur beneath the epithelium. The younger developmental stages lie near the host cell nuclei; as they grow they not only displace the nuclei but also destroy the host cell and come to lie free and partly beneath the epithelium. The meronts are spherical, 15–30 μm in diameter, and contain 15–35 merozoites. The macrogametes are usually ellipsoidal or irregularly spherical, uniformly coarsely granular, and 20–25 × 16–21 μm. The microgamonts are spherical or ellipsoidal and 28–36 × 23–31 μm. The prepatent period is 4–9 days.

PATHOGENESIS. *E. nocens* is moderately pathogenic. Intestinal hyperemia, mucus production, and tiny flakes of coagulated blood may be present.

Eimeria stigmosa Klimes, 1963

This species, found in the domestic goose, is quite common in Czechoslovakia. It occurs in the small intestine, usually in the anterior part but also the posterior part and even in the cecum and large intestine.

OOCYSTS. The oocysts are ovoid; 23 × 16–17 μm; with a punctate, radially striated, brown, 1-layered wall; with a micropyle 2–3 μm wide; with 1–2 polar granules; without a residuum. The sporocysts are ovoid, 11 × 8 μm, possibly with a Stieda body and a residuum. The sporozoites contain clear globules. The sporulation time is 2 days.

LIFE CYCLE. The endogenous stages occur in the epithelial cells of the villi. The macrogametes average 20 × 13 μm and the microgamonts 21 × 17 μm. The prepatent period is 4 days, the patent period 2–3 days.

PATHOGENESIS. Nonpathogenic.

Eimeria truncata (Railliet and Lucet, 1891) Wasielewski, 1904

SYNONYM. *Coccidium truncatum.*

This species is relatively common worldwide in the kidney tubules of the domestic goose. It has also been found in the graylag goose *Anser anser,* Ross' goose *A. rossi,* and Canada goose *Branta canadensis.*

OOCYSTS. The oocysts are ovoid; with a narrow, truncate small end; 14–27 × 12–22 μm; with a smooth, delicate wall; with a micropyle and micropylar cap; sometimes with a residuum. The sporocysts have a residuum. The sporulation time is 1–5 days.

LIFE CYCLE. The macrogametes are 12–18 × 11–15 μm. The microgamonts are 15–22 × 13–18 μm. The prepatent period is 5–14 days.

PATHOGENESIS. *E. truncata* is highly pathogenic for goslings, sometimes wiping out whole flocks within a few days. The disease is usually acute, lasting only 2–3 days. Affected birds are extremely weak and emaciated. Their kidneys are greatly enlarged, light-colored, with small, yellowish white nodules, streaks, and lines on the surface and throughout the parenchyma. Infected cells are destroyed, and adjacent, uninfected cells are also destroyed by pressure. The infected tubules are so filled with urates and oocysts that they are enlarged to 5–10 times the diameter of normal tubules.

Eimeria anatis Scholtyseck, 1955

This species has been found in the small intestine of the domestic duck and wild mallard in Germany and the USSR.

OOCYSTS. The oocysts are ovoid, 14–19 × 11–16 (mean 17 × 14) μm, with a smooth wall about 0.7–1.0 thick, with a thickened ring forming shoulders around the micropyle, without a polar granule or residuum. The sporocysts are elongate ovoid or ellipsoidal, with a slight thickening at the small end but not a true Stieda body, with a few residual granules.

Eimeria battakhi Dubey and Pande, 1963

This species is common in India in the feces of the domestic duck.

OOCYSTS. The oocysts are ovoid to subspherical; 19–24 × 16–21 μm; with a smooth, 2-layered wall 1–2 μm thick, the outer layer pale yellow to orange and the inner layer dark green; without a micropyle or a residuum; with a polar granule. The sporocysts are elongate ovoid, 11–13 × 6–8 μm, with a small Stieda body and a residuum. The sporozoites are elongate with one end broad and rounded and the other end narrower and pointed, lie lengthwise head to tail in the sporocysts, and are 10 × 2–3 μm, with 1 or 2 clear globules. The sporulation time is 1 day.

Eimeria danailova Gräfner, Graubmann, and Betke, 1965

This species was found in the small intestine of the domestic duck in East Germany and transmitted experimentally to the domestic goose.

OOCYSTS. The oocysts are ovoid and flattened at the small end; 19–23 × 11–15 μm; with a smooth, yellowish green or greenish, 2-layered wall 0.6–1.0 μm thick, the inner layer thinner and darker than the outer; with a broad micropyle and a polar granule; without a residuum. The sporocysts are ovoid, 10 × 5 μm, apparently with a Stieda body and a residuum. The sporozoites are elongate, lie lengthwise head to tail in the sporocysts, and apparently have a clear globule at the large end. The sporulation time is 4 days.

LIFE CYCLE. The meronts are in the epithelial cells of the mucosa in the crypts of Lieberkuehn, and extend down to the muscularis mucosae; most are above the host cell nuclei. They produce 8 merozoites 6–8 × 2 μm. The gamonts are infrequent in the duodenum but more common in the jejunum and ileum. They are near the muscularis mucosae in the crypts of Lieberkuehn. The microgamonts reach a size of 12 × 10 μm. The prepatent period (in geese) is 7 days.

PATHOGENESIS. Pathogenic for ducks (although usually subclinical) but not for geese. In ducks it can cause fatal diarrhea and dysentery. The mucosa of the duodenum, jejunum, and ileum is edematous and set with many fine petechiae. The intestinal contents are brownish red. The liver is engorged, and encephalitis and lymphocytic leptomeningitis are present. Histologically the mucosa of the duodenum and jejunum is infiltrated with inflammatory cells and its surface is desquamated. Catarrhohemorrhagic enteritis is present. The intestine contains epithelial cells, lymphocytes, neutrophils, histiocytes, numerous erythrocytes, and fecal particles. The blood vessels of the muscularis are hyperemic. The lesions are less pronounced in the ileum and are absent in the ceca and rectum. There may be merozoites in the brain (especially the cerebellum).

Eimeria saitamae Inoue, 1967

This species was found in the small intestine of the domestic duck in Japan.

OOCYSTS. The oocysts are ovoid; colorless; 17–21 × 13–15 (mean 19 × 13) μm; with a smooth, 2-layered wall 0.7–0.8 μm thick; with a micropyle, polar granule, and residuum. The sporulation time is 3 days.

LIFE CYCLE. All stages are in the epithelial cells (and occasionally in the lamina propria). First generation meronts are present 1 DAI. Their merozoites are banana-shaped, 6–10 × 1.5–3 (mean 8 × 2.2) μm. Second generation meronts are present 3 DAI. Their merozoites are 3–6 × 1–2 (mean 4.5 × 1.4) μm. Gamonts are present 3 and 4 DAI. The prepatent period is 3–4 days, the patent period 3 days.

PATHOGENESIS. Presumably highly pathogenic.

Eimeria schachdagica Musaev, Surkova, Elchiev, and Alieva, 1966

This species was found in the feces of the domestic duck in the USSR.

OOCYSTS. The oocysts are ovoid; 16–26 × 12–20 (mean 24 × 18) μm; with a smooth, colorless, 1-layered wall 1.6–2.0 μm thick; without a micropyle or a residuum; with 1 to several polar granules. The sporocysts are ellipsoidal or ovoid, 5–14 × 3–10 (mean 9 × 8) μm, without a Stieda body, with a residuum. The sporozoites were described as piriform but illustrated as elongate. The sporulation time is 3–4 days.

Eimeria columbae Mitra and Das Gupta, 1937

This species was found in the intestine of the pigeon in India.

OOCYSTS. The oocysts are subspherical, with a maximum size of 16 × 14 μm, and said to differ from those of *E. labbeana* in having a residuum.

Eimeria labbeana Pinto, 1928

SYNONYMS. *Coccidium pfeifferi, E. pfeifferi, E. columbarum.*

This species is common worldwide in the small and large intestines of the pigeon and various doves.

OOCYSTS. The oocysts are subspherical to spherical; 13–24 × 12–23 μm; with a smooth, colorless or slightly yellowish brown, 2-layered wall, the inner layer darker than the outer; without a micropyle or a residuum; with a polar granule. The sporocysts are elongate ovoid, with a Stieda body and residuum. The sporozoites are slightly crescent-shaped with one end wider than the other, lie lengthwise head to tail in the sporocysts, and have a clear globule at each end. The sporulation time is 4 days or less.

LIFE CYCLE. After the sporulated oocysts are ingested, the sporozoites are released and invade the epithelial cells of the intestine. Merogony is essentially ectopolygenous. First generation meronts containing 5–15 (mean 10) merozoites 4–6 × 2 (mean 4 × 2) μm are present 20–48 hr after inoculation in the epithelial cells of the anterior part of the ileum. Mature second generation meronts containing 8–19 (mean 14) merozoites 4–5 × 1.5–2 (mean 4 × 2) μm are present 96 hr after inoculation, and mature third generation meronts containing 6–16 (mean 7.5) merozoites 5–8 × 2 (mean 5 × 2) μm are present 144 hr after inoculation. The macrogametes are in the epithelial cells of the ileum and average 11 × 9 μm. The microgamonts form a large number of biflagellate microgametes about 3 μm long with flagella 10 μm long. The prepatent period is about 5 days.

PATHOGENESIS. *E. labbeana* is slightly to markedly pathogenic, depending on the strain of parasite and age of the birds. Adults are fairly resistant, although fatal infections have been seen. The birds become weak and emaciated, eat little but drink a great deal, and have a greenish diarrhea. The heaviest losses occur among squabs in the nest. A high percentage of the squabs may die, and those that recover are often somewhat stunted.

Eimeria tropicalis Malhotra and Ray, 1961

This species was found in the feces of the pigeon in India.

OOCYSTS. The oocysts are spherical or subspherical; 19–24 × 18–23 (mostly 20 in diameter) μm; with a smooth, transparent, slightly orange-pink wall with a thick "endocystic membrane," usually with a residuum. The sporocysts are ellipsoidal, 10 × 6 μm, with a Stieda body and residuum. The sporozoites are globular. The sporulation time is 40–48 hr.

LIFE CYCLE. The prepatent period is 7 days.

PATHOGENESIS. Presumably markedly pathogenic.

Eimeria dunsingi Farr, 1960

This species occurs presumably worldwide in the small intestine (primarily anterior part) of the parakeet.

OOCYSTS. The oocysts are ovoid; 25–39 × 22–28 (mean 33 × 24) μm; with a smooth or slightly rough, 2-layered wall, the outer layer colorless or light brown, the inner layer colorless or light blue; without a micropyle or a residuum; usually without a polar granule. The sporocysts are ovoid, 12–18 × 8–10 (mean 15 × 9) μm, with a prominent Stieda body and a residuum. The sporozoites are long and narrow, lie lengthwise head to tail in the sporocysts, and have a clear globule.

LIFE CYCLE. The endogenous stages are usually below the villi in areas of disorganized tissues adjacent to the circular muscle layer. The macrogametes are 9–22 × 8–15 μm. The microgamonts, 9–15 × 9–14 μm, produce many microgametes.

PATHOGENESIS. Markedly pathogenic.

Genus *Isospora* Schneider, 1881

In this genus the oocyst contains 2 sporocysts, each of which contains 4 sporozoites.

Isospora aksaica Bazanova, 1952

This species was found in the feces of the ox in the USSR. It may be a pseudoparasite of the ox.

OOCYSTS. The oocysts are spherical; 26 μm in diameter; with a smooth, 2-layered wall 1.6 μm thick; presumably without a micropyle or residuum; possibly with polar granules. The sporocysts are ellipsoidal or spherical, 22 × 15 μm, presumably without a residuum. The sporozoites, 15 × 11 μm, are spherical, bean-shaped, or ellipsoidal.

Isospora sp. Levine and Mohan, 1960

This species is rare, apparently worldwide, in the feces of the ox, ox-zebu hybrids, and presumably zebu and water buffalo. It is probably a pseudoparasite of the ox and resembles an *Isospora* of the house sparrow.

OOCYSTS. The oocysts are usually subspherical but occasionally spherical; 21–33 × 20–32 μm; with a smooth, colorless, 1-layered wall about 1 μm thick, sometimes apparently lined by a thin membrane; without a micropyle or residuum; with several irregular polar granules. The sporocysts are lemon-shaped; 14–20 × 10–12 μm; quite thick-walled; with a Stieda body, a substiedal body, an endostiedal body, and a residuum. The sporozoites are sausage-shaped, not arranged in any particular order in the sporocysts.

Isospora sp. Shah, 1963

This species is rare, presumably worldwide, in the feces of the domestic sheep and mouflon. It is probably a pseudoparasite and resembles an *Isospora* of the house sparrow.

OOCYSTS. The oocysts are usually subspherical but occasionally spherical; 20–25 × 20–24 μm; with a smooth, yellowish, 2-layered wall; without a micropyle or residuum; ordinarily with a polar granule. The sporocysts are lemon-shaped; 14–15 × 9–10 μm; quite thick-walled; with a Stieda body, a substiedal body, and a residuum. The sporozoites are sausage-shaped, not arranged in any particular order in the sporocysts. (See Fig. 7.12.)

Isospora orlovi Tsygankov, 1950

This species was found in the feces of the camel (species not stated) in the USSR. It may well be a pseudoparasite of the camel.

OOCYSTS. The oocysts are ellipsoidal, ovoid, piriform, cylindrical, or figure 8-shaped; 27–35 × 15–20 μm; with a smooth, 2-layered wall about 1 μm thick; without a micropyle, polar granule, or residuum. The sporocysts are ellipsoidal, ovoid, or spherical (the spherical ones 13–15 μm in diameter, the others 15–20 × 13–17 μm); without a Stieda body; with a residuum. The sporozoites are elongate ellipsoidal, 7–10 × 4–6 μm.

Isospora almataensis Paichuk, 1953

This species was found in the feces of the domestic pig in the USSR. It may well be a pseudoparasite.

OOCYSTS. The oocysts are ovoid or spherical (the spherical oocysts 26–34 μm in diameter, the ovoid ones 25–32 × 23–29 μm); with a smooth, dark brown, 3-layered wall up to 3 μm thick; without a residuum. The sporocysts are 12–19 × 9–12 μm, with a residuum. The sporozoites are short ovoid, 6 × 4 μm. The sporulation time is 5 days at room temperature.

Isospora neyrai Romero and Lizcano, 1971

This species was found in the feces of the domestic pig in Spain. It may be a pseudoparasite.

OOCYSTS. The oocysts are ovoid or ellipsoidal, 9–17 × 6–13 μm, with a 2-layered wall, without a micropyle, a polar granule said to be not present and a residuum said to be present (but the photomicrographs show what may be a polar granule and no residuum). The sporocysts are ovoid or subspherical, 4–8 × 2–6 μm, without a Stieda body, presumably with a residuum. The sporozoites are elongate ovoid, with a clear globule.

Isospora suis Biester, 1934

This species is quite common worldwide in the small intestine and sometimes colon of the domestic pig and wild boar.

OOCYSTS. The oocysts are spherical to subspherical; 17–25 × 16–21 μm; with a smooth, colorless, 1-layered wall 0.5–0.7 μm thick; without a micropyle, polar granule, or residuum. The sporocysts are ellipsoidal, 13–14 × 8–11 μm, often lying against each other with one side somewhat flattened, with a wall 0.2–0.4 μm thick, without a Stieda body, with a residuum. The sporozoites are sausage-shaped, with one end somewhat pointed, 9–11 × 3–4 μm. The sporulation time is 3–5 days. (See Fig. 7.13.)

LIFE CYCLE. The asexual stages are in the villar epithelial cells of the small intestine and sometimes of the colon, usually in the distal one-third of the villi and usually below the host cell nucleus. First generation meronts, present 2–3 DAI, are 10–19 × 5–10 μm in smears and contain 2–16 merozoites 11–18 × 4–9 μm. Second generation meronts, present 4 DAI, are 9–20 × 2.5–12 μm in smears and produce 4–16 crescent-shaped merozoites 7–12 × 2.5–5 μm.

Mature macrogametes, microgamonts, and oocysts are present 5 DAI. The microgametes have 2 flagella. The prepatent period is 5 days, the patent period 3–13 days.

PATHOGENESIS. This species causes diarrhea for some days in baby pigs and has recently been found to be a significant pathogen in young pigs.

Isospora sp. Shrivastava and Shah, 1968

This species was found in the feces of the domestic pig in India. It is probably a pseudoparasite of sparrow origin.

OOCYSTS. The oocysts are subspherical to spherical; 25–26 × 25 μm; with a 2-layered wall 1.2 μm thick; without a micropyle, polar granule, or residuum.

The sporocysts are lemon-shaped; 14 × 9 μm; with a Stieda body, substiedal body, and residuum. The sporozoites are more or less sausage-shaped, with a large clear globule at the broad end.

Isospora burrowsi Trayser and Todd, 1978

This species is apparently uncommon and apparently worldwide in the posterior small intestine and cecum of the dog. Transport hosts are the laboratory mouse and rat, in which hypnozoites are in the lymph nodes, liver, spleen, and occasionally skeletal muscles.

OOCYSTS. The oocysts are spherical to ellipsoidal; 17–24 × 15–22 μm; with a smooth, yellow-green, 1-layered wall about 1 μm thick; without a micropyle, polar granule, or residuum. The sporocysts are ovoid to ellipsoidal, 12–16 × 8–11 μm, without a Stieda body, with a residuum. The sporozoites are elongate, about 6–8 × 4–5 μm, not lying in any particular order in the sporocysts.

LIFE CYCLE. The endogenous stages are above the host cell nuclei in the epithelial cells of the tips of the villi and in the lamina propria cells. First generation meronts are mature on day 4, at which time they are 11–18 × 9–18 μm in tissue sections and are tightly packed with merozoites 8–14 × about 3–3.5 μm. Immature macrogametes appear on day 6. When mature they are ovoid to spherical, 11–25 × 8–18 μm. Immature microgamonts also appear on day 6. When mature they are 13–27 × 10–21 μm and contain a large number of microgametes 4–5 × 0.4 μm. The prepatent period is 6–9 days, the patent period 4–12 days.

Isospora canis Nemeséri, 1959

SYNONYMS. *Diplospora bigemina* in part, *I. felis* from the domestic dog, *I. bigemina* in part, *Levinea canis, Cystoisospora canis.*

This species is quite common worldwide in the small intestine (mostly the posterior third) and also to some extent in the large intestine of the dog and coyote. Transport hosts are the house mouse, Norway rat, hamster, newborn kitten, and dog, in which hypnozoites are in the lymph nodes, liver, spleen, mesentery, and occasionally skeletal muscles.

OOCYSTS. The oocysts are ellipsoidal to slightly ovoid; 34–42 × 23–36 μm; with a smooth, very pale tan to light green, 1-layered wall 1.3–1.5 μm thick sometimes appearing to be lined by a very thin membrane; without a micropyle, polar granule, or residuum, but with a tiny blob adherent to the inside of the oocyst wall at the broad end. The sporocysts are ellipsoidal; 18–28 × 15–19 μm; with a smooth, colorless, 2-layered wall 0.4 μm thick; without a Stieda body; with a prominent residuum. The sporozoites are sausage-shaped, usually oriented more or less lengthwise in the sporocysts, with a subcentral clear globule. The sporulation time is 2 days at 20 C, 16 hr at 30 or 35 C, and not complete after 16 days at 10 C. (See Fig. 7.18.)

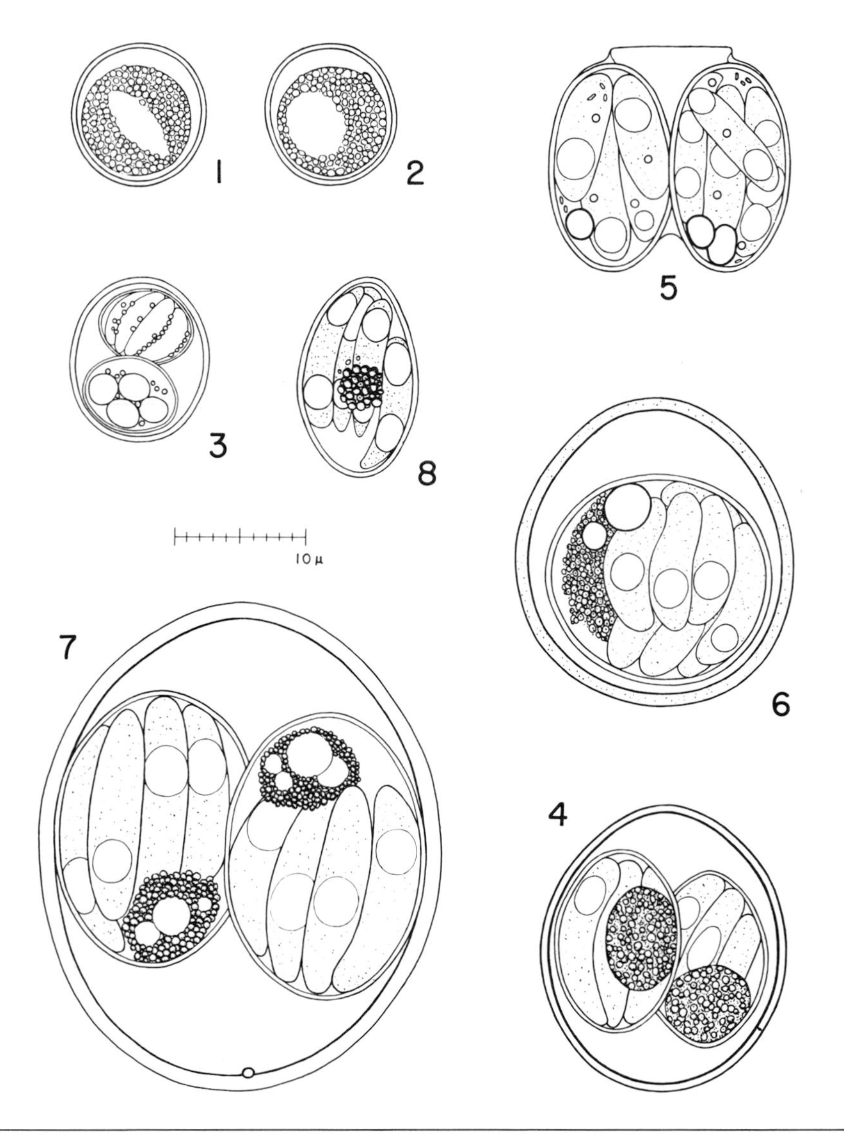

Fig. 7.18. Oocysts of coccidia of the dog: 1 and 2. unsporulated oocysts of *Toxoplasma bahiensis* in the process of development; 3. sporulated oocyst of *T. bahiensis;* 4 and 5. sporulated oocysts of *I. neorivolta* or *I. ohioensis;* 6. *Caryospora*-like abnormal sporulated oocyst of *I. neorivolta* or *I. ohioensis;* 7. sporulated occyst of *I. canis;* 8. sporulated sporocyst of *Sarcocystis* sp. (*S. cruzi*, *S. miescheriana*, *S. tenella*, or *S. capracanis*). (From Levine and Ivens 1965b)

LIFE CYCLE. The meronts are mostly just beneath the epithelium, but a few are deeper in the lamina propria; none occur in the epithelial cells. The macrogametes and microgamonts are in the epithelial cells, subepithelial connective tissue of the villi of the small intestine, and the mucosa of the large intestine. There are three meront generations. The first generation meronts mature in 5–7 days, are ellipsoidal to spherical, 16–38 × 11–23 μm, and contain 4–24 merozoites 8–11 × 3–5 μm. The second generation meronts appear on day 6 and become mature in about a day. They are spherical to slightly ovoid, 12–13 × 8–13 μm, and contain 3–12 merozoites 11–12 × 3–5 μm. The third generation meronts become mature on days 7–8. They are 13–38 × 8–24 μm and contain 6–72 merozoites 8–13 × 1.5–3 μm.

The earliest macrogametes and microgamonts can be seen on day 7. The mature macrogametes are ovoid, 22–29 × 14–23 μm. The mature microgamonts are 20–38 × 14–26 μm and produce numerous microgametes about 5 × 0.8 μm and a central residuum. The prepatent period is 9–11 days, the patent period about 4 wk.

PATHOGENESIS. There appear to be strain differences in pathogenicity, but on the whole this species is only mildly pathogenic.

Isospora neorivolta Dubey and Mahrt, 1979

SYNONYM. *I. rivolta* in part.

This species occurs in the posterior half of the small intestine and rarely the cecum and colon of the dog. It has so far been found only in Illinois.

OOCYSTS. Similar to those of *I. ohioensis* (see below). The sporulation time is 1–2 days.

LIFE CYCLE. The endogenous stages are in the distal one-third of the villi, mostly in the subepithelial cells of the lamina propria but occasionally in the epithelial cells. There are at least four types of meront. Type I meronts occur 3 DAI, are 6–10 × 2–3 μm, and contain 1–8 merozoites. Type II meronts are found 4–5 DAI, are 9–18 × 7–18 μm, and contain 2–12 short, broad merozoites 8–12 × 3–5 μm. Type III meronts occur mostly 4 DAI, are 7–25 × 4–16 μm, and contain 2–20 merozoites 8–14 × 2–3 μm. Type IV meronts occur mostly 5–6 DAI, are 9–18 × 7–18 μm, and contain 4–30 short, thin merozoites 7–8 × 1–1.5 μm.

Mature macrogametes and microgamonts are present 6 DAI. The mature microgamonts are 13 × 9 μm and contain 50–70 biflagellate microgametes 6 × 0.6 μm. The prepatent period is 6 days, the patent period 13–23 days.

PATHOGENESIS. Apparently only slightly pathogenic.

Isospora ohioensis Dubey, 1975

SYNONYMS. *Diplospora bigemina* in part, *I. rivolta* from the dog in part, *Lucetina rivolta* from the dog in part, *Cystoisospora ohioensis, Levinea ohioensis.*

This species is quite common worldwide in the epithelial cells throughout the small intestine and also in the cecum and colon of the dog, coyote, and possibly dingo, red fox, and raccoon dog. Experimental transport hosts are the house mouse, Norway rat, golden hamster, kitten, and dog, in which hypnozoites are in the lymph nodes, liver, spleen, mesentery, and occasionally skeletal muscles.

OOCYSTS. The oocysts are ellipsoidal to ovoid; 20–27 × 15–24 μm; with a smooth, colorless to pale yellow, 1-layered wall about 0.8 μm thick lined by a thin membrane; without a micropyle, polar granule, or residuum. The sporocysts are ellipsoidal, 12–19 × 9–13 μm, with a wall about 0.4 μm thick, without a Stieda body, with a residuum. The sporozoites are about 9–13 × 2.5–5 μm, with 1 or more clear globules. The sporulation time is less than 1 wk.

LIFE CYCLE. All stages are in the epithelial cells of the small intestine; the gamonts and gametes are also in the cecum and colon to some extent. The first generation meronts are in the jejunum. When mature at 3 days the meronts are 9–19 × 2.5–4 μm (in sections) and contain 2–8 merozoites 9–12 × 2.5–4 μm. There are one or two additional generations.

Gamonts appear in the surface epithelial cells of the small intestine, cecum, and colon 4–5 DAI. The macrogametes are 13–17 × 11–12 μm in sections and 21–26 × 17–25 μm in smears. The microgamonts are 13–17 × 8–15 μm in sections and 24–30 × 15–24 μm in smears; they contain up to 50 microgametes.

The prepatent period is about 4.5 days after ingestion of oocysts or 3.5 days after ingestion of mouse transport hosts. The patent period is 3–5 wk in pups inoculated at 6–10 days of age, or 1–2 wk in pups inoculated at 4–384 days of age.

PATHOGENESIS. Usually nonpathogenic.

Histologic lesions are necrosis and desquamation of the tips of the villi of the ileum and atrophy of the villi.

Isospora sp. Dubey, Weisbrode, and Rogers, 1978

This species was found in the distal half of the ileum, cecum, and colon of the dog in Ohio.

OOCYSTS. The oocysts are similar to those of *I. ohioensis,* 16–23 × 14–20 μm in smears. Sporulation occurs outside the host's body.

LIFE CYCLE. There are two or more generations of meront and at least three sizes of uninucleate free merozoites in smears. The meronts are generally spindle-shaped and contain up to 7 nuclei. Binucleate ones are 15–16 × 2–3 μm and 4–6-nucleate ones are 18–19 × 4–6 μm. Type A merozoites are 8–10 × 1.5–2 μm, Type B merozoites 12–15 × 1.5–2 μm, and Type C merozoites 12–16 × 4–6 μm.

PATHOGENESIS. Apparently markedly pathogenic for pups.

Isospora felis Wenyon, 1923

SYNONYMS. *Diplospora bigemina* in part, *I. bigemina* in part, *I. rivolta* in part, *I. cati, Lucetina felis, Cystoisospora felis, Levinea felis.*

This species is common worldwide in the small intestine, sometimes cecum, and occasionally colon of the domestic and wild cat, ocelot, serval, tiger, lion, jaguar, and lynx. Experimental transport hosts are the house mouse, laboratory rat, golden hamster, calf, ox, cat, and dog, in which hypnozoites are in the lymph nodes, liver, spleen, mesentery, and occasionally skeletal muscles.

OOCYSTS. The oocysts are ovoid; 32–53 × 26–43 μm; with a smooth, yellowish to pale brown, 1-layered wall about 1.3 μm thick apparently lined by a thin membrane; without a micropyle or a residuum; generally without a polar granule. The sporocysts are ellipsoidal; 20–27 × 17–22 μm; with a smooth, colorless wall 0.4 μm thick; without a Stieda body; with a residuum. The sporozoites are sausage-shaped with one end slightly narrowed, 10–15 μm long, generally lie lengthwise in the sporocysts, with a subcentral clear globule. The sporulation time is 2 days or less.

LIFE CYCLE. All the endogenous stages occur above the host cell nucleus in the epithelial cells of the distal parts of the villi of the ileum and occasionally of the duodenum and jejunum. There are three asexual generations. The mature first generation meronts are 11–30 × 10–23 μm and contain 16–17 merozoites 11–15 × 3–5 μm. They are mature in 4 or sometimes 5 days. The second generation meronts form 2–10 spindle-shaped second generation merozoites. At 6 days multinucleate third generation meronts are present; they are 12–16 × 4–5 μm. Each forms up to 6 third generation merozoites 6–8 × 1–2 μm within the "cyst" of the second generation meronts; they are not all formed at the same time, so that "cysts" may be found containing both fully formed third and second generation merozoites. Mature meronts containing only third generation merozoites are found 6–9 DAI and are most abundant 7 DAI. They contain 36–70 or more merozoites. The third generation meronts and merozoites develop within the same host cell and parasitophorous vacuole as the second generation meronts and merozoites.

Macrogametes and microgamonts are first seen 6 DAI and reach their maximum number at 8–9 days. Mature macrogametes are 16–22 × 8–13 μm. Mature microgamonts are 24–72 × 18–32 μm and contain a large number of biflagellate microgametes 5–7 × 0.8 μm and a central residuum. The prepatent period is 7–10 days, the patent period 10–20 days.

Hypnozoites in the lymph nodes of transport hosts have a crystalloid body like that of *I. canis.* (See Fig. 7.19.)

PATHOGENESIS. Slightly if at all pathogenic under normal circumstances.

Isospora rivolta (Grassi, 1879) Wenyon, 1923

SYNONYMS. *Coccidium rivolta, Diplospora bigemina* in part; *Lucetina rivolta, I. rivoltae, I. rivoltai, I. novocati, Cystoisospora rivolta, Levinea rivolta.*

This species is common worldwide in the small intestine, cecum, and colon of the domestic and wild cat, jungle cat, tiger, and leopard. Experimen-

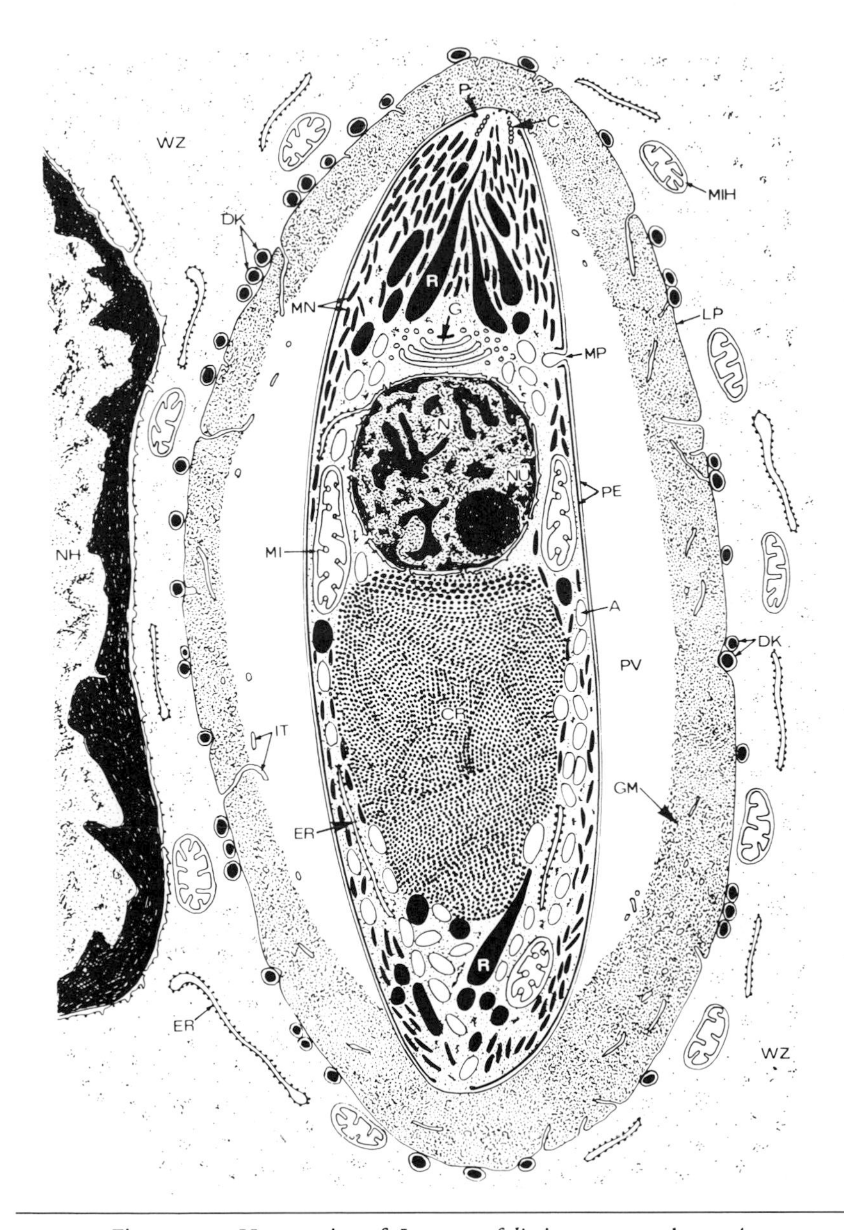

Fig. 7.19. Hypnozoite of *Isospora felis* in mouse spleen: *A* = polysaccharide granule, *C* = conoid, *CR* = crystalloid body, *DK* = dark granule, *ER* = endoplasmic reticulum, *G* = Golgi apparatus, *GM* = electron-dense granules in parasitophorous vacuole, *IT* = intravacuolar tubule, *LP* = limiting membrane of parasitophorous vacuole, *MI* = mitochondrion, *MIH* = mitochondrion of host, *MN* = microneme, *MP* = micropore, *N* = nucleus, *NH* ⇒ host cell nucleus, *NU* = nucleolus, *P* = polar ring *PE* = pellicle, PV = parasitophorous vacuole, *R* = rhoptry, *WZ* = host cell. (From Mehlhorn and Markus 1976)

tal transport hosts are the house mouse, laboratory rat, golden hamster, calf, domestic cat, and dog, in which hypnozoites are in the lymph nodes, liver, spleen, mesentery, and occasionally skeletal muscles.

OOCYSTS. The oocysts are ellipsoidal to somewhat ovoid; 21–29 × 18–26 μm; with a smooth, colorless to very pale brown, 1-layered wall about 0.5 μm thick apparently lined by a membrane; without a micropyle, polar granule, or residuum. The sporocysts are broadly ellipsoidal; 14–16 × 10–13 μm; with a smooth, colorless wall about 0.3 μm thick; without a Stieda body; with a residuum. The sporozoites are elongated, usually oriented more or less lengthwise in the sporocysts, with a subcentral clear globule. The sporulation time is 1–2 days.

LIFE CYCLE. Asexual stages are in the epithelial cells of the villi and glands of Lieberkuehn of the small intestine, cecum, and colon. There are at least three structurally different meronts. Type I meronts are seen 12–48 hr after feeding mice containing hypnozoites to cats. They are 6–13 × 3–6 μm in sections and contain 2–8 merozoites. Type II meronts occur 2–7 DAI. They are in parasitophorous vacuoles 9–18 × 9–13 μm in sections that contain 1–5 merozoite-shaped meronts 7–13 × 3–5 μm in sections or 10–16 × 5–6 μm in smears. Type III meronts occur 3–8 DAI. They are 7–24 × 4–21 μm in sections and contain 2–30 slender merozoites 5.5 × 1 μm. The endogenous stages in cats fed sporulated oocysts are essentially similar, but merogony occurs 12–48 hr later than in the mouse-induced cycle.

Gamonts and gametes of the mouse-induced cycle occur 3–4 DAI. The macrogametes are 11–18 × 5–13 μm in sections and 18 × 16 μm in smears. The microgamonts are 9–15 × 6–9 μm in sections and up to 21.5 × 14 μm in smears; they contain up to 70 microgametes. Gamogony occurs in cats fed sporulated oocysts ½–2 days later than in the mouse-induced cycle.

The prepatent period is 4–7 days; oocysts may be passed intermittently for at least 66 days.

PATHOGENESIS. Slightly if at all pathogenic for cats.

Isospora belli Wenyon, 1923

This species is fairly common worldwide in the small intestine of man.

OOCYSTS. The oocysts are ellipsoidal; 20–33 × 10–19 μm; with a smooth, thin, colorless wall; sometimes with a very small micropyle visible; with a polar granule at first that quickly disappears; with a residuum. The sporocysts are subspherical to ellipsoidal, 9–14 × 7–12 μm, without a Stieda body, with a residuum. The sporozoites are slender and somewhat crescent-shaped. The sporulation time is up to 5 days.

PATHOGENESIS. Most infections appear to be subclinical and self-limiting. Mild symptoms may occur in some cases.

REMARKS. For further information, see any textbook of human parasitology.

Isospora natalensis Elsdon-Dew, 1953

This species was found in the feces of man in Africa (Natal).

OOCYSTS. The oocysts are subspherical; 25–30 × 21–24 μm; with a smooth, thin wall; without a micropyle, polar granule, or residuum. The sporocysts are ellipsoidal, 17 × 12 μm, without a Stieda body, with a residuum. The sporulation time is 1 day.

Isospora gallinae Scholtyseck, 1954

This species was found in the feces of chickens in Germany. A similar form has been found in India and the USSR. It may be a pseudoparasite of the chicken.

OOCYSTS. The oocysts are ellipsoidal, 19–27 × 15–23 μm, without a micropyle or a residuum, with polar granules. The sporocysts are piriform.

Isospora heissini Svanbaev, 1955

This species was found in the feces of the turkey in the USSR. It may be a pseudoparasite of the turkey.

OOCYSTS. The oocysts are spherical or rarely broadly ovoid; 25–33 μm in diameter; with a greenish, smooth, double-contoured wall 1.5–1.7 μm thick; without a micropyle or a residuum; with a polar granule. The sporocysts are spherical or ovoid, 15 × 10 μm, without a residuum. The sporozoites are ovoid, 7–9 × 4–5 μm. The sporulation time is 16–20 hr.

Isospora mayuri Patnaik, 1966

SYNONYM. *I. pellerdyi.*

This species has been found in the feces of the peafowl in India, England, and Germany. It is probably a pseudoparasite of the peafowl.

OOCYSTS. The oocysts are subspherical or spherical; 20–27 × 18–24 μm; with a smooth, yellowish brown, 2-layered wall 1.5 μm thick; without a micropyle; with a residuum. The sporocysts are piriform, 14–16 × 10–11 μm, with a Stieda body and a residuum. The sporozoites are 6–7 × 1.5–2 μm.

Isospora sp. Anpilogova, 1965

This form was found in the feces of the duck and goose in the USSR. It is probably a pseudoparasite of these birds.

OOCYSTS. The oocysts are spherical or ellipsoidal, 25–31 × 23–31 μm. The sporocysts are ovoid, 17 × 11 μm, with a Stieda body.

Genus *Wenyonella* Hoare, 1933

In this genus the oocyst contains 4 sporocysts, each of which contains 4 sporozoites.

Wenyonella gallinae Ray, 1945

This species occurs uncommonly in India in the terminal part of the intestine of the chicken.

OOCYSTS. The oocysts are ovoid, rough, punctate, 29–34 × 20–23 μm. The sporocysts are flask-shaped, 19 × 8 μm. The sporulation time is 4–6 days.

PATHOGENESIS. This species may cause diarrhea, with blackish green, semisolid excreta. The terminal part of the intestine is thickened and congested, and there are pinpoint hemorrhages in the mucosa.

Wenyonella columbae Haldar and Ray-Choudhury, 1974

This species occurs in the small intestine of the pigeon in India.

OOCYSTS. The oocysts are spherical or slightly ovoid; 21–27 × 21–26 μm; with a 2-layered wall, the outer layer slightly thinner than the inner one; without a micropyle, polar granule, or residuum. The sporocysts are ovoid, 10–13 × 6–9 μm, with a Stieda body and a residuum. The sporozoites are elongate, 4.5 μm long, with a clear globule at the broad end.

Wenyonella anatis Pande, Bhatia, and Srivastava, 1965

This species was found in the feces of the duck in India.

OOCYSTS. The oocysts are ovoid with the micropylar end somewhat neck-shaped; 11–17 × 7–10 μm; with a colorless, punctate, 2-layered wall 0.7–1 μm thick; with a depressed micropyle 4.5–6 μm wide, a polar granule, and a residuum. The sporocysts are 6–7 × 4–5 μm, apparently without a Stieda body, with a residuum. The sporozoites are ovoid, 3 × 2 μm, with a prominent globule at one end. The sporulation time is 2 days.

Genus *Tyzzeria* Allen, 1936

In this genus the oocysts contain 8 naked sporozoites and no sporocysts.

Tyzzeria perniciosa Allen, 1936

This species is uncommon, presumably worldwide, in the small intestine, especially the upper half, of the domestic duck, pintail, and possibly the lesser scaup duck and diving duck *Aythya erythropus*.

OOCYSTS. The oocysts are ellipsoidal; 10–13 × 9–11 μm; with a colorless, 2-layered wall, the outer layer thin and transparent and the inner layer thicker; without a micropyle; with a residuum. The sporozoites are curved, about 10 × 3.5 μm, wider at one end than the other. The sporulation time is 1 day.

LIFE CYCLE. The endogenous stages are in the mucosal and submucosal cells. There are at least three asexual generations. The first generation meronts are about 12 × 8 μm and contain relatively few, small merozoites. The later meronts are about 15–16 × 14–15 μm and contain more and larger merozoites than the first generation ones. Merogony continues long after the formation of gametes

Microgamonts appear 2 DAI. They are about 7.5 × 6 μm and produce a large number of tiny microgametes. The macrogametes are somewhat irregular in shape. The prepatent period is 5 days.

PATHOGENESIS. *T. perniciosa* is highly pathogenic for ducklings. They stop eating, lose weight, become weak, and cry continually. At necropsy, inflammation and hemorrhagic areas are seen throughout the small intestine and especially in its upper half. The intestinal wall is thickened; round, white spots are visible through its serosal surface. In severe cases the lumen is filled with blood and often contains a cheesy exudate. The intestinal epithelium sloughs off in long pieces, sometimes forming a tube that can easily be lifted out.

Tyzzeria parvula (Kotlan, 1933) Klimes, 1963

SYNONYMS. *Eimeria anseris* in part, *E. parvula, Tyzzeria anseris.*

This species occurs apparently worldwide in the small intestine of the domestic goose, white-fronted goose, snow goose, Ross' goose, Atlantic brant, Canada goose, and possibly whistling swan. It is rare in domestic geese in the United States but common in Europe.

OOCYSTS. The oocysts are ellipsoidal, subspherical, or spherical; 10–16 × 9–14 μm; with a smooth, colorless, 1- or 2-layered wall about 0.6 μm thick; without a micropyle; with a residuum. The sporozoites are banana-shaped.

LIFE CYCLE. The prepatent period is 4 days.

PATHOGENESIS. Apparently not or only slightly pathogenic.

FAMILY CRYPTOSPORIDIIDAE

Members of this family are homoxenous, but they differ from the Eimeriidae in that they develop just under the surface membrane of the host cell or within its brush border rather than in the cell proper. In addition, their meronts have a knoblike attachment organelle, the oocysts may or may not contain sporocysts, and the microgametes lack flagella. Levine (1984) reviewed the genus.

Genus *Cryptosporidium* Tyzzer, 1907

With the characters of the family.

Cryptosporidium muris Tyzzer, 1907

SYNONYMS. *C. agni, C. bovis, C. cuniculus, C. felis, C. garnhami, C. parvum, C. rhesi, C. wrairi.*

This species occurs in the small and also large intestines and occasionally stomach of the house mouse (type host) and has also been found in man, the rhesus monkey, ox, sheep, goat, red deer, mule deer, pig, horse, dog, cat, raccoon, domestic rabbit, field mouse, red-backed mouse, guinea pig, and gray squirrel and has been transmitted experimentally to the laboratory rat, golden hamster, and probably chicken. It is found worldwide and is apparently not rare.

OOCYSTS. The oocysts are spherical, 1.5–5 μm in diameter, and are passed in the feces. Sporulation apparently occurs outside the host. There are 4 elongate sporozoites, which some workers say are enclosed in a sporocyst and others believe are naked in the oocyst. The sporozoites contain rhoptries and micronemes, among other structures.

LIFE CYCLE. The meronts are up to 5 μm in diameter and occur in the brush border of the intestinal epithelial cells. There are apparently two generations. First generation meronts are present 2 DAI. They contain 8 merozoites with an apical complex, a mitochondrion, and 28 subpellicular microtubules. Second generation meronts contain 4 or 8 merozoites (depending on the authority). Macrogametes and microgamonts up to 5 μm in diameter are present 3 DAI. The latter produce 16 nonflagellate microgametes 1–2 μm long. After fertilization a wall is laid down around the zygote, and the resultant oocyst enters the lumen of the intestine.

PATHOGENESIS. This species is known to cause a mild to moderate diarrhea in young calves, sheep, pigs, and rats, and in both immunodeficient and nonimmunodeficient human beings. Affected animals have edema and mild reticuloendothelial cell hyperplasia of the mesenteric lymph nodes. The intestinal microvilli at the site of attachment are absent or disintegrated, and the terminal web is disorganized.

REMARKS. A number of human cases of cryptosporidial diarrhea have been reported in persons who have had contact with infected calves. There is no doubt that bovine cryptosporidiosis is a zoonosis, and the infection in other animals probably is also. Recognition of this fact is so recent that much remains to be learned about the organism and its effects. Among the many recent papers on it are those of Pohlenz et al. (1978a, 1978b), Schmitz and Smith (1975), Barker and Carbonell (1974), Snodgrass et al. (1980), Tzipori et al. (1980), Reese et al. (1982), and Pivont and Antoine (1982).

Cryptosporidium meleagridis Slavin, 1955

SYNONYMS. *C. anserinum, C. tyzzeri.*

This species has been found in the turkey (type host), chicken, peafowl, quail *Coturnix coturnix,* goose, and red-lored parrot *Amazona autumnalis.* It occurs in the brush border of the gastrointestinal and/or respiratory tracts and is presumably worldwide.

OOCYSTS. The oocysts, which resemble those of *C. muris,* are more or less spherical, up to 5 μm in diameter, and contain 4 elongate sporozoites. Sporulation takes place outside the host's body.

LIFE CYCLE. The meronts, gamonts, and gametes resemble those of *C. muris.* The meronts are about 4 × 5 μm and produce 8 slender, falciform merozoites 5 × 1 μm with one end blunter than the other. The number of meront generations is unknown. The macrogametes are roughly ovoid, 4–5 × 3–4 μm. The microgamonts are spherical or ovoid, generally about 5 μm in their greatest diameter, and produce 16 rodlike, nonflagellate microgametes 1 × 0.3 μm.

PATHOGENESIS. This species can be either nonpathogenic or mildly or moderately pathogenic. It has been associated not only with diarrhea but also with respiratory disease in chickens and young peafowl.

REMARKS. The relationship of *C. meleagridis* to *C. muris* remains to be determined. Among recent reports of it are those of Proctor and Kemp (1974), Hoerr et al. (1978), and Dhillon et al. (1981).

COCCIDIOSIS

Coccidiosis in Cattle

EPIDEMIOLOGY. Infections with a single species of coccidium are rare in nature; mixed infections are the rule. *Eimeria zuernii* and *E. bovis* are the most pathogenic species, but *E. auburnensis* and the other species may contribute to the total disease picture; some of them may even cause marked signs by themselves if present in large enough numbers.

Bovine coccidiosis is primarily a disease of young animals. It ordinarily occurs in calves 3 wk to 6 mo old. Older calves and even adult animals may be affected under conditions of gross contamination, but they are usually symptomless carriers.

Calves become infected by ingesting oocysts along with their feed or water. The severity of the disease depends on the number of oocysts they receive. If they get only a few, there are no signs; repeated infections produce immunity without disease. If they get more, the disease may be mild and immunity may also develop. If they get a large number, severe disease and even death may result.

Crowding and lack of sanitation greatly increase the disease hazard. Successive passage of coccidia from one animal to another often builds infection up to a pathogenic level, since in each passage the recipient receives more oocysts than in the previous one. This is the reason for the common observation that calves placed in a lot where others are already present may suffer more from coccidiosis than those that were there from the first. This successive passage from a carrier to a symptomless multiplier to a subclinical case to a fatal case was described by Boughton (1945) as typical of the transmission of bovine coccidiosis. In addition, it is likely that recycling by repeated infections of a single individual may also play an important part. (It must be said that the role of light infections in developing immunity is opposed to this idea; perhaps quantitative factors determine the eventual result.)

Winter coccidiosis is the most puzzling type of coccidiosis in cattle. Hammond (1964) reviewed our knowledge of it and our ideas of its cause. It occurs primarily in calves during or following cold or stormy weather in the winter months. The most common cause is *E. zuernii.* Since winter is the period when the temperature is too low for oocyst sporulation, other factors must be operative. One idea is that there is enough heat in the bedding to permit sporulation. Roderick (1928) suggested that accessory or predisposing factors must be present. Marsh (1938) thought that coccidia must be normally present in the intestine without causing appreciable damage until the host's resistance was decreased by factors such as exposure to cold and change in feed. Marquardt (1962) gave evidence that seemed to support this idea; he found that little or no transmission occurred during the winter in Montana, although he suggested that some oocysts might sporulate in the feces that accumulated on the body surface, if the body's heat warmed them sufficiently, even in cold weather. Hammond et al. (1965) obtained similar results in Utah. However, the factors involved in winter coccidiosis still remain undetermined since the disease cannot be produced at will. The life cycles of almost all coccidia that have been studied carefully appear to be self-limiting. Is this true of the bovine coccidia that cause winter coccidiosis? Is the continual presence of oocysts in the feces due to repeated reinfections? Or is it due to hidden merogony in some internal organs? Further research is needed to elucidate the matter.

The difficulty of understanding the factors responsible for winter coccidiosis is well illustrated by the careful study made by Fitzgerald (1962) from June 1958 to March 1960 of coccidia in a herd of Hereford calves in Utah that had had a history of winter coccidiosis for several years. The disease had occurred in the fall or winter 2–3 mo after weaning. The herd was kept on a 35,000-acre mountain range during the grazing season and then brought together in late October. The animals were weaned in the course of 10 days and then moved to feedlots. In the first year, 100 of them were kept in a 90-acre mountain ranch feedlot, while 60 others were kept in a 2-acre desert ranch feedlot. About 3 wk after weaning, coccidiosis appeared in both groups. This coincided with a marked increase in the percentage of fecal samples containing coccidia. *E. bovis* was the most prevalent species, and *E. zuernii* was second. Other species present were *E. auburnensis, E. canadensis, E. cylindrica, E. ellipsoidalis,* and *E. subspherica.* However, severe manifestations occurred only in calves that were discharging 100,000 or more oocysts of *E. zuernii* per g feces; the highest oocyst count was 324,000 *E. zuernii* and 400 *E. bovis* per g. Several calves died with severe dysentery.

The next year Fitzgerald followed the appearance of oocysts from birth on. The incidence of infection increased from 0% at birth to 100% about 2 mo after weaning. *E. zuernii* and *E. bovis* appeared later than the other species. When the animals were brought in from the summer pasture the second year, he tried without success to prevent coccidiosis in them by use of a tranquilizer to decrease the stress of weaning.

Marquardt (1962) carried out a series of careful studies on small numbers of cattle during the winter months in Montana in an effort to elucidate the factors responsible for transmission of coccidia and for winter coccidiosis in that region. In one experiment he followed oocyst production during the fall and winter in a group of weaned beef calves from a ranch where coccidiosis was occurring. *E. bovis, E. canadensis,* and *E. ellipsoidalis* accounted for 78% of the oocysts produced by these calves, while *E. zuernii* accounted for only 9%. In another experiment, calves that had been raised under conditions of minimal exposure to coccidia were placed in contact with older calves that were shedding oocysts. The receptor calves were 2–3 mo old at the beginning of the experiment, which was conducted during cold weather. They were almost completely free of coccidia when placed with the donors, and they remained free for 79 days. After this time, infections appeared to have been established in the receptors in contact with the donors, but not in isolated calves.

Once infected, the calves continued to produce oocysts for long periods under conditions that would appear to preclude transmission. This raises the probability that coccidial infections in cattle may not be self-limiting.

Marquardt's findings and those of others give evidence that the type of coccidiosis that appears in calves during the fall and winter in the northern range states is not due to exposure of the host to large numbers of infective oocysts but to other factors that help bring on the disease. Coccidiosis seems to follow weaning, shifting calves from range to feedlots, severe winter weather, and changes in feed. Affected calves are usually 6 mo or more old. Heavy contamination with sporulated oocysts in these areas of low carrying capacity and in cold weather seems unlikely.

Davis, Herlich, and Bowman (1959a, 1959b, 1960a, 1960b) found that concurrent infections of cattle with the nematodes *Trichostrongylus colubriformis* or *Cooperia punctata* exacerbated the effects of coccidia in calves, but that *Ostertagia ostertagi* and *Strongyloides papillosus* had no such effect.

DIAGNOSIS. Bovine coccidiosis can be diagnosed from a combination of history, signs, gross lesions at necropsy, and microscopic examination of scrapings of the intestinal mucosa and of feces. Diarrhea or dysentery accompanied by anemia, weakness, emaciation, and inappetence are suggestive of coccidiosis in calves. Secondary pneumonia is often present. The lesions found at necropsy have already been described.

Microscopic examination is necessary to determine whether the lesions are due to coccidia or to some other agent. However, diagnoses will often be missed if one relies only on finding oocysts in the feces. There may be none there at all in the acute stage of *zuernii* coccidiosis. Similarly, the mere presence of oocysts in the feces is not proof that coccidiosis is present; it may be coccidiasis. To be sure of a diagnosis, scrapings should be made from the affected intestinal mucosa and examined under the microscope. It is not enough to look for oocysts, however; meronts, merozoites, and gametes should be recognized.

TREATMENT. The sulfonamides have some value against bovine coccidia, but they are only partially effective. Most other types of compound used in avian coccidiosis are too toxic for calves; these include nicarbazin, furacin, and nitrofurazone. Other compounds that have received particular attention include amprolium, monensin, lasalocid, salinomycin, lincomycin, and decoquinate. None, however, is completely satisfactory.

Monensin is perhaps the most effective. McDougald (1978) found that

16.5 g monensin per metric ton of feed prevented clinical signs of coccidiosis in young calves if it was started 3 days before experimental inoculation with *E. bovis* and continued for 31 days. Stockdale and Yates (1978) found that 1 mg monensin per kg feed daily beginning 10–20 days after artificial inoculation with *E. zuernii* was effective against the coccidium and that calves so treated became resistant to reinfection.

Fitzgerald and Mansfield (1979) found that about 3 mg lasalocid per kg body weight added to a complete feed was effective in controlling clinical coccidiosis in calves inoculated with 70% *E. bovis,* 25% *E. zuernii,* and 5% of three other species.

PREVENTION. Sanitation and isolation are effective in preventing coccidiosis. Beef calves should be dropped and kept on clean, well-drained pastures. Overstocking and crowding should be avoided. Feed and water containers should be high enough to prevent fecal contamination. Feedlots should be kept dry and cleaned as often as possible. Concrete or small gravel are preferable to dirt.

Dairy calves should be isolated within 24 hr after birth and kept separately. Individual box stalls that are cleaned daily may be used. Slat-bottom pens are also effective and require less cleaning. Allen and Duffee (1958) described a simple, homemade, raised, wooden stall with a 4 × 2.5-foot slatted floor in which dairy calves can be raised separately for the first 3 mo. Davis (1949) and Davis and Bowman (1952) described a 5 × 10 × 3-foot outdoor portable pen that can be moved to a fresh site once a week, thus eliminating the need for cleaning. It is made primarily of net wire and one-by-four lumber, with a removable roof and siding at one end. The pens should not be returned to the same ground for a year.

These methods will not eliminate all coccidia but will prevent the calves from picking up enough oocysts to harm them. In addition, they will greatly reduce lice, helminth parasites, pneumonia, white scours, and other diseases.

The unsporulated oocysts of *E. zuernii* are killed by sunlight in 4 hr or by drying at 25% relative humidity or lower in several days. They are not harmed by freezing at −7 to −8 C for as long as 2 mo, and half of them survive as long as 5 mo; at −30 C, however, only 5% survive 1 day. The oocysts are killed by 10^{-6} M mercuric chloride, 0.05 M phenol, 0.25 M formaldehyde, 1.25% sodium hypochlorite, or 0.5% cresol (Marquardt, Senger, and Seghetti 1960).

Coccidiosis in Sheep and Goats

EPIDEMIOLOGY. Mixed infections with coccidia are the rule. For instance, Gregory et al. (1980) found a single species of *Eimeria* in only 3% of 765 ovine outbreaks in England and Wales in 1978–1979.

Coccidiosis in sheep is primarily a disease of feedlot lambs. It begins 12 days to 3 wk after the lambs arrive in the feedlot. Diarrhea, depression, and inappetence appear, followed by weakness and loss of weight. The diarrhea continues for several days up to about 2 wk, and some lambs may die during this period. Most, however, recover. The mortality varies but is seldom more than 10%. Even if there are no deaths, there may be loss of weight or reduced weight gains.

When lambs are brought into the feedlot, they are usually shedding small numbers of coccidian oocysts. As a result of crowding, and under conditions that promote fecal contamination of the feed, the coccidial infections build up. The number of oocysts in the feces rises for about 1 mo, remains stationary for 1–3 wk, and then decreases rather rapidly, only a few oocysts being present at the end of the feeding period. Whether or not disease will appear depends on the number and species of oocysts that the lambs ingest during the crucial first or second week. By the end of the first month, there is little danger of coccidiosis. The lambs have been infected, but the exposing dose of oocysts has been small enough to permit immunity to develop. In other words, there has been coccidiasis but no coccidiosis.

Feeding of chopped feed in open troughs low enough to be contaminated with feces promotes coccidiosis.

In breeding flocks, the ewes are the source of infection, and the lambs become infected by ingesting sporulated oocysts from the bedding. The first oocysts appear when the lambs are a few weeks old; they build up to a peak that lasts 1–4 wk and then decline.

Temperature affects oocyst sporulation. The first sporulated oocysts appear in the bedding when the mean temperature is about 9 C. The optimum temperature for the oocysts of *E. ovina* to sporulate is 20–25 C, the sporulation time being 2–3 days at this temperature in the presence of oxygen. The oocysts survive less than 4 mo in fecal sediment at this temperature. Sporulation is slow at 0–5 C, although oocysts remain alive for at least 10 mo in fecal sediment or moist pellets. No sporulation takes place at 40 C, and the oocysts are killed within 4 days. If the fecal sediment is allowed to putrefy, however, no sporulation takes place at any temperature.

The oocysts of *E. ovina, E. ovinoidalis,* and *E. parva* do not survive 24 hr in sheep pellets when frozen directly to −30 C and survive less than 2 days when conditioned at −19 C prior to freezing to −30 C. They survive without essential mortality when frozen directly to −25 C for 7 days, but only about half the first two species and a quarter of the *E. parva* oocysts will survive 14 days. Repeated freezing and thawing at −19 or −25 C up to six or seven times has no significant effect on survival. In an average winter at Laramie, Wyo., the minimum soil surface temperature would probably be between −15 and −20 C and unsporulated oocysts would not normally be killed by such temperatures (Landers 1953).

DIAGNOSIS. Coccidiosis in sheep and goats can be diagnosed from a combination of history, signs, gross lesions at necropsy, and microscopic examination of the intestinal mucosa and feces. Recognition of coccidia in the lesions at necropsy is necessary for positive diagnosis. The mere presence of oocysts in the feces does not necessarily mean that the disease is due to coccidia. On the other hand, acute coccidiosis may be present before any oocysts appear.

TREATMENT. Essentially the same things can be said about treatment of coccidiosis in sheep and goats as in cattle. Among recent studies, Horton and Stockdale (1981) found that 12.5–100 mg lasalocid per kg feed, and to a somewhat lesser extent 22 mg monensin per kg feed, were effective against *E. ovinoidalis* and *E. ahsata* in naturally infected lambs. They improved the performance of early weaned lambs under feedlot conditions. McDougald and Dunn (1978) found that 5, 10, or 20 ppm monensin in the feed protected lambs against death, impaired body weight gain, and diarrhea due to *E. ovinoidalis* and *E. ahsata;* 10 ppm reduced the number of oocysts passed, and 20 ppm almost completely eliminated them. Horton and Stockdale (1979) found that 11 mg monensin per kg feed, and to a lesser extent 10 mg amprolium per kg feed, were effective against *E. ahsata, E. ovinoidalis,* and naturally occurring coccidiosis in general in feedlot lambs; they gave the medicated feed for 106 days. Bergstrom and Maki (1976) found that 10–30 g monensin per metric ton feed produced satisfactory weight gains and feed conversion during an 84-day experimental period in lambs infected with 18,000 sporulated oocysts of *E. ovinoidalis,* a number sufficient to produce diarrhea and rapid weight loss in controls. Leek et al. (1976) found that 1 mg monensin per kg body weight daily by mouth beginning before experimental infection with *E. ovinoidalis* prevented diarrhea and reduced oocyst production, while 2 mg per kg daily by mouth beginning at the time signs of infection appeared reduced oocyst production but did not eliminate diarrhea. In both cases, the treated lambs did not gain weight as well as the untreated, uninfected controls. Fitzgerald and Mansfield (1978) found that 10 or 20 mg monensin per kg in the ration during the feeding period somewhat improved the performance of lambs naturally or artificially infected with *E. ovinoidalis.* Foreyt et al. (1979) found that 100 mg lasalocid, or 17–33 mg monensin, per kg in pelleted feed fed continuously from weaning to market weight caused somewhat more weight gain and lower feed consumption in lambs experimentally infected with a mixture of 24,000 *E. ovina, E. crandalis, E. ovinoidalis, E. parva,* and *E. pallida* oocysts. Foreyt (1981) found that 25 mg lasalocid per kg feed was somewhat effective against a mixture of *E. ovina, E. ovinoidalis, E. parva, E. pallida,* and *E. intricata.*

I find it impossible to be enthusiastic about any of these drugs for use in sheep or goat coccidiosis.

PREVENTION. Good sanitation will largely prevent coccidiosis in lambs. Coccidiosis is not a problem in suckling lambs on the western range but appears when the animals are brought together in the feedlot. Feedlots should be kept dry and clean. Clean water and feed should be supplied, and feed troughs should be so constructed that they cannot be contaminated with feces.

Coccidiosis is a potential hazard if lambing takes place in a barn or restricted area. The bedding is the most common source of infection. If lambs are raised with their ewes, the animals should be kept in concrete pens with straw bedding and the pens cleaned twice a week.

Raising lambs in slatted-floor pens as described by Watson (1962), Mansfield (1964), and Mansfield et al. (1967) will prevent coccidiosis.

Coccidiosis in Swine

EPIDEMIOLOGY. Coccidia are common in swine, but we know little about the prevalence and importance of the disease coccidiosis. Enteritis is so common in young pigs and caused by so many different agents that the importance of some of them has not been properly assessed. Coccidia are among the least known of these agents. However, in recent years *Isospora suis* has been found to be a significant cause of loss in baby pigs (Stuart et al. 1980).

Coccidiosis is primarily a disease of young pigs. Adults are carriers. Pigs become infected by ingesting sporulated oocysts along with their feed or water. The presence or severity of the disease depends on the number of oocysts they receive. Crowding and lack of sanitation greatly increase the disease hazard.

The oocysts of *Eimeria debliecki* and *E. scabra* can survive and remain infective in the soil for 15 mo when the soil surface temperature varies between −4.5 and 40 C. Unsporulated oocysts withstand continuous freezing at −2 to −7 C or alternate freezing and thawing at 0.5 and −3 C for at least 26 days, although subsequent sporulation is somewhat decreased.

As with other coccidia, the coccidia of swine are not transmissible to other farm animals, and pigs cannot be infected with the latter's coccidia.

DIAGNOSIS. Coccidiosis in swine can be diagnosed by finding the endogenous stages in lesions in the intestine. As with other animals, the presence of oocysts in the feces does not necessarily mean that coccidiosis is present, nor does their absence necessarily mean that it is absent. Oocysts may not be produced until 2 or 3 days after the first signs of disease appear.

TREATMENT. Little is known about treatment of coccidiosis in pigs. The drugs used for the disease in cattle and sheep might be tried.

PREVENTION AND CONTROL. Sanitation will prevent coccidiosis in swine. Pens should be cleaned frequently, overcrowding should be avoided, and pigs should be raised under conditions that will prevent them from eating many infective oocysts.

Coccidiosis in Horses, Asses, and Mules

Coccidiosis is such a rarity in horses, asses, and mules that little can be said about it. The same measures that are effective in cattle should control coccidiosis in equids.

Coccidiosis in Domestic Rabbits

EPIDEMIOLOGY. The most important species of rabbit coccidium is *Eimeria stiedai,* which occurs in the liver. All the other species are found in the intestine. Of these, the most important are probably *E. intestinalis, E. irresidua, E. magna,* and *E. media* (Kheisin 1957).

Coccidiosis is primarily a disease of young rabbits; adults are carriers.

Rabbits become infected by ingesting oocysts along with their feed or water. The severity of the disease depends on the number of oocysts ingested and the species and strain involved. Mixed infections are the rule; infections with a single species are usually seen only under laboratory conditions. Crowding and lack of sanitation greatly increase the disease hazard.

Some of the coccidia of the domestic rabbit *Oryctolagus cuniculus* also occur in cottontails *Sylvilagus* spp. Some have also been reported from jackrabbits and hares *Lepus* spp., but it is questionable if any of them actually occur in *Lepus*.

DIAGNOSIS. Liver coccidiosis can be diagnosed by finding the characteristic lesions containing coccidia. Intestinal coccidiosis can be diagnosed by finding the coccidia on microscopic examination. However, the mere presence of these parasites in a case of enteritis does not mean that they caused it. Many rabbits carry a few coccidia without suffering any noticeable effects.

TREATMENT. Some of the sulfonamides have been found helpful in preventing coccidiosis if given continuously in the feed or drinking water. Catchpole and Norton (1975) found that sulfaquinoxaline in the feed was partially effective against *E. intestinalis* but that clopidol was not; sulfadimidine in the drinking water was effective but amprolium-ethopabate was not. Fitzgerald (1976) found that three polycyclic, polyether monocarboxylic acid antibiotics extracted from a strain of *Streptococcus albus* were all effective in preventing *E. stiedai* and intestinal coccidial infections; young rabbits ate them readily in pelleted feed, but they decreased the rabbits' weight gains. Sambeth and Raether (1980) found that continuous feeding of salinomycin or monensin in the feed markedly reduced oocyst excretion; lasalocid was inferior.

Peeters et al. (1981) found that clopidol was better than sulfaquinoxaline/pyrimethamine (which was better than sulfadimidine/robenidine) in suppressing oocyst production.

Long-term, continuous feeding of such drugs is not particularly desirable, nor is it usually necessary. It has been the usual experience with poultry, and the same is true with rabbits, that if the hosts are exposed to coccidiosis during the drug-feeding period (as they usually are), an aborted infection occurs that is sufficient to induce immunity. The drug can then be safely stopped.

PREVENTION. Coccidiosis can be prevented by proper management. Feeders and waterers should be designed so that they do not become contaminated with droppings and should be kept clean. Hutch floors should be self-cleaning or should be cleaned frequently and kept dry. Manure should be removed frequently. The animals should be handled as little as possible, and care should be taken not to contaminate either the animals themselves or their feed, utensils, or equipment. In addition, the rabbitry should be kept as free as possible of insects, rodents, and other pests. See McPherson et al. (1962) for a practical method of eliminating coccidia from a rabbitry.

Coccidiosis in Man

Coccidiosis due to homoxenous species is quite rare in man; coccidiosis due to heteroxenous species (discussed in Chapter 8) is somewhat commoner, especially in countries where raw meat is eaten.

In addition to the species that produce infections in man, a number of other coccidia have been found in human feces and mistaken for parasites of man. Perhaps the most famous of these were *Eimeria wenyoni, E. oxyspora,* and *E. snijdersi,* which Dobell (1919) described as human parasites. The first turned out to be *E. clupearum,* a coccidium of herring, sprats, and mackerel; the second two were both *E. sardinae,* a parasite of sardines, herring, and sprats. In addition, oocysts of *E. stiedai* of the rabbit have been found in a mental hospital patient who liked to eat raw rabbit livers, and oocysts of *E. debliecki* of the pig have been found in other persons who probably acquired them from sausage casings.

Coccidiosis in Chickens

EPIDEMIOLOGY. Infections with a single species of coccidium are rare, and mixed infections are the rule. *Eimeria tenella* is the most pathogenic and important species. In recent years, control of this species with coccidiostats has revealed more and more coccidiosis due to *E. necatrix.* Improvement of coccidiostats has reduced this species, too, so that the other species are becoming more important. *E. brunetti* is markedly pathogenic but uncommon. *E. maxima, E. mivati,* and *E. acervulina* are slightly to moderately pathogenic, depending on the strain; all three are common.

Coccidiosis is primarily a disease of young birds. Older birds are carriers. Birds become infected by ingesting oocysts along with their feed or water. Under farm conditions, and even in the laboratory unless extreme precautions are taken, it is practically impossible to prevent exposure to at least a few oocysts.

The disease picture depends on the number of oocysts and the strain of each species that the birds ingest. If they get only a few oocysts, there are no signs and repeated infections produce immunity without disease. If the birds get more, the disease may be mild and the birds will become immune. Only if they get a large number of oocysts of a pathogenic strain do severe disease and death result.

Crowding and lack of sanitation greatly increase the disease hazard. As the oocysts accumulate, the birds receive heavier and heavier exposures and the disease becomes increasingly severe in each successive batch of birds placed in contaminated surroundings.

DIAGNOSIS. Avian coccidiosis can be diagnosed by finding lesions containing coccidia at necropsy. Diarrhea with or without blood in the droppings, inappetence, and emaciation are suggestive, but scrapings of the affected intestinal mucosa must be examined microscopically to determine whether coccidia are present. It is not enough to look for oocysts; meronts, merozoites, gamonts, and gametes should be recognized also.

White bands in the upper small intestine are suggestive of *E. acervulina* infection; rounded lesions in the same place of *E. mivati* infection; petechiae, catarrh, enteritis, blood, and coagulation necrosis of the lower small intestine of *E. brunetti* infection; large meronts (up to 66 μm in diameter), thickening of the mucosa, and absence of oocysts from the small intestine of *E. necatrix* infection; the presence of gamonts deep in the small intestine epithelium of *E. maxima* infection; and cecal enteritis, blood, and cecal plugs of *E. tenella* infection (Long 1964; Reid 1964; Long and Reid 1982). However, microscopic examination is necessary to confirm the diagnosis. Unfortunately most species cannot be differentiated from the size of their oocysts. Those of *E. maxima* average 30.5 × 21 μm, those of *E. mivati* 16 × 13 μm, those of *E. mitis* 16 × 16 μm, and those of all the other species 16–25 × 15–19 μm (Reid 1964; Long and Reid 1982).

Coccidiasis is much more common than coccidiosis; hence the mere presence of oocysts in the feces cannot be relied on for diagnosis. Conversely, the absence of oocysts does not necessarily mean that coccidiosis is not present, since the disease may be in too early a stage to produce oocysts.

Since some species of coccidia are highly pathogenic for the chicken while others are practically nonpathogenic, the species present must be identified to establish a diagnosis. This can often be done in a rough way from the type and location of the lesions (see above).

TREATMENT. Many hundreds of papers have been written on the treatment of coccidiosis in chickens, and space in this chapter allows only a brief discussion. A multitude of compounds has been tested and many have been found effective; their trade names are more numerous than their chemical ones. By far the greatest part of the research has been done on *E. tenella.* Reid (1964) said that $25 million was spent annually for coccidiostats in the United States and that over $10 million worth of coccidiostats was exported annually. These figures would be considerably greater today.

Among the first compounds found effective against coccidia were borax and sulfur (Herrick and Holmes 1936; Hardcastle and Foster 1944). Both were too toxic to be satisfactory.

P. P. Levine (1939) found that sulfanilamide was active, and this discovery opened up the field. Many (probably several hundred) different sulfonamides were tested, and a number of them were found of practical value. Among them were sulfaguanidine, sulfaquinoxaline, sulfadimidine, and sulfadimethoxine. Several organic arsenic compounds are also effective against coccidia, as are a number of diphenylmethane, diphenyldisulfide, nitrofuran, pyrimidine, imidazole, benzamide, quinoline, guanidine and other derivatives, and carbanilide complexes. Latterly there has been a trend toward antibiotics, such as monensin and lasalocid.

The great majority of these compounds are coccidiostatic rather than coccidiocidal, that is, they prevent the coccidia from developing but will not cure coccidiosis once signs of disease have appeared. When given continuously in the feed, they abort the disease. Very few of them affect the sporozoites, so these may enter the intestinal cells and stimulate the development of immunity. However, if too much is given, immunity may not develop.

Coccidiostats have been mixed in poultry feed for so many years that it was inevitable that drug-resistant strains of coccidia would develop. The first report of this was by Waletzky, Neal, and Hable (1954). Cuckler and Malanga (1955) found additional cases. Among more recent papers on the subject are those of McLoughlin and Chute (1974, 1978, 1979a,b), Jeffers (1974), Chapman (1975, 1976a, 1976b, 1976c, 1978), and Denek and Strossova (1977).

Drug resistance is becoming increasingly common. As a consequence, there is a race between the discovery of new coccidiostats and development by the parasites of resistance against the older ones. Thirteen compounds or formulations had been abandoned by 1967 (Reid 1969), and the number is much higher today. The effective life of a coccidiostat is generally only a few years, and we are in danger of running out of new ones one of these days.

General papers on anticoccidial drugs were written by Reid (1974) and McDougald (1982). Among papers on specific drugs are those on robenidine by Reid et al. (1970), Manuel and Lacuata (1972), Kennett et al. (1974), and McLoughlin and Chute (1978); methyl benzoquate by Ryley and Wilson (1972), Long and Millard (1973), and Chapman (1975); arprinocid by Ruff et al. (1978), Ruff et al. (1979), Kutzer et al. (1979), and McManus et al. (1981); monensin by Reid et al. (1972), McLoughlin and Chute (1974), Ruff et al. (1976), Chapman (1976a,b,c, 1978), Reid et al. (1977), Ruff and Jensen (1977), Long and Millard (1978), and Long et al.(1979); lasalocid by Reid et al. (1975); and salinomycin by Danforth et al. (1977a,b) and Chappel (1979).

Karlsson and Reid (1978) found that when administered in the feed at the levels indicated for 15 consecutive days while the chickens were at the same time fed *E. tenella* oocysts daily, 121 ppm monensin, 80 ppm salinomycin, or 75 ppm lasalocid strongly suppressed the coccidia; 100 ppm monensin, 30 ppm decoquinate, 125 ppm clopidol, or 80 ppm narasin induced moderate suppression; 70 ppm arprinocid, 125 ppm nicarbazin, or 125 ppm amprolium plus 4 ppm ethopabate induced slight suppression; while 33 ppm robenidine, 125 zoalene, or 250 ppm aklomide had no effect. It must be remembered, of course, that these results were obtained against one strain of one species of coccidium; the results might differ with other strains.

Chappel (1982) reviewed the contemporary poultry anticoccidials salinomycin, monensin, lasalocid, arprinocid, and halofuginone. Probably the best was 60 ppm arprinocid in the feed, but they all work well enough. No drug resistance had developed at the time of this writing.

There is a question as to the practical value of coccidiostats in the United States today. Since clinical coccidiosis occurs only occasionally under modern management conditions and since most flocks are naturally exposed to most species of coccidia and become immune regardless of whether they have received coccidiostats, it may not be worthwhile to use them under these conditions. It should be noted that replacement flocks are used for egg-laying and the birds in them are older than broilers, which are killed early. Development of immunity is important in replacement birds but not necessarily in broilers.

PREVENTION AND CONTROL. Coccidian oocysts are extremely resistant to environmental conditions. They may remain alive in the soil for 1 yr or more. They will not sporulate in the absence of oxygen, and they are killed in time by subfreezing temperatures.

Ordinary antiseptics and disinfectants are ineffective agents against coccidian oocysts. They are not harmed by 5% formalin, 5% phenol, 5% copper sulfate, 10% sulfuric acid, 5% potassium hydroxide, or 5% potassium iodide. Indeed, the standard storage solutions for live coccidian oocysts are 2.5% potassium dichromate and 1% chromic acid solutions.

The oocysts may be destroyed by ultraviolet light, heat, desiccation, or bacterial action in the absence of oxygen. Exposure to a temperature of 52 C for 15 min kills the oocysts of *E. tenella* and *E. maxima.* However, even a blowtorch will not kill all the oocysts on the floors of poultry houses unless it is applied long enough to make the wood start to char (Horton-Smith and Taylor 1939). The problem is reaching and maintaining a lethal temperature at the spot where the oocysts are located.

Ammonia fumigation is of practical value (Horton-Smith, Taylor, and Turtle 1940). Poultry houses can be fumigated successfully with 3 oz ammonia gas per 10 ft^3. For satisfactory results, the houses should be sealed so that the gas does not leak out.

Methyl bromide is also an effective fumigant. It should be applied to the litter or soil at the rate of approximately 1 lb per 1,000 ft^2 (0.3 ml per ft^2). When used as a space fumigant at the rate of 2 lb per 1,000 ft^3 it prevents infection in brooder houses using artificially contaminated cane pulp litter on wooden floors (Boney 1948).

Since it is practically impossible under farm conditions to prevent chickens from picking up at least a few oocysts, prevention of coccidiosis depends on preventing a heavy enough infection to produce disease while at the same time permitting a symptomless infection (coccidiasis) to develop and to produce immunity. This can be accomplished by proper sanitation and management. Strict sanitation is effective alone but may be supplemented by use of a coccidiostatic drug.

Young chickens should be raised apart from older birds, which are a source of infection. If birds are raised on the floor, each new brood of chicks should be placed in a clean house containing clean, new litter. The litter should be kept dry, stirred frequently, and removed when wet. The feeders and waterers should be washed in boiling water before use and should be cleaned at least weekly with hot water and detergent. The waterers should be placed on wire platforms over floor drains, and the feeders should be raised high enough to prevent their being fouled. Enough feeders should be provided so that all the birds can feed at once without crowding.

Chicks raised on wire have much less chance of contamination than those raised on the floor. However, the wire should be cleaned regularly. Flies, rats, and mice around the poultry houses and yards should be eliminated, since they may carry coccidia mechanically. Damp areas around the poultry house should be filled in or drained.

Feeding a coccidiostat during times when the birds are especially susceptible may also be helpful. The drug may be fed until the birds are 8–9 wk old, after which they have ordinarily become immune. In addition, it is often recommended that a coccidiostat be fed to pullets for the first 2–3 wk after they have been moved into laying houses.

If an outbreak of coccidiosis occurs, all sick birds should be removed from the flock and placed in a separate pen. They should be given ample feed and water, but it is useless to attempt to treat them. The remaining, apparently healthy birds should be treated with a coccidiostat in the dosage recommended by the manufacturer. Birds that become ill should be removed. The litter should be kept dry and stirred frequently.

All dead birds should be burned. The litter should also be burned or put where chickens will never have access to it. Care should be taken not to track coccidia from sick birds to healthy ones. Special rubbers or overshoes should be put on before entering pens containing sick birds and should be cleaned thoroughly after each use. Veterinarians going from one farm to another should clean and disinfect their boots before leaving each farm.

The use of old, built-up, deep floor litter, which was recommended to reduce losses from coccidiosis, has not been found satisfactory.

Edgar (1955) developed a coccidiosis "vaccine" (Coccivac) that is said to be highly successful in immunizing chicks. It contains fully virulent oocysts of *E. acervulina, E. brunetti, E. hagani, E. maxima, E. mivati, E. praecox, E. necatrix,* and *E. tenella.* Rose and Long (1980) said of it,

> It is usually administered in the drinking water at 4 to 10 days of age but can be given in moist food and its efficacy depends very largely on adequate but not excessive reinfection from the litter by oocysts passed as a result of the immunizing dose. Its use is consequently only applicable to birds kept on litter, and careful management of the litter moisture is necessary to provide the conditions necessary for sporulation. For many years it was recommended that the vaccine be given in conjunction with low concentrations of anticoccidial drugs in the food in order to reduce the potentially pathogenic effects of the inoculum on broiler chickens (lowered body weight gain). This is no longer done and the vaccine is now used mainly for replacement breeding and egg producing stock due, in part, to the introduction of new and improved anticoccidial drugs for the protection of broiler flocks.

Coccidiosis in Turkeys

EPIDEMIOLOGY. The most important coccidia of turkeys are *Eimeria meleagrimitis* and *E. adenoeides.* The former affects the jejunum and the latter the lower ileum, ceca, and rectum. The USDA (1965) estimated that coccidiosis caused an annual loss of $11,866,000 from 1951–1960, and the figure is probably higher now.

Coccidiosis is primarily a disease of young birds. Older birds are carriers. Poults become infected by ingesting oocysts along with their feed or water. The severity of the disease depends on the number of oocysts they receive. If they ingest relatively few they may develop immunity without ever showing signs of illness, while if they ingest large numbers they may become seriously ill or die. Crowding and lack of sanitation greatly increase the disease hazard.

DIAGNOSIS. Coccidiosis of turkeys can be diagnosed in the same way as coccidiosis of chickens, by finding endogenous stages of the coccidia in scrapings of the affected parts of the intestinal tract of birds that show signs of the disease. The mere presence of coccidia in the absence of disease cannot be relied on. Since several species of turkey coccidia (in particular *E. innocua, E. subrotunda,* and *E. meleagridis*) are nonpathogenic or nearly so, they must be differentiated from the pathogenic ones. The sporulated oocysts of *E. meleagrimitis* and *E. adenoeides* have polar bodies, which differentiates them from all but *E. meleagridis* and *E. gallopavonis*. The oocysts of *E. meleagrimitis* are ellipsoidal, but apparently only pathogenesis and absence of cross-immunity differentiate *E. adenoeides* from the other two. This last is hardly a practical diagnostic test, since it requires a colony of turkeys immunized against the various species.

TREATMENT. Some of the drugs used against coccidiosis in chickens are also effective in turkeys, but others are not. Among effective drugs are the sulfonamides, clopidol, and monensin (Horton-Smith and Long 1961; Anderson et al. 1976; Eckman et al. 1976; Long and Millard 1977; Reid et al. 1978).

PREVENTION AND CONTROL. The same measures should be used in turkeys as in chickens.

Coccidiosis in Dogs and Cats

EPIDEMIOLOGY. Coccidiosis, common in dogs and cats, is a cause of diarrhea and even death in puppies and kittens. Crowding and lack of sanitation promote its spread. Coccidia sometimes seed a breeding kennel, boarding kennel, or veterinarian's wards so heavily that most of the puppies born or brought there become infected.

DIAGNOSIS. Coccidiosis can be disagnosed at necropsy by finding coccidia in the intestinal lesions. It can be diagnosed in affected animals by finding oocysts in association with diarrhea or dysentery. However, care must be taken to differentiate coccidiosis from coccidiasis, since many animals may be shedding oocysts without suffering from disease. Other disease agents should be searched for and their absence confirmed. The presence of a wave of oocysts during and shortly after an attack of enteritis and their marked diminution or disappearance soon thereafter would suggest that coccidia had caused the attack.

The fact that some cases of coccidiosis in dogs and cats may be caused by heteroxenous genera (*Sarcocystis* in dogs and *Sarcocystis* or *Toxoplasma* in cats) must be borne in mind continually. (Information about these genera is given in Chapter 8.)

TREATMENT. There is no good treatment for coccidiosis in dogs or cats once the signs of disease have appeared. All the coccidiostatic agents on the market are preventive rather than curative in action. The fact that coccidiosis is a self-limiting disease has often led to the belief that some ineffective drug, administered at the time natural recovery was due to begin, was responsible for the cure. Uncontrolled studies on coccidiosis therapy are worse than useless, since they may lead to false conclusions regarding a drug's value.

Supportive treatments, and particularly the use of antibiotics such as chlortetracycline and oxytetracycline to control secondary infections, may be helpful even though they do not act on the coccidia themselves.

PREVENTION. Sanitation and isolation are effective in preventing coccidiosis. Animal quarters should be cleaned daily. Runways should be concrete. Ordinary disinfectants are ineffective against coccidian oocysts, but water that is still boiling when it reaches the oocysts will kill them. Prevention of infection with heteroxenous coccidia is discussed in Chapter 8.

Coccidiosis in Geese and Ducks

Our knowledge of the coccidia of ducks and geese is quite deficient. Except for renal coccidiosis of the goose caused by *E. truncata,* coccidiosis appears to be of relatively little importance in these birds, and homoxenous coccidia have seldom been reported from them. Some outbreaks of intestinal coccidiosis have been reported, however, in ducks in Europe.

TREATMENT. Little is known of the treatment of coccidiosis in ducks and geese. *E. truncata* infections of geese seemed to respond to sodium sulfamethazine, and the urinary excretion of sulfonamides in general would suggest that they should be particularly effective against this species.

PREVENTION AND CONTROL. The same measures should be used as with chickens.

LITERATURE CITED

For citations before 1965, see the *Index-Catalogue of Medical and Veterinary Zoology, Zoological Record, Biological Abstracts, Protozoological Abstracts,* and *Veterinary Bulletin.*

Anderson, W. I., W. M. Reid, and L. R. McDougald. 1976. Avian Dis. 20:387–94.
Andreassen, J., and O. Behnke. 1968. J. Parasitol. 54:150–63.
Barker, I. K., and P. L. Carbonell. 1974. Z. Parasitenkd. 44:289–98.
Bergstrom, R. C., and L. R. Maki. 1976. Am. J. Vet. Res. 37:79–81.
Catchpole, J., and C. C. Norton. 1975. J. Protozool. 22:49A.
Chapman, H. D. 1975. Parasitology 71:41–49.
______. 1976a. Avian Pathol. 5:283–90.
______. 1976b. Parasitology 73:265–73.
______. 1976c. Parasitology 73:275–82.
______. 1978. Parasitology 76:177–83.

Chapman, H. D. 1982. In The Biology of the Coccidia, 2d ed., ed. P. L. Long, 429–52. Baltimore: University Park Press.
Chappel, L. R. 1979. J. Parasitol. 65:137–43.
———. 1982. In Parasites: Their World and Ours, ed. D. F. Mettrick and S. S. Desser, 327–29. Amsterdam: Elsevier.
Chobotar, B., and E. Scholtyseck. 1982. In The Biology of the Coccidia, 2d ed., ed. P. L. Long, 101–65. Baltimore: University Park Press.
Danforth, H. D., M. D. Ruff, W. M. Reid, and J. Johnson. 1977a. Poult. Sci. 56:933–38.
Danforth, H. D., M. D. Ruff, W. M. Reid, and R. L. Miller. 1977b. Poult Sci. 56:926–32.
Denek, J., and Z. Strossova. 1977. J. Protozool. 24:14A.
Desser, S. S. 1978. J. Parasitol. 64:933–35.
Dhillon, A. S., H. L. Thacker, A. V. Dietzel, and R. W. Winterfield. 1981. Avian Dis. 25:747–51.
Doran, D. J. 1982. In The Biology of the Coccidia, 2d ed., ed. P. L. Long, 229–85. Baltimore: University Park Press.
Eckman, M. K., R. D. Vatne, E. V. Reed, and B. L. Rachunek. 1976. Down to Earth 32(3):27–32.
Fayer, R., and W. M. Reid. 1982. In The Biology of the Coccidia, 2d ed., ed. P. L. Long, 453–87. Baltimore: University Park Press.
Fernando, M. A. 1982. In The Biology of the Coccidia, 2d ed., ed. P. L. Long, 287–327. Baltimore: University Park Press.
Fitzgerald, P. R. 1975. Bovine Pract. 11:28–33.
———. 1976. J. Protozool. 23:150–54.
Fitzgerald, P. R., and M. E. Mansfield. 1978. Am. J. Vet. Res. 39:7–10.
———. 1979. J. Parasitol. 65:824–25.
Foreyt, W. J., N. L. Gates, and R. B. Wescott. 1979. Am. J. Vet. Res. 40:97–100.
Foreyt, W. J., S. M. Parish, and K. M. Foreyt. 1981. Am. J. Vet. Res. 42:57–65.
Gregory, M. W., L. P. Joyner, J. Catchpole, and C. C. Norton. 1982. Vet. Res. Commun. 5:307–25.
Hammond, D. M., F. Sayin, and M. L. Miner. 1965. Am. J. Vet. Res. 26:83–89.
Hammond, D. M., E. Scholtyseck, and M. L. Miner. 1967. J. Parasitol. 53:235–47.
Hoerr, F. J., F. M. Ranck, and T. F. Hastings. 1978. J. Am. Vet. Med. Assoc. 173:1591–93.
Horton, G. M. J., and P. H. G. Stockdale. 1979. Am. J. Vet. Res. 40:966–70.
———. 1981. Am. J. Vet. Res. 42:433–36.
Jeffers, T. K. 1974. J. Parasitol. 60:900–904.
Jeffers, T. K., and M. W. Shirley. 1982. In The Biology of the Coccidia, 2d ed., ed. P. L. Long, 63–100. Baltimore: University Park Press.
Joyner, L. P. 1982. In The Biology of the Coccidia, 2d ed., ed. P. L. Long, 35–62. Baltimore: University Park Press.
Karlsson, T., and W. M. Reid. 1978. Avian Dis. 22:487–95.
Kennett, R. L., S. Kantor, and A. Gallo. 1974. Poult. Sci. 53:978–86.
Kutzer, E., J. Leibetseder, H. Prosl, and A. Mitterlehner. 1979. Wien. Tieraerztl. Monatsschr. 66:197–202.
Leek, R. G., R. Fayer, and D. K. McLoughlin. 1976. Am. J. Vet. Res. 37:339–41.
Levine, N. D. 1982a. In The Biology of the Coccidia, 2d ed., ed. P. L. Long, 1–33. Baltimore: University Park Press.
Levine, N. D., and V. Ivens. 1965a. Ill. Biol. Monogr. 33. Urbana: University of Illinois Press.
———. 1965b. J. Parasitol. 51:859–64.

______. 1967. J. Protozool. 14:351–60.
______. 1970. Ill. Biol. Monogr. 44. Urbana: University of Illinois Press.
______. 1981. Ill. Biol. Monogr. 51. Urbana: University of Illinois Press.
______. 1985. The Coccidia of Artiodactyls. Urbana: University of Illinois Press. (In press)
Levine, N. D., et al. 1980. J. Protozool. 27:37–58.
Lima, J. D. 1980. J. Protozool. 27:153–54.
Long, P. L., ed. 1982. The Biology of the Coccidia. 2d ed. Baltimore: University Park Press.
Long, P. L., and B. J. Millard. 1973. Avian Pathol. 2:111–25.
______. 1977. Avian Pathol. 6:227–33.
______. 1978. Avian Pathol. 7:373–81.
Long, P. L., and W. M. Reid. 1982. Univ. Ga. Coll. Agric. Exp. Stn. res. rep. 404.
Long, P. L., B. J. Millard, and K. M. Smith. 1979. Avian Pathol. 8:453–67.
McDougald, L. R. 1978. Am. J. Vet. Res. 39:1748–49.
______. 1982. In The Biology of the Coccidia, 2d ed., ed. P. L. Long, 373–427. Baltimore: University Park Press.
McDougald, L. R., and W. J. Dunn. 1978. Am. J. Vet. Res. 39:1459–62.
McLoughlin, D. K., and M. B. Chute. 1974. J. Parasitol. 60:835–37.
______. 1978. J. Parasitol. 64:874–77.
______. 1979a. Proc. Helm. Soc. Wash. 46:138–41.
______. 1979b. Proc. Helm. Soc. Wash. 46:265–69.
McManus, E. C., G. Olson, and J. D. Pulliam. 1981. J. Parasitol. 66:765–70.
Mansfield, M. E., J. M. Lewis, and G. E. McKibben. 1967. J. Am. Vet. Med. Assoc. 151:1182–85.
Manuel, M. F., and A. Q. Lacuata. 1972. Philipp. J. Vet. Med. 10:81–100.
Melhorn, H., and M. B. Markus. 1976. Z. Parasitenkd. 51:15–24.
Peeters, J. E., R. Geeroms, R. Froyman, and P. Halen. 1981. Res. Vet. Sci. 30:328–34.
Pellérdy, L. 1974. Coccidia and Coccidiosis. 2d ed., Budapest: Akad. Kiado; West Berlin: Paul Parey.
Pivont, P., and H. Antoine. 1982. Ann. Med. Vet. 126:189–203.
Pohlenz, J., W. M. Bemrick, H. W. Moon, and N. F. Cheville. 1978a. Vet. Pathol. 15:417–27.
Pohlenz, J., H. W. Moon, N. F. Cheville, and W. M. Bemrick. 1978b. J. Am. Vet. Med. Assoc. 172:452–57.
Proctor, S. J., and R. L. Kemp. 1974. J. Protozool. 21:664–66.
Reese, N. C., W. L. Current, J. V. Ernst, and W. S. Bailey. 1982. Am. J. Trop. Med. Hyg. 31:226–29.
Reid, W. M. 1969. Publ. 1679. Natl. Acad. Sci., Washington, D.C., 87–99.
______. 1974. Proc. Symp. Coccidia Relat. Organ., Guelph, Ontario, 119–34.
Reid, W. M., L. M. Kowalski, E. M. Taylor, and J. Johnson. 1970. Avian Dis. 14:788–79.
Reid, W. M., L. Kowalski, and J. Rice. 1972. Poult. Sci. 51:139–46.
Reid, W. M., J. Johnson, and J. Dick. 1975. Avian Dis. 19:12–18.
Reid, W. M., J. Dick, J. Rice, and F. Stino. 1977. Poult. Sci. 56:66–71.
Reid, W. M., W. I. Anderson, and L. R. McDougald. 1978. Avian Pathol. 7:569–76.
Rose, M. E. 1967. J. Parasitol. 53:924–29.
______. 1982. In The Biology of the Coccidia, 2d ed., ed. P. L. Long. Baltimore: University Park Press.
Rose, M. E., and P. L. Long. 1980. Symp. Br. Soc. Parasitol. 18:57–74.
Ruff, M. D., and L. S. Jensen. 1977. Poult. Sci. 56:1956–59.
Ruff, M. D., W. M. Reid, and A. P. Rahn. 1976. Am. J. Vet. Res. 37:963–67.

Ruff, M. D., W. I. Anderson, and W. M. Reid. 1978. J. Parasitol. 64:306-11.
Ruff, M. D., W. M. Reid, J. K. Johnson, and W. A. Anderson. 1979. Poult. Sci. 58:298-303.
Ryley, J. F. 1982. In Parasites: Their World and Ours, ed. D. F. Mettrick and S. S. Desser, 319-26. Amsterdam: Elsevier, 319-26.
Ryley, J. F., and R. G. Wilson. 1972. J. Parasitol. 58:664-68.
Sambeth, W., and W. Raether. 1980. Zentralbl. Veterinaermed. [B] 27:446-58.
Schmitz, J. A., and D. H. Smith. 1975. J. Am. Vet. Med. Assoc. 167:731-32.
Scholtyseck, E. 1964. Z. Zellforsch. 64:688-707.
______. 1965. Z. Zellforsch. 66:625-42.
______. 1979. Fine Structure of Parasitic Protozoa. Berlin: Springer-Verlag.
Scholtyseck, E., A. Rommel, and G. Heller. 1969. Z. Parasitol. 31:289-98.
Sheffield, H. G., and D. M. Hammond. 1967. J. Parasitol. 53:831-40.
Snodgrass, D. R., K. W. Angus, E. W. Gray, W. A. Keir, and L. W. Clerihew. 1980. Vet. Rec. 106:458-60.
Stockdale, P. H. G., and W. D. G. Yates. 1978. Vet. Parasitol. 4:209-14.
Stuart, B. P., D. S. Lindsay, J. V. Ernst, and H. S. Gosser. 1980. Vet. Pathol. 17:84-93.
Todd, K. S., Jr., and D. M. Hammond. 1971. In Infectious and Parasitic Diseases of Wild Birds, ed. J. W. Davis et al., 234-81. Ames: Iowa State University Press.
Tzipori, S., K. W. Angus, E. W. Gray, and I. Campbell. 1980. N. Engl. J. Med. 303:818.
USDA. 1965. Agric. Handb. 291.
Vetterling, J. M. 1965. J. Parasitol. 51:897-912.
______. 1966. J. Protozool. 13:290-300.
Wąng, C. C. 1982. In The Biology of the Coccidia, 2d ed., ed. P. L. Long, 167-228. Baltimore: University Park Press.

8
Apicomplexa: *Sarcocystis, Toxoplasma,* and Related Protozoa

Sarcocystis, Toxoplasma, and related protozoa belong to the family Sarcocystidae. This family belongs to the order Eimeriorina, in which the macrogamete and microgamont develop independently, there is no syzygy, and the microgamonts typically produce many flagellated microgametes. The Sarcocystidae differ from the Eimeriidae in that they are heteroxenous, requiring one type of host for the sexual stages and another type for the asexual stages. So far as is known, all the hosts (both definitive and intermediate) are vertebrates.

FAMILY SARCOCYSTIDAE

In the family Sarcocystidae the meronts and merozoites are in a prey animal and the oocysts in a predator, which can be a mammal, bird, or snake, depending on the species. So far as is known, all members of the family produce oocysts containing 2 sporocysts, each with 4 sporozoites. Oocysts are produced in the predator's intestinal cells, while the asexual stages are in the tissues of the prey host.

Genus *Sarcocystis* Lankester, 1882

In this genus, the last generation meronts (sarcocysts) are found in the striated and heart muscles (and rarely in the brain) and are ordinarily divided into compartments by thin internal septa. The sarcocysts may be either microscopic or visible to the naked eye, depending on the species and stage of development. They contain *metrocytes* (rounded premerozoites) at first, and then, when mature, bradyzoites. The oocysts sporulate in the predator host; sporulated sporocysts and a few sporulated oocysts are passed in the feces. Synonyms of this name are *Miescheria* and *Balbiania.* This genus has been reviewed by Tadros and Laarman (1976), Markus (1978), and Rommel et al. (1979).

Sarcocystis is common in many species of domestic and wild animals.

Levine and Tadros (1980) prepared a list of 93 named species, and since that time several others have been added. At the time of writing, both the definitive and intermediate hosts were known for 35 of them. The species occurring in carnivores have been reviewed by Levine and Ivens (1981), those in artiodactyls by Levine and Ivens (1985).

STRUCTURE. The oocysts and sporocysts in the definitive host resemble those of *Isospora,* with 2 sporocysts, each containing 4 sporozoites, but the sporocysts lack a Stieda body. They sporulate in the intestinal cells and are fully sporulated when passed in the feces. Indeed, the sporocysts usually break out of the oocysts before passage, so that most of the cysts in the feces are sporocysts. The sporocysts have often been mistaken for oocysts of *Cryptosporidium.*

The asexual stages in the intermediate hosts are meronts of two types. In perhaps all species there are one or two preliminary generations (all, so far as is known, in or under endothelial cells of the smaller blood vessels) that produce tachyzoites without a "cyst" wall. These are followed by the typical sarcocysts in the muscles. The sarcocysts are actually meronts. They are enclosed in a wall, which is responsible for their being called cysts. Some are easily visible to the naked eye; others are microscopic. The large ones are known as Miescher's tubules. They are usually cylindroid or spindle-shaped and run lengthwise in the muscles, but they may also be ellipsoidal or rather irregular. The ellipsoidal cysts of *S. gigantea* in the sheep may reach 1 cm in diameter, but considerably smaller ones are the rule. Those of *S. rileyi* in the duck are 1–2 μm in diameter and 1 cm or more long. In some species, such as *S. tenella* of the sheep, the sarcocysts are microscopic.

The sarcocyst wall varies in appearance with the species. It is composed of a primary (outer) layer and a more or less homogeneous secondary layer beneath it. The primary layer may be smooth and thin, or it may be thick and contain radial spines, protrusions, or villi, with or without fibrils within them; these spines or protrusions are known as *cytophaneres.* The secondary layer has thin extensions (trabeculae or septa) that extend into the sarcocyst itself, dividing it into compartments. The young sarcocysts contain only metrocytes at first; they are more or less ellipsoidal, and divide repeatedly by endodyogeny, giving rise to intermediate forms and then to bradyzoites, so that when the sarcocyst is completely mature it contains only bradyzoites.

The tachyzoites and bradyzoites are elongate and resemble the merozoites of the Eimeriidae and other coccidia, having an apical complex consisting of conoid, polar ring(s), rhoptries, micronemes and subpellicular microtubules, one or more micropores, a nucleus with a nucleolus, one or more mitochondria, amylopectin granules, various vacuoles, endoplasmic reticulum, and a 3-layered pellicle. The metrocytes are similar but lack rhoptries and micronemes. (See Figs. 8.1 and 8.2.)

LIFE CYCLE. *Sarcocystis* is heteroxenous. The asexual stages occur in a prey animal, the sexual stages in a predator. Oocysts develop in the lamina propria of the small intestine of the predator. They sporulate there, forming 2 sporocysts, each with 4 sporozoites. The oocysts have a smooth wall, lack a micro-

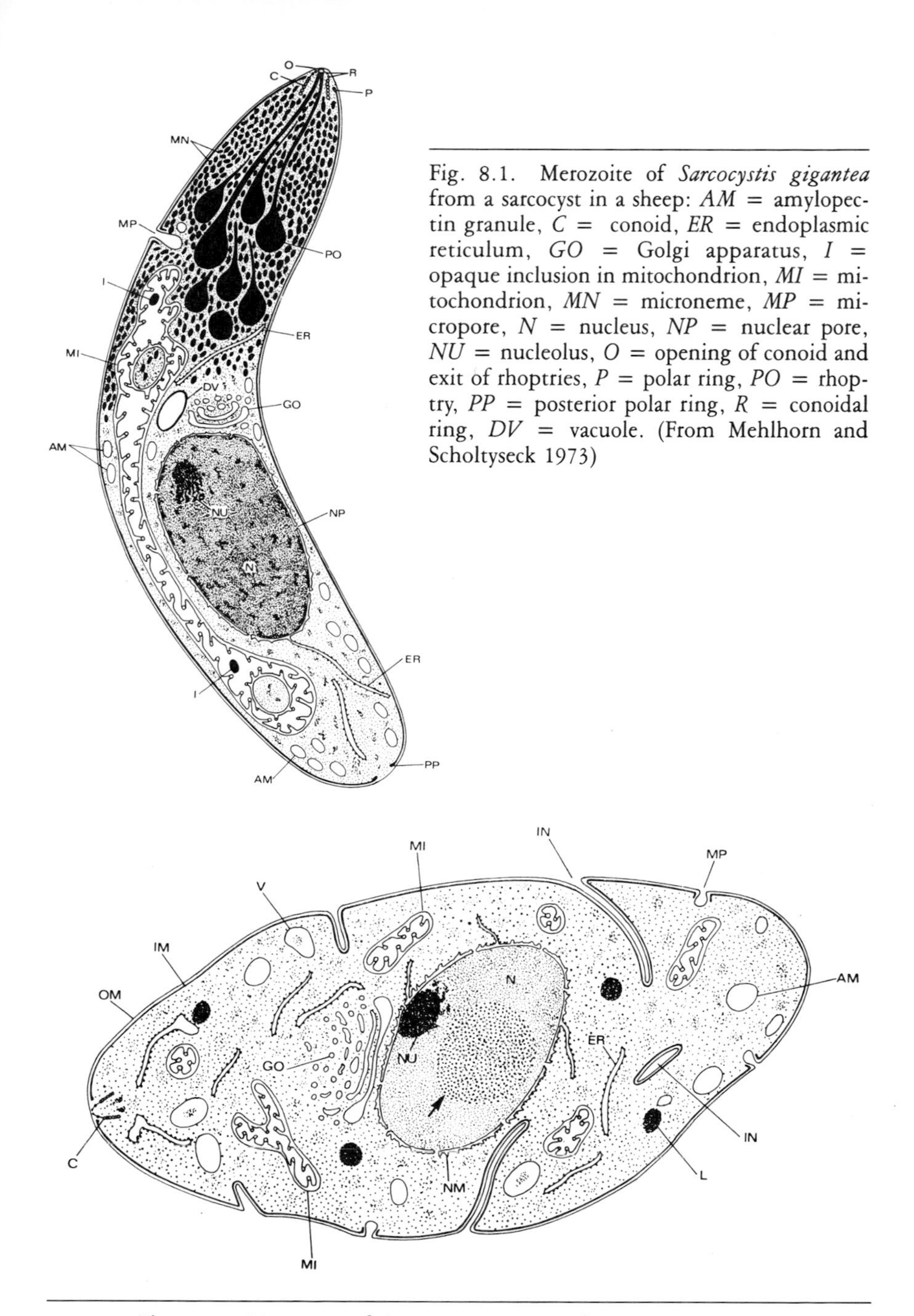

Fig. 8.1. Merozoite of *Sarcocystis gigantea* from a sarcocyst in a sheep: *AM* = amylopectin granule, *C* = conoid, *ER* = endoplasmic reticulum, *GO* = Golgi apparatus, *I* = opaque inclusion in mitochondrion, *MI* = mitochondrion, *MN* = microneme, *MP* = micropore, *N* = nucleus, *NP* = nuclear pore, *NU* = nucleolus, *O* = opening of conoid and exit of rhoptries, *P* = polar ring, *PO* = rhoptry, *PP* = posterior polar ring, *R* = conoidal ring, *DV* = vacuole. (From Mehlhorn and Scholtyseck 1973)

Fig. 8.2. Metrocyte of *Sarcocystis gigantea* from a sarcocyst in a sheep: *AM* = amylopectin granule, *C* = conoid, *ER* = endoplasmic reticulum, *IM* = inner membrane of pellicle, *IN* = invagination, *L* = lipid, *MI* = mitochondrion, *MP* = micropore, *N* = nucleus, *NM* = nuclear membrane, *NU* = nucleolus, *OM* = outer membrane of pellicle, *V* = vacuole, *GO* = Golgi apparatus. (From Mehlhorn and Scholtyseck 1973)

pyle, polar granule, and residuum. The sporocysts are ellipsoidal; with a relatively thick, smooth wall; without a Stieda body, but with a residuum. The sporozoites are elongate. The sporocysts generally break out of the oocysts in the predator's intestine so that sporulated sporocysts (which have often been confused with *Cryptosporidium* oocysts) and a few sporulated oocysts are found in the feces. These oocysts are usually dumbbell-shaped, with a very thin wall sunken between the sporocysts. No meronts or merozoites are produced in the definitive host. (See Fig. 8.3.)

The intermediate host can be infected by ingesting sporocysts. While they have not been recognized in all species of *Sarcocystis,* there appear to be one or two generations of meront in the endothelial cells of the small blood vessels in various organs, followed by another and final generation in the striated muscles. The early generations do not have a cyst wall around them. They produce a variable number (probably not over 200) of tachyzoites; the last generation of

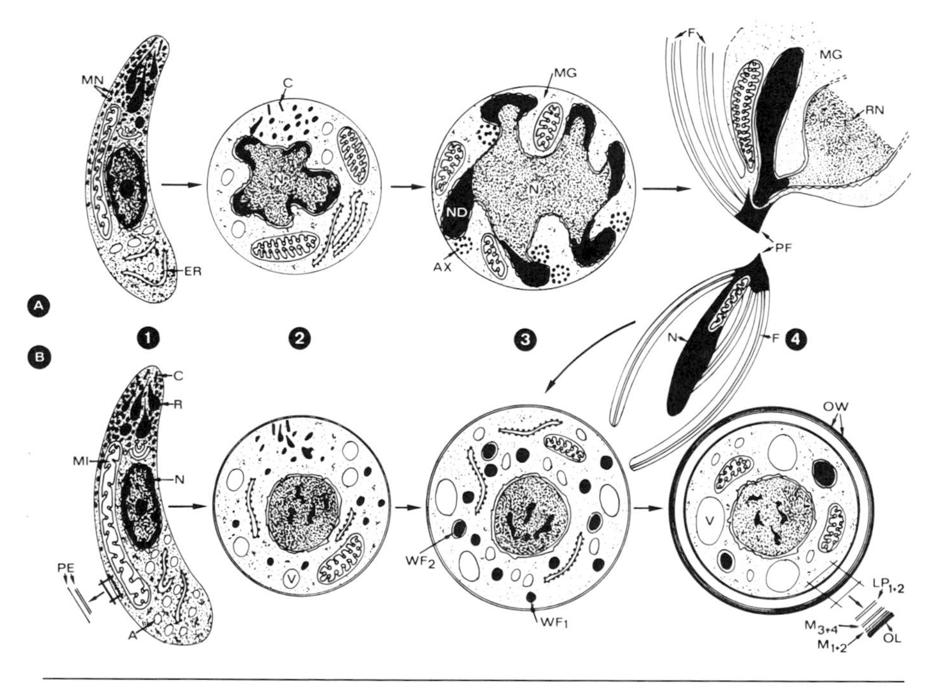

Fig. 8.3. Formation of sexual stages of *Sarcocystis:* *A* = amylopectin granule, *AX* = axoneme, *C* = conoid, *ER* = endoplasmic reticulum, *F* = flagellum, *LP* = inner layers of oocyst wall, *M* = middle layers of oocyst wall, *MG* = microgamont, *MI* = mitochondrion, *OL* = outer layer of oocyst wall, *PE* = pellicle, *PF* = perforatorium, *R* = rhoptry, *RN* = residual nucleus, *WF* = wall-forming bodies, *MN* = microneme, *ND* = nuclear diverticulum, *N* = nucleus, *OW* = outer wall, *V* = vacuole. From Becker et al., 1979, Gustav Fischer Verlag)

the latter enter the striated muscles and form the last meront (sarcocyst) generation, as described above. It has a definite wall, which, depending on the species, may be smooth and relatively thin or may contain cytophaneres and be relatively thick. It forms metrocytes within itself; these divide repeatedly by endodyogeny to form intermediate stages and then bradyzoites, which come to live in compartments divided by septa in the mature sarcocyst. The tachyzoites are able to infect other intermediate hosts, but the bradyzoites cannot.

The definitive host becomes infected by ingesting mature (but not immature) sarcocysts containing bradyzoites, usually by eating the prey animal. The metrocytes are not infective for the predator host.

PATHOGENESIS. Most species of *Sarcocystis* do not appear to be particularly pathogenic, but some very definitely are, causing various signs or symptoms and even death. The pathogenic effects of each species, if known, are described in the individual species section.

The effects in the intermediate hosts (prey animals) are caused by the tachyzoites and meronts in the endothelial cells. The intact sarcocysts and bradyzoites do not appear to be pathogenic, and if an animal does not die before they appear, it recovers spontaneously. While the intact sarcocysts are considered nonpathogenic, they may contain a powerful endotoxin known as sarcocystin, which is highly toxic for rabbits, mice, and sparrows but probably less so for rats, sheep, and some other animals.

In a few species the gamonts, gametes, and oocysts developing in the intestine may cause diarrhea or even more serious effects. In most species, however, they appear to be quite harmless.

IMMUNITY. Predator animals apparently do not become immune to reinfection. Oocysts and sporocysts develop in them each time they are reinfected. A certain degree of resistance may develop, however, against the meronts in prey animals.

DIAGNOSIS. Sarcocystosis is diagnosed in the intermediate host by histologic examination. However, because the sarcocysts cause no recognizable signs, the infection is rarely diagnosed until after death. Infections in the definitive host can be found by fecal examination for sporocysts. However, these may often be present without any indications of disease.

CULTIVATION. Both merogony and gamogony have been obtained in culture (Fayer 1970, 1972; Vetterling et al. 1973; Dubremetz et al. 1975; Fayer and Thompson 1975; Becker et al. 1979; Mehlhorn and Heydorn 1979; Vershinin et al. 1979).

TREATMENT. Little success has been obtained in treating sarcocystosis.

PREVENTION AND CONTROL. Intermediate hosts of *Sarcocystis* acquire their infections by ingesting the sporocysts in grazing or otherwise. Hence, prevention of infection in them depends on keeping predators and prey animals separate and in keeping the prey animals away from areas where predators have defecated. Disposal of the definitive host's feces in such a way that intermediate hosts cannot get at them is relatively easy if the definitive host is man, but practically impossible if it is some other animal.

Definitive hosts acquire their infections by eating prey animals or their meat. Freezing kills the bradyzoites, but they may still be present and infective in uncooked or poorly cooked meat.

S. hominis appears to be much more common in man than is generally believed. To avoid it, meat should be eaten well done.

The species of *Sarcocystis* discussed below are arranged by intermediate host and alphabetically under them.

Sarcocystis cruzi (Hasselmann, 1923) Wenyon, 1926

SYNONYMS. *Miescheria cruzi, S. fusiformis* in part, *S. iturbei, S. marcovi* in part, *S. bovicanis, Isospora rivolta* in part, *Endorimospora hirsuta, Cryptosporidium vulpis* (?).

This species is common throughout the world. The intermediate host is the ox and definitive hosts are the dog, coyote, wolf, red fox, and raccoon.

OOCYSTS. The oocysts in the feces are fully sporulated and dumbbell-shaped, with the thin oocyst wall sunken between the sporocysts. They are 19–21 × 15–18 μm, without a micropyle, polar granule, or residuum. The sporocysts are ellipsoidal, with one side flatter than the other; 13–22 × 6–15 (mean 15–17 × 8–11) μm; with a smooth, colorless to very pale yellow wall about 0.5 μm thick; without a Stieda body; with a residuum. The sporozoites are banana-shaped, with one end rounded and the other bluntly pointed; about 10–11 × 2–3 μm; usually with a clear globule near the broad end; and lie lengthwise in the sporocysts.

LIFE CYCLE. The first generation meronts are in the endothelial cells of small- and medium-sized arteries in the cecum, large intestine, kidney, pancreas, cerebrum, and mesenteric lymph nodes of the ox; they are 17–52 × 7–28 μm and contain 8–200 nuclei at 15–16 days. The second generation meronts occur at 26–33 days in the endothelial cells of the capillaries of many organs, especially the kidney glomeruli; they are 8–27 × 4–13 μm and contain many tachyzoites 7–8 × 2–3 μm. The last generation meronts (the sarcocysts) can be seen beginning 45 days after inoculation (DAI). They are in the striated muscles; when mature they are up to 1 cm or more long, are compartmented, and contain thousands of bradyzoites about 10 μm long. Most numerous in the tongue, they are also found in many other organs. They have a thin wall less than 1 μm thick bearing a layer of folded-over protrusions that can be seen with the electron microscope (under the light microscope the wall appears smooth). They are mature and infective for dogs in about 3 mo. In the domestic dog the prepatent period is 8–33 days, the patent period 3–70 days or more.

PATHOGENESIS. So far as is known, this species is not pathogenic for carnivores.

However, it is highly pathogenic for the ox, causing anorexia, cachexia, weight loss, anemia, accelerated heart rates, and death in calves 23–54 days after experimental inoculation, and abortion and even death in experimentally infected adult cattle. The second generation meronts are the pathogenic stage; the first generation meronts cause only mild fever, and the sarcocysts are apparently nonpathogenic. The condition characterized by abortion and death called Dalmeny disease reported in Canadian cattle by Corner et al. (1963) was undoubtedly due to *Sarcocystis* (Markus et al. 1974).

Sarcocystis hirsuta Moulé, 1888

SYNONYMS. *Miescheria cruzi* in part, *S. fusiformis* in part, *S. marcovi* in part, *S. bovifelis, Isospora bigemina* in part, *Endorimospora cruzi.*

This species is common throughout the world. The intermediate host is the ox, and definitive hosts are domestic and wild cats.

OOCYSTS. The oocysts in the lamina propria and villar epithelial tips of the cat are 12–18 × 11–14 μm. When passed in the feces they have a thin, smooth, colorless, 1-layered wall that is stretched between the 2 sporocysts, producing a dumbbell-like appearance; without a micropyle, polar granule, or residuum. The sporocysts are ellipsoidal, 11–14 × 7–9 μm, without a Stieda body, with a residuum. The sporozoites are 7.5–9 × 1.5–2 μm.

LIFE CYCLE. The first generation meronts are found 7–23 DAI in arteries associated with the mesenteric lymph nodes and intestine and in the mesenteric lymph nodes themselves. When mature they are 37 × 22 μm and contain more than 100 tachyzoites 5 × 1 μm. Second generation meronts in the capillaries of the heart and also of the thigh, diaphragm, tongue, and eye muscles can be found 15–23 DAI. When mature they are 14 × 6.5 μm and contain 3–35 tachyzoites 4 × 1.5 μm. Sarcocysts form 25–75 DAI in the striated muscles but not in the heart; they are most common in the esophageal muscles. When mature they are up to 800 μm long and have a striated wall up to 6 μm thick. They become infective for cats 75–120 DAI. They are compartmented; the primary wall contains palisadelike protrusions about 5 μm long and 1.5 μm in diameter that contain 200–300 parallel fibrils running into the interior of the sarcocyst. The metrocytes are about 12–14 × 5–7 μm, the bradyzoites about 13–17 × 2.5–3 μm.

The oocysts develop extracellularly in the lamina propria of the cat small intestine. The prepatent period in the cat is 7–9 (or more) days, the patent period 6–17 (or more) days.

PATHOGENESIS. This species is not or only slightly pathogenic for calves and apparently not at all pathogenic for cats.

Sarcocystis hominis (Railliet and Lucet, 1891) Dubey, 1976

SYNONYMS. *Coccidium bigeminum* var. *hominis* in part, *Miescheria cruzi* in part, *Isospora hominis* in part, *Lucetina hominis* in part, *S. fusiformis* in part, *S. bovihominis, Endorimospora hominis.*

This species is common throughout the world. The intermediate host is the ox (and presumably the zebu); the known definitive hosts are man, the rhesus monkey, baboon, and presumably the chimpanzee.

OOCYSTS. The oocyst wall is very thin, stretched around the sporocysts, usually constricted between them, and sometimes not visible; the oocysts in the feces are dumbbell-shaped. They are about 20 × 15 μm, without a micropyle, polar granule, or residuum. The sporocysts are ellipsoidal or with one side flattened; about 14–15 × 10–11 μm; without a Stieda body; with a residuum. The sporozoites are elongate.

LIFE CYCLE. Early meronts, if any, are unknown. Sarcocysts are present in the striated muscles of the ox by 98 days after sporocysts from man have been ingested. They are compartmented, with a wall about 6 μm thick that appears radially striated by light microscopy. The striations are due to protrusions in the primary wall. The sarcocysts contain globular metrocytes at first, which divide repeatedly by endodyogeny, finally becoming elongate bradyzoites about 13–17 × 2.5–3 μm that resemble those of *S. cruzi* and *S. hirsuta* in other respects. These bradyzoites are infective for man. The prepatent period is 9–10 days, the patent period more than 6 wk.

PATHOGENESIS. This species is slightly if at all pathogenic for calves. It may cause diarrhea in man.

Sarcocystis gigantea (Railliet, 1886) Ashford, 1977

SYNONYMS. *Balbiania gigantea, S. ovifelis, S. tenella* in part, *Endorimospora tenella.*

This species is common throughout most of the world. Intermediate hosts are the domestic sheep and Rocky Mountain bighorn sheep; definitive hosts are the cat and red fox.

OOCYSTS. The oocysts are ellipsoidal at first; about 15–18 × 10–14 μm; with a smooth, colorless, 1-layered wall about 0.25 μm thick that decreases to 0.1 μm during sporulation. Sporulated oocysts are dumbbell-shaped, without a micropyle, polar granule, or residuum. The sporocysts are ellipsoidal, 10–14 × 8–10 (mean 12 × 8–9) μm, without a Stieda body, with a residuum. The sporozoites are sausage-shaped, with a clear globule at one end.

LIFE CYCLE. Early meronts, if any, are unknown. Sarcocysts are rather ellipsoidal, up to 1 cm long, compartmented, with a primary wall about 25 nm thick with many cauliflowerlike protrusions containing microtubules; they are compartmented. Globular metrocytes are present at first. These divide repeatedly to form bradyzoites much like those of *S. cruzi.* The prepatent period in the cat is 10–14 days, the patent period 4–43 days.

PATHOGENESIS. This species is apparently only slightly if at all pathogenic for either lambs or cats.

Sarcocystis medusiformis Collins, Atkinson, and Charleston, 1979

This species has been found so far only in New Zealand. The intermediate host is the domestic sheep, the definitive host the cat.

OOCYSTS. Oocysts are unknown. The sporocysts are 10–13 × 7–9 (mean 12 × 8) μm.

LIFE CYCLE. The sarcocyst wall is thin (in contrast with the thick wall of *S. gigantea*) and has rounded villi. There are apparently no septa. The prepatent period in cats is 11–20 days.

Sarcocystis odoi Dubey and Lozier, 1983

This species occurs in North America. The intermediate host is the white-tailed deer, the definitive host the cat.

OOCYSTS. Sporulated oocysts are 16–21 × 12.5–16 (mean 18 × 14) μm, without a residuum or polar granule. Sporulated sporocysts are ellipsoidal, 11–15 × 9–11 (mean 13 × 10) μm, without a Stieda body, with a residuum and 4 sporozoites 8 × 1.5 μm. Sporocysts in cat feces are 11–13 × 7–11 (mean 12 × 9) μm.

Sarcocystis tenella (Railliet, 1886) Moulé, 1886

SYNONYMS. *Miescheria tenella, S. ovicanis, Isospora rivolta* in part, *I. bigemina* in part, probably *Hoareosporidium pellerdyi, Endorimospora ovicanis.*

This species is common throughout the world. The intermediate host is the domestic sheep, and definitive hosts are the dog, coyote, red fox, and probably dingo.

OOCYSTS. The oocysts are spherical before sporulation but become dumbbell-shaped, with a thin wall stretched between the 2 sporocysts after sporulation, without a micropyle, polar granule, or residuum. The sporocysts are ellipsoidal, with one side flatter than the other; 13–16 × 8–11 (mean 14–15 × 9–10) μm; with a smooth, colorless to very pale yellow wall about 0.5 μm thick; without a Stieda body; with a residuum. The sporozoites are banana-shaped with one end rounded and the other bluntly pointed, lie lengthwise in the sporocysts, and usually have a clear globule near the broad end.

LIFE CYCLE. There are three meront generations in sheep. The first generation meronts are small and found in or beneath the endothelium of the arteries and arterioles of many organs (but not the brain) 6–21 days after ingestion of sporocysts from the dog or coyote. Second generation, small meronts and merozoites are in the capillary endothelium of many organs, especially the kidney glomeruli and convoluted tubules, 16–40 days after ingestion of sporocysts. First generation meronts are 19–29 × 7.5–24 μm and contain 120–280 merozoites 5–7.5 × 1–2 μm; second generation meronts are 8–15 × 7–13.5 μm and contain 30–80 merozoites 6 × 1–2 μm.

The sarcocysts, which are always microscopic (about 500 × 60–100 μm), can be found in the striated muscle fibers as early as 35 days after sporocyst ingestion. They are compartmented and have a wall up to 25 nm thick folded regularly to form palisadelike protrusions about 3.5 μm long containing neither microfibrils nor microtubules; with light microscopy the combined protrusions look like a radially striated thick wall. The earliest sarcocysts contain only metrocytes. Bradyzoites appear 52–66 DAI, and the sarcocysts are infective for coyotes 75 DAI.

There is no merogony in the definitive host. Oocysts are formed in a day and sporulation is complete within 8 days. The prepatent period in dogs and coyotes is usually 8–14 days but may be 36 days.

PATHOGENESIS. This species is highly pathogenic for lambs, causing fever, inappetence, anemia, and even death. It is perhaps mildly pathogenic for the dog and is not pathogenic for the red fox.

Sarcocystis capracanis Fischer, 1979

This species is common throughout the world. The intermediate host is the goat, the definitive host the dog.

OOCYSTS. The oocysts have not been described. The sporocysts are ellipsoidal, 12–15 × 8–10 μm, without a Stieda body, with a residuum.

LIFE CYCLE. Immature first generation meronts and many merozoites 6–8 × 2–4 μm are in endothelial cells of the small blood vessels of many organs 12 DAI. Second generation meronts are in the endothelial cells of small blood vessels 20–23 DAI. Sarcocysts containing metrocytes alone are in the heart muscle, striated muscles, and brain of the goat by day 43. Bradyzoites appear on day 65; they are about 14 × 4 μm when alive and 12–16 × 3–5 μm in fixed, stained smears. By day 92 the sarcocysts are 130–800 × 50–70 μm; sarcocysts in naturally infected goats are up to 3 mm long. The sarcocyst wall averages 2.6 μm in width, appears striated, and has fingerlike protrusions.

The prepatent period in the dog is 7–8 days, the patent period several weeks.

PATHOGENESIS. This species is not pathogenic for the dog, but the first and second generation meronts are pathogenic for the goat and can even cause death. Pregnant goats may abort without clinical signs.

Sarcocystis moulei Neveu-Lemaire, 1912

This species is apparently common throughout the world. The intermediate host is the goat, and the definitive host is unknown. The sarcocysts are said to be larger than those of *S. gigantea.* This species is presumably not pathogenic.

Sarcocystis fusiformis (Railliet, 1897) Bernard and Bauche, 1912

SYNONYMS. *Balbiania fusiformis, S. blanchardi, S. siamensis, S. bubali.*

This species occurs commonly in Africa and Asia. The intermediate host is the water buffalo, the definitive host the cat.

OOCYSTS. The oocysts are 17–18 × 13 μm, the sporocysts 8–14 × 6–10 μm.

LIFE CYCLE. Early meronts, if any, are unknown. Sarcocysts in the muscle fibers of the water buffalo are compartmented, whitish, fusiform bodies up to 5–15 mm long and 2–4 mm in diameter, with cauliflowerlike cytophaneres. The prepatent period in the cat is 10–33 days, the patent period more than 4 wk.

Sarcocystis levinei Dissanaike and Kan, 1978

SYNONYM. Large form of *Isospora bigemina* in part.

This species occurs commonly in Asia. The intermediate host is the water buffalo, the definitive host the dog.

OOCYSTS. Oocysts scraped from the intestinal wall are 21 × 15 μm; with a smooth, thin, apparently 1-layered wall; without a micropyle, residuum, or (apparently) polar granule. Sporocysts in dog feces are ellipsoidal, 11–16 × 8–10 μm, without a Stieda body, with a residuum. Sporozoites in sections are 8 × 1.5 μm.

LIFE CYCLE. Early meronts, if any, are unknown. Sarcocysts in water buffalo muscles are compartmented, spindle-shaped, 0.8–1.15 × 0.09–0.14 mm, with a thin wall, with a few peripheral metrocytes and many banana-shaped bradyzoites 17–18 × 4–5 μm. The sarcocyst wall is 33–44 nm thick, with invaginations 18–33 nm deep and 56–89 nm apart, with sloping cytophaneres with a highly folded, wavy wall 6–10 μm high. The prepatent period in dogs is 17–20 days, the patent period more than 21 wk.

Sarcocystis cameli Mason, 1910

This species occurs commonly in North Africa (Egypt). The intermediate host is the dromedary camel, the definitive host the dog.

OOCYSTS. Sporocysts in dog feces average 12 × 9 μm and contain four sporozoites and a coarse residuum.

LIFE CYCLE. The sarcocysts are compartmented, up to 12 mm long and 2 mm in diameter, with striated walls. The bradyzoites are banana-shaped, 15–20 × 4–6 μm. Metrocytes are apparently also present.

Hilali and Mohamed (1980) said that the sarcocysts in the camel esophagus and diaphragm muscles are microscopic, 35–389 × 22–33 μm, with striated walls 1–2 μm thick. The prepatent period in dogs is 10–14 days, the patent period 69–73 days.

Sarcocystis capreoli Levchenko, 1963

SYNONYM. *S. capreolicanis.*

This species is common in Europe and the USSR. Its intermediate host is the roe deer, its definitive hosts the dog and red fox.

REMARKS. The number, names, and definitive hosts of the species of *Sarcocystis* in the roe deer have yet to be determined. Four species have been named; in chronologic order, they are *S. gracilis, S. sibirica, S. capreoli,* and *S. capreolicanis.* However, some of the later authors were not aware of the earlier names.

Sarcocystis hemionilatrantis Hudkins and Kistner, 1977

This species occurs in North America. The intermediate host is the mule deer, definitive hosts the coyote and dog.

OOCYSTS. The sporocysts are ellipsoidal, 14–16 × 9–12 μm, without a Stieda body, with a residuum.

LIFE CYCLE. There appear to be three types of meront in mule deer. Microscopic meronts have been found 27–39 days after ingestion of sporocysts in macrophages between muscle fibers, and near blood vessels in the esophagus, heart, striated muscles, diaphragm, and tongue. Sarcocysts occur in the muscles 60 DAI. In the coyote the prepatent period is 9–12 days, the patent period 12–36 days.

PATHOGENESIS. This species is apparently not pathogenic for the coyote or dog but may kill fawns.

Sarcocystis bertrami Doflein, 1901

SYNONYMS. *Isospora rivolta* sporocysts in part, large form of *I. bigemina* in part, *Endorimospora bertrami, S. equicanis,* perhaps *Hoareosporidium pellerdyi.*

This species is presumably common throughout the world. Intermediate hosts are the horse and ass, the definitive host the dog.

OOCYSTS. The sporocysts are ellipsoidal, 15–16 × 9–11 μm, without a Stieda body, with a residuum.

LIFE CYCLE. The sarcocysts in horse muscle are compartmented, up to 10 mm long, and have a layer of cytophaneres on their outside. The metrocytes are 5–8 × 3–5 μm and the bradyzoites 8–10 × 2.5–3.5 μm, with 22 subpellicular microtubules and the usual apical complex.

PATHOGENESIS. This species is apparently not pathogenic for either the horse or the dog.

REMARKS. The relationship of this species to *S. fayeri* is not clear.

Sarcocystis fayeri Dubey, Streitel, Stromberg, and Toussant, 1977

This species has apparently been found only in North America (Ohio). Its intermediate host is the horse, its definitive host the dog. Its relationship to *S. bertrami* remains to be determined.

OOCYSTS. The sporocysts are ellipsoidal, 11–13 × 7–9 μm, without a Stieda body, with a residuum. In fixed sections they are 10–13 × 8.5 μm. The sporozoites are 5–7 × 1–1.5 μm.

LIFE CYCLE. The sarcocysts are up to 900 × 70 μm, without obvious compartments, and with a striated wall 1–2 μm thick. The bradyzoites in the sarcocysts are banana-shaped, with one end pointed, and 15–20 × 2–3 μm when fixed and stained. The prepatent period is 11–14 days.

Sarcocystis miescheriana (Kühn, 1865) Labbé, 1899

SYNONYMS. *Mieschieria utriculosa* in part, *Synchrytium miescherianum, S. miescheri, Sarcocystis suicanis, Coccidium bigeminum* in part, *Lucetina bigemina* probably, *Cryptosporidium vulpis, Endorimospora miescheriana, S. bigemina.*

This species is common throughout the world. The intermediate host is the pig, the definitive hosts the dog, wolf, red fox, and raccoon.

OOCYSTS. The sporulated oocysts are 12–18 × 7–13 μm, ellipsoidal, without a Stieda body, with a residuum. The sporozoites are 7–8 × 3–4 μm.

LIFE CYCLE. The first generation meronts are in the endothelial cells of the venules of the liver up to the 7th DAI. The second generation meronts are in the endothelial cells of capillaries in all organs, especially the heart muscle, 8–13 (or more) DAI. The sarcocysts in porcine muscle are compartmented, up to 0.5–4 mm long and up to 3 mm in diameter, with a striated wall. The prepatent period is 8–9 days.

PATHOGENESIS. This species is apparently not pathogenic for the dog. Small numbers of sporocysts do not cause significant changes in the pig, but moderate numbers depress growth and large numbers may even kill young pigs.

Sarcocystis porcifelis Dubey, 1976

SYNONYM. *Miescheria utriculosa* in part, *S. miescheriana* in part.

This species is apparently uncommon. Its intermediate host is the pig, its definitive host the cat.

OOCYSTS. The sporocysts average 13.5 × 8 μm, without a Stieda body, with a residuum. The sporozoites average 9.5 × 4 μm.

LIFE CYCLE. The prepatent period is 4–9 days, the patent period 10–15 days.

Sarcocystis suihominis (Tadros and Laarman, 1976) Heydorn, 1977

SYNONYMS. *Miescheria utriculosa* in part, *Coccidium bigeminum* var. *hominis in part, Isospora hominis* in part, *Lucetina hominis* in part, *S. porcihominis, Endorimospora suihominis.*

This species is apparently quite common, presumably worldwide. Its intermediate host is the pig, its definitive hosts are man, the chimpanzee, rhesus monkey, and cynomolgus monkey.

OOCYSTS. The oocysts in the feces are 18–20 × 12–15 μm, without a micropyle, polar granule, or residuum. The sporocysts are ellipsoidal, 11–14 × 10–11 μm, without a Stieda body, with a residuum.

LIFE CYCLE. There are two presarcocyst generations of meront. The first generation meronts occur in the endothelial cells of the veins of the liver 6 DAI. The second generation meronts occur in the endothelial cells of the veins of all organs about 7–20 DAI. Sarcocysts can be found in the striated muscles, heart, and even brain. At first they contain only metrocytes. These divide repeatedly, and bradyzoites can be found by day 56. Mature sarcocysts on day 142 are compartmented, up to 1.5 mm long, and have protrusions up to 13 μm long folded closely on the surface, so that the sarcocysts appear to be relatively thin walled. The metrocytes are globular and about 10 μm long, the bradyzoites up to 13 μm long.

The macrogametes and microgamonts in the primate host are about 10 μm in diameter. The latter produce 20–30 triflagellate microgametes about 4–6 μm long. The prepatent period in the chimpanzee, rhesus, and cynomolgus monkeys is 12–14 days, the patent period at least 18 days.

PATHOGENESIS. This species is not pathogenic for nonhuman primates but is extremely pathogenic for pigs and man.

Sarcocystis cuniculi Brumpt, 1913

This species is quite common, presumably throughout the world. Its intermediate host is the domestic rabbit, the definitive host the cat.

OOCYSTS. The sporocysts are ellipsoidal, 13 × 10 μm, without a Stieda body, with a residuum. The sporocysts are sausage-shaped.

LIFE CYCLE. The sarcocysts are elongate, compartmented, up to several mm long and 0.5 mm in diameter, with a wall containing many fine projections up to 11 μm high that are tightly packed to form a luxuriant pile. The metrocytes are 4–5 μm in diameter, the bradyzoites 11–16 × 4–6 μm. The sarcocysts may or may not be macroscopically visible. It takes 93 or more DAI for the sarcocysts in rabbits to become infective for cats. The prepatent period in the cat is 9 days.

Sarcocystis horvathi Ratz, 1908

SYNONYM. *Sarcocystis gallinarum.*

This species is uncommon; it presumably occurs throughout the world. Its intermediate host is the chicken, the definitive host the dog and perhaps the cat.

OOCYSTS. The sporocysts in dog feces are 10–13 × 7–9 μm, with a residuum.

LIFE CYCLE. The sarcocysts in chicken muscles have a striated wall. The prepatent period is 6–10 days, the patent period in the dog apparently 21–23 days.

PATHOGENESIS. This species may cause severe myositis in chickens.

Sarcocystis rileyi (Stiles, 1893) Minchin, 1903

SYNONYMS. *Balbiania rileyi, Sarcocystis anatina.*

This species is quite common in dabbling ducks, rather uncommon in diving ducks, and rare in domestic ducks throughout the world. Its intermediate hosts are domestic and wild ducks, the definitive hosts the striped skunk, possibly the opossum, and perhaps the dog and cat.

OOCYSTS. The sporocysts in skunk feces are 10–14 × 5–10 μm, with a granular residuum.

LIFE CYCLE. Microscopic sarcocysts can be found in the skeletal muscles 85 DAI and small macroscopic sarcocysts 154 DAI. They are 2–8 × 1 mm, with irregular cauliflowerlike projections on the primary wall. Later the sarcocysts become several centimeters long and may be extremely abundant. The prepatent period in the skunk is 17–21 days, the patent period 29–44 days.

PATHOGENESIS. The sarcocysts in duck muscles are apparently not pathogenic.

Other Species of Sarcocystis

S. muris (Railliet, 1886) Labbé, 1899, has sarcocysts in the house mouse and sexual stages in the cat. Other species with sarcocysts in the house mouse are *S. dispersa* Černá and Sénaud, 1977, *S. scotti* Levine and Tadros, 1980, and *S. sebeki* (Tadros and Laarman, 1976) Levine, 1978. The definitive hosts of *S. dispersa* are the barn owl *Tyto alba,* masked owl *T. novaehollandiae,* and long-eared owl *Asio otus.* The definitive host of *S. scotti* and *S. sebeki* is the tawny owl *Strix aluco.*

S. cymruensis Ashford, 1978, has sarcocysts in the Norway rat and sexual stages in the cat. *S. murinotechis* Munday and Mason, 1980, has sarcocysts in the Norway rat and sexual stages in the tiger snake *Notechis ater. S. singaporensis* Zaman and Colley, 1976, *S. villivillosi* Beaver and Maleckar, 1981, and *S. zamani* Beaver and Maleckar, 1981, have sarcocysts in the Norway rat and sexual stages in the reticulated python *Python reticulatus.*

S. caviae de Almeida, 1928, has sarcocysts in the guinea pig; the definitive host is unknown. *S. kortei* Castellani and Chalmers, 1910, has sarcocysts in rhesus and other monkeys; the definitive host is unknown. *S. nesbitti* Mandour, 1969, has sarcocysts in the rhesus and other monkeys; the definitive host is unknown. *S. leporum* Crawley, 1914, has sarcocysts in cottontails and sexual stages in the raccoon and, to a lesser extent, in the cat.

Genus *Arthrocystis* Levine, Beamer, and Simon, 1970

In this genus the merozoites are spherical and the meront septa thick. The meronts appear jointed like bamboo, with one to four or more compartments. The exact taxonomic position of this genus is unknown. Some workers believe it to be *Leucocytozoon* in a foreign host.

Arthrocystis galli Levine, Beamer, and Simon, 1970

This species was found in the muscles of chickens in Uttar Pradesh, India. It formed *Sarcocystis*-like meronts up to 1,400 μm long and 126 μm in diameter in the skeletal and heart muscles. About 100 hens in three flocks died.

Genus *Toxoplasma* Nicolle and Manceaux, 1909

In this genus merogony occurs in both the intermediate and definitive hosts. There are no metrocytes. Oocysts are produced in the intestinal cells of the definitive host and are not sporulated when passed. Transmission may be either by sporulated oocysts or by merozoites. About seven species have been named.

Because of its importance as a cause of human disease, *T. gondii* is the only member of the genus of major research interest. The literature on it is vast. Jira and Kozojed (1970) published a two-volume bibliography of papers from 1908–1967 that contained 7,763 titles; many more papers have been published since then. Various aspects of *Toxoplasma* and toxoplasmosis have been reviewed since 1970 by Jacobs (1970), Frenkel (1973, 1974a, 1974b), Jones (1973), Siim (1974), Dubey (1977), Scott (1978), Zasukhin (1980), Beverley (1976), Overdulve (1978), and Beyer et al. (1979), among others.

The generic name *Hammondia* Frenkel, 1974, is a synonym of *Toxoplasma*.

Toxoplasma gondii (Nicolle and Manceaux, 1908) Nicolle and Manceaux, 1909

SYNONYMS. See Levine (1977) for a list of 18 synonyms. To these should be added the small form of *Isospora bigemina,* in which unsporulated oocysts are passed in the feces.

This species is common throughout the world. About 25–30% of the human population of the world has antibodies against it, and the disease has been found in practically all domestic animals.

INTERMEDIATE HOSTS. *T. gondii* was first found in the gondi *Ctenodactylus gundi,* a North African rodent, and has since been found in over 200 species of mammals and birds. Its host list includes rodents, lagomorphs, insectivores, carnivores, marsupials, primates (including man), and many species of birds, including the chicken, pigeon, and canary. Turtles, lizards, geckos, chameleons, and skinks can be infected experimentally. However, most of the organisms reported as *Toxoplasma* from the blood of various wild birds are probably *Atoxoplasma.*

DEFINITIVE HOSTS. Only members of the carnivore family Felidae have been found to be definitive hosts of *T. gondii.* Those recognized so far are the cat, jaguarundi, ocelot, mountain lion, Asian leopard cat, bobcat, and probably cheetah.

LOCATION. Several types of meront and sexual stages are in the epithelial cells of the villi of the small intestine of the cat and other felids. Meronts and merozoites are in many types of cell of the intermediate host, including neurons, microglia, endothelium, liver parenchyma cells, lung and glandular epithelial cells, cardiac and skeletal muscle cells, fetal membranes, and leukocytes. In acute infections, merozoites may be found free in the blood and peritoneal exudate.

OOCYSTS. Oocysts are produced in the epithelial cells of the small intestine villi of cats and other felids, but not, so far as is known, in other animals. The oocysts in the feces are unsporulated and spherical at first. After sporulation they are subspherical; 11–14 × 9–11 (mean 12.5 × 11) μm; without a micropyle, residuum, or polar granule; and contain 2 ellipsoidal sporocysts about 8.5 × 6 μm, without a Stieda body, with a residuum. The sporozoites are about 8 × 2 μm. (See Fig. 8.4.)

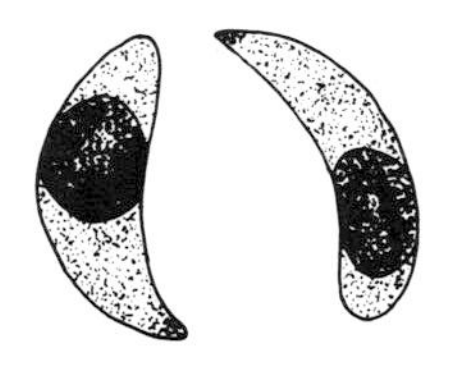

Fig. 8.4. *Toxoplasma gondii* tachyzoites from mouse peritoneal exudate. Giemsa stain, light microscope. (×2,800)

LIFE CYCLE. Intermediate hosts become infected by ingesting sporulated oocysts or infected meat or animals, or congenitally via the placenta. (Definitive hosts can also be intermediate hosts.) Congenital toxoplasmosis of the newborn resulting from infection of the mother while she is pregnant is probably the most common form in man and perhaps sheep but is not nearly so important in cats. Mice, at least, can be infected congenitally for generation after generation. Experimental infections can be established by intravenous, intraperitoneal, or any other type of parenteral inoculation or by feeding merozoites. Following experimental inoculation the protozoa proliferate for a time at the site of injection and then invade the blood stream and cause a generalized infection. Susceptible tissues all over the body are invaded, and the parasites multiply in them, causing local necrosis. The parasitemia continues for some time, until antibodies appear in the plasma, after which the parasites disappear from the blood and more slowly from the tissues. They finally remain only in meronts ("cysts"), and only in the most receptive tissues. In general, the spleen, lungs, and liver are cleared of parasites relatively rapidly, the heart somewhat more slowly, and the brain much more slowly. Residual infections may persist for a number of years. (See Fig. 8.5.)

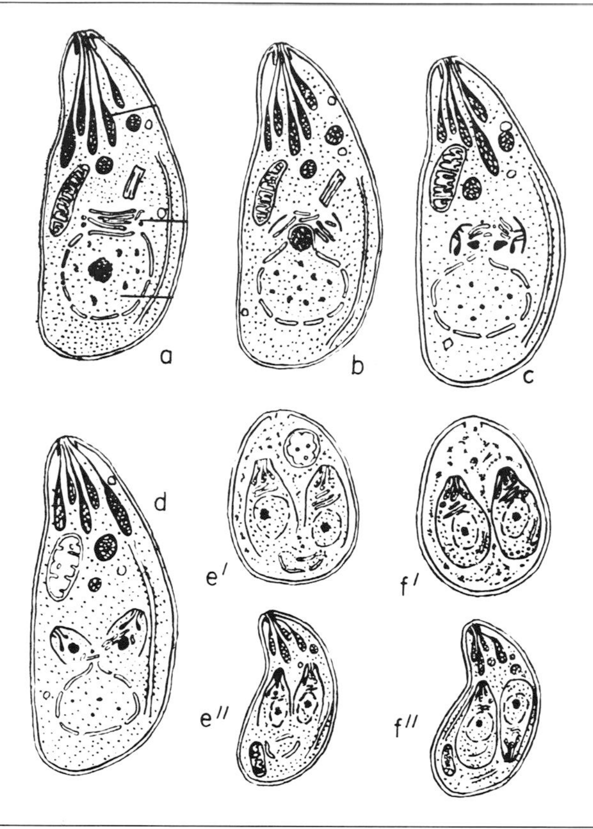

Fig. 8.5. Endodyogeny in *Toxoplasma gondii:* a. formation of ball of DNA in nucleus; b. ball of DNA leaves nucleus; c and d. development of daughter cells; e′and f′. final phase of endodyogeny within a cyst; e″ and f″. final phase of endodyogeny during the proliferation stage. (From van der Zypen and Piekarski 1968)

When the sporulated oocyst is ingested by a susceptible animal, the sporozoites emerge and pass to the parenteral tissues via the blood and lymph; any type of cell may be invaded. Here they multiply by endodyogeny. The stage in which this occurs has been called a pseudocyst, terminal colony, colony aggregate stage, or (preferably) group stage. The merozoites within it are tachyzoites, which multiply by endodyogeny. The group stage with its tachyzoites is the stage found in the leukocytes in peritoneal exudate, but it also occurs in other parenteral locations, such as the liver, lungs, and submucosa; this is the stage occurring in acute toxoplasmosis. (See Fig. 8.6.)

There is an indefinite number of tachyzoite generations. Eventually they enter other cells and induce the host cell to form a wall around them, forming a pseudocyst (a meront). Within it a large number of bradyzoites is formed by endodyogeny. These meronts and bradyzoites are much more resistant to trypsin and pepsin than are the tachyzoites, and they may remain viable in the tissues for years. This stage, occurring commonly in the brain, also occurs in other tissues such as muscle. It is the stage found in chronic infections.

The above meront is the end of the life cycle in all animals except felids, so far as is known. *T. gondii* can be transmitted from one intermediate host to another by means of tachyzoites, bradyzoites, or meronts. The merozoites are 5–8 × 1–2 μm and have a 3-membraned complex at the surface, each membrane consisting of 2 electron-dense layers separated by electron-light material. They have an apical complex consisting of 2 polar rings at the anterior end (and a similar ring at the posterior end); a short, truncate hollow conoid 0.2–0.36 ×0.15 × 0.36 μm composed of 6–7 microtubules spirally coiled at an angle of 45°–50°; 20 to perhaps 30 cylindrical or club-shaped rhoptries of variable length, which apparently open to the outside at the anterior end after passing through the conoid; about 50 curved, rodlike micronemes anterior to the nucleus; and 22 longitudinal subpellicular microtubules arising from a ring at the level of the conoid and running posteriorly for about one-fifth to two-thirds of the body length. Just in front of the nucleus is the Golgi apparatus. There are one or more micropores in the pellicle. The cytoplasm is somewhat vacuolated and contains numerous ribosomes, rough endoplasmic reticulum, and one to several mitochondria. The nucleus is usually spherical or ovoid, about 1–2 μm in diameter, and contains a large nucleolus.

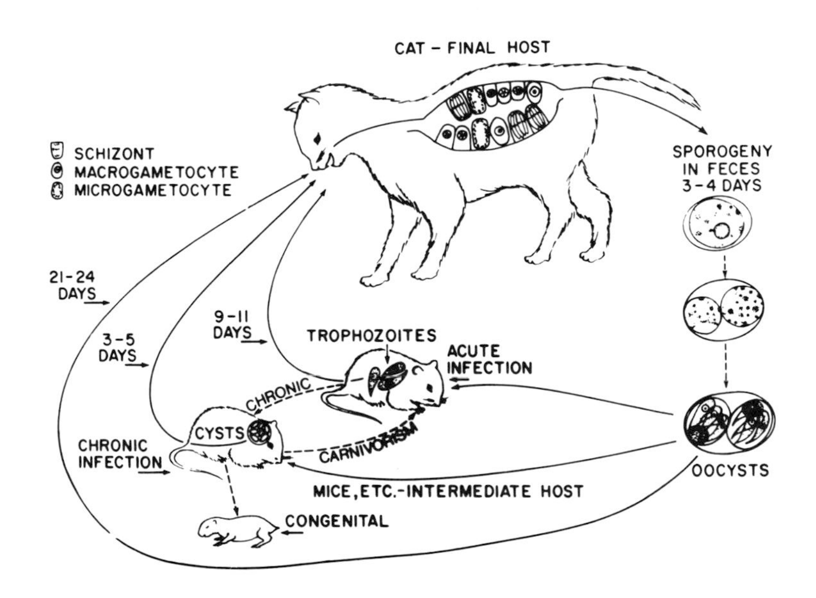

Fig. 8.6. Life cycle of *Toxoplasma gondii*. (From Frenkel, Dubey, and Miller 1970, Science 167 [6 Feb.]:893–96, © 1970, American Association for the Advancement of Science)

In the cat and other felids the bradyzoites enter the intestinal tissues and multiply. Five types have been recognized. It is not known to how many generations they belong.

The macrogametes and microgamonts are in the intestinal epithelial cells. The microgamonts produce 12–32 slender, crescentic microgametes about 3 μm long; they have 2 flagella plus a rudiment of a third. The macrogametes are presumably haploid and simply grow. Fertilization takes place; the resultant zygotes form walls around themselves, become oocysts, and are released into the intestinal lumen.

The prepatent period in cats is 2–7 days after feeding parenteral meronts, 7–10 days after feeding merozoites, and 1–3 wk after feeding fecal oocysts. The patent period in cats is 1–2 wk.

CULTIVATION. *T. gondii* is readily cultivated in tissue culture and chicken embryos.

PATHOGENESIS. The sexual stages of *T. gondii* are apparently not pathogenic, or perhaps only slightly so, for felids. The parenteral stages may or may not cause symptoms. Toxoplasmosis may vary from an inapparent infection to an acutely fatal one. Asymptomatic toxoplasmiasis is the commonest type.

In man a common form of the disease is the congenital type found in newborn infants. It is characterized by encephalitis, rash, jaundice, and hepatomegaly, usually associated with chorioretinitis, hydrocephalus, and microcephaly; the mortality rate is high. Congenital toxoplasmosis occurs in different countries in 0.25–7.0 per 1,000 live births.

Acquired (i.e., noncongenital) human toxoplasmosis has many different manifestations. The most common type is characterized by lymphadenopathy; it may be febrile, nonfebrile, or subclinical. The onset may be acute, with chills and fever, or gradual. The temperature may last 2–4 wk or even longer. The lymph nodes are enlarged, the throat is often sore, and the patients suffer from malaise. Fatigue may persist for some time following recovery, and the lymph nodes remain enlarged for months.

The second type of acquired human toxoplasmosis is a typhuslike, exanthematous disease. In addition to the exanthema, there may be atypical pneumonia, myocarditis, and meningoencephalitis, and the termination is often fatal. Lymphadenopathy may or may not be present.

The third type is a cerebrospinal form, characterized by fever, encephalitis, convulsions, delirium, lymphadenopathy, and mononuclear pleocytosis, followed by death. This form is quite rare.

The fourth type is an ophthalmic form, characterized by chronic chorioretinitis. It is presumably more common than neonatal toxoplasmosis in man.

The disease in domestic animals is similar to that in man. It has been found especially in dogs, cats, and sheep. It may perhaps be associated with distemper in dogs. In sheep it often causes abortion or neonatal death. It also occurs naturally in the pig, cattle, goat, rhesus monkey, and even chicken.

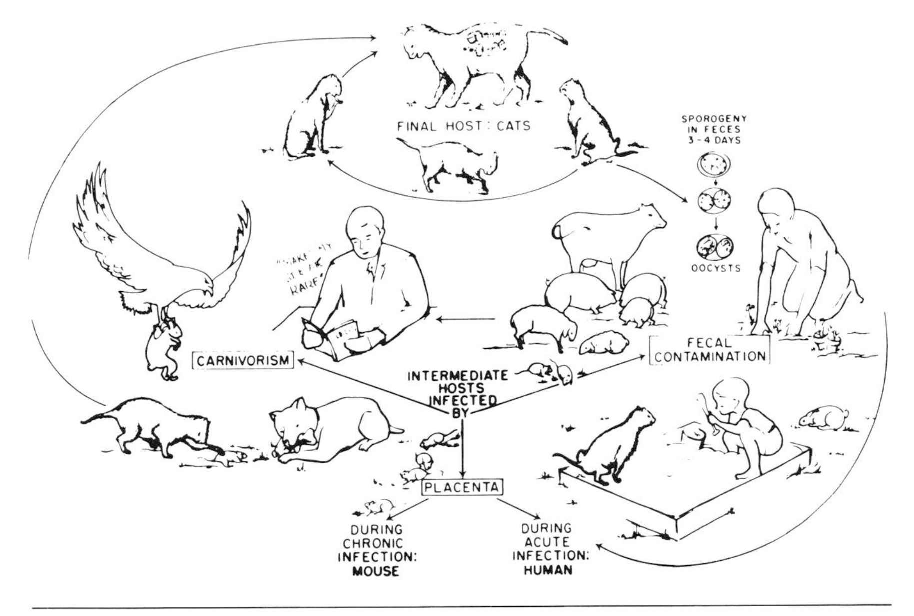

Fig. 8.7. Transmission of toxoplasmosis. (From Frenkel 1971, in *Protozoan and Helminthic Diseases of Man*, ed. R. A. Marcial-Rojas, Williams & Wilkins, Baltimore)

EPIDEMIOLOGY. Transplacental transmission from mother to offspring is probably the commonest cause of frank toxoplasmosis. Apparently the mother must become infected while she is pregnant. She generally has no signs or symptoms but transmits the parasite to the fetus. (See Fig. 8.7.)

There is an obvious relationship between the presence of cats in the household or the farmyard and the presence of *T. gondii* in the people or livestock. Following dissemination to the general public of knowledge of the association between *T. gondii* oocysts in cat feces and toxoplasmosis in people in households containing cats, there was a rash of inquiries about the danger of human infections. All that can be said is that the danger exists unless proper precautions are taken.

Common sources of human toxoplasmosis are raw meat juice fed to babies, wild rabbits, cat feces containing oocysts, and insufficiently cooked meat.

IMMUNITY. There is a definite age immunity. Congenital infections are the commonest, and the mothers usually do not show signs of disease themselves. Young animals are more susceptible than adults. Nude mice (which lack a thymus gland) do not develop immunity.

DIAGNOSIS. The dye test is based on the fact that both the cytoplasm and nucleus of *T. gondii* merozoites stain deeply with alkaline methylene blue after incubation with normal serum, but after incubation with antibody-containing serum only the nuclear endosome will stain. The merozoites used in the dye test can be obtained from peritoneal exudate or tissue culture. Dye test antibodies usually persist for a number of years, probably for more than a decade, although their titer declines slowly.

The complement fixation test is also used. Complement-fixing antibodies rarely appear earlier than 1 mo after infection and decrease relatively rapidly with time.

A number of other serologic tests for *T. gondii* have been reported. These include the hemagglutination test, the skin test, the latex agglutination test, a fluorescent antibody test, the enzyme-linked immunosorbent assay (ELISA), the India ink immunoreaction (IIIR), and a fluorescent antibody test using antigen fixed to cellulose acetate disks on a dipstick (fluoroimmunoassay or FIAX).

The most certain method of diagnosis of toxoplasmosis is by isolation of the parasites themselves by inoculation of experimental animals: mice, ground squirrels, multimammate rats, hamsters, or guinea pigs.

T. gondii can be found in stained smears and sections of tissues and in exudates. It must be differentiated from other organisms, including *Sarcocystis, Besnoitia,* and *Encephalitozoon,* which is not always possible on structural grounds alone.

The dye test still appears to be the most satisfactory serologic test available at present. Reviewing serologic tests, Kagan (1980a, 1980b) gave 149 references and tabulated 31 sources of commercial reagents. He said that the dye test was probably the best but that the indirect hemagglutination and indirect immunofluorescence tests are used routinely by most laboratories, and that they are very good. He recommended the IF (indirect immunofluorescence) test but said the ELISA and FIAX techniques were just as good and also measured IgM antibody.

TREATMENT. No really satisfactory treatment for toxoplasmosis is known. Perhaps the best results have been obtained by the use of pyrimethamine and a sulfonamide simultaneously; the two drugs act synergistically.

PREVENTION AND CONTROL. The measures customarily employed to control infectious diseases should be used against toxoplasmosis. In addition, contact with cats and probably many wild mammals, which are apparently reservoir hosts, should be avoided and rodents should be controlled. Raw meat or meat juice should not be eaten or fed, especially from sheep, rabbits, or pigs.

All meat should be cooked thoroughly before being eaten by man or cats. Pet cats should be fed dry, canned, or boiled food, and they should be kept indoors and trained to defecate in a litter pan. Contact with cats that have an uncontrolled diet should be avoided. Litter pans should be handled with disposable gloves. Children's sandboxes should be covered when not in use. Contact with contaminated soil should be avoided, and the hands should be washed with soap if such contact is unavoidable. Stray cats should be avoided. Pregnant women, and especially those who are serologically negative, should not handle or take care of cats unless they wear plastic gloves. Serologically positive persons are probably quite immune.

Toxoplasma hammondi (Frenkel, 1974) Levine, 1977

SYNONYM. *Isospora hammondi, I.* (*Toxoplasma*) *datusi, Hammondia hammondi.*

This species presumably occurs throughout the world but is uncommon. The definitive host is the cat, and known intermediate hosts (all experimental) are the house mouse, laboratory rat, multimammate rat, golden hamster, guinea pig, dog, and marmoset.

OOCYSTS. Unsporulated oocysts are spherical to subspherical; 11–13 × 10–13 (mean 11 × 11) μm; with a colorless, 2-layered wall about 0.5 μm thick. After sporulation, the oocysts are subspherical to ellipsoidal; 13–14 × 10–11 (mean 13 × 11) μm; with a colorless, 2-layered wall about 0.5 μm thick; without a micropyle, polar granule, or residuum. The sporocysts are ellipsoidal, 8–11 × 6–8 (mean 10 × 6.5) μm, without a Stieda body, with a residuum. The sporozoites are elongate and curved, about 7 × 2 μm, with a nucleus near the center.

LIFE CYCLE. The oocysts are unsporulated when shed. They sporulate on the ground in about 3 days. Intermediate hosts are infected by ingestion of sporulated oocysts. The meronts in the striated muscles are 100–340 × 40–95 μm. They are similar to the sarcocysts of *Sarcocystis muris* in many ways. However, they have a relatively smooth, intensely folded primary wall, beneath which is a zone of granular ground substance. They have no secondary cyst wall, septa, or metrocytes. In all these respects they are similar to the meronts of *Toxoplasma gondii* in striated muscles. The merozoites within these meronts are about 4–5 μm long.

Definitive hosts are infected by eating intermediate hosts or tissues containing mature meronts. After cats ingest these meronts, there is multiplication in the small intestine epithelium followed by gamogony there. The prepatent time in the cat is 5–16 days and the initial patent period 1–2 wk, after which oocysts may be shed intermittently for as long as 136 DAI.

PATHOGENESIS. So far as is known, this species is not pathogenic for the cat, but it may cause severe anorexia with apathia after 7–8 days and sometimes death in experimentally infected mice.

IMMUNITY. There is considerable cross-immunity between this species and *T. gondii.*

Toxoplasma bahiensis (de Moura Costa, 1956) Levine, 1983

SYNONYMS. *Isospora bigemina* in part, *I. bahiensis, I. heydorni, Hammondia heydorni, T. heydorni.*

This species occurs fairly commonly throughout the world. Intermediate hosts are the ox, goat, sheep, moose, guinea pig, and dog. Definitive hosts are the dog, dingo, coyote, and probably red fox, mink, and guinea pig.

OOCYSTS. The oocysts are 13 × 11 μm, without a micropyle or residuum, and are dumbbell-shaped after sporulation. The sporocysts have no Stieda body but have a residuum.

LIFE CYCLE. The oocysts sporulate outside the definitive host in about three days. Meronts and merozoites in the muscles of the intermediate hosts have apparently not been described. Direct dog-to-dog transmission does not occur. The prepatent period in definitive hosts that have been fed infected guinea pigs or dogs is 6–7 days.

PATHOGENESIS. There is no evidence of pathogenicity in lightly infected dogs, coyotes, sheep, goats, or moose.

Genus *Besnoitia* Henry, 1913

Members of this genus are heteroxenous, producing unsporulated oocysts in felids and multiplying by merogony in a variety of prey animals. Sporulation occurs outside the host. Meronts and merozoites can transmit the infection from one intermediate host to another (except for *B. wallacei*). This genus differs from *Toxoplasma* in that the meront wall is thick and contains a number of flattened, giant host nuclei.

The merozoites are banana-shaped, crescentic, or elongate ovoid, and slightly pointed at one end. They move by body flexion. They reproduce by binary fission or endodyogeny; multiple fission has also been described.

This genus is not well known. Species have been found in cattle, horses, reindeer, caribou, rodents, opossums, primates, and reptiles.

Besnoitia besnoiti (Marotel, 1913) Henry, 1913

SYNONYMS. *Sarcocystis besnoiti, Gastrocystis besnoiti, Globidium besnoiti, Isospora besnoiti.*

This species is quite common in the subtropics and tropics, especially in South Africa, Kazakhstan, and Israel. The definitive hosts are domestic and wild cats. Naturally infected intermediate hosts are the ox, zebu, goats, wildebeest, impala, and kudu; experimental intermediate hosts are the rabbit, sheep, gerbil, golden hamster, ground squirrels, house mouse, and marmot.

OOCYSTS. The oocysts are similar to those of *Toxoplasma*. They are ovoid and unsporulated, 14–16 × 12–14 μm, apparently without a micropyle when passed. After sporulation they contain 2 sporocysts, each with 4 sporozoites.

LIFE CYCLE. The oocyst sporulation time is presumably 7–8 days or less. First generation meronts are in the endothelial cells of the blood vessels. The merozoites produced there enter cells of the cutis, subcutis, scleral conjunctiva, connective tissue, fascia, serosae, nasal mucosa, larynx, trachea, and other tissues of intermediate hosts. There they become large nonseptate meronts (pseudocysts or "cysts") about 100–500 μm in diameter, with a thick wall containing several host cell nuclei, and form thousands of merozoites but no metrocytes. Gamogony has apparently not been studied.

The prepatent period in cats is 4–25 days, the patent period 3–15 days.

PATHOGENESIS. *B. besnoiti* is apparently not pathogenic for the cat definitive host and does not affect antelopes severely. However, it may cause serious disease in cattle. Most infected cattle have low-grade, chronic infections without skin lesions. The typical signs in clinically affected cattle are anasarca followed by scleroderma, alopecia, and seborrhea. In the first (febrile) stage, a temperature as high as 41.7 C (107 F) develops, photophobia is present, anasarca develops, lachrymation and hyperemia of the sclera are present, and the cornea and nasal mucosae are studded with whitish, elevated specks (*Besnoitia* pseudocysts). This stage lasts 5–10 days. It then subsides and the second (depilatory) stage begins. In this stage the pathologic and clinical pictures are dominated by skin lesions. The hair falls out over the swollen parts. The skin becomes greatly thickened and loses its elasticity, its surface cracks, and a serosanguinous fluid oozes out. Necrosis develops on the parts in contact with the ground when the animal lies down, and hard sitfasts develop on the stifles, briskets, and elbows. The anasarca subsides, leaving the skin with typical, broad wrinkles along the lower line; the photophobia decreases, and grazing is resumed. The animal may die during this stage, but if not the stage lasts 2 wk to about 1 mo and gradually passes into the third (seborrhea sicca) stage. The hair does not come back, and the denuded parts are covered by a thick, scurfy layer. The sitfasts crack away, fissures remain in the flexor surfaces, the skin hardens, deep scars show plainly, and the hide resembles that of an elephant; the animal looks as though it has mange. Its lymph nodes are permanently enlarged, it is listless and debilitated, and the protozoan pseudocysts remain. In light infections in which there has been little hair loss, the animals appear practically normal, but in severe cases the remaining hair forms patterns resembling the markings on a giraffe.

IMMUNITY. There is no cross-reaction with *T. gondii.* Cattle and rabbits can be immunized against *B. besnoiti* by use of a live vaccine grown in tissue culture and prepared from a strain from the blue wildebeest.

DIAGNOSIS. Besnoitiosis can be diagnosed by biopsy examination of affected skin or other areas. The spherical, encapsulated pseudocysts are pathognomonic. Merozoites are often found in blood smears, sometimes in large numbers, but most of them are introduced when a pseudocyst is cut in obtaining blood. The best method of diagnosis is examination of the scleral conjunctiva; the pseudocysts can be seen with the naked eye, revealing many chronic cases without clinical signs.

PREVENTION AND CONTROL. Separation of cattle from cats, and of domestic cattle from wild ruminants, should help prevent transmission of this parasite, as should the elimination or isolation of infected animals.

Besnoitia bennetti Babudieri, 1932

This species is relatively uncommon. It occurs in Africa, southern Europe, Mexico, and the southern United States. Intermediate hosts are the horse and ass. The definitive host is unknown.

LIFE CYCLE. The oocysts are unknown. The pseudocysts (meronts) in the horse are more or less spherical, without septa, and about 100–1,000 μm in diameter. Their wall is about 10–15 μm thick and contains several flattened, giant host nuclei. The bradyzoites in the pseudocysts are 5–10 × 1–4 μm.

PATHOGENESIS. The disease is a chronic one, running a course of many months. Affected horses are weak and dejected, although their appetite is good. The skin is scurfy and thickened and contains many scabs and whitish scars. The hair may be destroyed by the lesions. The conjunctiva is a peculiar brick red color, with a few petechiae. The temperature is slightly elevated. The muscles in advanced cases are pale brown and friable but contain no parasites. Pseudocysts are abundant in the skin and may also be found in the mucous membrane covering the larynx, nostrils, soft palate, etc. Infected burros are apparently unharmed.

Besnoitia darlingi (Brumpt, 1913) Mandour, 1965

SYNONYMS. *Sarcocystis darlingi, B. panamensis, B. sauriana.*

This species occurs in Central and North America. Its definitive host is the cat. It is common in its opossum intermediate host and has also been found in basilisk and "borroguero" lizards. Experimental infections have been produced in the marmoset, house mouse, golden hamster, and free-tailed bat.

OOCYSTS. Unsporulated oocysts in cat feces are spherical, 11–13 (mean 12) μm in diameter. Sporulated oocysts are 11–13 × 10–11 (mean 12 × 10) μm, without a micropyle or residuum. The sporocysts are ellipsoidal, 6–9 × 5–6 (mean 8 × 5) μm, without a Stieda body, with a residuum. The sporozoites are about 5 μm long.

LIFE CYCLE. The sporulation time is 2–3 days. Meronts 66–300 × 62–156 μm containing hundreds to thousands of merozoites occur in the tongue, ears, mesenteries, gastrointestinal tract, adrenal glands, kidneys, abdominal muscles, and uterus of the opossum; they are most numerous in the adrenal glands. Gamogony has not been described. The prepatent and patent periods in the cat are both 10 days.

PATHOGENESIS. This species is presumably not pathogenic for the cat, but it may cause necrosis, fibrosis, and commensatory hyperplasia of the adrenal glands in some naturally infected opossums.

Besnoitia wallacei (Tadros and Laarman, 1976) Dubey, 1977

SYNONYM. *Isospora wallacei.*

This species presumably occurs throughout the world. Its definitive host is the cat, and its natural intermediate host is the Norway rat; experimental intermediate hosts are the Polynesian rat, Mongolian gerbil, vole, house mouse, and laboratory rabbit.

OOCYSTS. Unsporulated oocysts are subspherical; 16–19 × 10–13 (mean 14–12) μm; with a 2-layered, smooth, light brown to pink wall 0.5 μm thick. Sporulated oocysts are ellipsoidal to subspherical; 15–18 × 12–15 (mean 16 × 13) μm; without a micropyle, polar granule, or residuum. The sporocysts are ellipsoidal, 11 × 7–9 (mean 11 × 8) μm, without a Stieda body, with residual granules. The sporozoites are about 10 × 2 μm and lie lengthwise in sporocysts.

LIFE CYCLE. The oocyst sporulation time is 64–96 hr. Meronts up to 200 μm in diameter with a wall up to 30 μm thick can be found in the heart, mesentery, small intestine wall, liver, and lungs of intermediate hosts 30–60 DAI. They may persist 393 days or more and contain hundreds to thousands of bradyzoites.

There is at least one asexual generation with meronts 500–800 μm in diameter in the lamina propria of the small intestine of newborn kittens 4–14 days after ingestion of meronts containing merozoites from mice. They generally extend into a blood vessel, appearing to be in endothelial or intimal cells. They may also occur in the liver after 10 days.

Macrogametes 10–13 μm in diameter occur in goblet cells in 13–16 days; microgamonts up to 11 μm in diameter occur on day 13.

The prepatent period in kittens is 12–15 days, the patent period 5–12 days.

PATHOGENESIS. There is no evidence that this species is pathogenic, either for cats or rodents.

LITERATURE CITED

For citations before 1965, see the *Index-Catalogue of Medical and Veterinary Zoology, Zoological Record, Biological Abstracts, Protozoological Abstracts,* and *Veterinary Bulletin.*

Becker, R., H. Mehlhorn, and A. O. Heydorn. 1979. Zentralbl. Bakteriol. [Orig. A.] 244:394–404.

Beverley, J. K. A. 1976. Vet. Rec. 99:123–27.

Beyer, T. V., N. A. Bezukladnikova, I. G. Galuzo, S. I. Konovalova, and S. M. Pak, eds. 1979. *Toksoplazmidy.* [*The Toxoplasmids.*] Akad. Nauk SSSR, Vses. Obshch. Protozool., vol. 4, 116 pp.

Dubey, J. P. 1977. In Parasitic Protozoa, vol. 3, ed. J. P. Kreier, 101–237. New York: Academic.

Dubremetz, J. F., E. Porchet-Henneré, and M. D. Parenty. 1975. C. R. Acad. Sci.[D] 280:1793–95.

Fayer, R. 1970. Science 168:1104–5.

______. 1972. Science 175:65–67.

Fayer, R., and D. E. Thompson. 1975. J. Parasitol. 61:466–75.

Frenkel, J. K. 1971. In Pathology of Protozoan and Helminthic Diseases with Clinical Correlation, ed. R. A. Marcial-Rojas. Baltimore: Williams & Wilkins.
______. 1973. In The Coccidia, ed. D. M. Hammond with P. L. Long, 343–410. Baltimore: University Park Press.
______. 1974a. Bull. N.Y. Acad. Med. [Ser. 2] 50:182–91.
______. 1974b. Z. Parasitenkd. 45:125–62.
Gregory, M. W., L. P. Joyner, J. Catchpole, and C. C. Norton. 1980. Vet. Rec. 106:461–62.
Hilali, M., and A. Mohamed. 1980. Tropenmed. Parasitol. 31:213–14.
Jacobs, L. 1970. J. Wildl. Dis. 6:305–12.
Jira, J., and V. Kozojed. 1970. Toxoplasmose—1908–1967. 2 vols. Stuttgart. G. Fischer.
Jones, S. R. 1973. J. Am. Vet. Med. Assoc. 163:1038–42.
Kagan, I. G. 1980a. In Manual of Clinical Microbiology, 3d ed., ed. E. H. Lennette et al. 724–50. Washington, D.C.: American Society for Microbiology.
______. 1980b. In Manual of Clinical Immunology, 2d ed., ed. N. R. Rose and H. Friedman, 573–604. Washington, D.C.: American Society for Microbiology.
Levine, N. D. 1977. J. Protozool. 24:36–41.
Levine, N. D., and V. Ivens. 1981. The Coccidian Parasites (Protozoa, Apicomplexa) of Carnivores. Ill. Biol. Monogr. 51. Urbana: University of Illinois Press.
______. 1985. The Coccidian Parasites (Protozoa, Apicomplexa) of Artiodactyls. Ill. Biol. Monogr. (in press). Urbana: University of Illinois Press.
Levine, N. D., & W. Tadros. 1980. Syst. Parasitol. 2:41–59.
Markus, M. B. 1978. Adv. Vet. Sci. Comp. Med. 22:159–93.
Markus, M. B., R. Killick-Kendrick, and P. C. C. Garnham. 1974. J. Trop. Med. Hyg. 77:248–59.
Mehlhorn, H., and A. O. Heydorn. 1979. Z. Parasitenkd. 58:97–113.
Mehlhorn, H., and E. Scholtyseck. 1973. Z. Parasitenkd. 41:291–310.
Overdulve, J. P. 1978. Prudish Parasites: *Isospora* (*Toxoplasma*) *gondii*. Utrecht, Netherlands: Rijksuniversiteit.
Rommel, M., A. O. Heydorn, and M. Erber. 1979. Berl. Münch. Tieraerztl. Wochenschr. 92:457–64.
Scott, R. J. 1978. Trop. Dis. Bull. 75:809–27.
Siim, J. C. 1974. Actual. Protozool. 1:203–12.
Tadros, W., and J. J. Laarman. 1976. Acta Leiden. 44:1–107.
Van der Zypen, E., and G. Piekarski. 1968. Bol. Chil. Parasitol. 23:90–94.
Vershinin, I. I., V. I. Petrenko, and F. F. Leont'eva. 1979. In Toksoplazmidy, Akad. Nauk SSSR, ed. Beyer et al. 115–16. Protozoologiya, vol. 4.
Vetterling, J. M., N. D. Pacheco, and R. Fayer. 1973. J. Protozool. 20:613–21.
Zasukhin, D. N., ed. 1980. Problema Toksoplazmoza. Moskva, USSR: "Meditsina."

9

Apicomplexa: *Klossiella* and *Hepatozoon*

THESE genera belong to the suborder Adeleorina, which is differentiated from the Eimeriorina and Haemospororina by the fact that the macrogamete and microgamont are presumably associated in syzygy (i.e., they are attached to each other) during development. Correlated with this is the fact that the microgamonts produce very few microgametes. The zygote may or may not be motile, and the sporozoites are enclosed in an envelope.

The great majority of the Adeleorina are parasites of lower vertebrates and invertebrates, but a few occur in domestic and laboratory animals.

Genus *Klossiella* Smith and Johnson, 1902

This genus is homoxenous, with gametogony and merogony in different locations in the same host; generally the meronts and merozoites are in the Bowman's capsules and the gamonts and gametes in the tubules of the kidney, although other organs may be infected in some species. The zygote is inactive. A typical oocyst is not formed; a number of sporocysts, each with many sporozoites, develops within a membrane, which is perhaps laid down by the host cell.

Infection takes place by ingestion of sporulated sporocysts. The sporozoites pass into the blood stream and enter the endothelial cells of the capillaries and arterioles of the kidneys, lungs, spleen, and other organs. Here they turn into meronts, and these then produce merozoites. There are probably several asexual generations.

Eventually some merozoites enter the epithelial cells of the convoluted tubules of the kidney, where they become gamonts and where gamogony and sporogony take place. A macrogamete and microgamont are found together in syzygy within a vacuole in the host cell. The microgamont divides to form 2–10 nonflagellate microgametes, one of which fertilizes the macrogamete. The resultant zygote divides by multiple fission to form a number of sporoblasts. Each of these develops into a sporocyst containing 8–25 or more sporozoites. The sporocysts are enclosed within a membrane, but all authorities do not agree whether it is a true oocyst wall or simply the remnant of the host cell.

The sporocysts, released into the lumen of the kidney tubules by rupture of the host cell, pass out in the urine.

Klossiella equi Baumann, 1946

This species is apparently quite common throughout the world but is seldom seen. It occurs in the kidneys of the horse, ass, and zebra.

LIFE CYCLE. Meronts 8–12 μm in diameter with 20–30 nuclei occur in endothelial cells of Bowman's capsule in the kidneys. Second generation meronts 15–23 μm in diameter containing 15–20 merozoites 8 μm long occur in epithelial cells of the proximal convoluted tubules. Gamogony and sporogony occur in the epithelial cells of the thick limb of Henle's loop. The microgamonts form 4–10 microgametes.

The microgametes are 2–3 μm in diameter, the macrogametes 8–15 μm. The zygotes increase in size, becoming sporonts 20–23 μm in diameter. These grow, and about 40 buds form on their periphery when they are 35–45 μm in diameter. Each bud becomes a sporoblast, and a residual mass is left. Each sporoblast divides by multiple fission until 10–15 or more nuclei are formed. These condense and come to lie along the periphery of the sporoblast. The sporocysts each contain 10–15 sporozoites and are themselves contained in a sac formed by the host cell. Infection is presumably by ingestion of sporocysts (Vetterling and Thompson 1972).

PATHOGENESIS. Apparently nonpathogenic.

Other Species of *Klossiella*

Klossiella muris Smith and Johnson, 1902, is apparently fairly common in laboratory mice throughout the world and has been reported twice in wild house mice. It is ordinarily nonpathogenic, although in heavy infections the kidneys may have minute grayish necrotic foci over their entire surface and the epithelium of the infected kidney tubules may be destroyed.

Klossiella cobayae Seidelin, 1914, occurs sporadically in the guinea pig throughout the world. It is apparently nonpathogenic.

Genus *Hepatozoon* Miller, 1908

In this genus merogony takes place in the viscera of a vertebrate, and the gamonts are either in the leukocytes or erythrocytes, depending on the species. Fertilization and sporogony occur in a tick, mite, louse, tsetse fly, mosquito, or other blood-sucking invertebrate, again depending on the species. The microgamont forms two to four microgametes. A synonym of this generic name is *Leucocytogregarina.* Mammalian forms once assigned to the genus *Leucocytozoon* probably all belong to this genus also.

Species of *Hepatozoon* have been described from mammals, reptiles, and birds. Among mammals they are especially common in rodents and wild carnivores.

The vertebrate hosts become infected by eating the invertebrate hosts.

The sporozoites are released in the intestine, penetrate its wall, and pass via the bloodstream to the liver, lungs, spleen, or bone marrow; different species prefer different organs. The sporozoites enter the tissue cells and become meronts, which divide by multiple fission to produce a number of merozoites. There are several asexual generations in the visceral cells, but their number is known in only a few cases. There may be two types of meront; macromeronts produce a few macromerozoites, and micromeronts produce many micromerozoites. (So far as I can determine, both types of meront are about the same size; their names derive from the size of the merozoites that they produce.) The last generation merozoites enter the blood cells and become gamonts. These look alike; presumably the female is a macrogamete and the male a microgamont, but no evidence is available on this point.

No further development takes place until the parasites reach the alimentary tract of the vector. The gamonts then leave their host cells, associate in syzygy, and the microgamont forms two to four microgametes, which may or may not be flagellated and are relatively large but smaller than macrogametes. One of them fertilizes the macrogamete; the resultant ookinete penetrates the intestinal wall and comes to lie in the haemocoel. Here it grows considerably and becomes an oocyst. Several nuclear divisions take place in the sporont within the oocyst wall. The daughter nuclei migrate to its periphery, and each one buds off to form a sporoblast, leaving a large residual mass. Each sporoblast then forms a wall around itself, becoming a sporocyst. Sporozoites develop in the sporocysts, their number depending on the species. When the vertebrate host ingests the invertebrate one, the oocysts and sporocysts rupture in its intestine and release the sporozoites.

Hepatozoon canis (James, 1905) Wenyon, 1926

SYNONYMS. *Leucocytozoon canis, Haemogregarina canis.*

This species occurs in the dog throughout the world (but rarely in the United States or other parts of the temperate zones). It has also been reported from the jackal, hyena, palm civet, and cat, but these hosts possibly harbor different species.

LIFE CYCLE. Merogony takes place in the spleen, bone marrow, and to a lesser extent the liver. There are several types of meront. Macromeronts produce a small number of macromerozoites (usually three), micromeronts produce a large number of micromerozoites, and intermediate types produce merozoites of intermediate numbers and sizes.

The micromerozoites enter the polymorphonuclear leukocytes to form gamonts. These are elongate rectangular bodies with rounded ends, about 8–12 × 3–6 μm, with a central, compact nucleus. Their cytoplasm stains pale blue and their nucleus dark reddish with Giemsa stain. They are surrounded by a delicate capsule. They may emerge from the leukocytes and capsule and lie free in citrated blood, becoming difficult to distinguish from platelets.

The vector is the brown dog tick *Rhipicephalus sanguineus.* Both the nymph and adult can transmit the infection, but there is no transovarial transmission. The oocysts are found in the haemocoel. They are about 100 μm in longest diameter and contain 30–50 sporocysts 15–16 μm long, each containing about 16 banana-shaped sporozoites and a residuum. Dogs become infected by eating infected ticks.

PATHOGENESIS. *H. canis* has often been found in apparently healthy dogs but can also cause serious disease and death. The principal signs are irregular fever, progressive emaciation, anemia, and splenomegaly. Lumbar paralysis has also been reported. Affected dogs may die in 4–8 wk.

TREATMENT. Unknown.

PREVENTION AND CONTROL. Since *H. canis* is transmitted by the brown dog tick, elimination of ticks will eliminate the disease.

Hepatozoon felis Patton, 1908

SYNONYM. *Hepatozoon felisdomesticae.*

HOST. This apparently rare species has been found in the cat in India and Israel.

Other Species of *Hepatozoon*

H. muris (Balfour, 1905) Miller, 1908, occurs in the wild and laboratory Norway rat, black rat, and possibly other rats throughout the world. Merogony takes place in the parenchymal cells of the liver, and the gamonts are found in the lymphocytes and rarely in the polymorphonuclear leukocytes. The vector is the spiny rat mite *Echinolaelaps echidninus.*

H. epsteini Kakabadze and Zasukhin, 1969, was described from the Norway rat in Sukhumi, USSR. Merogony occurs in monocyte-type cells in the liver.

H. musculi (Porter, 1908) was reported from the white mouse in England. It differs from *H. muris* in that merogony takes place only in the bone marrow.

H. cuniculi (Sangiorgi, 1914) was reported from the domestic rabbit in Italy. Its gamonts are found in the leukocytes and its meronts in the spleen.

H. procyonis Richards, 1961, occurs commonly in raccoons in the southern United States and in Panama.

LITERATURE CITED

Vetterling, J. M., and D. E. Thompson. 1972. J. Parasitol. 58:589–94.

10

Apicomplexa: *Plasmodium*, *Haemoproteus*, *Leucocytozoon*, and Related Protozoa

THESE genera belong to the suborder Haemospororina, which is differentiated from the Eimeriorina and Adeleorina by the facts that the microgamont produces a moderate number (generally 8) of flagellate microgametes, and the sporozoites and merozoites lack a conoid. The gamonts are similar and develop independently. The zygote is motile (i.e., it is an ookinete) and turns into an oocyst. All species are heteroxenous; merogony takes place in a vertebrate host, sporogony in an invertebrate. There are no sporocysts; the sporozoites lie "naked" in the oocysts.

The sporozoites and merozoites have a polar ring, rhoptries, micronemes, subpellicular microtubules, and a micropore, but lack a conoid. Most of these structures are lost by dedifferentiation when the trophozoite is formed. The microgametes have 1 or 2 (usually 1) flagella, compared with the 2 or 3 present in cocccidia.

If the erythrocytes are invaded, pigment (hemozoin) may be formed from the host cell hemoglobin. It is a heme-containing protein different from hemoglobin and hematin.

Garnham (1966) reviewed the blood protozoa of vertebrates. The International Reference Centre for Avian Haematozoa, established at the Memorial University of Newfoundland in 1968 under the auspices of the World Health Organization, issued a 123-page bibliography of the avian blood-inhabiting protozoa in 1976 (Herman et al. 1976) and a 243-page host-parasite catalog in 1982 (Bennett et al. 1982). By 1982 it had records and/or samples of blood parasites from about 2,590 of the more than 8,000 known species of birds. Greiner and Bennett (1975) and Greiner et al. (1975) gave color photomicrographs of some of the species.

Genus *Plasmodium* Marchiafava and Celli, 1885

The gamonts occur in the erythrocytes. Merogony takes place in the erythrocytes and also in various other tissues, depending on the species. The exoerythrocytic ("e.e") meronts are solid or, at the most, vacuolated bodies. Members of this genus are parasites of mammals, birds, and reptiles. They are normally transmitted by mosquitoes; *Anopheles* transmits the mammalian species, and culicines or sometimes *Anopheles* the avian and reptilian ones.

Members of this genus cause malaria, which is still the most important disease of man. They also cause a similar disease in birds. The subgenera used by Garnham (1966) and modified by Sarkar and Ray (1969), Greiner et al. (1975), Ayala (1977), and Killick-Kendrick (1978) are accepted (at least tentatively) in this book. According to this classification, there are three subgenera in mammals, four in birds, and three in reptiles. The mammalian subgenera are *Plasmodium* (type species *P. malariae*), with large erythrocytic meronts and round gamonts, occurring in primates; *Vinckeia* (type species *P. bubalis*), with small erythrocytic meronts and round gamonts, occurring in antelopes, rodents, lemurs, and other mammals; and *Laverania* (type species *P. falciparum*), with large erythrocytic meronts and elongate (crescentic) gamonts, occurring in primates. The avian subgenera are *Haemamoeba* (type species *P. relictum*), with large erythrocytic meronts and round gamonts; *Giovannolaia* (type species *P. circumflexum*), with large erythrocytic meronts and elongate gamonts; *Novyella* (type species *P. vaughani*), with small erythrocytic meronts and elongate or oval gamonts; and *Huffia* (type species *P. elongatum*), with large erythrocytic meronts, elongate gamonts, and marked attraction to the immature hemopoietic system. The reptilian subgenera are *Sauramoeba* (type species *P. agamae*), with large erythrocytic meronts; *Carinamoeba* (type species *P. minasense*), with small erythrocytic meronts; and *Ophidiella* (type species *P. wenyoni*) in snakes.

At present, I have records of 156 presumably valid named species: 50 in mammals, 41 in birds, and 65 in reptiles. Man has 4 species, higher anthropoids (chimpanzee, gorilla, orangutan) 4, gibbons 4, Asian monkeys 7, African monkeys 1, New World monkeys 2 (which are probably human species), and lemurs 2. Their relations are given in Table 10.1.

TABLE 10.1. Species of *Plasmodium* in primates

Type[a]	Man	Higher Anthropoids	Gibbons	Asian Monkeys	African Monkeys	New World Monkeys	Lemurs
Quotidian	(*knowlesi*)[b]			*knowlesi*			
Tertian	*falciparum*	*reichenowi*		*coatneyi*[c]			
	vivax	*schwetzi*	*youngi*	*cynomolgi*			
		pitheci	*eyelesi*	*fragile*	*gonderi*		*lemuris*
		silvaticum	*jefferyi*				
	ovale			*simiovale*		*simium*[e]	
				fieldi[d]			
Quartan	*malariae*	*malariae*	*hylobati*	*inui*		*brasilianum*[f]	*girardi*

[a]*P. falciparum* and *P. reichenowi* belong to the subgenus *Laverania* and *P. lemuris* and *P. girardi* to the subgenus *Vinckeia;* all others belong to the subgenus *Plasmodium.*

[b]A single natural case is known.

[c]Resembles *P. falciparum,* but gamonts are round.

[d]Resemblance to *P. ovale* is superficial.

[e]Resembles *P. ovale* and also *P. vivax* in some respects; probably human *P. ovale* or *P. vivax* in New World monkeys.

[f]Probably human *P. malariae* in New World monkeys.

Only the species occurring in primates and domestic animals are taken up in this book. The most comprehensive book on *Plasmodium* is by Garnham (1966). The species in primates were reviewed by Coatney et al. (1971). More recent reviews were by Rieckmann and Silverman (1977), Bruce-Chwatt (1980), and the authors of the three-volume treatise edited by Kreier (1980) for man, Collins and Aikawa (1977) for nonhuman primates, Carter and Diggs (1977) and Killick-Kendrick and Peters (1978) for rodents, Seed and Manwell (1977) for birds, and Ayala (1977, 1978) for reptiles.

LIFE CYCLE. The life cycle of *Plasmodium vivax* of man is representative (Fig. 10.1). The sporozoites enter the blood through a mosquito bite. They stay in the bloodstream less than an hour, quickly entering liver parenchyma cells. Here they become primary exoerythrocytic meronts (schizonts; known also as cryptozoites because of their "hidden" location). These enlarge and divide by multiple fission to form merozoites (also known as metacryptozoites). These merozoites enter new liver parenchyma cells, become secondary exoerythrocytic meronts (schizonts), undergo multiple fission, and form new metacryptozoites. This process may go on indefinitely in *P. vivax,* but in another human species, *P. falciparum,* there is only a single generation of metacryptozoites (i.e., only primary exoerythrocytic meronts are formed); this is apparently true of the whole subgenus *Laverania.*

The metacryptozoites produced by exoerythrocytic merogony break out of the liver cells, pass into the bloodstream, and enter the erythrocytes about 1 wk to 10 DAI. Here they round up and develop a large vacuole in their center. They are called ring stages because in Romanowsky-stained smears they resemble a signet ring, with a red nucleus at one edge and a thin ring of blue cytoplasm around the vacuole. These grow and are then called meronts (schizonts) or trophozoites. They form food vacuoles containing host cell cytoplasm by invagination, pinching off portions of the cytoplasm. The hemozoin pigment granules within these food vacuoles are formed by digestion of the hemoglobin.

The trophozoites undergo merogony to produce merozoites, the number depending on the species. These break out of the erythrocytes and enter new ones, repeating the cycle indefinitely.

The length of each cycle depends on the parasite species: 2 days in *P. vivax* and *P. falciparum;* 3 days in another human species, *P. malariae;* and 1 day in *P. knowlesi.* Practically all the parasites are generally in the same stage of the cycle at the same time, rupturing their host cells all together. Along with them go the hemozoin granules and other waste products produced by the parasites' metabolism. These are toxic and cause a violent reaction or paroxysm in the host—the chills and fever characteristic of malaria.

After the infection has been present for some time and after an indefinite number of asexual generations, some merozoites entering the erythrocytes develop into macrogametes and others develop into microgamonts (microgametocytes). The former are customarily called macrogametocytes, but this name is incorrect since they are haploid from the start. They remain in this stage until the blood is ingested by a mosquito.

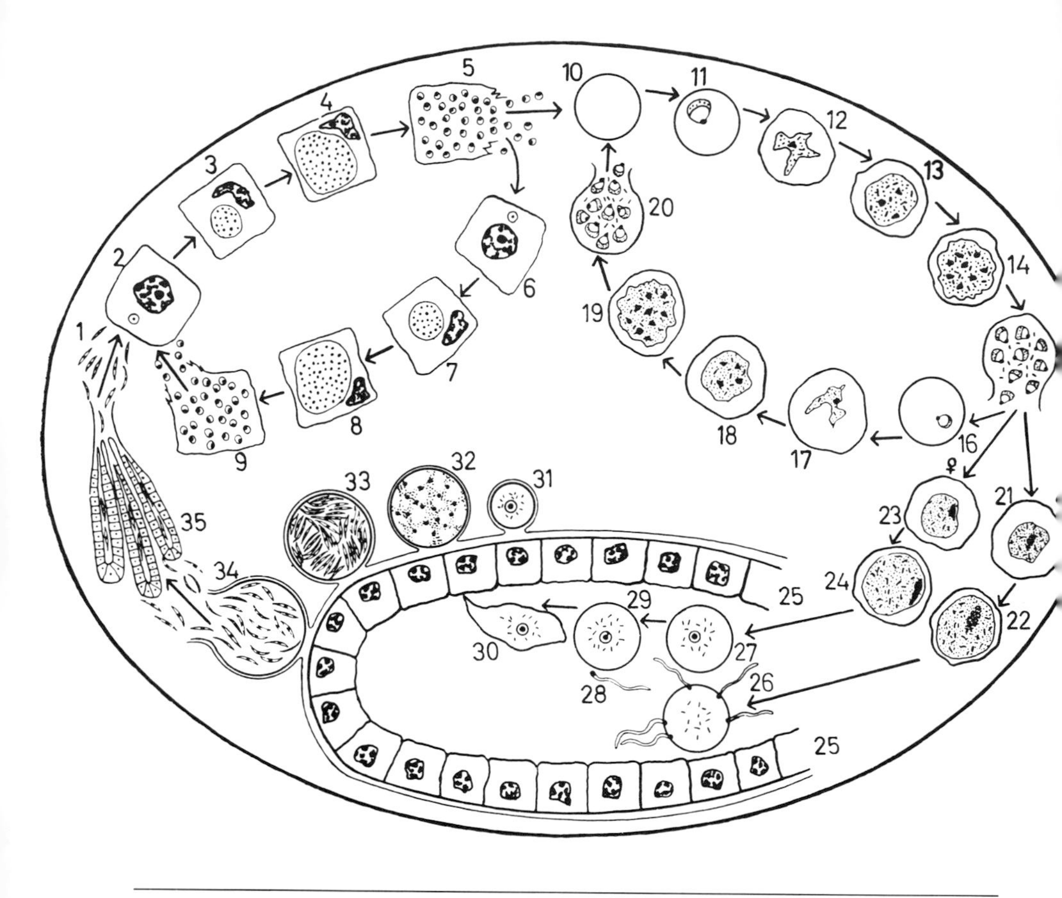

Fig. 10.1. Life cycle of *Plasmodium* sp.: 1. sporozoites from mosquito salivary glands entering host liver cells; 2–4. primary exoerythrocytic meront developing in host liver cell; 5. exoerythrocytic merozoites escaping from host liver cell; 6–9. further generations of exoerythrocytic meronts and merozoites developing in host liver cells; 10. host erythrocyte; 11. ring stage in host erythrocyte; 12. trophozoite in host erythrocyte; 13 and 14. meront in host erythrocyte; 15. merozoites escaping from host erythrocyte; 16–20. further generations of erythrocytic meronts and merozoites; 21 and 22. formation of microgamonts in host erythrocytes; 23 and 24. formation of macrogamete in host erythrocyte; 25. mosquito midgut ("stomach") wall; 26. formation of microgametes from microgamont (exflagellation); 27. macrogamete in mosquito midgut; 28 and 29. fertilization of macrogamete by microgamete; 30. zygote (ookinete) in mosquito midgut; 31. oocyst on outside of mosquito midgut wall; 32 and 33. development of sporozoites within oocyst; 34. release of sporozoites into mosquito hemocoel; 35. sporozoites in mosquito salivary glands. (From Garnham 1966)

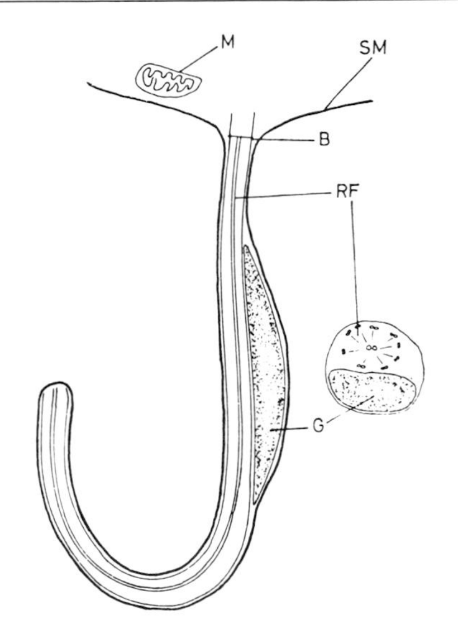

Fig. 10.2. Microgamete of *Plasmodium berghei* being formed from microgamont: *B* = basal body, *G* = nucleus, *M* = mitochondrion, *RF* = reflexed (attached) flagellum, *SM* = surface membrane of microgamont. (From Garnham 1966)

In the midgut of the mosquito, microgametes are produced (Fig. 10.2). The changes in the microgamonts are striking. Within 10–15 min the nucleus divides, and six to eight long, heavy flagellumlike microgametes are extruded in a process known as exflagellation. The microgametes break off and swim freely. They are composed essentially of a nucleus, a single flagellum, and an outer cell membrane. When they find a macrogamete, fertilization takes place, and a motile zygote (ookinete) is formed.

The ookinete penetrates into the midgut (stomach) wall and grows into an oocyst, which forms a ball about 40–55 μm in diameter on the outer surface of the midgut. The oocyst nucleus divides repeatedly, and a number of "sporoblasts" is formed. The nucleus of each "sporoblast" then divides repeatedly, and eventually each oocyst comes to contain 10,000 or more slender, spindle-shaped sporozoites about 15 μm long with a nucleus in the center. These break out of the oocyst into the body cavity and migrate to the salivary glands. They are then injected into a new host when the mosquito bites again. The process of sporozoite development takes 10 days to 3 wk or longer, depending on the species of *Plasmodium* and the temperature. Once infected, a mosquito remains infected for life and can transmit the parasites every time it bites.

Transmission from man to man without the intervention of mosquitoes occurs on rare occasions. Blood transfusion from an infected person is the commonest means. Congenital malaria acquired by a newborn infant by passage of the parasites through the placenta from an infected mother may also occur. A third means of transmission is the use of contaminated syringes or other instruments, especially by drug addicts.

Relapses occur in most types of malaria. They are of two types (WHO 1963), recrudescences and recurrences. In recrudescences, parasites that were present in small numbers in the erythrocytes begin to multiply again and increase to such numbers that they produce disease again; this type of relapse is characteristic of the avian malarias. In recurrences, there are no parasites in the blood between attacks; exoerythrocytic stages in the liver parenchyma cells (or elsewhere) continue to multiply until such time as the body's defenses have decreased sufficiently so that the parasites can again invade the blood. This type of relapse occurs in vivax and malariae malarias in man.

In the avian species, exoerythrocytic merogony takes place not in the liver parenchyma but in either the endothelial cells or (largely) in the hemopoietic cells. In bird malaria also, but apparently not in mammalian malaria, some of the merozoites that have been formed in the erythrocytes are able to enter tissue cells and develop exoerythrocytically. Known as phanerozoites, they do not differ in structure from the forms derived from sporozoites.

Plasmodium is haploid throughout its life cycle except for a brief period following fertilization and zygote formation.

IMMUNITY. A great deal of research has been done and a large number of review papers written on immunity against malaria, especially with the human and rodent species. Briefly, there are many strains of each species of *Plasmodium*. Infection with each generally produces immunity against it but generally not against other strains. As a consequence, an individual who develops considerable immunity to malaria in one geographic region as a result of repeated infections may not be immune at all to strains in other regions.

Immunity can be induced by injection of various antigenic fractions. It can be transferred passively. It may be developed independently against sporozoites (and hence against mosquito transmission), merozoites (and hence against blood transmission), and gamonts (and hence against vertebrate-to-invertebrate transmission). Because of disappointment in the results of efforts to eradicate human malaria by chemotherapy and mosquito control, attention has turned toward immunization. Perhaps effective, broad-spectrum immunization procedures will be developed sometime in the future. A workshop on immunology against malaria was held in October 1978 in Bethesda, Md., under the auspices of the U.S. Naval Medical Research Institute, U.S. Agency for International Development, and World Health Organization; 38 papers were given, and 87 persons from 16 countries participated. Work on production, purification, cultivation, characterization, storage, use in vaccination, diagnosis, assessment of immune response, and ethical questions related to use of the vaccines were considered. A great deal of work has been done on these subjects in recent years but much more remains to be done (WHO 1979a).

CULTIVATION. Various species of *Plasmodium* have been cultivated in fluid media and in avian embryos and tissue culture, but continuous cultivation was not achieved until recently (see Trager and Jensen 1976, and other studies). This success may open the door to development of an effective vaccine.

Malaria

Human Malaria

The following discussion of human malaria is necessarily brief. Further details and references can be found in any textbook of human parasitology and, in more detail, in the reviews cited earlier in this chapter.

Man has four recognized species of *Plasmodium*. *P.* (*Laverania*) *falciparum* (Welch, 1897) Schaudinn, 1902, is the cause of malignant tertian, aestivoautumnal, or falciparum malaria. This is by far the commonest species in sub-Saharan Africa. Paroxysms of chills and fever occur every other day (i.e., on days 1 and 3, which accounts for the name *tertian*). The ring forms are about one-sixth to one-fifth the diameter of a red blood cell. The meronts and merozoites (segmenters) rarely occur in the peripheral circulation but are found in clumped erythrocytes in the viscera. The meronts are usually compact and rounded, with coarse, blackish pigment. The segmenters occupy two-thirds to three-fourths of the host cell and form 8–32 merozoites. The host erythrocyte is not enlarged but contains reddish clefts known as Maurer's dots and may also have bluish stippling. The gamonts are crescent- or bean-shaped, with pigment granules clustered around a central nucleus or scattered except at the poles. The microgamonts have pale blue cytoplasm and a relatively large, pink nucleus when stained with Giemsa stain. The macrogametes have darker blue cytoplasm and a more compact, red nucleus.

P. (*Plasmodium*) *vivax* (Grassi and Feletti, 1890) Labbé, 1899, is the cause of benign tertian or vivax malaria. Paroxysms occur every other day as in falciparum malaria. The ring forms are about one-third to one-half the diameter of the host cell. The meronts are highly active and sprawled out irregularly over the host cell, with small, brown pigment granules usually collected in a single mass. The host cell is enlarged, pale, and contains red dots known as Schüffner's dots. The segmenters nearly fill the host cell and produce 15–20 or occasionally up to 32 irregularly arranged merozoites. The gamonts are rounded, 10–14 μm in diameter (i.e., larger than normal erythrocytes), and have fine, brown, evenly distributed pigment granules. The microgamonts have pale blue cytoplasm and a relatively large, pink nucleus when stained with Giemsa stain. The macrogametes are slightly larger, with darker blue cytoplasm and a small, red nucleus.

P. (*Plasmodium*) *malariae* (Laveran, 1881) Grassi and Feletti, 1890, is the cause of quartan or malariae malaria. Paroxysms occur every 3 days (i.e., on days 1 and 4). The ring forms are similar to those of *P. vivax*. The host cell is not enlarged and does not contain Schüffner's dots. The segmenters nearly fill the host cell and produce 6–12 (usually 8–9) merozoites arranged in a rosette. The gamonts are rounded and smaller than those of *P. vivax*. They do not quite fill the host cell and contain blacker and coarser pigment granules than *P. vivax*.

P. (*Plasmodium*) *ovale* Stephens,.1922, is a rather uncommon species that causes a tertian type of malaria. It is found in tropical Africa and also in the Philippines, New Guinea, and Vietnam. Its ring forms are similar to those of *P. vivax*. Its meronts are usually round, with brownish, coarse, somewhat scattered pigment granules. The host cell is oval, often fimbriated, not much enlarged, and contains Schüffner's dots. The segmenters occupy three-fourths of the host cell and produce 8–10 merozoites in a grapelike cluster. The gamonts are rounded, occupy three-fourths of the host cell, and have coarse, black, evenly distributed pigment granules.

The human species of *Plasmodium* are transmissible to lower primates; the owl monkey is a popular experimental host.

P. brasilianum of New World monkeys is probably *P. malariae* of man gone wild in the monkeys. Further, six species of nonhuman primate *Plasmodium* have been transmitted experimentally to man, and accidental laboratory infections of man with *P. cynomolgi* and natural ones with *P. knowlesi* and *P. simium* of monkeys have been recorded. These findings and the presence of *P. malariae* in chimpanzees make it clear that malaria can be a zoonosis.

PATHOGENESIS. The malarial paroxysm is highly characteristic. It begins with a severe chill. The patient shivers uncontrollably, with teeth chattering and gooseflesh, although his or her temperature is actually above normal. The chill is followed by a burning fever, headache, and sweating. This gradually subsides, the temperature falls, and after 6–10 hr the patient feels much better—until the next paroxysm. The destruction of erythrocytes causes anemia.

After a certain number of paroxysms, the malaria attack subsides. Relapses may occur over a period of years in vivax and malariae malaria.

In general, mortality from malaria is higher in children than in adults in endemic areas; by the time the children become adult they have had repeated attacks and those who have survived have developed a good deal of immunity. Therefore to determine the prevalence of malaria in a region it is better to examine children than adults.

A highly fatal, cerebral form of malaria may occur in falciparum infections. It is due to clogging of the capillaries of the brain by agglutinated (sludged), infected erythrocytes. If enough clogging takes place in the viscera, a severe gastrointestinal disease resembling typhoid, cholera, or dysentery may occur. Another complication of falciparum malaria is blackwater fever, which gets its name from the color of the urine. There is tremendous destruction of the erythrocytes (60–80% may be destroyed in 24 hr), accompanied by fever, intense jaundice, and hemoglobinuria. Severe attacks are usually fatal. The cause of blackwater fever is not known, but it probably involves an immunologic reaction (perhaps an autoimmune reaction) that hemolyzes the erythrocytes.

EPIDEMIOLOGY. Human malaria is transmitted by *Anopheles* mosquitoes. There are perhaps 400 species in this genus, but only about 60 are good vectors, and the epidemiology of the disease in any particular locality depends on the particular vectors present and their breeding habits, food preferences, susceptibility to infection, etc. The subject is an extremely complex one.

Nowadays malaria is primarily a disease of warmer climates, but at one time it was common in the temperate zone. Nevertheless, it is still the most important human disease from a global standpoint. Previously there were 200 million cases of malaria in the world, with 2 million deaths every year, but malaria control measures have decreased these figures. Bruce-Chwatt (1978) estimated that now there are only "at least" 100 million cases and 1 million malaria deaths a year. According to the United Nations Development Program/World Bank/World Health Organization (1982), 107 countries were affected by malaria in 1981, with about 1.8 billion people exposed to infection. There were about 215 million people (most in tropical Africa) with chronic malaria, and about 150 million new cases a year. They expected the global situation to worsen over the next few years.

DIAGNOSIS. Malaria can be diagnosed with certainty only by finding and identifying the causative organisms in the blood. This is done by microscopic examination of smears stained with one of the Romanowsky stains; Giemsa stain is best. A fluorescent dye such as acridine orange can also be used. At one time thin smears were used almost entirely, but thick, laked smears are much better, since they permit a much larger amount of blood to be examined in a given time. An excellent guide with outstanding colored illustrations is that of Wilcox (1960, Manual for the Microscopical Diagnosis of Malaria in Man, U.S. Public Health Service); her illustrations have been reproduced in many texts and reference works.

TREATMENT. A number of drugs have been used in treating malaria. The first one was quinine, the most active ingredient of cinchona bark, which was identified in 1820 by Pelletier and Caventou. It is both suppressive and curative but does not prevent relapses. Chemically, it is 6-methoxy-1-(5-vinyl-2-quinuclidyl-4-quinoline-methanol).

Quinacrine (Atebrin, Atabrine, mepacrine) is 2-chloro-5-diethylamino-isopentylamino-7-methoxyacridine dihydrochloride. It is prophylactic against falciparum malaria and suppressive against vivax and malariae malarias. It cures attacks of the disease but does not prevent relapses. One disadvantage is that it is a dye and stains the skin yellow.

Chloroquine is 7-chloro-4,4-dimethylamino-1-methylbutylaminoquinoline. Aralen is its diphosphate. It is effective in the treatment and suppression of all types of human malaria. The recommended human therapeutic dose is 1.5 g chloroquine base (2.5 g chloroquine phosphate) in 3 days. Following its use, fever subsides in a day, and the parasites disappear from the blood in 2–3 days. The suppressive dose is 0.3 g per wk.

Primaquine is 8-(4-amino-1-methylbutyl-amino)-6-methoxyquinoline. It is truly curative against vivax malaria and is best used in combination with chloroquine if the patient is having an attack but can be used alone in between relapses to prevent further relapses. The dosage is 15 mg (26.4 mg primaquine diphosphate) daily for 14 days.

Amodiaquine (Camoquine, Flavoquine, Basoquin) is closely related to primaquine. It is used to prevent acquisition of malaria. The dosage is 520 mg (400 mg base) once a week, continued for 6 wk after the last exposure in a malarious area.

Pyrimethamine (Daraprim, Malocide) is 2,4-diamino-5-*p*-chlorophenyl-6-

ethylpyrimidine. It is a good suppressive drug but not recommended for the treatment of malarial attacks. In single weekly doses of 25 mg it completely suppresses all *Plasmodium* species and is prophylactic against *P. falciparum* and some strains of *P. vivax.*

At present, the drugs generally recommended are chloroquine, primaquine, and pyrimethamine. There are, unfortunately, increasing numbers of drug-resistant strains. Strains of *Plasmodium* resistant to chloroquine have been found commonly in southeast Asia, rather less commonly in South America, and recently in Africa; strains resistant to pyrimethamine have been found in Africa. Bruce-Chwatt et al. (1981) reviewed the chemotherapy of human malaria.

Chloroquine or amodiaquine is recommended to prevent acquisition of malaria in areas without known chloroquine-resistant malaria. Pyrimethamine-sulfadoxine (Fansidar, Falcidar, Antemal, Methipox) is recommended at 50 mg pyrimethamine and 1.0 g sulfadoxine every other week and continued for 6 wk after last exposure in a malarious area to prevent acquisition of malaria in areas with known chloroquine-resistant strains of *P. falciparum.* Primaquine is recommended to prevent relapses of vivax and ovale malaria.

Vaccination against malaria has been the subject of considerable research. No practical vaccine has been developed as yet, but the situation appears encouraging (USAID/WHO 1983).

PREVENTION AND CONTROL. Malaria control has eliminated or almost eliminated malaria from many parts of the world. According to Bruce-Chwatt (1978): "Out of some 2000 million people inhabiting the originally malarious areas of the world, some 1200 million are now living in areas where malaria has disappeared or where the risk of infection is very low. Another 460 million people live in areas under a reasonably satisfactory degree of malaria control especially in urban areas. But there is no denying the grim fact that nearly 400 million people of the developing world in the tropics inhabit regions where malaria is still endemic."

Control has been achieved largely by use of residual spraying with DDT and other insecticides. In 1981 1,103 persons were known to have become ill with malaria in the United States. Only 13 cases were acquired in the United States, of which 10 were congenital, 2 from blood transfusions, and 1 of undetermined origin in California (CDC 1982).

Malaria eradication is an eventual goal, but now the talk is only of malaria control. Mosquitoes are becoming resistant to insecticides, especially to DDT and dieldrin, which are relatively cheap and which have been the principal ones used in residual spraying (WHO 1979b). Many strains of *Plasmodium* are becoming resistant to antimalarial drugs, including pyrimethamine and especially chloroquine. A reservoir of malaria has been found in lower primates, so that zoonotic malaria has entered the picture.

The World Health Organization is directing the malaria eradication campaign. The campaign consists of five sequential phases: (1) a preeradication survey, (2) a preparatory phase, (3) an attack phase, (4) a consolidation phase, and (5) a maintenance phase.

Nonhuman Primate Malaria

Interest in nonhuman primate malaria has grown considerably in recent years. Seven species of *Plasmodium* (*P. brasilianum, P. cynomolgi, P. eylesi, P. inui, P. knowlesi, P. schwetzi,* and *P. simium*) have been transmitted to man, and some of these have been found occurring naturally in man. Table 10.1 gives the names of those in primates. Two of them (in New World monkeys) are probably human species gone wild in these monkeys, despite their names. *Plasmodium* probably originated in southeast Asia as a parasite of monkeys. For further information, see Garnham (1966), Coatney et al. (1971), Collins (1974), and Collins and Aikawa (1977).

P. (*Plasmodium*) *brasilianum* Gonder and von Berenberg-Gossler, 1908, is the most common malarial parasite of South and Central American monkeys, being found frequently in howler monkeys, spider monkeys, capuchins, squirrel monkeys, and ukaris. It can be transmitted experimentally to man, in whom it produces a low-level parasitemia. It causes a quartan type of malaria that may be fatal in some monkeys. It is likely that this species is actually *P. malariae* of man introduced into the New World by the early explorers and gone wild in monkeys. Various species of *Anopheles* can transmit it, including especially *A. aztecus* and *A. freeborni.*

P. cynomolgi Mayer, 1907, is common in the cynomolgus monkey and other macaques. It can be transmitted to man, and several accidental laboratory infections have been reported. It causes a tertian type of malaria in both monkeys and man. Its known natural vectors are *A. hackeri, A. balabacensis,* and *A. maculatus.*

P. (*Plasmodium*) *eylesi* Warren, Bennett, Sandosham, and Coatney, 1965, occurs in the white-handed gibbon *Hylobates lar* in Malaya and has been transmitted experimentally to man. It appears to be nonpathogenic in gibbons. Its natural vector is unknown.

P. (*Plasmodium*) *inui* Halberstaedter and von Prowakek, 1907 (syns., *P. osmaniae, P. shortii*), is common in various macaques in southern and southeast Asia, Indonesia, the Philippines, and Taiwan. It has been experimentally transmitted to man. It causes a mildly pathogenic type of quartan malaria in monkeys. Its known vectors are *A. hackeri* and *A. leucosphryus;* many other *Anopheles* species can be infected.

P. (*Plasmodium*) *knowlesi* Sinton and Mulligan, 1932, occurs commonly in macaques and leaf monkeys in the Malay peninsula, Indonesia, the Philippines, and Taiwan. A single natural human case has been found (Chin et al. 1963). It is easily transmissible to the rhesus monkey and from it to man. It causes a quotidian type of disease that is mild in the natural hosts but acute and often fatal in experimentally infected rhesus and other monkeys. The known natural vector is *A. hackeri.*

P. (*Plasmodium*) *schwetzi* Brumpt, 1939, occurs in the chimpanzee and gorilla in Africa and has been transmitted experimentally to man. It may actually be *P. vivax* or *P. ovale.* It causes a mild tertian malaria in chimpanzees and man. Its natural vector is unknown.

P. (*Plasmodium*) *simium* da Fonseca, 1951, is common in the howler monkey in parts of Brazil. A single human case is known. It causes a tertian type of malaria. Presumably it is actually *P. vivax* gone wild in New World monkeys. Its natural vectors are unknown.

Although the rhesus monkey is a favorite laboratory host for monkey malaria parasites, it is rarely infected in nature.

Bird Malaria

A tremendous amount of work has been done on the bird malarias. The avian species of *Plasmodium* lend themselves well to experimentation, and until the discovery of *P. berghei* in 1978, birds were the only experimental animals in which malaria could be conveniently studied. Around 50 species of *Plasmodium* have been described from birds, but only about 40 are accepted as valid. The International Reference Centre for Avian Haematozoa is actively engaged in studying the species of *Plasmodium* in birds. In 1980 the Centre had type material for 31 species (Bennett et al., 1980). Greiner et al. (1975) gave color photomicrographs of the erythrocytic stages of 24 species and a checklist of the 35 species accepted at that time and of 30 species no longer considered valid. Reviews of bird malaria have been written by Garnham (1966) and Seed and Manwell (1977). Many wild birds are commonly infected.

Bird malaria is not of great veterinary importance, but it may occasionally cause losses, especially in pigeons. Many (most?) of the species are not strongly host-specific and can infect several species of birds.

The avian species of *Plasmodium* fall into two groups, depending on whether their gamonts are round or elongate. Among those with round gamonts are *P. cathemerium, P. gallinaceum, P. matutinum,* and *P. relictum.* Among those with elongate gamonts are *P. circumflexum, P. elongatum,* and *P. rouxi. P. lophurae* is somewhat different; its gamonts are elongate at first but continue to grow and come to fill up the whole host cell except for the nucleus. The exoerythrocytic stages of all these species except *P. elongatum* occur in endothelial cells and cells of the lymphoid-macrophage system; those of *P. elongatum* occur in blood-forming cells.

DIAGNOSIS. Bird malaria can be diagnosed by finding and identifying the protozoa in stained blood smears. If meronts or merozoites are present, it is easy to differentiate *Plasmodium* from *Haemoproteus,* since these stages are not found in the peripheral blood in the latter. However, if elongate gamonts alone are found, differentiation is usually impossible.

TREATMENT. The bird malarias respond to treatment with quinacrine, chloroquine, and other antimalarial drugs.

PREVENTION AND CONTROL. Since bird malaria is carried by mosquitoes, prevention depends on mosquito control. Residual spraying of poultry houses with insecticides such as DDT or lindane should be effective. Birds can also be raised in screened quarters where mosquitoes cannot get to them. Mosquito control will also control such virus diseases as equine encephalomyelitis.

Plasmodium (Haemamoeba) cathemerium Hartman, 1927

This species was first found in the house sparrow. It is common in passerine birds, including the canary.

STRUCTURE. The gamonts and meronts are more or less round, displacing and often expelling the host cell nucleus. The pigment granules in the gamonts are coarse, often elongate and rodlike. The meronts produce 6–24 merozoites.

LIFE CYCLE. The life cycle of this species has been studied extensively. The asexual cycle takes 24 hr, and synchronicity is high. Huff (1965) listed 9 species of *Culex,* 2 of *Aedes,* 3 of *Anopheles,* and 1 of *Culiseta* that can act as vectors. He also listed 14 nonsusceptible species.

PATHOGENESIS. *P. cathemerium* causes a highly fatal disease in canaries. Affected canaries have subcutaneous hemorrhages, anemia, splenomegaly, and hepatomegaly.

Plasmodium (Giovannolaia) circumflexum Kikuth, 1931

This species is quite common in a wide variety of hosts. The type host is a German thrush *Turdus pilaris,* and passerine birds are most commonly infected, but it also occurs in the ruffed grouse, woodcock, Canada goose, wild ducks, and other birds. It is primarily a temperate zone parasite.

STRUCTURE. The gamonts and trophozoites are elongate. They tend to encircle the host cell nucleus, are generally in contact with it, and do not displace it. The meronts produce 12–30 (mean 19) merozoites.

LIFE CYCLE. The life cycle is similar to those of other avian species of *Plasmodium.* The known vector mosquitoes are *Culex tarsalis, Culiseta annulata, C. melaneura,* and *Mansonia crassipes.* Two other species of *Culex* and five of *Aedes* have been found insusceptible (Huff 1954, 1965).

Plasmodium (Giovannolaia) durae Herman, 1941

The natural host of this species is unknown but is probably some wild African bird. It occurs in the turkey and the peafowl and has been transmitted to baby chicks and ducks. It has been found in Africa and Florida.

STRUCTURE. The gamonts are elongate, at the end or side of the host cell, and often displace the host cell nucleus. The pigment granules are usually large, round, and black. The host cell is not enlarged. The trophozoites are more or less ameboid. The mature meronts rarely displace the host cell nucleus. There are 6–14 (usually 8) merozoites. The host cell is not distorted.

LIFE CYCLE. Merogony in the erythrocytes apparently takes 24 hr. The vectors are unknown.

Plasmodium (*Giovannolaia*) *fallax* Schwetz, 1930

This species occurs in the guinea fowl and other birds in tropical Africa. Many birds can be infected experimentally.

STRUCTURE. The gamonts are elongate and usually have one end hooked around the host cell nucleus. The meronts are also elongate and produce about 12–18 large merozoites.

LIFE CYCLE. The life cycle is highly asynchronous. The natural mosquito vector is unknown, but 4 species of *Aedes* and 1 of *Culex* can be infected, as can *Anopheles quadrimaculatus*.

PATHOGENESIS. This species is apparently only slightly pathogenic in its natural hosts.

Plasmodium (*Haemamoeba*) *gallinaceum* Brumpt, 1945

The natural hosts are jungle fowls. It occurs in chickens in southern and southest Asia and Indonesia, and many other gallinaceous birds can be infected.

STRUCTURE. The gamonts and meronts are round or irregular. The host cell nucleus is displaced but seldom expelled. The pigment granules in the gamonts are rather large and not very numerous. The meronts produce 8–30 merozoites. (See Fig. 10.3.)

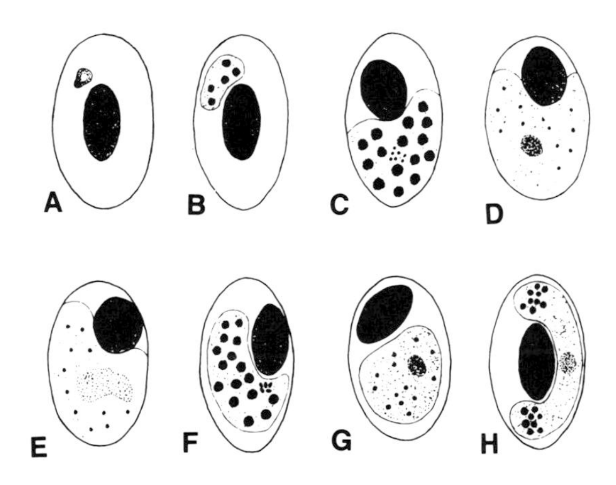

Fig. 10.3. Avian *Plasmodium* and *Haemoproteus* in host erythrocytes: A. *Plasmodium gallinaceum* young trophozoite (ring stage); B. *P. gallinaceum* older trophozoite (meront); C. *P. gallinaceum* mature meront (trophozoite, segmenter); D. *P. gallinaceum* macrogamete; E. *P. gallinaceum* microgamont; F. *P. relictum* mature meront (segmenter); G. *P. relictum* macrogamete; H. *Haemoproteus columbae* macrogamete. (×2,800)

LIFE CYCLE. The life cycle is similar to those of other *Plasmodium* species. The natural vector is the mosquito *Mansonia crassipes.* Various other mosquitoes can be infected or are potential vectors. Huff (1965) listed 37 susceptible species; 24 are *Aedes,* 6 *Armigeres,* 2 *Culex,* 3 *Anopheles,* 1 *Culiseta,* and 1 *Mansonia;* 15 are not susceptible.

PATHOGENESIS. Jungle fowls are relatively resistant, but outbreaks of disease occur in domestic chickens, in which it causes high mortality. The body temperature fluctuates, and anemia and splenomegaly are present. The birds may become paralyzed and die due to blocking of the brain capillaries by the exoerythrocytic stages.

REMARKS. Hundreds of papers have been written on *P. gallinaceum* because it does so well in the laboratory.

Plasmodium (Huffia) hermani Telford and Forrester, 1975

This species occurs in the turkey in Florida and can infect the bobwhite but is not infective for chicks and anserorid or passerorid birds.

STRUCTURE. The meronts are oval or round, with 6–14 merozoites, typically in immature erythrocytes. The gamonts are elongate, always in mature erythrocytes.

LIFE CYCLE. The primary natural vector is apparently *Culex nigripalpus. Wyeomyia vanduzeei* is also a satisfactory vector.

PATHOGENESIS. *P. hermani* causes anemia, splenomegaly, and decreased growth rate in both domestic and wild turkey poults but does not kill them.

Plasmodium (Novyella) juxtanucleare Versiani and Gomes, 1941

SYNONYM. *P. japonicum.*

This species occurs in the chicken and wild gallinaceous birds in South and Central America, Asia, and East Africa. Anserorid and passerorid birds, pigeons, and various wild birds cannot be infected.

STRUCTURE. The meronts are round, ovoid, or irregular; relatively small; and usually in contact with the red cell nucleus. They produce 2–7 (usually 4) merozoites. The gamonts are round, ovoid, irregular, or elongate piriform. The host cell is often distorted.

LIFE CYCLE. Merogony takes 24 hr. Synchronicity is low. The known natural vectors are *Culex sitiens* and *C. annulus,* but other species of *Culex* can be infected experimentally; species of *Aedes, Mansonia,* and *Anopheles* are refractory.

PATHOGENESIS. The New World strains are apparently highly pathogenic, causing enlargement of the spleen, pericardial effusions, usually a high parasitemia, and distortion of the host red cell. Affected birds do not have any marked signs. Shortly before death they appear listless and weak, with pale combs. Their temperature is not elevated. There are deposits of pigment in the liver and spleen.

Plasmodium (Haemamoeba) matutinum Huff, 1937

SYNONYM. *P. relictum matutinum.*

This species is presumably common in wild birds throughout the world. It was first found in a robin in Kansas, Ill.

STRUCTURE. This species resembles *P. relictum* (see below). Its gamonts and meronts are round or irregular. The meronts produce 10–22 (usually 16) merozoites. The merozoites and young meronts have characteristic "punched out," circular vacuoles.

LIFE CYCLE. The life cycle is synchronous in canaries; asexual merogony takes exactly 24 hr. The natural vectors appear to be unknown.

PATHOGENESIS. This species is highly pathogenic for canaries, almost always causing death. It is less pathogenic for columborid birds, which generally acquire a high degree of immunity in nature.

Plasmodium (Haemamoeba) relictum (Grassi and Feletti, 1891)

SYNONYM. *P. praecox* (a nomen nudum)

This species was first described from the house sparrow. It is common in wild passerine birds throughout the world and occurs in pigeons, doves, ducks, passerines, and other wild birds. Garnham (1966) listed 141 natural avian host species.

STRUCTURE. The gamonts and meronts are round or irregular. The host cell nucleus is displaced and often expelled by the larger forms. The pigment granules of the gamonts are relatively fine and dotlike. The meronts produce 8–32 merozoites, the number depending on the particular strain.

LIFE CYCLE. The asexual cycle has been reported to take from 12 to 36 hr in different strains; some have a very high and others a low degree of synchronicity. Many species of mosquitoes can act as vectors, various species of *Culex* apparently being the preferred ones. Huff (1965) listed 12 of *Culex,* 5 of *Anopheles,* 4 of *Aedes,* and 2 of *Culiseta,* as well as 3 species that were not susceptible.

PATHOGENESIS. *P. relictum* is highly pathogenic for the pigeon but less so for the mourning dove and canary. Affected squabs become weak and anemic, with enlarged and heavily pigmented spleens and livers. Pigment may also be deposited in the fat. Anemia is the principal cause of death.

Genus *Haemoproteus* Kruse, 1890

The gamonts occur in the erythrocytes and are usually described as halter-shaped (hence the synonym *Halteridium*); they are elongate and curve around the host cell nucleus. Merogony takes place in the endothelial cells of the blood vessels, especially in the lungs, and not in the erythrocytes as in *Plasmodium*. The known vectors are louse-flies (Hippoboscidae) and midges (*Culicoides*), but the vectors of the great majority of species are unknown. Members of this genus are parasites of birds, reptiles, and a few amphibia. They are extremely common in wild birds and also occur in domestic pigeons, ducks, and turkeys. They are not an important cause of disease. At present, I have records of 154 presumably valid species: 130 in birds, 20 in reptiles, and 4 in amphibia. The International Reference Centre for Avian Haematozoa is actively engaged in studying the species in birds. In 1980, the Centre had type material for 41 species (Bennett et al. 1980). Greiner and Bennett (1975) gave color photomicrographs of the gamonts in the blood of 19 species of avian *Haemoproteus*.

LIFE CYCLE. The life cycle of *Haemoproteus* is similar to that of *Plasmodium* except (1) merogony takes place not in the erythrocytes but in the endothelial cells of the blood vessels and (2) the vectors are not mosquitoes but hippoboscid flies, midges, or *Chrysops*.

DIAGNOSIS. *Haemoproteus* infections can be diagnosed by finding and identifying the protozoa in stained blood smears. However, some infections in which gamonts alone are found are *Plasmodium*, not *Haemoproteus*, infections.

TREATMENT. Little is known about treatment of *Haemoproteus* infections. However, in view of their slight pathogenicity treatment does not seem necessary.

PREVENTION AND CONTROL. Prevention of *Haemoproteus* infections depends on control of their vectors, or in preventing the hosts from being bitten.

Haemoproteus (*Haemoproteus*) *columbae* Kruse, 1890

This species occurs commonly in the domestic pigeon and possibly also wild pigeons, doves, and other columborid birds throughout the world.

STRUCTURE. The mature macrogametes and microgamonts, which are elongate and sausage-shaped, partially encircle the host cell nucleus and may displace it to some extent. They contain a variable number of dark brown pigment granules. The host cell is not enlarged. Gametogenesis resembles that of *Plasmodium*. The microgametes apparently have 2 flagella.

LIFE CYCLE. Birds become infected when bitten by the dipteran vector, the hippoboscid fly *Pseudolynchia canariensis*. Merogony occurs in the endothelial cells of the blood vessels of the lungs, liver, and spleen. There may be several generations. Following merogony, the merozoites enter red blood cells and become macrogametes and microgamonts. These appear 28–30 DAI.

In the midgut of the hippoboscid vector, the microgamonts produce 4 or more snakelike microgametes by exflagellation. These fertilize the macrogametes; the resultant zygotes are ookinetes, which crawl to the midgut wall and form oocysts on its outer surface. Within them very large numbers of slender, falciform sporozoites up to 10 μm long are formed in 10–12 days. These enter the body cavity and pass to the salivary gland, where they accumulate and are injected into a new host when the fly bites it.

PATHOGENESIS. *H. columbae* is not or is only slightly pathogenic.

Haemoproteus meleagridis Levine, 1961

This species occurs in domestic and wild turkeys in North America. Its life cycle and pathogenesis are unknown.

STRUCTURE. The macrogametes and microgamonts are elongate, sausage-shaped, curve around the host cell nucleus, and occupy about one-half to three-fourths of the host cell.

Haemoproteus (Parahaemoproteus) nettionis (Johnston and Cleland, 1909) Coatney, 1936

SYNONYMS. *Halteridium nettionis, Haemoproteus anatis, H. hermani, H. anseris.*

This species occurs commonly in the domestic duck, domestic goose, and about 50 other species of wild ducks, geese, and swans throughout the world. It is essentially a parasite of wild waterfowl and may infect domestic birds in heavily endemic regions.

STRUCTURE. The mature macrogametes and microgamonts are elongate and sausage-shaped, partially encircling the host cell nucleus and often displacing it. They contain a few to 30 or more pigment granules, which are usually coarse and round and tend to be grouped at the ends of the cell. The host cell is not enlarged.

LIFE CYCLE. The vectors of *H. nettionis* are the biting midge *Culicoides downesi* and other species of *Culicoides.* The prepatent period in experimentally infected ducks is 21 days.

PATHOGENESIS. *H. nettionis* is only slightly if at all pathogenic.

Haemoproteus (Haemoproteus) sacharovi Novy and MacNeal, 1904

This species occurs in the domestic pigeon and various doves throughout the world. Common in mourning doves, it is primarily a parasite of wild doves and may also infect pigeons.

STRUCTURE. The mature macrogametes and microgamonts completely fill the host cell, enlarging and distorting it and often pushing the host cell nucleus to the edge of the cell. They contain very little pigment.

LIFE CYCLE. The hippoboscid fly *Pseudolynchia canariensis* is an experimental vector, but the natural vectors are still unknown.

PATHOGENESIS. *H. sacharovi* is only slightly if at all pathogenic in the mourning dove.

Genus *Leucocytozoon* Sambon, 1908

The macrogametes and microgamonts are in the leukocytes or, in some species, in the erythrocytes. Pigment granules (hemozoin) are not formed from hemoglobin. Merogony takes place in the parenchyma of the liver, heart, kidney, or other organs, the meronts forming large bodies divided into cytomeres. There is no merogony in the erythrocytes or leukocytes. The vectors are blackflies (*Simulium*) or midges (*Culicoides*). Members of this genus are parasites of birds.

Leucocytozoon is common in many wild birds and also causes disease in ducks, geese, turkeys, and chickens. Bennett et al. (1982) listed 91 named species. Greiner and Bennett (1975) gave color photomicrographs of the gamonts in the blood of 10 species. Fallis and Desser (1977) reviewed the genus somewhat.

Leucocytozoon andrewsi Atchley, 1951

Atchley (1951) found this species in 15% of 400 domestic chickens in South Carolina. The macrogametes are 12–14 μm in diameter with a nucleus generally 3–4 μm in diameter; the microgamonts are 10–12 μm in diameter with a nucleus 10–12 μm in diameter occupying most of the cell. The host cell is 13–17 μm in diameter. The gamonts are round. There appear to be six microgametes. This form has apparently not been seen again.

Leucocytozoon caulleryi Mathis and Leger, 1909

SYNONYMS. *L. schueffneri* in part, *Akiba caulleryi.*

This species occurs in the chicken and also guinea fowl in Asia. The gamonts are in erythrocytes. See, for example, Morii et al. (1981).

STRUCTURE. The mature gamonts are round, 15.5 × 15.0 μm. The host cell is round also, about 20 μm in diameter. The host cell nucleus forms a narrow, dark band extending about one-third of the way around the parasite.

LIFE CYCLE. The vectors in Japan are *Culicoides arakawa* and also *C. circumscriptus* and *C. odibilis.* Other species of *Culicoides* can also be infected. The zygote is round and about 14 μm in diameter; it elongates into an ookinete about 21 μm long, which passes through the midgut wall to form subspherical oocysts 4–13 × 5–14 μm on the midgut outer wall. Sporozoites

7–11 × 1–2 μm are formed, occur in the salivary gland, and are introduced into new hosts when the midges bite them. See Akiba (1960, 1964, 1970).

Exoerythrocytic meronts occur especially in the kidneys, liver, and lungs, and also in the blood spaces or tissues of the heart, spleen, pancreas, thymus, muscles, intestine, trachea, ovaries, adrenals, and brain. They are spherical or lobed and divide at first into cytomeres, which eventually fuse, forming megalomeronts. These are 26–300 μm in diameter and produce a great number of spherical merozoites.

Gamonts appear in the peripheral blood 14 DAI. They are found in erythrocytes or sometimes erythroblasts, which they distort as they grow. When mature, they may break out of their host cell to lie free in the plasma.

PATHOGENESIS. Some strains are nonpathogenic; others are highly pathogenic, killing a high percentage of the chickens in a flock. The "Bangkok hemorrhagic disease" described by Campbell (1954) in Thailand was probably caused by *L. caulleryi.*

Affected chickens are anemic, listless, diarrheic, have pallid combs and wattles, and have hemorrhages in their lungs, liver, and kidneys. The signs are due primarily to the exoerythrocytic megalomeronts, which cause hemorrhage when they rupture. There may be gross hemorrhage from the kidney lesions into the peritoneal cavity.

EPIDEMIOLOGY. Akiba (1970) said that *C. arakawa* develops in the surface mud of rice paddies, so that outbreaks occur frequently in chickens in Japan in June when the paddy fields are ready for rice planting. They become sporadic in October and later months.

TREATMENT. One ppm pyrimethamine, 50 ppm sulfonamides, or 125 ppm clopidol in the feed prevent infection in chickens (Akiba 1975). These drugs do not affect the late meronts or gamonts. Thirty to 40 ppm sulfamonomethoxine in the feed for 29 days beginning 2 days before sporozoite inoculation prevent infection (Nakamura et al. 1979).

PREVENTION AND CONTROL. These depend on elimination of *Culicoides.*

Leucocytozoon marchouxi Mathis and Leger, 1910

SYNONYM. *L. turtur.*

This species occurs in at least 17 species of columborid birds, including the turtledove and mourning dove, throughout the world. There is only one report of it in domestic pigeons, by Jansen (1952) in South Africa.

STRUCTURE. The macrogametes are rounded or elliptical, 8–15 × 7–11 (mean 12 × 9) μm; they stain dark blue with Giemsa stain and have a compact, reddish nucleus. The microgamonts are often distorted or ruptured by the smearing process, but if not badly damaged are 8–15 × 5–11 (mean 11 × 8) μm. The host cell nucleus forms a darkly staining band along about one-third of the periphery of the parasite.

PATHOGENESIS. This species is apparently only slightly if at all pathogenic.

Leucocytozoon sabrazesi Mathis and Leger, 1910

SYNONYMS. *L. schueffneri* in part, *L. sabrazi.*

This species occurs rather uncommonly in the chicken and jungle fowls in south and southeast Asia and Indonesia.

STRUCTURE. The mature gamonts are elongate and about 22–24 × 4–7 μm. The host cells are spindle-shaped, with long, cytoplasmic "horns" extending beyond the parasites, and are about 67 × 6 μm. The host cell nucleus forms a narrow, darkly staining band along one side of the parasite.

PATHOGENESIS. This species causes a disease in chickens characterized by anemia, pyrexia, diarrhea, paralysis of the legs, and a ropy discharge from the mouth.

Leucocytozoon simondi Mathis and Leger, 1910

SYNONYMS. *L. anatis, L. anseris.*

This species occurs commonly in domestic ducks and geese and many wild anserorid birds in northern United States, Canada, Europe, and Vietnam, in mountainous or hilly areas where cold, rapid streams permit suitable blackfly vectors to breed.

STRUCTURE. The mature macrogametes and microgamonts are more or less elongate, 14–22 μm long, and sometimes round. Their host cells are ordinarily elongate, up to 45–55 μm long, with their nucleus forming a very long, thin, dark band along one side and with pale cytoplasmic "horns" extending out beyond the parasite and the nucleus (Desser 1967; Desser et al. 1978). Both leukocytes and erythrocytes may be infected (Desser 1967; Desser et al. 1970). (See Fig. 10.4.)

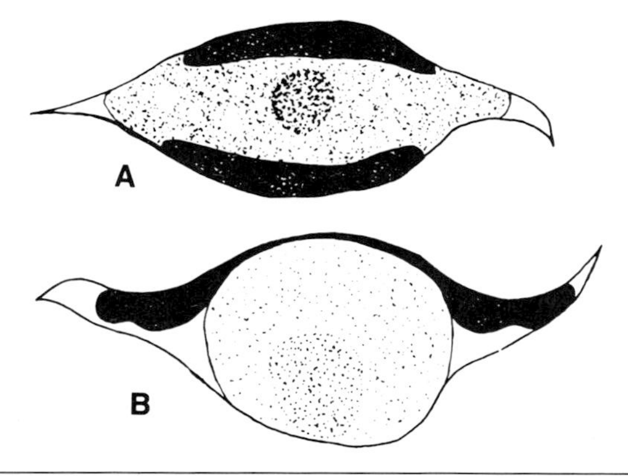

Fig. 10.4. Species of *Leucocytozoon* in avian blood cells: A. *L. smithi* macrogamete from turkey; B. *L. simondi* microgamont from duck. (×1,400)

LIFE CYCLE. Herman et al. (1976) illustrated the life cycle in 48 color photomicrographs on a microfiche. Birds become infected when bitten by a blackfly vector. The sporozoites enter the blood stream, invade various tissue cells, round up, and become meronts (Fig. 10.5).

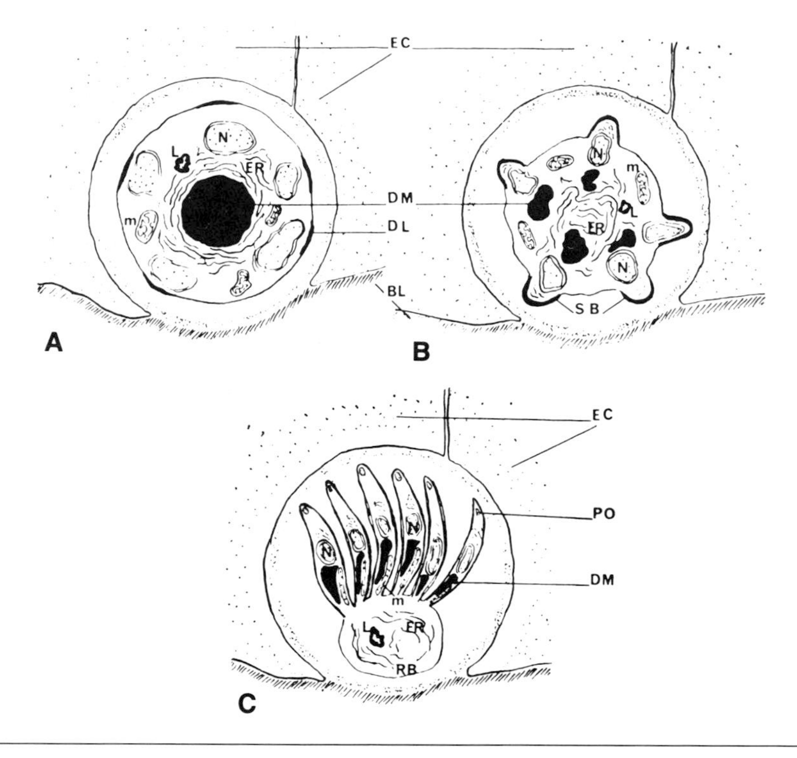

Fig. 10.5. Diagram of development of sporozoites within oocysts of *Leucocytozoon simondi: BL* = basal lamina of midgut, *DL* = dense line, *DM* = dense materials, *EC* = epithelial cells, *ER* = endoplasmic reticulum, *L* = lipid, *m* = mitochondrion, *N* = nucleus, *PO* = rhoptries (paired organelles), *RB* = residual body, *SB* = sporozoite bud. In oocyst capsule relatively few sporozoites are shown. (From Desser and Wright 1968)

Two types of meront occur in the duck. Hepatic meronts 11–18 μm in diameter occur in the liver cells; they form a number of cytomeres, which in turn form small merozoites by multiple fission.

Megalomeronts 6–164 μm in diameter when mature are found in the brain, lungs, liver, heart, kidney, gizzard, intestine, and lymphoid tissues 4–6 days after exposure. They are more common than the hepatic meronts. Each megalomeront produces many thousands of bipolar merozoites.

The merozoites enter blood cells and form gamonts. Merogony continues in the internal organs for an indefinite, long time, although at a much reduced rate. There are about 1,000 fewer gamonts in the relapse phase than in the primary infection, and these adult birds are not seriously affected. However, they are the source of infection for the new crop of ducklings.

The vectors of *L. simondi* are various species of blackflies (*Simulium* and other simuliids). *S. anatinum* and *S. innocens* (which may be the same species) are most important early in the season, and *S. rugglesi* later (Desser et al. 1978); *Cnephia ornithophilia* is also a vector (Tarshis 1976), as is *S. parnassum*.

In the blackfly's midgut 4–8 microgametes are formed within a few minutes by exflagellation from the microgamonts. These fertilize the macrogametes to form a motile zygote or ookinete about 33 μm long and 5 μm wide. Ookinetes are present in the blackfly midgut 2–6 hr after ingestion of infected blood. They develop into oocysts both in the midgut wall and in the midgut itself.

The oocysts are 10–13 μm in diameter. They can be found 2–3 DAI and complete their development 2.5–4 DAI. They produce relatively few sporozoites compared with *Plasmodium*. The sporozoites are 5–10 μm long, slender, with one end rounded and the other pointed. They break out of the oocysts and pass to the salivary glands, where they accumulate. Viable sporozoites can be found for at least 18 days after an infective feeding.

PATHOGENESIS. *L. simondi* is markedly pathogenic for ducks and geese. The heaviest losses occur among young birds.

The outstanding feature of an outbreak of leucocytozoonosis is its suddenness of onset. A flock of ducklings may appear normal in the morning, become ill in the afternoon, and be dead by the next morning. Acutely affected ducklings are listless and do not eat. Their breathing is rapid and labored due to obstruction of the lung capillaries with meronts. They may go through a short period of nervous excitement just before death. Adult birds are more chronically affected. They are thin and listless, and the disease develops more slowly in them. If they die at all, it is seldom in fewer than 4 days after the appearance of signs. Death usually occurs as the peripheral parasitism approaches its peak, 10–12 days after infection (Desser 1967). Ducklings that recover often fail to grow normally, and recovered birds remain carriers.

The principal lesions of leucocytozoonosis are splenomegaly and liver hypertrophy and degeneration. Anemia and leukocytosis are present, and the blood clots poorly.

DIAGNOSIS. Leucocytozoonosis can be diagnosed by finding and identifying the gamonts in stained blood smears or the meronts in tissue sections.

TREATMENT. No effective treatment is known.

PREVENTION AND CONTROL. Prevention depends on blackfly control—ordinarily a difficult task—or on raising ducks and geese under conditions that prevent them from being bitten by blackflies. In blackfly areas this means raising them in screened quarters. Blackflies pass readily through ordinary, 16-

mesh-per-in. window screening, so 32- or 36-mesh screening is necessary. Since this type of screening is expensive, a good grade of cheesecloth has been recommended for a single season's use.

Since wild ducks and geese are reservoirs of infection for domestic birds, the latter should not be raised close to places where wild waterfowl congregate. And since blackflies are the vectors, ducks and geese should preferably be raised in regions where these do not occur in significant numbers.

Leucocytozoon smithi Laveran and Lucet, 1905

This species occurs in domestic and wild turkeys in North America and Europe in mountainous or hilly areas where suitable blackfly vectors breed.

STRUCTURE. The mature gamonts are rounded at first but later become elongate, averaging 20–22 μm in length. Their host cells are elongate, averaging 45 × 14 μm, with pale cytoplasmic "horns" extending out beyond the enclosed parasite. The host cell nucleus is elongate, forming a long, thin, dark band along one side of the parasite; often it splits to form a band on each side of the parasite.

LIFE CYCLE. The life cycle of *L. smithi* is similar to that of *L. simondi* but is not known in nearly so much detail. The prepatent period is about 9 days. Hepatic meronts in the hepatocytes of infected turkeys are 10–20 × 7–14 (mean 13.5 × 10.5) μm. The earliest stage contains round and crescent-shaped, basophilic cytomeres, which develop into masses of deeply staining merozoites that completely fill the host cell cytoplasm. Megalomeronts have not been seen.

The vectors of *L. smithi* are the blackflies *Simulium meridionale* (syn., *S. occidentale*), *S. slossonae, S. jenningsi, S. congareenarum,* and *Prosimulium hirtipes.*

PATHOGENESIS. *L. smithi* is markedly pathogenic for turkeys, and extremely heavy losses have been reported. Adult birds are less seriously affected than poults, and the disease runs a slower course in them, but even they may die.

Affected poults fail to eat, appear droopy, and tend to sit. They move with difficulty when disturbed; in the later stages there may be incoordination, and the birds may suddenly fall over, gasp, become comatose, and die. If the birds do not die 2–3 days after signs of disease appear, they recover.

Recovered birds continue to carry parasites in their blood. Some may have no serious aftereffects, but others may develop a chronic type of the disease. They never again regain their vigor, and the males pay little attention to the females and rarely strut. They often have moist rales and cough and repeatedly try to clear their throats when disturbed. They may die suddenly under stress caused by undue excitement or handling.

The spleen and liver of affected birds are enlarged, the duodenum more or less inflamed. This enteritis may sometimes extend throughout the small intestine. The birds are anemic and emaciated, their flesh is flabby, and their muscles may be brownish. There are no gross lesions in adult carriers, but the liver may occasionally be icteric, enlarged, and cirrhotic.

DIAGNOSIS, TREATMENT, PREVENTION, AND CONTROL. These are the same as for *L. simondi*.

LITERATURE CITED

For citations before 1965, see the *Index-Catalogue of Medical and Veterinary Zoology, Zoological Record, Biological Abstracts, Protozoological Abstracts,* and *Veterinary Bulletin.*

Akiba, K. 1970. Natl. Inst. Anim. Health Q. Tokyo 10(Suppl.):131–47.

______. 1975. Proc. 3rd Int. Congr. Parasitol. 3:1322.

Ayala, S. C. 1977. In Parasitic Protozoa, vol. 3, ed. J. P. Kreier, 267–309. New York: Academic.

______. 1978. J. Protozool. 25:87–100.

Bennett, G. F., E. M. White, N. A. Williams, and P. R. Grandy. 1980. J. Parasitol. 66:162–65.

Bennett, G. F., M. Whiteway, and C. B. Woodworth-Lynas. 1982. A Host-Parasite Catalogue of the Avian Haematozoa. Mem. Univ. Newfoundland occas. pap. biol. 5.

Bruce-Chwatt, L. J. 1978. In Rodent Malaria, ed. R. Killick-Kendrick and W. Peters, xi–xxv. New York: Academic.

______. 1980. Essential Malariology. London: William Heinemann.

Bruce-Chwatt, L. J., R. H. Black, C. J. Canfield, D. F. Clyde, W. Peters, and W. H. Wernsdorfer, eds. 1981. Chemotherapy of Malaria. 2d ed. Geneva: World Health Organization.

Carter, R., and C. L. Diggs. 1977. In Parasitic Protozoa, vol. 3, ed. J. P. Kreier, 359–465. New York: Academic.

CDC (Center for Disease Control). 1982. Malaria Surveillance Annual Summary 1981. Atlanta: U.S. Department of Health and Human Services, Center for Disease Control.

Chin, W., P. G. Contacos, G. R. Coatney, and H. R. Kimball. 1963. Science 149:865.

Coatney, G. R., W. E. Collins, M. Warren, and P. G. Contacos. 1971. The Primate Malarias. Washington, D.C.: GPO.

Collins, W. E. 1974. Adv. Vet. Sci. Comp. Med. 18:1–23.

Collins, W. E., and M. Aikawa. 1977. In Parasitic Protozoa, vol 3, ed. J. P. Kreier, 467–92. New York: Academic.

Desser, S. S. 1967. J. Protozool. 14:244–54.

Desser, S. S., and K. A. Wright. 1968. Can. J. Zool. 46:303–7.

Desser, S. S., J. R. Baker, and P. Lake. 1970. Can. J. Zool. 48:331–36.

Desser, S. S., J. Stuht, and A. M. Fallis. 1978. J. Wildl. Dis. 14:124–31.

Fallis, A. M., and S. S. Desser. 1977. In Parasitic Protozoa, vol. 3, ed. J. P. Kreier, 239–66. New York: Academic.

Garnham, P. C. C. 1966. Malaria Parasites and Other Haemosporidia. Oxford: Blackwell.

Greiner, E. C., and G. F. Bennett. 1975. Wildl. Dis. 66. Microfiche.

Greiner, E. C., G. F. Bennett, M. Laird, and C. M. Herman. 1975. Wildl. Dis. 68. Microfiche.

Herman, C. M., E. C. Greiner, G. F. Bennett, and M. Laird. 1976. Bibliography of the Avian Blood-inhabiting Protozoa. St. John's, Newfl., Canada: Memorial University of Newfoundland.
Huff, C. G. 1965. Exp. Parasitol. 16:107–32.
Killick-Kendrick, R. and W. Peters, eds. 1978. Rodent Malaria. New York: Academic.
Kreier, J. P., ed. 1980. Malaria. 3 vols. New York: Academic.
Morii, T., T. Shihara, Y. C. Lee, M. F. Manuel, K. Nakamura, T. Iijima, and K. Hooji. 1981. Int. J. Parasitol. 11:187–90.
Nakamura, K., T. Morii, and T. Iijima. 1979. Jpn. J. Parasitol. 28:377–83.
Rieckmann, K. H., and P. H. Silverman. 1977. In Parasitic Protozoa, vol. 3, ed. J. P. Kreier, 493–527. New York: Academic.
Sarker, A. C., and H. N. Ray. 1969. Prog. Protozool. 3:353–54.
Seed, T. M., and R. D. Manwell. 1977. In Parastic Protozoa, vol. 3, ed. J. P. Kreier, 311–57. New York: Academic.
Tarshis, L. B. 1976. J. Med. Entomol. 13:337–41.
Trager, W., and J. B. Jensen. 1976. Science. 193:673–75.
United Nations Development Program/World Bank/World Health Organization. 1982. Newsletter: Special program for research and training in tropical diseases, 5–8.
USAID (U.S. Agency for International Development)/WHO (World Health Organization). 1983. Bull. WHO 61:81–92.
WHO (World Health Organization). 1979a. Bull. WHO 57 (Suppl. 1).
———. 1979b. WHO Expert Committee on Malaria. Seventeenth Report. WHO Tech. Rep. Ser. 640. Geneva, Switzerland: World Health Organization.

11

Apicomplexa: The Piroplasms

THE piroplasms are blood cell parasites of vertebrates. They are smaller than the Plasmodiidae, from which they probably originated. They are piriform, round, ameboid, or rod-shaped, depending in part on the genus. They occur in the erythrocytes; some genera occur in the leukocytes or other blood system cells as well. At some stage they have an apical complex, but they do not have all the structures that are present in other apicomplexans. The Theileriidae have rhoptries (and micronemes and subpellicular microtubules in certain stages), while the Babesiidae have polar rings, subpellicular microtubules, and possibly micronemes as well. There is a micropore at some stage. Pigment (hemozoin) is not formed from the host cell hemoglobin. No spores or oocysts are formed, and no flagella or cilia are present. Locomotion is by body flexion or gliding. Reproduction in the vertebrate host is asexual, by binary fission or merogony. Budding has also been said to occur, but the processes described under this name are more probably endodyogeny or endopolygeny with the formation of 2 or 4 daughter cells. Sexual reproduction probably occurs in the vector. The piroplasms are heteroxenous; the known vectors are ixodid or argasid ticks.

The single order, Piroplasmorida, contains two families (Babesiidae and Theileriidae), which contain parasites of domestic animals. Their taxonomy has been reviewed by Levine (1971) and Krylov (1981). Ristic (1970) reviewed immunity to both families.

Genus *Babesia* Starcovici, 1893

SYNONYMS. *Piroplasma, Achromaticus, Nicollia, Nuttallia, Smithia, Rossiella, Microbabesia, Babesiella, Francaiella, Luhsia, Pattonella, Rangelia, Sogdianella, Entopolypoides,* and *Gonderia* in part.

This is the most important genus in the family Babesiidae. Its species cause babesiosis, a group of highly fatal and economically important diseases of livestock.

Babesia and babesiosis occur in most parts of the world where there are ticks. They are most important in the tropics, where, along with the trypanosomes, they often dominate the livestock disease picture. They also occur, however, in the temperate zone. *B. microti* of rodents, a favorite research organism, has been found causing disease in man in the northeastern United States.

Babesiosis was once an extremely important disease of cattle in the United States. It was associated with Texas fever, which was once thought to be due to *Babesia bigemina* but is now believed to have been due to a combination of *B. bigemina, B. bovis,* and the rickettsia *Anaplasma marginale.* Texas fever has now been eliminated from the United States, and the only domestic animal species left in this country are *B. canis* of dogs and the horse species *B. equi* and *B. caballi.*

LIFE CYCLE. The merozoites of *Babesia* occur in the erythrocytes of vertebrates, where they multiply by binary fission, endodyogeny, or endopolygeny ("budding") or schizogony. In some species 2 merozoites are formed, which break out of the erythrocytes and enter new red cells, while in others tetrads composed of 4 merozoites are formed. The above asexual cycle continues indefinitely, the animals sometimes remaining infected for life. (See Figs. 11.1 and 11.2.)

Babesia is transmitted by ticks. The discovery of this fact—by Smith and

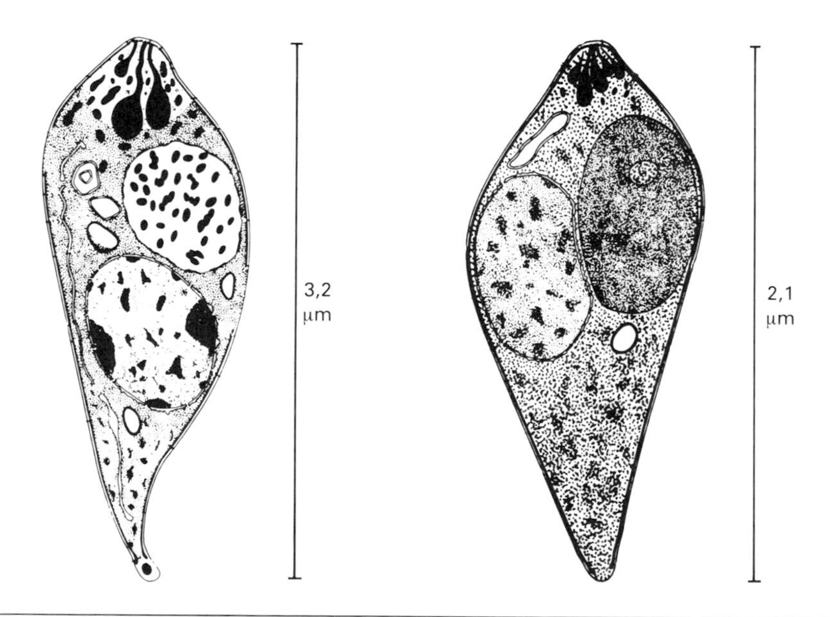

Fig. 11.1. *Babesia bigemina* merozoite in ox erythrocyte. (From Friedhoff and Liebisch 1978)

Fig. 11.2. *Babesia ovis* merozoite. (From Friedhoff and Liebisch 1978)

Kilborne (1893) for *B. bigemina* of cattle—was a milestone in the history of parasitology, since it was the first proof that an arthropod was the vector of any disease.

Whether there is sex in the life cycle of *Babesia* is still undetermined, although it seems probable. Mehlhorn et al. (1981) saw two types of "Strahlenkörper" in dog blood in vitro, one with dense and the other with less dense cytoplasm; they were often paired and in close contact and appeared to fuse. Mehlhorn et al. (1981) considered them to be gametes and interpreted the fusion as initial syngamy. They multiply by a series of binary fissions, producing more than 1,000 individuals in 2–3 days. These become vermiform and pass into the body cavity. The vermiform stages are broadly rounded at the anterior end and pointed posteriorly, about 16 μm long, and have a gliding motion. They enter the ovary, where they penetrate the eggs (Fig. 11.3). Here they round up and divide a few times, forming very small, round individuals. They do not develop further in the larval tick that hatches from the egg, but when it molts they enter the salivary gland and continue their development, undergoing a series of binary fissions and entering the cells of the salivary gland acini. Here they multiply further, becoming smaller and filling the whole host cell until it finally contains thousands of minute parasites. These become vermiform, break out of the host cell, come to live in the lumen of the gland, and are injected into the host when the tick sucks blood. The developmental process in the salivary glands takes 2–3 days. (See Fig. 11.4.) For other studies of the life cycle in ticks, see Shortt (1936), Muratov and Kheisin (1959), Polyanskii and Kheisin (1959), Holbrook et al. (1968a), Friedhoff (1969), Friedhoff and Büscher (1976), Rudzinska et al. (1979), Spielman et al. (1979), Mehlhorn et al. (1980), and Friedhoff and Smith (1981).

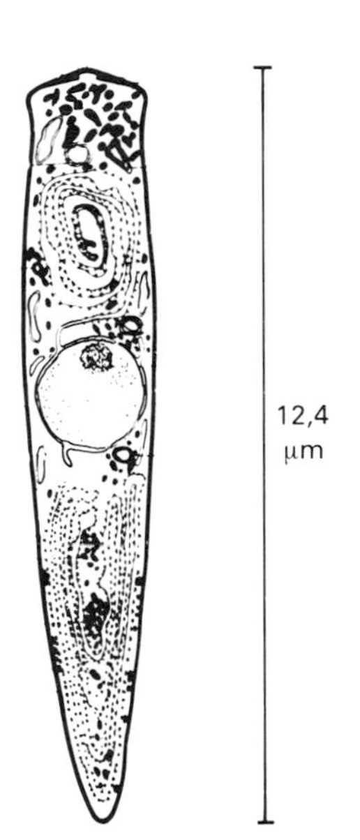

Fig. 11.3. *Babesia bigemina* sporokinete in tick ovary. (From Friedhoff and Liebisch 1978)

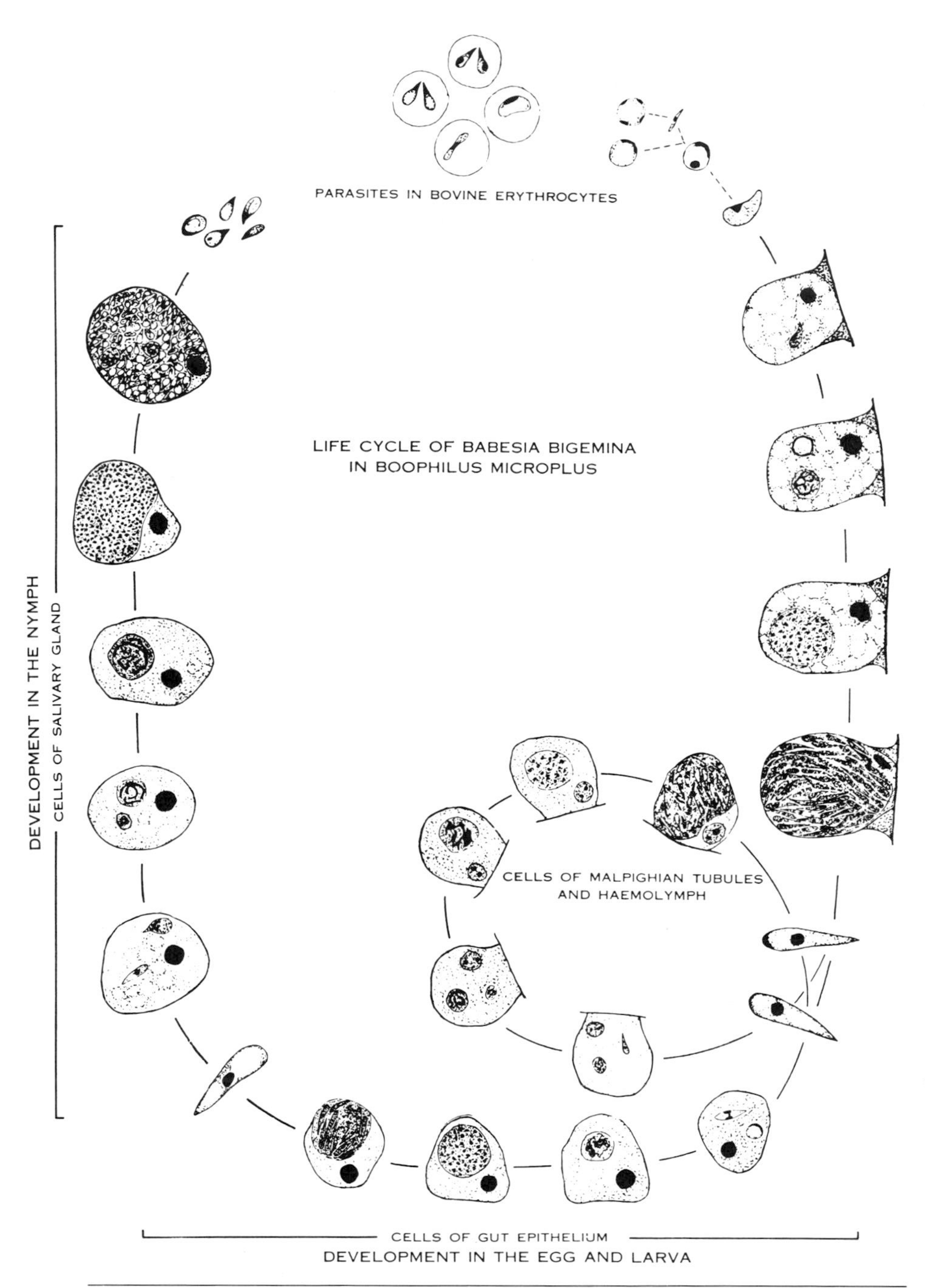

Fig. 11.4. Life cycle of *Babesia bigemina* (omitting sexual stages, if any), with the tick *Boophilus microplus* as intermediate host. (From Riek 1964, Aust. J. Agric. Res. 15:802–21)

PATHOGENESIS. Babesiosis, a highly pathogenic disease in most hosts, is unusual in that the death rate is much higher in adults than in young animals. In most cases there are fever, malaise, and listlessness. Affected animals do not eat, or eat little. There is severe anemia, and destruction of the erythrocytes is accompanied by hemoglobinuria. The mucous membranes become pale, and icterus develops. The spleen is greatly enlarged, with soft, dark red pulp and prominent splenic corpuscles. The liver is enlarged and yellowish brown. The lung may be slightly edematous. There may be diarrhea or constipation, and the feces are yellow except in very early or peracute cases. Affected animals lose condition, become emaciated, and often die. The signs of babesiosis may vary markedly from this typical picture, however.

Death, if it occurs, is due to organic failure which, in turn, is due not only to the destruction of erythrocytes with resultant anemia, edema, and icterus but also to the clogging of the capillaries of various organs by parasitized cells and free parasites. The stasis resulting from this sludging causes degeneration of the endothelial cells of the small blood vessels, anoxia, accumulation of toxic metabolic products, capillary fragility, and eventually perivascular escape of erythrocytes and macroscopic hemorrhage. There is a great similarity between babesiosis in domestic animals and malaria in man, although there are no periodic paroxysms. Reviews on the pathogenesis of *Babesia* have been given by Wright (1981) and Hildebrandt (1981).

IMMUNITY. Cattle that have recovered from an attack of babesiosis due to *B. bigemina* remain infected for life and are immune to reinfection. This type of immunity due to continuing low-grade infection is known as *premunition.* Premunition in cattle due to species other than *B. bigemina* and in sheep, swine, and dogs lasts up to 2 yr or more. Premunized animals may show signs of disease under stress of one sort or another. For instance, an attack of foot-and-mouth disease may reactivate babesiosis in cattle; distemper may do the same in dogs.

The spleen plays an important role in maintaining immunity; splenectomy is often followed by a severe or fatal relapse in premunized animals. In addition, splenectomized animals are much more susceptible to infection with *Babesia* and much more seriously affected than are intact ones.

Calves, foals, young pigs, and kids are much less seriously affected by babesiosis than are adult animals. This is the reason that cattle can often be raised in highly endemic areas without being seriously affected, whereas imported animals usually die; the native cattle are infected as calves and premunized. Lambs and puppies, however, are highly susceptible.

There is no cross-immunity between the different species of *Babesia.*

TREATMENT. The acridine derivative acriflavine (trypaflavine, Gonacrine, flavin, euflavin; a mixture of 2,8-diamino-10-methylacridinium chloride with a small amount of 2,8-diaminoacridinium chloride) and the quinoline derivative acaprin (Acapron, Pirevan, Babesan, Piroparv, Zothelone, Piroplasmin; 6,6′-di[N-methylquinolyl] urea dimethosulfate) are still being used in some areas, although they have been superseded by better drugs.

A number of aromatic diamidines are effective against *Babesia*, among them stilbamidine (4,4′-diamidinostilbene), propamidine (4,4′-diamidino-1,3-diphenoxypropane), pentamidine (Lomidine; 4,4′-diamidino-1,5-diphenoxypentane), phenamidine (4,4′-diamidinodiphenyl ether) and diminazene (Ganaseg, Berenil; 4,4′-diamidinodiazoaminobenzene diacetate). They are injected subcutaneously or intramuscularly, depending on the compound. Many of them tend to cause a fall in blood pressure, but it soon returns to normal. Subcutaneous injection of concentrated solutions may cause irritation. Transitory anaphylactic swelling of the face and lips sometimes occurs with phenamidine.

Amicarbalide (M & B 5062A, Diampron; 3,3′-diamidinocarbanilide diethionate) is probably the drug of choice for cattle. It is also effective against *B. caballi* in horses. Joyner & Brocklesby (1973) and Kuttler (1980, 1981), among others, reviewed the chemotherapy of babesiosis. In practice, 3–5 mg/kg diminazene aceturate intramuscularly, 5–10 mg/kg amicarbalide intramuscularly, or 1–3 mg/kg imidocarb intramuscularly are most often used, but none of these compounds is approved in the United States. Larger amounts are usually required to treat babesiosis in horses and dogs.

PREVENTION AND CONTROL. Since babesiosis is transmitted by ticks, prevention and control depend primarily on tick elimination. This can be done by regular dipping, which should be carried out at least on an area basis for livestock. Dogs and riding horses can be treated individually.

Artificial premunization of young animals has been practiced with a good deal of success, especially in North Africa. A mild strain of the organism is ordinarily used. The practice is not necessary if the animals are raised in an endemic area where they will all become naturally infected at an early age, but it is worthwhile in areas where only a certain proportion of the animals become infected or for animals destined to be shipped to endemic areas later on.

Vaccination of adult animals against babesiosis is not yet commercially practical but should be in not too many years (Callow 1976; Purnell 1980; Mahoney 1981 a,b; Ristic 1981). The cultivation of *Babesia* in erythrocytes or tick tissues (Erp et al. 1978, 1980; Levy and Ristic 1980; Levy et al. 1981; Ristic and Kreier 1981) provides a convenient source of antigen.

REMARKS. This genus has been reviewed by, among others, Mahoney (1977), Ristic and Lewis (1977), Ristic and Kreier (1981), and McCoster (1981).

Babesia bigemina (Smith and Kilborne, 1893)

SYNONYMS.. *Pyrosoma bigeminum, Apiosoma bigeminum, Piroplasma bigeminum, P. australe, P. bubali, B. hudsonius bovis, B. bubali, Luhsia bigemina.*

DISEASE. Bovine babesiosis, piroplasmosis, redwater, Texas fever, tick fever.

This species occurs in the ox, zebu, water buffalo, African buffalo, and certain deer in Africa, Australia, Europe, the Middle East, South America, Central America, and Mexico. It has been essentially eliminated from the

United States. It is common in the tropics and subtropics, where it causes one of the most important diseases in cattle.

STRUCTURE. The merozoites in the erythrocytes are piriform, round, oval, or irregularly shaped. The piriform merozoites occur characteristically in pairs, a feature that gives the species its name. *B. bigemina* is relatively large. The round forms are 2–3 μm in diameter, the elongate ones 4–5 μm long.

The erythrocytic stages lack a conoid, micropores, and typical mitochondria. They have mitochondrionlike vesicles without prominent tubules or cristae and may contain an anterior spheroid body of obscure function. They have an anterior and posterior polar ring and typically 2 rhoptries. There is no typical Golgi apparatus. (See Fig. 11.5.)

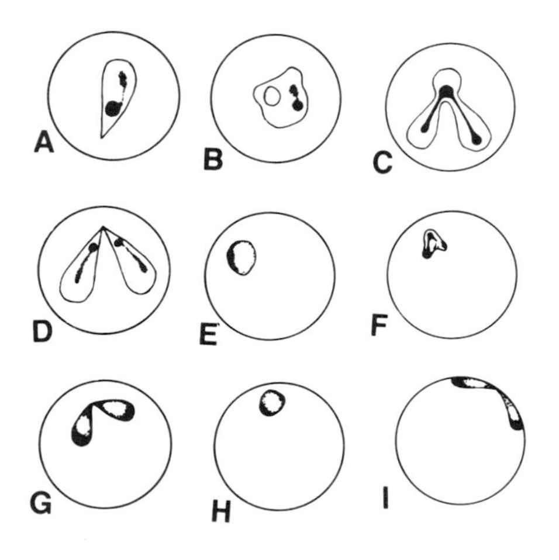

Fig. 11.5. Bovine species of *Babesia* in erythrocytes, as seen with light microscope: A–D. *Babesia bigemina;* E–G. *B. bovis.* H and I. *B. divergens.* (A–D after Nuttall and Graham-Smith 1908, in Parasitology, published by Cambridge University Press; E–I after Davies, Joyner, and Kendall 1958, in *Annals of Tropical Medicine and Hygiene*, published by Liverpool School of Tropical Medicine)

The spherical forms in the ovary of *Boophilus microplus* and *B. decoloratus* occur within a parasitophorous vacuole (formed by the host cell), where they are transformed into elongate, folded kinetes. They have many micronemes, 32 subpellicular ribs, 32 subpellicular microtubules, a polar ring, and rhoptries but apparently no Golgi complex, mitochondria, or conoid. The kinetes are 11–12 μm long.

The sporozoites in the salivary glands of *B. microplus* are longish piriform, with a 3-membraned pellicle, a nucleus, apical micronemes, an uncertain number of rhoptries, a mitochondrion, and 1 or more spheroid bodies (rather electron-dense, spongiform bodies with a single, rather inconspicuous membrane) (Friedhoff and Scholtyseck 1969; Weber and Friedhoff 1979; Weber 1980).

LIFE CYCLE. The tick vectors are *Boophilus annulatus* in North America, *B. microplus* in South and Central America, *B. australis* and *B. microplus* in Australia, *B. calcaratus* in North Africa and the USSR, *B. decoloratus* in South Africa, *Haemaphysalis punctata* in Europe, *Rhipicephalus appendiculatus* and *R. evertsi* in South Africa, and *R. bursa* in North Africa. Transmission takes place through the egg in all species; stage-to-stage transmission also occurs in *Haemaphysalis* and *Rhipicephalus*. Intrauterine transmission may also take place (Enigk 1942).

PATHOGENESIS. *B. bigemina* is highly pathogenic for adult animals but much less so for calves. The incubation period is 8–15 days or less. The first sign of disease is a rise in temperature to 106–108 F. The temperature persists for 1 wk or more. Affected animals are dull and listless, fail to eat, and stop ruminating. The feces are yellowish brown. Severe anemia is caused by invasion and destruction of the erythrocytes; up to 75% of them may be destroyed. Hemoglobinuria is ordinarily present but may be absent. Affected animals become thin, emaciated, and icteric. In chronic cases the temperature is not very high and there is usually no hemoglobinuria, but diarrhea or constipation with hard, yellowish feces is present.

Death may occur in 4–8 days in acute cases. The mortality is as high as 50–90% in untreated cases; treatment reduces it markedly. Calves less than 1 yr old are seldom seriously affected.

Chronically affected animals lose condition quite rapidly and remain thin, weak, and emaciated for weeks before finally recovering.

The principal lesions are splenomegaly with soft, dark red splenic pulp and prominent splenic corpuscles. The liver is enlarged and yellowish brown. The gall bladder is distended with thick, dark bile. The mucosa of the abomasum and intestine is edematous and icteric, with patches of hemorrhage. The subcutaneous, subserous, and intramuscular connective tissues are edematous and icteric, and the fat is yellow and gelatinous. The blood is thin and watery, the plasma may be tinged with red, and the urine in the bladder is usually red.

Some strains are less pathogenic than others.

IMMUNITY. Recovered cattle are premunized, and premunition due to latent infection generally persists for life. There are strain differences, however, and sterile immunity rather than premunition may exist (Callow 1964). Cross-reactions between *B. bigemina* and other species of *Babesia* are slight, if they occur at all.

DIAGNOSIS. Fever associated with hemoglobinuria, anemia, and icterus is suggestive of babesiosis. The diagnosis can be confirmed by finding *B. bigemina* by microscopic examination of stained thin or thick blood smears. Serologic tests are unsatisfactory for diagnosis of disease, although they may be used to detect premunized animals.

TREATMENT. Reviewing chemotherapy of babesiosis in cattle, Kuttler (1980, 1981) said that in practice 3–5 mg/kg diminazene aceturate intramuscularly, 5–10 mg/kg amicarbalide intramuscularly, and 1–3 mg/kg imidocarb intramuscularly are most often used in bovine babesiosis.

PREVENTION AND CONTROL. Since *B. bigemina* is transmitted by ticks, infection can be prevented by tick control, dipping the cattle regularly. In this way Texas fever was eliminated from the United States.

Another measure that has been used is artificial premunization of young animals with a mild strain, especially before shipping them to endemic areas.

Babesia bovis (Babès, 1888) Starcovici, 1893

SYNONYMS. *Haematococcus bovis, Piroplasma bovis, P. argentinum, Babesiella bovis, B. berbera, Babesia argentina, B. berbera, B. colchica, Francaiella colchica, F. argentina, F. berbera, Microbabesia argentina, M. berbera, M. bovis, Luhsia bovis.*

DISEASE. Bovine babesiosis, piroplasmosis, redwater.

This species occurs in cattle, roe deer, and red deer throughout the world (but not the United States or Canada) and is common in southern Europe, the Middle East, and the USSR. In Australia and Mexico it is more important than *B. bigemina*.

STRUCTURE. The merozoites in the erythrocytes are piriform, round, or irregular. Vacuolated "signet-ring" forms are especially common. *B. bovis* has merozoites about 2.4 × 1.5 μm, usually in the center of the erythrocyte.

LIFE CYCLE. The tick vectors are *Ixodes persulcatus* in the USSR, *Boophilus calcaratus* and *Rhipicephalus bursa* in Europe and North Africa, *B. microplus* in South America, and *B. australis* and *B. microplus* in Australia. Transmission occurs through the egg, except that stage-to-stage transmission may occur in *Rhipicephalus*. Relapses may occur for 1.5–2 yr. Intrauterine transmission has also been reported (Neitz 1956).

PATHOGENESIS. The disease caused by *B. bovis* is similar to that caused by *B. bigemina*. The incubation period is 4–10 days, and the first sign is a temperature of 104–106 F that usually lasts 2–3 days. Hemoglobinuria, anemia, icterus, diarrhea, and rapid heartbeat are present; affected animals may die. Strains vary in virulence; *B. bovis* is more pathogenic than *B. bigemina* in Australia (Pierce 1956) and Mexico (Smith et al. 1980).

IMMUNITY. Same as for *B. bigemina*, except that premunition does not last more than about 2 yr, and the minimum time in which cattle regain susceptibility is 5–6 mo.

DIAGNOSIS. Diagnosis is the same as for *B. bigemina*. Cross-reactions between *B. bovis* and other species of *Babesia* are apparently slight and not of importance in serologic tests.

TREATMENT. Treatment is the same as for *B. bigemina.*

PREVENTION AND CONTROL. Methods are the same as for *B. bigemina.* A culture-derived vaccine gives promise of effectiveness (Kuttler et al. 1983).

Babesia caballi (Nuttall and Strickland, 1910)

SYNONYM. *Piroplasma caballi.*

This species occurs in the horse, mule, donkey, and zebra in southern Europe through Asia, USSR, Africa, South America, Central America, and southern United States. Introduced into Florida from Cuba in 1961, it has spread as far north as Tennessee and New Jersey.

STRUCTURE. This is a large species, resembling *B. bigemina.* The merozoites in the erythrocytes are piriform and 2–5 μm long, or round or oval and 1.5–3 μm in diameter. The piriform merozoites are often found in pairs at an acute angle to each other. They have rhoptries, microtubules, free ribosomes, and both smooth and rough endoplasmic reticulum.

LIFE CYCLE. The life cycle is similar to that of *B. bigemina.* The vectors in Europe and the USSR are *Dermacentor marginatus, D. pictus, D. silvarum, Hyalomma anatolicum, H. marginatum, H. volgense, Rhipicephalus bursa,* and *R. sanguineus.* That in North Africa is *H. dromedarii,* that in North America the tropical horse tick *D. nitens.* Transmission through the egg occurs in *D. marginatus, D. silvarum, D. nitens, H. marginatum, H. volgense, R. sanguineus,* and *H. dromedarii.* Stage-to-stage transmission occurs in *D. marginatus, D. pictus, H. anatolicum, H. marginatum, R. bursa,* and *R. sanguineus.*

Holbrook et al. (1968) described the development of *B. caballi* in *D. nitens*, giving 24 colored illustrations. After the ticks ingest infected blood, most of the parasites are destroyed. Small, spherical bodies 4–6 μm in diameter become evident in the gut contents; they apparently turn into clavate (club-shaped) bodies 10–14 × 4–6 μm, which in turn develop into large, round bodies 12–16 μm in diameter that segment into vermicules about 8–12 × 2–4 μm. The vermicules penetrate the gut wall, and some invade other cells of the tick. In the Malpighian tubule cells, hemolymph, and ovaries, the vermicules undergo multiple fission to form a new generation of vermicules similar to those occurring earlier in the gut. The vermicules that invade the ova also undergo similar multiple fission in the larval tick. The resultant vermicules invade the salivary glands, where they undergo multiple fission again to produce large numbers of small, oval, or piriform sporozoites 2.5–3 μm in maximum length. These forms are presumably injected into a horse by the nymphs when they suck blood. Holbrook et al. (1968) saw no forms that could be designated sexual.

B. caballi has also been found in fetuses.

PATHOGENESIS. The incubation period is 6–10 days. The signs of this disease vary markedly. The disease may be either acute or chronic; in either case it may be relatively mild or severe, ending in death. Hemglobinuria is rare, but fever, anemia, and icterus are present. Gastroenteritis is common. Locomotor signs are usually present, and posterior paralysis may occur. Most of the damage in peracute infections is caused by sludged blood in the capillaries. In fatal cases death occurs from 1 wk to about 1 mo after the appearance of signs. The mortality rate varies from 10% or less to 50% or more.

IMMUNITY. Young animals are less susceptible than old ones. There is cross-immunity between *B. caballi* and *B. equi*. Recovered horses may remain carriers for 10 mo to 4 yr.

DIAGNOSIS. Diagnosis depends on identification of the parasites in stained blood smears. They are most numerous in the blood during the first febrile attack. Diagnosis often requires serial passage from ticks to horses.

Various serologic tests have been suggested, including the complement fixation test, but none is as satisfactory as direct observation of the parasites themselves.

TREATMENT. Reviewing the literature, Kuttler (1981) said that 11 mg/kg diminazine aceturate, 8.8 mg/kg phenamidine isethionate, 8.8 mg/kg amicarbalide, or 2.2 mg/kg imidocarb, all given intramuscularly twice at 24-hr intervals, are effective against carrier infections of *B. caballi*. In treating acute infections, he said the drugs of choice are 9 mg/kg amicarbalide, 2 mg/kg imidocarb, or 11 mg/kg diminazine aceturate used in a single intramuscular injection.

PREVENTION AND CONTROL. Tick elimination is the key to prevention. This can be done by spraying the horses with an appropriate tickicide. Inspection, treatment, and spraying should be repeated every 3 wk in outbreak areas.

Babesia canis (Piana and Galli-Valerio, 1895)

SYNONYMS. *Piroplasma canis, P. rossi, P. vitalii, P. commune, Pyrosoma bigeminum* var. *canis, Achromaticus canis, Nicollia canis, Rossiella rossi, Rangelia vitallii, Plasmodium canis, B. rossi, B. vitalii, B. commune.*

DISEASE. Canine babesiosis, canine piroplasmosis, biliary fever, malignant jaundice, nambiuvu.

This species occurs in the dog, wolf, coyote, jackals, red fox, and raccoon dog in South, Central, and North America; Puerto Rico; southern Europe; the USSR; Asia; and Africa. It is common in many tropical regions. In the United States it has been found as far north as Pennsylvania.

STRUCTURE. This is a large form. The merozoites in the erythrocytes are piriform and 4–5 μm long or ameboid and 2–4 μm in diameter, and generally contain a vacuole.

Mature blood stages have a polar ring, subpellicular microtubules, rhoptries, free ribosomes, and endoplasmic reticulum, but apparently not a true conoid, micronemes, or a micropore (Büttner 1968).

There are many "spiky-rayed" stages, and a few ovoid ones without protrusions in the tick (*Dermacentor marginatus*) gut after lysis of infected dog erythrocytes. There are presumably gametes (Mehlhorn et al. 1980; Mehlhorn et al. 1981). The zygotes presumably enter the intestinal epithelial cells and form large numbers of club-shaped forms that enter the hemocoel and then pass to the salivary glands.

Kinetes reach the salivary glands during or before the second day after attachment. They are 15 × 2.5 μm. They lose their typical organelles, reduce their 3-layered pellicle to 1 layer and become spherical. Numerous binary fissions follow, and the parasites progressively become piriform sporozoites about 2.5 × 1.5 μm, with a 3-layered pellicle, rhoptries, a few micronemes, but never spherical bodies. This process takes about 2-3 days, so a dog could be infected by a tick that was still engorging (Schein et al. 1979).

LIFE CYCLE. The vectors are *Rhipicephalus sanguineus* throughout the world; *Dermacentor marginatus* (syn., *D. reticulatus*), *D. pictus,* and *D. venustus* in Europe; *D. pictus* and *Hyalomma marginatum* in the USSR; and *Haemaphysalis leachi* in South Africa. Transmission takes place through the egg in all but *D. pictus* and is stage-to-stage in *D. pictus, R. sanguineus,* and *H. leachi.*

PATHOGENESIS. The severity of infections with *B. canis* varies considerably with the strain. In some localities it is a comparatively mild disease, while in others it may be highly pathogenic. Both young and old dogs are susceptible.

The incubation period is 10–21 days in naturally infected dogs. In acute cases the first sign of disease is fever. This is quickly followed by marked anemia with icterus, inappetence, marked thirst, weakness, prostration, and often death. Hemoglobinuria is sometimes but not usually present.

In chronic cases the fever is not high and seldom lasts more than a few days; there is little icterus. Anemia is severe, and the dogs are listless and become very weak and emaciated.

Protean in its manifestations, canine babesiosis may take on many different clinical forms. Involvement of the circulatory system may produce edema, purpura, and ascites, there may be stomatitis and gastritis, and involvement of the respiratory system causes catarrh and dyspnea. Keratitis and iritis are seen if the eyes are affected, and myositis and rheumatic signs if the muscles are involved. Central nervous system involvement causes locomotor disturbances, paresis, epileptiform fits, and other CNS signs. A tendency to clog the capillaries is common to many species of *Babesia*. In cerebral babesiosis the signs may be confused with those of rabies.

In South America, the disease is called nambiuvu, meaning "bloody ears" in the Guarani language. As the name suggests, it is a hemorrhagic disease. There is bleeding from the edges of the ears and from the muzzle, particularly in young dogs in summer. There are also internal hemorrhages.

The spleen is enlarged, with dark red, soft pulp and prominent splenic corpuscles. The liver is enlarged and yellow, with pathologic changes ranging from congestion to centrilobular necrosis. The heart is pale and yellowish. The kidneys are yellowish; there is considerable nephrosis or nephritis. The muscles are pale and yellow, and the fat and mucous membranes may be yellowish. There may be a variable amount of fluid in the pleural, pericardial, and peritoneal cavities. Small hemorrhages are sometimes present on the heart, pleura, bronchi, and intestines. There is less icterus in chronic than in acute cases.

IMMUNITY. Recovered animals remain infected in a state of premunition. This persists for life if they are kept in an endemic area, but the parasites die out in a year or more in the absence of reinfection.

DIAGNOSIS. In endemic areas, signs of fever, anemia, and icterus, with or without hemoglobinuria, are suggestive of canine babesiosis. The diagnosis can ordinarily be confirmed by finding the parasites in stained blood smears.

TREATMENT. Reviewing the literature, Kuttler (1981) said that the same general groups of compounds that are active against bovine and equine babesiosis are active against canine babesiosis. Diminazene aceturate at a dose greater than 0.25–3.5 mg/kg has been found effective, and *B. canis* is susceptible to treatment with trypan blue and quinoline derivatives such as acaprin.

PREVENTION AND CONTROL. As for other *Babesia* infections, these depend on tick control.

Babesia cati Mudaliar, Achary, and Alvar, 1950

This species occurs in wild and domstic cats in India.

STRUCTURE. This is a large *Babesia*, in contrast to the small *B. felis*.

Babesia crassa Hashemi-Fesharki and Uilenberg, 1981

This species occurs in the erythrocytes of the domestic sheep in Iran. It is transmissible to the domestic goat.

STRUCTURE. This is a large species that differs structurally and immunologically from both *B. motasi* and *B. ovis*. There are frequently four organisms in a single erythrocyte; they result from quadruple division in some cases and from two successive binary fissions in others. They are 2–3 μm long.

Babesia divergens (M'Fadyean and Stockman, 1911)

SYNONYMS. *Piroplasma divergens, P. rupturaeliensis, Microbabesia divergens, M. occidentalis, Babesiella divergens, B. karelica, Babesia caucasica, B. karelica, B. occidentalis* in part, *Francaiella caucasica, F. occidentalis.*

This species occurs in the erythrocytes of cattle and rarely of man in Europe and the USSR. It is the common *Babesia* of cattle in Europe.

STRUCTURE. This species is smaller than *B. bovis.* The merozoites usually occur as paired, club-shaped organisms about 1.5 × 0.4 μm; the angle between the members of the pair is relatively large, so that they diverge more from each other than do the merozoites of *B. bovis*; in addition, they tend to lie along the circumference of the host erythrocyte (the so-called accolé position). Other merozoites are stouter and piriform (about 2 × 1 μm) or circular (up to 2 μm in diameter). The merozoites lack a conoid, micropores, a typical Golgi complex, and typical mitochondria. They have mitochondrionlike vesicles without prominent tubules or cristae, lack a spheroid body or subpellicular microtubules, and usually have 3 or more rhoptries.

LIFE CYCLE. The life cycle is the same as that of *B. bovis*. The vector tick is *Ixodes ricinus*. Transmission takes place through the egg and from stage to stage.

PATHOGENESIS. Same as for *B. bovis*.

IMMUNITY. Immunity factors are the same as for *B. bovis*. An irradiated vaccine appears effective in protecting cattle against natural infection.

DIAGNOSIS. Same as for *B. bovis*.

REMARKS. Five human cases have been reported, all in splenectomized persons, in Yugoslavia, Northern Ireland, Scotland, and the USSR (Skrabalo and Deanovic 1957; Fitzpatrick et al. 1968; Harambašić et al. 1976; Rabinovich et al. 1978; and Entrican et al. 1979a,b). It has been transmitted experimentally to the gerbil *Meriones unguiculatus* and to the splenectomized chimpanzee, rhesus monkey, mouflon, red deer, fallow deer, and roe deer but not to the splenectomized domestic rabbit, goat, or sheep (Garnham and Bray 1959; Enigk and Friedhoff 1962; Entrican et al. 1979a,b; Lewis and Young 1980).

Babesia equi (Laveran 1901)

SYNONYMS. *Piroplasma equi, Nuttallia equi, N. asini, N. minor, Achromaticus equi.*

This species occurs in the horse, mule, donkey, and zebras throughout the world. Splenectomized rabbits, mice, rats, guinea pigs, gerbils, hamsters, dogs, pigs, sheep, and cattle cannot be infected (Frerichs et al. 1969). It is generally more widely distributed than *B. caballi*, but relatively few cases have been reported in the United States.

STRUCTURE. *B. equi* is relatively small, only 2-3 μm long. The merozoites in the erythrocytes are rounded, ameboid, or most often piriform. The last are usually found in a group of 4 joined together in the form of a cross. There may be as many as eight *B. equi* in a single host cell.

LIFE CYCLE. Division is unlike that of *B. caballi* in that 4 daughter merozoites are formed at one time (Holbrook et al. 1968b).

The vectors are *Dermacentor marginatus* (syn., *D. reticulatus*), *D. pictus, Hyalomma marginatum* (syn., *H. detritum*), *H. uralense*, and *Rhipicephalus bursa* in the USSR; *H. anatolicum* (syn., *H. excavatum*) and *H. marginatum* in Greece; and *H. dromedarii* and *R. sanguineus* in Central Asia. The vector of *B. equi* in the United States is unknown; it will not develop in *D. nitens* (Thompson 1969). Transmission is through the egg in *H. anatolicum*, and stage to stage in all the others. Intrauterine transmission may also occur (Neitz 1956).

PATHOGENESIS. This species is more pathogenic than *B. caballi*. Mixed infections are not rare, however, so it is sometimes difficult to be sure which species is causing the signs. The incubation period following an infective tick bite is 10–21 days. The first sign of disease is a rise in temperature. This is followed by listlessness, depression, marked thirst, inappetence, watering of the eyes, and swelling of the eyelids. The most characteristic sign is icterus. There is marked anemia, more than half of the erythrocytes often being destroyed. Hemoglobinuria is present, but in contrast to *B. caballi* infections, posterior paralysis is absent. Edema of the head, legs, and ventral part of the body is sometimes present. Affected animals are constipated, passing small, hard balls of feces covered with yellow mucus; they lose condition fairly rapidly and may become extremely emaciated. Hemorrhages are present on the mucous membranes of the nasal passages, vagina, and third eyelid.

The disease usually lasts 7–12 days; but it may be peracute, with death occurring in 1–2 days, or it may be chronic and last for weeks. The mortality is generally not more than 10% but may sometimes reach 50%. Recovery is slow; it may be several weeks or even months before the animal returns to normal.

At necropsy, emaciation, icterus, anemia, and edema are present. There are accumulations of fluid in the pericardial sac and body cavities, and the fat is gelatinous and yellow. The spleen is enlarged, with soft, dark brown pulp. The lymyph nodes are swollen and sometimes inflamed. The liver is swollen, engorged, and brownish yellow; the hepatic lobules are yellow in the center and greenish yellow around the edges; the kidneys are pale yellow and may contain petechial hemorrhages. There are hemorrhages or red streaks on the mucosa of the intestine and stomach.

IMMUNITY. There is no cross-immunity between *B. equi* and *B. caballi*. Young animals are less seriously affected than adults.

DIAGNOSIS. Babesiosis can be diagnosed by identifying the parasites in stained blood smears. Examinations should be made as early as possible, since the parasites begin to disappear from the peripheral blood after the 5th day.

TREATMENT. Kuttler (1981) said that this species is more resistant to treatment than *B. caballi*. Amilicarb at 22 mg/kg (which approaches the toxic level) is effective only in some animals. Four mg/kg imidocarb intramuscularly 4 times at 3-day intervals is usually effective but might produce toxic signs.

Babesia felis Davis, 1929

SYNONYMS. *Babesiella felis, Nuttallia felis* var. *domestica.*

This species occurs in the erythrocytes of domestic and wild cats, the puma, lion, American lynx, and leopard in Africa and India. Its existence in North America is problematical.

STRUCTURE. This is a small form. Most of the merozoites are round or irregularly round and 1.5–2 μm in diameter. Some are elongate and 2–3 (or rarely 4) μm long. Piriform merozoites are rare. Division is quadruple, forming a cruciform meront, or binary.

LIFE CYCLE. The vectors are unknown, although *Haemaphysalis leachi* has been incriminated in South Africa.

PATHOGENESIS. Feline babesiosis is less severe than the canine disease, and affected animals usually recover without treatment. It is characterized by anemia, slow respiration, somnolence, listlessness, emaciation, constipation with yellow or orange feces, splenomegaly, and sometimes icterus and hemoglobinuria.

TREATMENT. Both trypan blue and acaprin are effective.

Babesia foliata Ray and Rhaghavachari, 1941

This species occurs in the erythrocytes of sheep in India. It resembles *B. ovis* but is leaf-shaped; it may nevertheless be a synonym. The vector is unknown.

Babesia gibsoni (Patton, 1910)

SYNONYMS. *Piroplasma gibsoni, P. tropicus, Achromaticus gibsoni, Babesiella gibsoni, Pattonella gibsoni, Nuttallia bauryi.*

This species occurs in the dog, jackal, wolf, Indian wild dog *Cuon dukhensis*, red fox, fennec fox, and probably the mongoose *Herpestes javanicus* and ferret-badger *Melogale personata* in Asia and occasionally North America and North Africa. The jackal is the natural host in India. It cannot be transmitted to the raccoon, golden hamster, or white-footed mouse (Anderson et al. 1979). Anderson et al. (1979) found it in a dog in Connecticut, the first finding in an American dog in North America. Farwell et al. (1982) felt that several hundred dogs arriving in the United States from Okinawa each year are carriers of *B. gibsoni*.

STRUCTURE. This species is smaller than *B. canis* and rarely has its characteristic paired, piriform merozoites. Its merozoites are usually annular or oval and not more than one-eighth of the diameter of the host erythrocyte. Occasionally, large ovoid forms half the diameter of the host cell or thin, elongate forms reaching almost across the cell may be found.

LIFE CYCLE. The life cycle is similar to that of *B. canis*. The vectors in India are *Haemaphysalis bispinosa* and *Rhipicephalus sanguineus*. Transmission is through the egg or stage to stage in the former, stage to stage in the latter.

PATHOGENESIS. This species is only slightly pathogenic for its natural host, the jackal, but is highly pathogenic for the dog, causing marked anemia, remittent fever, hemoglobinuria, constipation, marked splenomegaly, and hepatomegaly. The disease usually runs a chronic course, with remissions and relapses of fever, and death may not occur for many months. In imported dogs, however, death is said to occur in 3-4 wk.

IMMUNITY. There is no cross-immunity between *B. gibsoni* and *B. canis*.

TREATMENT. Diminazene aceturate in 1% solution intramuscularly in two doses 5 days apart, 11 mg/kg each, was the best of seven drugs that Fowler et al. (1972) tested. It reduced the death rate or prevented death.

Babesia major (Sergent, Donatien, Parrot, Lestoquard, and Plantureux, 1926)

SYNONYM. *Babesiella major.*

This species occurs in cattle and bison in North Africa, Europe, and the USSR. It is rare in England and Germany.

STRUCTURE. The merozoites in the blood resemble those of *B. bovis* but are larger. The piriform, paired forms are 2.6 × 1.5 μm, the round ones 1.8 μm in diameter. The parasites lie in the center of the host erythrocyte.

LIFE CYCLE. The life cycle is similar to that of *B. bovis*. The vector in the USSR is *Boophilus calcaratus*, that in England and Germany *Haemaphysalis punctata.*

PATHOGENESIS. This species is considerably less pathogenic than *B. bovis*, but strains vary considerably in pathogenicity.

IMMUNITY. This species is antigenically different from *B. bigemina, B. bovis,* and *B. divergens*.

TREATMENT. Same as for *B. bovis*.

Babesia microti (França, 1910) Reichenow, 1953

SYNONYMS. *Smithia microti, S. microtia, Nuttallia microti, N. colesi, B. rodhaini, B. bandicootia (?).*

This species occurs in voles (*Microtus* spp.), field mice (*Apodemus* spp.), deermice (*Peromyscus* spp.), house mice, ground squirrels, other rodents, cottontails, and man throughout the world. A variety of rodents and primates has been infected experimentally. It is quite common among rodents in certain areas.

STRUCTURE. This is a small species of *Babesia*. Multiplication within the erythrocytes takes place by binary or quadruple fission, the latter producing a cross-shaped structure composed of 4 merozoites. There are many strains, which differ in factors such as host specificity, isoenzymes, and DNA buoyancy. The erythrocytic merozoites have a single plasma membrane, micronemes, rhoptries, a spheroid body (which is possibly a kind of rhoptry), and Golgi apparatus. Large "food vacuoles" do not contain host cell cytoplasm, although it may appear so in sections; they are instead merely invaginations of the cytoplasm (Rudzinska 1976). The predominant forms in man are rings resembling those of *Plasmodium falciparum*, but they do not form pigment deposits in the parasitized erythrocytes as do mature *Plasmodium* stages (Healy and Ruebush 1980).

LIFE CYCLE. There are two types of organism in the erythrocytes. In one, the meront forms merozoites by endodyogeny or endopolygeny (budding). In the other, no merozoites are formed but rather what are probably gamonts. In ticks (*Ixodes dammini*) fed on infected hamsters, there are gamonts containing several arrowheadlike structures with a tail, microtubules, and a cytostome that form gametes. The microgametes (if that is what they are) are slender and spindle-shaped, about 2 × 0.8 μm, while the "macrogametes" are spheroids about 2.5 μm in diameter (Rudzinska et al. 1978; Rudzinska et al. 1979).

The vector in Europe is *I. ricinus* (Krampitz and Bäumler 1978), while that in the United States (at least on Nantucket Island) is *I. dammini* (Spielman et al. 1979), a species near *I. scapularis*. Transmission is apparently stage to stage.

PATHOGENESIS. This species does not appear to be particularly pathogenic in some wild rodents, but relatively little attention has been paid to this subject. Splenomegaly has been found in them. In unsplenectomized people, symptoms of fever, drenching sweats, shaking chills, myalgia, arthralgia, extreme fatigue, and mild to moderate hemolytic anemia have been reported. Even after successful treatment parasitemia and fatigue frequently persist for weeks to months (Ruebush et al. 1977). Human infections are self-limiting, and subclinical infections also occur.

DIAGNOSIS. Infections can be found by microscopic examination of stained blood smears. The splenectomized golden hamster may be inoculated to isolate *B. microti* from man or wild rodents (Brandt et al. 1977; Etkind et al. 1980).

TREATMENT. Oral chloroquine phosphate, presumably as used for malaria, is effective in man (Ruebush et al. 1977).

PREVENTION AND CONTROL. Elimination of ticks or avoidance of tick bites will prevent infection.

REMARKS. As of August 1978 at least 28 human cases of babesiosis had been reported in the literature from the United States, the USSR, Yugoslavia, France, Northern Ireland, and Scotland. The great majority were due to *B. microti*, and most of these were from Nantucket Island and Martha's Vineyard in the United States. Nine of the affected persons had been splenectomized some years before they became ill, but the cases due to *B. microti* were primarily in intact persons and were mild, whereas those due to *B. divergens* occurred only in splenectomized persons and were acute (fatal).

The principal reservoir on Nantucket Island is *Peromyscus leucopus*, but the vole, Norway rat, and eastern cottontail are also infected (Spielman et al. 1981).

Babesia motasi Wenyon, 1926

SYNONYM. *Haematococcus ovis* in part; *Piroplasma ovis* in part.

This species occurs in sheep and goats in Europe, the Middle East, the USSR, Vietnam, Africa, and parts of the tropics. It can be transmitted to the intact mouflon and splenectomized red and fallow deer, but not to the splenectomized calf, roe deer, and Soemmering's gazelle (Enigk, Friedhoff, and Wirahadiredja 1964). It has apparently not been found in sheep in Great Britain, although Lewis and Herbert (1980) found it in the tick *Haemaphysalis punctata* in Wales.

STRUCTURE. This is a large form, 2.5–4 × about 2 μm. The merozoites resemble those of *B. bigemina* and are usually piriform. They occur singly or in pairs; the angle between members of a pair is acute.

LIFE CYCLE. Similar to that of *B. bigemina*. The vector in Romania is *Rhipicephalus bursa*, that in Sardinia *Haemaphysalis punctata*, and those in the USSR *Dermacentor silvarum* and *Haemaphysalis otophila*. Transmission occurs both through the egg and stage to stage in *R. bursa.*

PATHOGENESIS. This species may cause either an acute or chronic disease. Fever, prostration, marked anemia, and hemoglobinuria are present in the acute disease; affected animals often die. There may be no characteristic signs in the chronic disease.

IMMUNITY. Sheep that are immune to *B. motasi* are not immune to *B. ovis* and vice versa.

DIAGNOSIS. Same as for *B. bigemina.*

TREATMENT. Same as for *Babesia* in cattle.

PREVENTION AND CONTROL. Same as for other species of *Babesia.*

Babesia ovis (Babes, 1892) Starcovici, 1893

SYNONYMS. *Amoebosporidium polyphagum, Haematococcus ovis* in part, *Piroplasma ovis, P. hirci, B. hirci, Babesiella ovis, Francaiella ovis, Microbabesia ovis.*

This species occurs in sheep, goat, mouflon, and argali in Europe, the USSR, the Middle East, and some tropical and subtropical regions. Splenectomized red and fallow deer are slightly susceptible (Enigk, Friedhoff, and Wirahadiredja 1964).

STRUCTURE. This is a small species, about 1–2.5 μm long. Most of the parasites are round; they usually lie in the margin of the host erythrocytes. The angle between the paired, piriform trophozoites is usually obtuse. The differentiated merozoites lack a conoid, micropores, and typical mitochondria but have mitochondrionlike vesicles without prominent tubules or cristae and have 2 spheroid bodies of obscure function. They have 5 or more rhoptries and an anterior and a posterior polar ring but apparently have no subpellicular microtubules (Friedhoff and Scholtyseck 1977).

The kinetes (vermicules) in the tick *Rhipicephalus bursa* have an apical complex consisting of an apical umbrella, a crown of microtubules beneath it, rhoptries, and micronemes. There are no typical subpellicular microtubules or conoid (Friedhoff and Scholtyseck 1968a; Weber 1980).

Large meronts are formed in the secretory glands of adult *R. bursa* ticks. They produce spindle-shaped sporozoites about 2.1 μm long and 0.9 μm wide, which have a "pellicular complex," anterior and posterior polar rings, rhoptries, micronemes, nucleus, mitochondrionlike vesicles, and a spheroid body, usually between the anterior end and the nucleus (Friedhoff et al. 1972).

LIFE CYCLE. The life cycle is similar to that of *B. bovis*. In *R. bursa*, infected erythrocytes are lysed in the gut, releasing the babesias, which lie extracellularly in the gut contents for about 3 days after the ticks have become replete. Most die, but a few survive and are phagocytized with the gut contents by the gut epithelial cells. They either develop directly into merozoitelike forms or pass through a still unknown multiplication and/or transformation process, ending with the production of differentiated forms (this phase is still not clear). The merozoites arising in this way enter the cytoplasm of the basal parts of the gut epithelial cells, where they turn into ameboid and then plasmodial forms; the latter divide to form many spheroid daughter individuals that elongate into merozoites that pass to the apical part of the host cell and thence (when the host cell breaks down) into either the gut lumen or the hemocoel. In the first case, they are phagocytized by other gut cells and repeat the above process; several such developmental cycles may take place. Most of the merozoites, however, go into the hemolymph where they enter the neutrophile hemocytes or granulocytes and turn into ameboid and then plasmodial forms; the latter produce many spheroid daughter cells that elongate into merozoites. These then enter new hemocytes and also muscle, excretory canal, and ovarian cells. The merozoites in the ovary multiply, especially in the peritracheal cells and nurse cells. Those in the nurse cells pass into the developing oocytes. After the eggs have been laid, further multiplication takes place and a new type of merozoite is formed in the egg. The daughter tick thus transmits the parasite (Friedhoff and Scholtyseck 1968b; Friedhoff 1969).

Transplacental infection may occur.

The vectors in the USSR are *R. bursa* and *Ixodes persulcatus*, and in Germany *R. bursa*.

PATHOGENESIS. This species is less pathogenic than *B. motasi* but may cause fever, anemia, and icterus. Usually not more than 0.6% of the erythrocytes are infected.

IMMUNITY. There is no cross-immunity between *B. ovis* and *B. motasi.*

TREATMENT. Two mg/kg imidocarb 3 times at 1-day intervals is highly successful (Kuttler 1981).

Babesia perroncitoi (Cerruti, 1939) Levine, 1971

SYNONYMS. *Babesiella perroncitoi, Babesia metalnikovi.*

This species occurs in the domestic and wild pig in Europe (Sardinia), the Sudan, and Vietnam.

STRUCTURE. This is a small form. It is usually annular, 0.7–2 μm in diameter, with a thin ring of cytoplasm surrounding a vacuole; but it may also be oval, quadrangular, lanceolate, or piriform. It measures 1–3 × 1–2 μm. The merozoites usually occur singly in the host cells, but sometimes 2 or more may be present.

LIFE CYCLE. The vector is unknown.

PATHOGENESIS. The disease caused by this species is similar to that caused by *B. trautmanni.*

TREATMENT. Treatment is the same as for *B. bovis*. Acaprin is effective against this species.

Babesia taylori (Sarwar, 1935)

SYNONYM. *Piroplasma taylori.*

This species occurs in the domestic goat in India.

STRUCTURE. This is a small species; the merozoites are about 2 × 1.5 μm when there is a single one per host cell and are 1 μm or less in diameter when there are several. The merozoites are mostly ovoid or round, rarely piriform. The host cell is enlarged.

LIFE CYCLE. Unknown.

PATHOGENESIS. This species is probably pathogenic.

Babesia trautmanni (Knuth and Du Toit, 1921)

SYNONYMS. *Piroplasma trautmanni, P. suis.*

This species occurs in the domestic pig in southern Europe, Africa, and the USSR.

STRUCTURE. This is a large form, the merozoites being 2.5–4 × 1.5–2 μm. They are oval, piriform, or (less commonly) round. They often occur in pairs. The host cells usually contain 1–4 or occasionally 5–6 parasites.

LIFE CYCLE. The vector is *Rhipicephalus sanguineus* (syn., *R. turanicus*). Transmission occurs through the egg. Other ticks have also been incriminated.

PATHOGENESIS. This species may cause either a mild disease or a fatal one with fever, listlessness, inappetence, anemia, hemoglobinuria, icterus, edema, and incoordination. Infected sows may abort. The spleen is enlarged and engorged; the liver is enlarged; there are pulmonary, renal, and gastrointestinal hyperemia and edema; petechiae are present on the serous membranes; and there are subepicardial and subendocardial hemorrhages.

TREATMENT. Same as for *B. bovis*.

Babesia vogeli Reichenow, 1937

SYNONYM. *B. major* in part.

This species occurs in the dog in southern Asia and North Africa.

STRUCTURE. This species is somewhat larger than *B. canis*.

LIFE CYCLE. The life cycle is similar to that of *B. canis*. The vector is *Rhipicephalus sanguineus*. Transmission occurs through the egg and stage to stage.

PATHOGENESIS. This species is less pathogenic than *B. canis*, but the disease it causes is otherwise similar.

IMMUNITY. Dogs infected with this species are not resistant to infection with *B. canis* transmitted by *Dermacentor*.

TREATMENT. Same as for *B. canis*.

Babesia moshkovskii (Schurenkova, 1938) Laird and Lari, 1957

SYNONYMS. *Sogdianella moshkovskii, B. ardeae, Nuttallia shortti, Aegyptianella moshkovskii*.

This species occurs in the chicken, possibly the turkey and pheasant, and perhaps the eagle and other wild birds in the United States, Asia, the USSR, and Africa. It is rare in the chicken and turkey.

STRUCTURE. The chicken form described by Henry (1939) was 0.2–2.5 μm in diameter, occurring as *Anaplasma*-like granules, as small rings, and as elongate bodies with a terminal dot of chromatin and a thin tail of cytoplasm. Four merozoites were usually produced.

The form described by McNeil and Hinshaw (1944) from turkey poults was roundish, oval, or piriform; 0.5–2 μm in diameter; and occurred singly or in pairs.

LIFE CYCLE. The life cycle is unknown. Croft and Kingston (1975) thought that the soft tick *Ornithodoros concanensis* was the most likely vector of their falcon form.

Genus *Theileria* Bettencourt, França, and Borges, 1907

Members of the family Theileriidae occur in mammals. They have small, round, ovoid, irregular, or bacilliform merozoites. The elements of the apical complex are much reduced, including only rhoptries; there are no polar rings or conoids. Micronemes and subpellicular microtubules are present only in certain stages. A micropore is present in the erythrocyte stage. Some meronts occur in vertebrate erythrocytes and others in vertebrate lymphocytes, histiocytes, erythroblasts, etc. Merogony occurs in the lymphocytes, histiocytes, erythroblasts, etc. and is followed by invasion of the erythrocytes. The forms in the erythrocytes may or may not reproduce; if they do, they divide into 2 or 4 merozoites. The vectors are ixodid ticks; binary fission and schizogony occur in the tick. Synonyms of this genus are *Gonderia, Cytauxzoon,* and *Haematoxenus*. Many of the named species occur in wild African ruminants. The genus has been reviewed by Barnett (1977).

Theileria annulata (Dschunkowsky and Luhs, 1904) Wenyon, 1926

SYNONYMS. *Piroplasma annulatum, T. dispar, T. turkestanica, T. sergenti, Gonderia annulata, G. dispar.*

DISEASE. Tropical theileriosis, tropical piroplasmosis, Egyptian fever, Mediterranean Coast fever.

This species occurs commonly in the lymphocytes and erythrocytes of the ox, zebu, and water buffalo in North Africa, southern Europe, southern USSR, and Asia, causing mortalities of 10–90%, depending on the area. In addition, an American bison in the Cairo zoo died of a natural infection (Carpano 1937). It causes one of the most important diseases of cattle in those regions.

STRUCTURE. The forms in the erythrocytes are predominantly (70–80%) round or oval but may also be rod-shaped, comma-shaped, or even *Anaplasma*-like. The round forms are 0.5–2.7 μm in diameter, the oval ones about 2.0 × 0.6 μm, the comma-shaped ones about 1.2 × 0.5 μm, and the *Anaplasma*-like ones 0.5 μm in diameter. Binary fission takes place, with the formation of 2 daughter individuals, or quadruple fission with the formation of 4 individuals in the form of a cross. Both the comma-shaped and round forms have a micropore, but neither has polar rings, conoid, or subpellicular microtubules (Schein et al. 1978).

So-called Koch bodies are in the lymphocytes of the spleen or lymph nodes, or free in these organs. They average 8 μm in diameter but range up to 15 or even 27 μm. Two types have been recognized, but actually one becomes the other. The so-called macromeronts or macroschizonts contain chromatin granules 0.4–1.9 μm in diameter; after their nuclei have divided further they become micromeronts, which contain chromatin granules 0.3–0.8 μm in diameter and produce merozoites 0.7–1.0 μm in diameter. (See Fig. 11.6.)

In the tick, spindle-shaped forms about 8–12 μm long and 0.8 μm in diameter at the middle, with a unit membrane, an electron-dense stilettolike

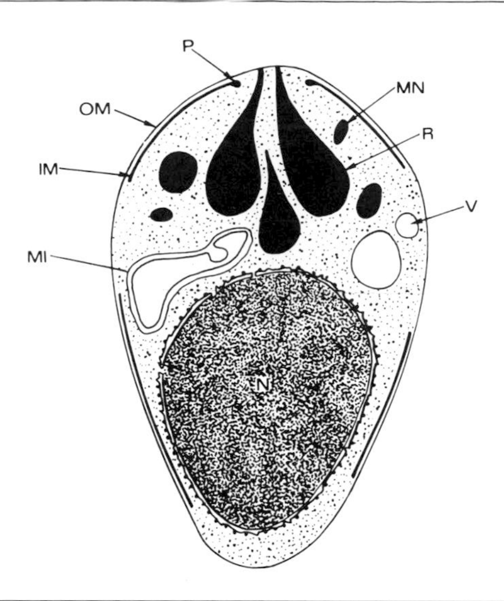

Fig. 11.6. Merozoite of *Theileria annulata*, *T. mutans*, and/or *T. parva*: *IM* = inner membrane of pellicle, *MI* = mitochondrion, *MN* = microneme, *N* = nucleus, *OM* = outer membrane of pellicle, *P* = polar ring, *R* = rhoptry, *V* = vacuole. (From Schein et al. 1978)

apex about 2–3 μm long, several flagellumlike protrusions 4–5 μm long, and a slender posterior pole, were interpreted by Schein et al. (1975), Mehlhorn et al. (1975), and Schein and Friedhoff (1978) as microgamonts. The stilettolike structure appears spongy at its base. The microgamonts form filiform bodies about 12 μm long with a central nucleus and several microtubules. These were interpreted as microgametes. Spherical bodies interpreted as macrogametes are present. Later (presumably after syngamy) kinetes up to 20 μm long and 4–5 μm in maximum diameter appear. They have a typical coccidian pellicle, with a characteristic surface striation. At the anterior end is a polar ring system consisting of a riblike structure, beneath which are 38–40 subpellicular microtubules, each beneath a rib. These terminate after a short distance. The apical third of the cell is completely filled with micronemes, and there are also smaller numbers of micronemes in the posterior regions. There are no rhoptries or crystalloid structures. Anterior to the nucleus is a zone of numerous double-membraned organelles 1.0 × 0.1–0.15 μm, which may be mitochondria although they contain no cristae or tubules and are essentially hollow spheres. There are also prominent membrane-bound vacuoles up to 4 μm in diameter and unknown granular bodies in the kinetes.

LIFE CYCLE. It had generally been supposed that sexual stages do not occur in *Theileria*. However, several workers (Schein et al. 1975; Mehlhorn and Schein 1977; Schein and Friedhoff 1978; and others) have stated that they do. According-ing to these workers, the erythrocytic merozoites develop into slender, spindle-shaped "microgamonts" in the gut of the tick about 1–4 days after repletion. The microgamonts then form up to 4 nuclei and several flagellumlike appendages, which break away from the microgamonts and presumably unite with spherical macrogametes. This union, however, has not been seen. The resultant zygotes are spherical and have a vacuolelike center. They enter the cells of the gut wall, slowly become elongate, and leave the gut cell, entering the hemolymph on days 14–17; they are then kinetes. They are fully differentiated by day 17–20. They then enter the salivary glands and transform into fission bodies, which grow up to about 20 μm in diameter. The ticks then molt and no further development occurs until activation. Shortly before infestation the salivary glands begin to proliferate, and the fission bodies grow rapidly. In adult ticks the primary fission bodies (sporonts) divide into numerous secondary fission bodies (primary sporoblasts), which then divide into tertiary fission bodies (secondary sporoblasts); these produce comma-shaped sporozoites, which are released into the saliva. The whole process of sporozoite formation in the salivary glands takes less than 2 days. Cattle are infected when the ticks suck blood. The sporozoites enter the lymphocytes and become meronts (Koch bodies); the early ones are macromeronts, which produce macromerozoites 2.0–2.5 μm in diameter and these in turn become micromeronts, which produce micromerozoites 0.7–01.0 μm in diameter. The number of asexual generations in the lymphocytes appears to be unknown. In due time micromerozoites enter the erythrocytes. After a few binary fissions, the comma-shaped parasites produce ovoid forms that are taken up by other ticks. (See Fig. 11.7.)

The vectors of *T. annulata* are *Hyalomma detritum* (syn., *H. mauretanicum*) in North Africa; *H. detritum* and *H. excavatum* (syn., *H. anatolicum*) in the USSR; *H. truncatum* in parts of Africa; *H. dromedarii* in Central Asia; *H. excavatum, H. turanicum* (syn., *H. rufipes glabrum*), and *H. marginatum* (syns., *H. savignyi, H. aegyptium*) in Asia Minor; *H. marginatum* in India; and *H. longicornis* (syns., *H. bispinosa, H. neumanni*) in Siberia and the Far East.

Transmission is stage to stage in all cases, and not through the egg. Either the tick larvae become infected and the nymphs pass the infection on to new cattle, or the nymphs become infected and the adults pass the infection on. The incubation period following tick transmission is 9–25 (mean 15) days.

Congenital infections occur occasionally in calves; every case occurring during the first week of life may be considered congenital.

PATHOGENESIS. Tropical theileriosis is similar to East Coast fever (caused by *T. parva*, discussed later in the chapter) in most respects. The mortality varies considerably, from 10% in some areas to 90% in others.

The disease lasts 4–20 (mean 10) days. Peracute, acute, subacute, mild, and chronic forms have been described. The acute form is the usual one. The first sign is fever, the body temperature rising to 104–107 F. The fever is continuous or intermittent and persists for 5–20 days. A few days after it begins, other signs appear, including inappetence, cessation of rumination,

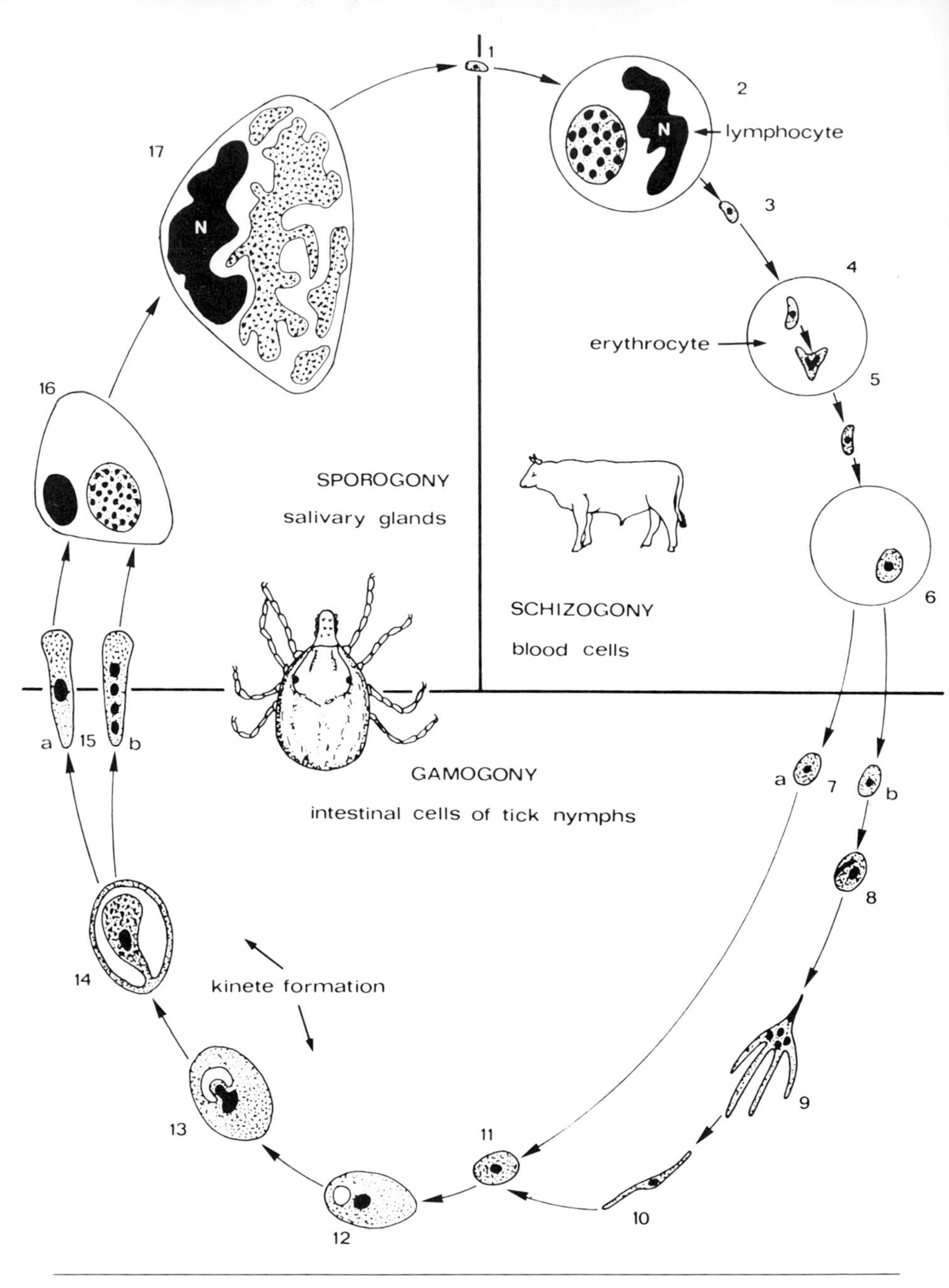

Fig. 11.7. Presumed life cycle of *Theileria annulata* and *T. parva:* 1. sporozoite; 2. merogony within lymphocyte; 3. merozoite; 4 and 5. intraerythrocytic merozoite dividing by endodyogeny; 6. intraerythrocytic merozoite becomes spherical gamont; 7. macrogamete (*a*) and early microgamont (*b*) in tick intestine; 8. developing microgamont; 9. mature microgamont; 10. microgamete; 11. fertilization; 12. presumed zygote; 13. beginning of kinete formation; 14. kinete formation nearly finished; 15. kinetes of *T. annulata* (mononuclear) (*a*) and *T. parva* (*b*), in which nuclear division has already started; 16 and 17. formation of large sporont containing thousands of small sporozoites in tick salivary gland cell. (From Mehlhorn et al. 1977)

rapid heartbeat, weakness, decreased milk production, and swelling of the superficial lymph nodes and of the eyelids. Marked anemia develops in a few days, and there may be hemoglobinuria. Bilirubinemia and bilirubinuria are always present. Diarrhea appears, and the feces contain blood and mucus. The conjunctiva is icteric and may bear petechial hemorrhages. Affected animals become greatly emaciated, and their erythrocyte count may drop below 1 million per mm^3. Death, if it comes, usually occurs 8–15 days after the onset.

In the peracute form of the disease, the animals may die in 3–4 days. In the subacute form, the fever is usually irregularly intermittent and lasts up to 10–15 days, after which the animals usually recover; pregnant animals sometimes abort. In the chronic form, intermittent fever, inappetence, marked emaciation, and some degree of anemia and icterus may persist for 4 wk or longer, but it may take 2 mo before the animals return to normal; in some cases the acute form may suddenly supervene and the animals may die in 1 or 2 days. In the mild form, little is seen but mild fever, inappetence, listlessness, slight digestive disturbances, and lachrymation lasting a few days. There may be moderate anemia.

The lymph nodes are often but not always swollen; the spleen is often much enlarged. The liver is usually enlarged, infarcts are usually present in the kidneys, the lungs are usually edematous, and characteristic ulcers are present in the abomasum and often in the small and large intestines.

Mixed infections with *Babesia* and/or *Anaplasma* are not uncommon; the resultant signs and lesions are then due to a combination of diseases and may differ from those described above.

IMMUNITY. Animals that recover from *T. annulata* infections are premunized. There is no cross-immunity between *T. annulata, T. mutans,* and *T. parva.*

DIAGNOSIS. Diagnosis is based on finding and identifying the parasites in the erythrocytes in stained blood smears or in stained smears made from the lymph nodes or spleen. Differential diagnosis between the species of *Theileria* is not easy, since they look alike; a serologic test may be useful in this case.

CULTIVATION. *T. annulata* has been cultivated by many workers in ox tissue culture. Tissue culture organisms have been used successfully for vaccination.

TREATMENT. The babesiacides are ineffective against *T. annulata*. However, the tetracyclines are active; 16 mg/kg chlortetracycline may be used for 4–16 days.

PREVENTION AND CONTROL. Tick control by regular, repeated dipping is the most important control measure. Quarantine measures, particularly with respect to importation of livestock from endemic areas into regions where suitable tick vectors exist, are also of great importance.

Immunization with a strain of low virulence or with a tissue culture vaccine has been used with success in North Africa and Israel.

Theileria camelensis Yakimoff, Schokhor, and Kozelkine, 1917

SYNONYM. *Gonderia camelensis.*

This name was given provisionally to a round organism that the authors thought was *Theileria;* it had been found in the erythrocytes of three camels (species not named) in Russian Turkestan. It was not described further, and what it was is questionable.

Theileria lestoquardi Morel and Uilenberg, 1981

SYNONYMS. *Gonderia hirci, T. ovis* in part, *T. hirci.*

This species is the cause of malignant ovine and caprine theileriosis in North Africa, southern Europe, the southern USSR, Asia Minor, and India. It is found in the lymphocytes and erythrocytes of sheep and goats.

STRUCTURE. The erythrocytic stages are about 80% round or oval, 18% rod-shaped, and 2% *Anaplasma*-like. The round forms are 0.6–2.0 μm in diameter, the more elongate ones about 1.6 μm long. Binary or quadruple fission takes place in the erythrocytes.

Koch bodies are common in the lymphocytes of the spleen and lymph node smears or free in these organs. They average 8 μm in diameter but may range up to 10 or even 20 μm. They contain 1–80 reddish purple granules 1–2 μm in diameter. Both macromeronts and micromeronts can be found; they produce merozoites 1–2 μm in diameter.

LIFE CYCLE. The vector is unknown but is possibly *Rhipicephalus bursa.*

PATHOGENESIS. This species is highly pathogenic for adult sheep and goats, with reported mortalities of 46–100%. The disease is relatively mild in young lambs and kids in endemic areas.

The incubation period is unknown. The disease itself lasts 5–42 days. Acute, subacute, and chronic forms have been described, the acute form being the usual one.

The disease resembles tropical bovine theileriosis in its manifestations. There is fever followed by listlessness, nasal discharge, atony of the rumen, and weakness. Affected animals are anemic, and icterus is frequently present. There is often a transitory hemoglobinuria. The lymph nodes are always swollen, the liver usually swollen, the spleen markedly enlarged, and the lungs edematous. Infarcts are often present in the kidneys, and there are petechiae on the mucosa of the abomasum and irregularly disseminated red patches on the intestinal mucosa, particularly in the cecum and colon.

IMMUNITY. Animals that recover from the disease are premune. There is no cross-immunity between *T. hirci* and *T. ovis.*

DIAGNOSIS. Diagnosis depends on identification of the parasites in stained blood, lymph node, or spleen smears. In contrast to *T. ovis*, the erythrocytic stages are usually present in relatively large numbers and Koch bodies are common in the lymph nodes and spleen. Inoculation of susceptible sheep or goats may also be resorted to.

TREATMENT. None known.

PREVENTION AND CONTROL. These depend on tick control.

Theileria mutans (Theiler, 1906) França, 1909

SYNONYMS. *Piroplasma mutans, T. buffeli, T. orientalis, Gonderia mutans, Babesia mutans.*

This species occurs in the erythrocytes and lymphocytes of the ox, zebu, and water buffalo, presumably worldwide. The African buffalo *Syncerus caffer* can be infected experimentally, but without causing death. *T. mutans* can survive and even multiply in splenectomized sheep for 9–10 mo. It causes benign bovine theileriosis, Tzaneen disease, Marico calf disease, and mild gallsickness. It is endemic throughout Africa, in much of Asia, and in many parts of the USSR and southern Europe. It is rare in North America.

STRUCTURE. The forms in the erythrocytes are round, oval, piriform, comma-shaped, or *Anaplasma*-like. About 55% are round or oval. The round forms are 1–2 μm in diameter, the oval ones about 1.5 × 0.6 μm. Binary and quadruple fission occur in the erythrocytes.

There are relatively few Koch bodies in the lymphocytes of the spleen and lymph nodes or free in these organs. They average 8 μm in diameter but may range up to 20 μm. They contain 1–80 chromatin granules 1–2 μm in diameter, and are practically all of the macromeront type. They become micromeronts, which produce merozoites.

LIFE CYCLE. Sexual stages occur in the gut of tick nymphs 5–7 days after repletion, zygotes in the gut beginning on day 29, and kinetes beginning on day 30 (i.e., 3–4 days after the nymphs have molted to adults). Beginning on day 34, kinetes are in the hemolymph.

The proven vectors of *T. mutans s. str.* are *Amblyomma variegatum, A. cohaerens,* and *Haemaphysalis hebraeum;* the low-pathogenic species of *Theileria* transmitted by *Rhipicephalus appendiculatus* can no longer be considered *T. mutans. H. bispinosa* and *H. bancrofti* are probably the vectors in Australia. *H. neumanni, H. japonica,* and *H. concinna* are vectors in the USSR. *H. punctata* is an experimental vector. Transmission is stage to stage in all these ticks.

PATHOGENESIS. *T. mutans* is seldom more than slightly if at all pathogenic, although an acute form of the disease may develop in cattle imported into an endemic area and exposed to massive tick infestation. The mortality is generally less than 1%. Some strains have recently been found to be pathogenic and to cause East Coast fever.

The signs, course of the disease, and lesions resemble those of mild *T. annulata* infections. Anemia is the main feature; if present, it is usually slight. Icterus is sometimes present, and the lymph nodes are moderately swollen. In acute cases the spleen and liver are swollen, the lungs may be edematous, there are characteristic ulcers in the abomasum, and infarcts may be present in the kidneys. Hematuria is absent.

The incubation period following tick transmission is 10–20 (mean 15) days. The disease lasts 3–10 (mean 5) days.

IMMUNITY. Animals that have once been infected with *T. mutans* are premunized. There is no cross-immunity between *T. mutans, T. annulata,* and *T. parva.*

DIAGNOSIS. Same as for other species of *Theileria.*

TREATMENT. None known.

PREVENTION AND CONTROL. These depend on tick control.

REMARKS. According to Morel and Uilenberg (1981), *T. mutans* is the benign *Theileria* of cattle only in sub-Saharan Africa, Madagascar, Reunion, Mauritius, Guadeloupe, and Antigua. It is transmitted only by *Amblyomma* spp. ticks. They said that the benign *Theileria* of cattle in Eurasia, North Africa, and Australia is *T. orientalis* (syns., *T. mutans* in part, *T. sergenti* in part, and *Gonderia orientalis*). They did not name its vectors and said that the name of the benign *Theileria* of cattle in other places remains to be determined. It is transmitted by *Haemaphysalis longicornis* in oriental Asia and Australia, by *H. punctata* in Europe, and probably also in North Africa, North America, and Cuba.

Theileria ovis Rodhain, 1916

SYNONYMS. *Babesia sergenti, Gonderia ovis, T. recondita, T. sergenti, Haematoxenus separatus.*

This species is the cause of benign ovine and caprine theileriosis. Much more widely distributed than *T. hirci*, it occurs in the erythrocytes and lymphocytes of the sheep and goat in Africa, Europe, USSR, India, Sri Lanka, and western Asia. The mouflon and splenectomized red deer, but not the splenectomized ox, can be infected experimentally.

STRUCTURE. The erythrocytic stages resemble those of *T. hirci* in shape and size but are much sparser in infected animals, with less than 2% of the erythrocytes infected in nonsplenectomized animals. The Koch bodies resemble those of *T. hirci* but have been found only in the lymph nodes and then only after prolonged search.

LIFE CYCLE. The vectors are *Rhipicephalus bursa* in the USSR and Asia, and this species and *R. evertsi* in Africa. Some species of *Haemaphysalis* may be the vector on Sri Lanka, where neither *R. bursa* nor *R. evertsi* occurs.

Structures considered gamonts by Mehlhorn et al. (1979) develop in the gut of *R. evertsi* nymphs. After the nymphs molt, spherical and ovoid parasites (perhaps zygotes) occur in the intestinal cells. They transform in 3 days into motile kinetes, with a fine structure similar to that of hemosporidian ookinetes. The kinetes enter the salivary gland cells where they produce sporozoites that are infective for sheep.

PATHOGENESIS. This species is nonpathogenic or almost so. The incubation period following tick transmission is 9–13 days, and the disease lasts 5–16 days. The only signs are fever, swelling of the lymph nodes in the region of tick attachment, and slight anemia. These would normally be overlooked in the field.

IMMUNITY. Animals that have been infected are premune. There is no cross-immunity between *T. ovis* and *T. hirci.*

DIAGNOSIS. Diagnosis depends on identification of the parasites in stained blood or lymph node smears. *T. ovis* is structurally indistinguishable from *T. hirci*, but the small number of parasites present and their lack of pathogenicity may help to differentiate them. Cross-immunity tests may be carried out if desired.

REMARKS. According to Morel and Uilenberg (1981) this is the benign *Theileria* species of small domestic ruminants. They said that synonyms are *Gonderia ovis, G. hirci* in part, *Babesia sergenti, T. sergenti* in part, *T. recondita,* and *T. musimoni.*

Theileria parva (Theiler, 1904) Bettencourt, França, and Borges, 1907

SYNONYMS. *Piroplasma kochi, P. parvum, P. bacilliformis, Gonderia bovis, G. lawrencei, T. kochi, T. bovis, T. lawrencei.*

This species causes East Coast fever, bovine theilerosis, African Coast fever, Rhodesian tick fever, Rhodesian redwater disease, or corridor disease in the erythroctyes and lymphocytes of the ox, zebu, water buffalo, and African buffalo *Syncerus caffer*. This is one of the most important diseases in east and central Africa. Corridor disease (named for the corridor between the Hluhluwe and Umfolozi Game Reserves) is widely distributed in cattle and African buffalo in Zimbabwe. A highly fatal encephalitis of cattle in East Africa known as turning sickness, *kizengerera,* or *muthiuko* may be caused by it.

STRUCTURE. The forms in the erythrocytes are predominantly (over 80%) rod-shaped and measure about 1.5–2.0 × 0.5–1.0 μm. Round, oval, and comma-shaped forms also occur. About 55% of those of "*T. lawrencei*" are round or oval.

The merozoites within the erythrocytes have a single outer unit membrane, a micropore with an inner diameter of 80 nm, and ingest food by pinocytosis through this micropore. The exoerythrocytic merozoites (which are produced by the meronts in the lymphoid cells) have an additional membrane within the outer one, a variable number of rhoptries, apparently a few micronemes, perhaps some subpellicular microtubules, but no micropore or conoid. Both types of merozoite have 2 vacuoles, free ribosomes, and a vesicular nucleus without nucleolus. They are somewhat smaller than the merozoites of *T. mutans.*

The multiplying forms, known as Koch's bodies or Koch's blue bodies, occur in the lymphocytes and occasionally in the endothelial cells. They are found especially in the lymph nodes and spleen, where they are usually very numerous. They are circular or irregular bodies averaging 8 μm in diameter and ranging up to 12 μm or more. They may be intracellular or free in the gland or spleen juice.

Two types of these meronts are recognized. Macromeronts (macroschizonts) contain chromatin granules 0.4–2.0 (mean 1.2) μm in diameter and are believed to produce macromerozoites 2.0–2.5 μm in diameter. Micromeronts (microschizonts) contain chromatin granules 0.3–0.8 (mean 0.5) μm in diameter and are believed to produce micromerozoites 0.7–1.0 μm in diameter. Practically all the meronts of *T. lawrencei* are macromeronts.

Apparently the same meront appears both ways in succession. During the phase of nuclear reproduction it looks like a macromeront under the light microscope, while during the phase of merozoite formation it looks like a micromeront.

LIFE CYCLE. The most important vector is *Rhipicephalus appendiculatus*. Other vectors are *R. ayrei, R. capensis, R. evertsi, R. jeanelli, R. neavei, R. simus, Hyalomma excavatum, H. dromedarii,* and *H. truncatum*. Transmission is stage to stage in all cases, and not through the egg. *R. appendiculatus*, for instance, acquires the infection as a larva and transmits it as a nymph, or acquires the infection as a nymph and transmits it as an adult. The parasite will not survive in the ticks through more than one molt.

When the tick takes a blood meal, the great majority of the parasites die in its intestine. A few become stages that Mehlhorn and Schein (1976) interpreted as microgamonts in the tick intestine 2–4 days after repletion (Fig. 11.8). These organisms are about 10.5 μm long with a maximum diameter of about 2.1 μm in the middle and are shaped like a spearhead. They have a unit membrane on the outside, an electron-dense stilettolike apex, up to 4 flagellumlike protrusions about 3 μm long originating near the base of the stilettolike structure, and usually 2 slender posterior protrusions. Each flagellumlike protrusion contains up to 6 microtubules, but their number is reduced to 2 at their free ends. More or less spherical macrogametes are also apparently present.

The protrusions break off to become what Mehlhorn and Schein (1976) identified as microgametes. They are filiform, 9–11 μm long, and contain several microtubules. No one has seen fertilization (syngamy), but it is assumed that it takes place to form a motile zygote. This enters a gut epithelial cell and rounds up, forming a comma-shaped kinete about 19 × 5.5 μm. It has a polar ring to which about 40 subpellicular microtubules are attached, many micronemes, and a micropore, but no conoid or rhoptries. Most of the kinetes have left the gut cells and started to penetrate the salivary glands by day 25. In the salivary glands, the kinetes round up to form sporonts. They enter the cells of the alveoli and multiply in them. The host cell becomes greatly enlarged and is filled with some 30,000 tiny sporozoites. These are about 1.5 μm in diameter and contain rhoptries and micronemes. Mature sporozoites are expelled into the lumen of the alveolus and are injected with the saliva into a new animal when the tick sucks blood.

The prepatent period in cattle is 4–12 days. After injection into cattle,

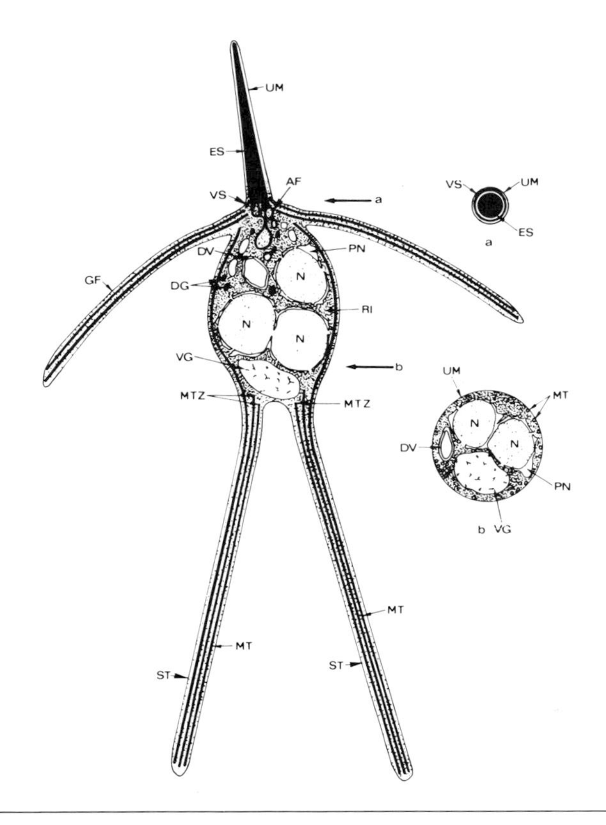

Fig. 11.8. Microgamont of *Theileria parva* from *Hyalomma anatolicum excavatum* gut: *a* and *b* = cross sections from areas indicated by arrows, *AF* = relaxed filamentous part of the electron-dense spike, *DG* = osmiophilic granule, *DV* = double-membraned structures, *ES* = electron-dense spike, *GF* = flagellumlike process, *MT* = microtubule, *MTZ* = microtubule (interrupted by drawing technique), *N* = nucleus, *PN* = perinuclear space, *RI* = ribosome, *ST* = taillike process, *UM* = unit membrane, *VG* = vacuole with granular contents, *VS* = reinforcing ring. (From Mehlhorn and Schein 1976, Tropenmed. Parasitol. 27(1976):182–91, Georg Thieme Verlag, Stuttgart)

the sporozoites enter the lymphocytes and become meronts. These produce merozoites, which enter other lymphocytes, become meronts, and produce another crop of merozoites. After several such generations, the reproducing meronts can be found in the prescapular lymph nodes. At first the meronts are macromeronts, but after about six generations the macromeront nuclei divide without cell division to form micromeronts. These produce micromerozoites, which enter the erythrocytes (Jarrett 1969). (See Fig. 11.9.)

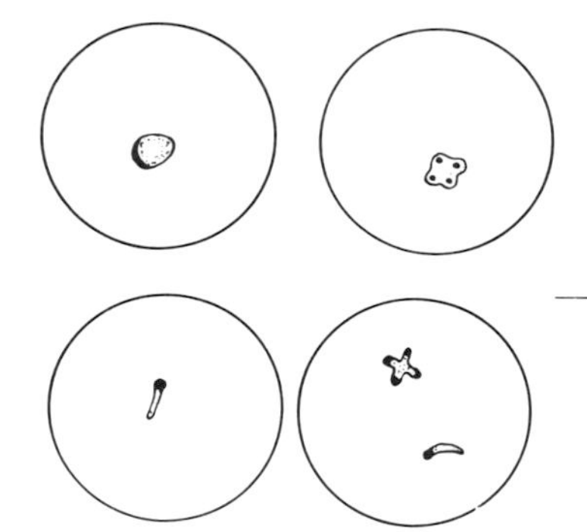

Fig. 11.9. *Theileria parva* in bovine erythrocytes as seen by light microscope. (×2,800) (After Nuttall 1913 in *Parasitology,* published by Cambridge University Press)

PATHOGENESIS. *T. parva* is highly pathogenic; 90–100% of affected cattle die, although the mortality is lower in endemic areas. In East Africa, for instance, immature cattle are more resistant than adults, and the mortality among calves is 5–50%.

The incubation period following tick transmission is 8–24 (mean 13) days. The disease itself lasts 10–23 (mean 15) days. Acute, subacute, mild, and inapparent forms have been described; the acute type is the usual one.

In the acute form, the first sign is fever. The body temperature is 104–107 F; it may continue high or it may decrease after 7–11 days and then increase again. Other clinical signs usually appear a few days after the initial rise in temperature. The animals cease to ruminate and to eat. Other signs are a serous nasal discharge; lachrymation; swelling of the superficial lymph nodes; sometimes swelling of the eyelids, ears, and jowl region; rapid heartbeat; general weakness; decreased milk production; diarrhea, frequently with blood and mucus in the feces; emaciation; coughing; and sometimes icterus. Breathing becomes rapid, and dyspnea is pronounced just before death. An oligocythemic anemia is present, but there is no hematuria in uncomplicated cases.

In the subacute form the signs resemble those in the acute form but are not so pronounced. Affected animals may recover, but it takes them several weeks to return to normal.

In the mild form, little is seen but a relatively mild fever lasting 3–7 days, listlessness, and swelling of the superficial lymph nodes.

The lymph nodes are usually markedly swollen, with a variable degree of hyperemia. The spleen is usually enlarged, with soft pulp and prominent Malpighian corpuscles. The liver is enlarged, friable, brownish yellow to lemon yellow, with parenchymatous degeneration. The kidneys are either congested or pale brown, with a variable number of hemorrhagic "infarcts" or grayish white lymphomatomata. The meninges may be slightly congested. The heart is flabby, with petechiae on the epicardium and endocardium. The lungs are often congested and edematous. There may be hydrothorax and hydropericardium, and the kidney capsule may contain a large amount of serous fluid. There may be petechiae in the visceral and parietal pleura, adrenal cortex, urinary bladder, and mediastinum. There are characteristic ulcers 2–5 mm or more in diameter in the abomasum; similar ulcers together with red streaks or patches may be present throughout the small and large intestines. These ulcers consist of a central, red or brown, necrotic area surrounded by a hemorrhagic zone. The Peyer's patches are swollen, and the intestinal contents yellowish.

IMMUNITY. Animals that recover from *T. parva* infection are solidly immune. The parasites disappear completely, and there is generally no premunition. There is no cross-immunity between *T. parva* and *T. mutans*.

DIAGNOSIS. Diagnosis is based on finding the parasites in the erythrocytes in stained blood smears or in stained smears made from the lymph nodes or spleen.

CULTIVATION. Tsur-Tchernomoretz, Neitz, and Pols (1957) were apparently the first to cultivate *T. parva*; they did so in bovine tissue cultures. Many others have done so since then.

TREATMENT. No drug is known to be effective against *T. parva* once signs of the disease have appeared. However, chlortetracycline and oxytetracycline seem to prevent clinical disease if given repeatedly during the incubation and reaction periods; treated animals become solidly immune.

PREVENTION AND CONTROL. These depend on tick control and quarantine measures. Repeated regular dipping of cattle in tickicide solutions has been found effective. Quarantine measures are effective in preventing the spread of East Coast fever. In isolated outbreaks the whole herd may be slaughtered and the farm kept free of cattle for 18 mo before restocking.

Theileria tarandi Levine, 1971

This species has been found in reindeer, but little is known about its manifestations.

Theileria velifera (Uilenberg, 1964) Uilenberg, 1977

SYNONYM. *Haematoxenus veliferus*.

This species occurs in the erythrocytes and presumably lymphocytes of the ox and zebu in Africa.

STRUCTURE. The forms in the erythrocytes are pleomorphic but are most often small rods. Most are 1–2 μm long. The great majority have a rectangular "veil" 1.0–3.5 μm long extending out from the side. This veil is actually crystallized hemoglobin and consequently Uilenberg sank the generic name *Haematoxenus* as a synonym of *Theileria*.

LIFE CYCLE. Known vectors are *Amblyomma variegatum, A. lepidum,* and *A. hebraeum*.

PATHOGENESIS. This species is apparently nonpathogenic.

IMMUNITY. There is no cross-immunity between this species and *T. mutans* or *T. parva.*

Theileria felis (Kier, Wagner, and Morehouse, 1982) Levine, 1982

SYNONYM. *Cytauxzoon* sp. Wagner 1976; *Cytauxzoon felis* Kier, Wagner, and Morehouse, 1982.

This species has been found in the erythrocytes and lymph nodes of cats and the bobcat *Lynx rufus* in several states in the United States. It has been transmitted to the laboratory rabbit, sheep, goat, and ox.

STRUCTURE. The erythrocytes contain signet-ring-like forms 1–1.2 μm in diameter; histiocytes in the spleen, lymph nodes, lungs, liver, kidneys, and sometimes veins of the heart, urinary bladder, and bone marrow may contain hundreds of merozoites 0.1–0.2 μm in diameter or indistinct Koch bodies.

PATHOGENESIS. *T. felis* may cause abortion or a fatal disease. The principal clinical signs in affected cats are pyrexia, anorexia, dehydration, and depression. Icterus and fever may also be present. The cats are markedly dehydrated, with generalized paleness and icterus and numerous petechiae and ecchymoses of the epicardium and serosal membranes of the abdominal organs, as well as beneath the visceral pleura of the lungs and in the urinary bladder mucosa. The lymph nodes are enlarged, congested, or hemorrhagic and edematous. The spleen is markedly enlarged. The liver is orange-brown. The parenchymatous organs appear somewhat swollen, and the pericardial sac is distended with icteric fluid.

The organism is not pathogenic for the nonpregnant rabbit, sheep, goat, or ox, but it causes pregnant animals of these species to abort or their offspring to be stillborn.

REMARKS. Further information on this parasite was given, among others, by Wagner (1976), Wagner et al. (1980), Kier et al. (1982), and Ferris and Dardiri (1978).

LITERATURE CITED

For citations before 1965, see the *Index-Catalogue of Medical and Veterinary Zoology, Zoological Record, Biological Abstracts, Protozoological Abstracts,* and *Veterinary Bulletin.*

Anderson, J. F., L. A. Magnarelli, C. S. Donner, A. Spielman, and J. Piesman. 1979. Science 204:1431–32.

Barnett, S. F. 1977. In Parasitic Protozoa, vol. 4, ed. J. P. Kreier, 77–113. New York: Academic.

Brandt, F., G. H. Healy, and M. Welch. 1977. J. Parasitol. 63:934–37.

Callow, L. L. 1976. World Anim. Rev. 18:9–15.

Croft, R. E. and N. Kingston. 1975. J. Wildl. Dis. 11:229–233.

Entrican, J. J., H. Williams, I. A. Cook, W. M. Lancaster, J. C. Clark, L. P. Joyner, and D. Lewis. 1979a. Br. Med. J. 1979(2):474.

______. 1979b. J. Infect. Dis. 1:227–34.

Erp, E. E., S. M. Gravely, R. D. Smith, M. Ristic, B. M. Osorno, and C. A. Carson. 1978. Am. J. Trop. Med. Hyg. 27:1061–64.
Erp, E. E., R. D. Smith, M. Ristic, and B. M. Osorno. 1980. Am. J. Vet. Res. 41:1141–42.
Etkind, P., J. Piesman, T. K. Ruebush II, A. Spielman, and D. D. Juranek. 1980. J. Parasitol. 66:107–10.
Farwell, G. E., E. K. LeGrand, and C. C. Cobb. 1982. J. Am Vet. Med. Assoc. 180:507–11.
Ferris, D. H., and A. H. Dardiri. 1978. Short Commun. 4th Int. Congr. Parasitol. H:71.
Fitzpatrick, J. E. P., C. C. Kennedy, M. G. McGeown, D. G. Oreopoulos, J. H. Robertson, and M. A. O. Soyannwo. 1968. Nature 217:861–62.
Fowler, J. L., M. D. Ruff, R. C. Fernau, and Y. Furusho. 1972. Am. J. Vet. Res. 33:1109–14.
Frerichs, W. M., A. J. Johnson, and A. A. Holbrook. 1969. Am. J. Vet. Res. 30:1333–36.
Friedhoff, K. 1969. Z. Parasitenkd. 32:191–219.
Friedhoff, K. T., and G. Büscher. 1976. Z. Parasitenkd. 50:345–47.
Friedhoff, K. T., and A. Liebisch. 1978. Tieraerztl. Prax. 6:125–39.
Friedhoff, K. T., and E. Scholtyseck. 1968a. J. Parasitol. 54:1246–50.
______. 1968b. Z. Parasitenkd. 30:347–59.
______. 1969. Z. Parasitenkd. 32:266–83.
______. 1977. Protistologica 13:195–204.
Friedhoff, K. T., and R. D. Smith. 1981. In Babesiosis, ed. M. Ristic and J. P. Kreier, 267–321. New York: Academic.
Friedhoff, K., E. Scholtyseck, and G. Weber. 1972. Z. Parasitenkd. 38:132–40.
Harambašić, H., B. Kršnjavi, and A. Urbanke. 1976. Acta Parasitol. Iugosl. 7:37–41.
Healy, G. R., and T. K. Ruebush II. 1980. Am. J. Clin. Pathol. 73:107–9.
Hildebrandt, P. K. 1981. In Babesiosis, ed. M. Ristic and J. P. Kreier, 459–73. New York: Academic.
Holbrook, A. A., D. W. Anthony, and A. J. Johnson. 1968a. J. Protozool. 15:391–96.
Holbrook, A. A., A. J. Johnson, and P. A. Madden. 1968b. Am. J. Vet. Res. 29:297–303.
Jarrett, W. F. H., G. W. Crighton, and H. M. Pirie. 1969. Exp. Parasitol. 24:9–25.
Joyner, L. P., and D. W. Brocklesby. 1973. In Advances in Pharmacology and Chemotherapy, vol. II, ed. S. Garattini et al., 321–55. New York: Academic.
Kier, A. B., J. E. Wagner, and L. G. Morehouse. 1982. Am. J. Vet. Res. 43:97–101.
Krampitz, H. E., and W. Bäumler. 1978. Z. Parasitenkd. 58:15–33.
Krylov, M. V. 1981. Piroplazmidy. Leningrad, USSR: Izdat. "Nauka."
Kuttler, K. L. 1980. J. Am. Vet. Med. Assoc. 176:1103–8.
______. 1981. In Babesiosis, ed. M. Ristic and J. P. Kreier, 65–85. New York: Academic.
Kuttler, K. L., M. G. Levy, and M. Ristic. 1983. Am. J. Vet. Res. 44:1456–59.
Levine, N. D. 1971. Trans. Am. Microsc. Soc. 90:2–33.
______. 1982. In Review of Advances in Parasitology, ed. W. Slusarski, 283–90. Warsaw: Panst. Wydawn. Nauk.
Levy, M. G., and M. Ristic. 1980. Science 207:1218–20.
Levy, M. G., E. Erp, and M. Ristic. 1981. In Babesiosis, ed. M. Ristic and J. P. Kreier, 207–23. New York: Academic.
Lewis, D., and I. Herbert. 1980. Vet. Rec. 107:352–53.
Lewis, D., and E. R. Young. 1980. J. Parasitol. 66:359–60.
McCoster, P. J. 1981. In Babesiosis, ed. M. Ristic and J. P. Kreier, 1–24. New York: Academic.

Mahoney, D. F. 1977. In Parasitic Protozoa, vol. 4, ed. J. P. Kreier, 1–52. New York: Academic.
______. 1981a. In Babesiosis, ed. M. Ristic and J. P. Kreier, 475–83. New York: Academic.
______. 1981b. In Babesiosis, ed. M. Ristic and J. P. Kreier, 555–62. New York: Academic.
Mehlhorn, H., and E. Schein. 1976. Tropenmed. Parasitol. 27:182–91.
______. 1977. Abstr. 5th Int. Congr. Protozool., 22.
Mehlhorn, H., G. Weber, E. Schein, and G. Büscher. 1975. Z. Parasitenkd. 48:137–50.
Mehlhorn, H., E. Schein, and M. Warnecke. 1979. J. Protozool. 26:377–85.
Mehlhorn, H., A. O. Heydorn, and E. Schein. 1977. Proc. 1st Jpn.-Ger. Symp., 91–101. Japan: Gakujutsu Tosho, Ltd.
Mehlhorn, H., E. Schein, and W. P. Voight. 1980. J. Parasitol. 66:220–28.
Mehlhorn, H., U. Moltmann, E. Schein, and W. P. Voight. 1981. Zentralbl. Bakteriol. [Orig. A.] 250:248–55.
Morel, P. C., and G. Uilenberg. 1981. Rev. Elev. Med. Vet. Pays Trop. 34:139–43.
Purnell, R. E. 1980. In Vaccines against Parasites, vol. 18, ed. A. E. R. Taylor and R. Muller, 25–55. Oxford, Eng.: British Society for Parasitology.
Rabinovich, S. A., Z. K. Voronina, N. I. Stepanova, G. M. Maruashvili, T. L. Bakradze, M. S. Obishariya, and N. I. Gvasaliya. 1978. US NAMRU 3, Cairo, Egypt. Transl. 1395.
Ristic, M. 1970. In Immunity to Parasitic Animals, vol. 2, ed. G. J. Jackson et al. 831–70. New York: Appleton-Century-Crofts.
______. 1981. In Babesiosis, ed. M. Ristic and J. P. Kreier, 563–71. New York: Academic.
Ristic, M., and J. P. Kreier, eds. 1981. Babesiosis. New York: Academic.
Ristic, M., and G. E. Lewis, Jr. 1977. In Parasitic Protozoa, vol. 4, ed. J. P. Kreier, 53–76. New York: Academic.
Rudzinska, M. A. 1976. J. Protozool. 23:224–33.
Rudzinska, M. A., R. F. Riek, and S. K. Lewengrub. 1978. J. Protozool. 25:32A–33A.
Rudzinska, M. A., A. Spielman, R. F. Riek, S. J. Lewengrub, and J. Piesman. 1979. Can. J. Zool. 57:424–34.
Ruebush, T. K., II, P. B. Cassaday, H. J. Marsh, S. A. Lisker, D. B. Voorhees, E. B. Mahoney, and G. R. Healy. 1977. Ann. Intern. Med. 86:6–9.
Schein, E., and K. T. Friedhoff. 1978. Z. Parasitenkd. 56:287–303.
Schein, E., G. Büscher, and K. T. Friedhoff. 1975. Z. Parasitenkd. 48:123–36.
Schein, E., H. Mehlhorn, and M. Warnecke. 1978. Protistologica 14:337–48.
Schein, E., H. Mehlhorn, and W. P. Voight. 1979. Acta Trop. 36:229–41.
Smith, R. D., E. Molinar, F. Larios, J. Monroy, F. Trigo, and M. Ristic. 1980. Am. J. Vet. Res. 41:1957–65.
Spielman, A., C. M. Clifford, J. Piesman, and M. D. Corwin. 1979. J. Med. Entomol. 15:218–34.
Spielman, A., P. Etkind, J. Piesman, T. K. Ruebush II, D. D. Juranek, and M. S. Jacobs. 1981. Am. J. Trop. Med. Hyg. 30:560–65.
Wagner, J. E. 1976. J. Am. Vet. Med. Assoc. 168:586–88.
Wagner, J. E., D. H. Ferris, A. B. Kier, S. R. Wightman, E. Maring, L. G. Morehouse, and R. D. Hansen. 1980. Vet Parasitol. 6:305–11.
Weber, G. 1980. J. Protozool. 27:59–71.
Weber, G., and K. T. Friedhoff. 1979. Z. Parasitenkd. 58:191–94.
Wright, I. G. 1981. In Babesiosis, ed. M. Ristic and J. P. Kreier, 171–205. New York: Academic.

12

Microspora and Myxozoa

MEMBERS of the phylum Microspora are all parasitic. They all produce spores and all have a vesicular nucleus. The spores are unicellular and contain a uninucleate or dinucleate sporoplasm and an extrusion apparatus that always has a polar tube (polar filament) and a polar cap. They do not have mitochondria. (See Fig. 12.1.) Members of this phylum may be of value for the biological control of vectors of mammalian or avian diseases (Laird 1979; WHO 1980).

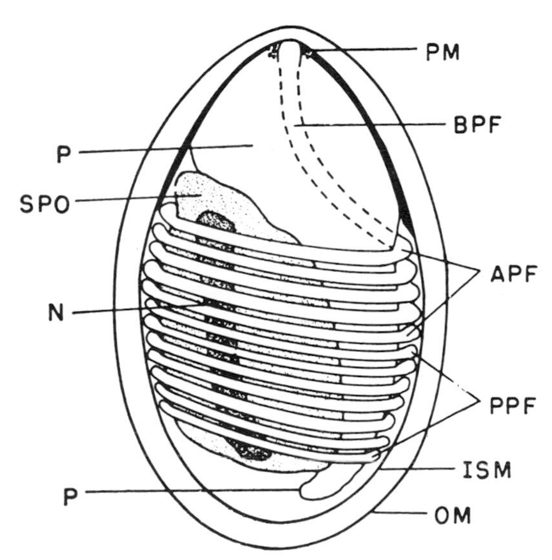

Fig. 12.1. Microsporidan spore (*Thelohania californica* from *Culex tarsalis*) as seen with the electron microscope: *APF* = anterior polar filament, *BPF* = basal part of filament, *ISM* = inner surface membrane, *N* = nucleus, *OM* = outer surface membrane, *P* = polaroplast, *PM* = polar mass and polaroplast membrane, *PPF* = posterior polar filament, *SPO* = sporoplasm. (From Kudo and Daniels 1963, J. Protozool. 10:112–20)

Another phylum, the Myxozoa, occurs in lower vertebrates, especially fish, and will not be mentioned further. The Microspora are mostly parasites of invertebrates, but two genera, *Encephalitozoon* and *Nosema,* occur in vertebrates, and another genus, *Thelohania,* has been found in invertebrates and a field mouse (*Apodemus sylvaticus*) in France.

Genus *Encephalitozoon* Levaditi, Nicolau, and Schoen, 1923

This genus belongs to the family Nosematidae, in which the spores are ellipsoidal, ovoid, or more or less piriform, and sometimes subcylindrical. The spores consist essentially of an external wall, sporoplasm, a coiled polar tube, and a polar capsule. The sporoplasm (sporont) divides to produce sporoblasts; in further division the nuclei, including the spores, are unpaired during most of the life cycle (Sprague and Vernick 1971).

Both polar tube and polar capsule are difficult to see with ordinary stains; in the past this has led investigators to think that *Encephalitozoon* was similar to *Toxoplasma.* Aside from the polar tube, the lack of an apical complex, and the presence of a capsule, *Encephalitozoon* differs from *Toxoplasma* in its trophozoites, which are smaller and rodlike; the clusters of trophozoites in the brain are not surrounded by an argyrophilic cyst wall, although they are said to form a pseudocyst; *Encephalitozoon* stains very poorly with hematoxylin and eosin but stains dark red with Wright and Craighead's carbol fuchsin-methylene blue stain (decolorized with 37% formaldehyde solution), whereas *Toxoplasma* stains well with hematoxylin and eosin and stains blue with carbol fuchsin-methylene blue; *Encephalitozoon* stains black with Weigert's iron hematoxylin, whereas *Toxoplasma* does not; and *Encephalitozoon* survives rapid freezing and storage at -70 C or storage in 50% buffered glycerol at 4 C, while *Toxoplasma* meronts do not (Perrin 1943; Frenkel 1956).

Several species of mammalian *Encephalitozoon* have been named, and the name has also been mistakenly given to the Negri bodies of rabies. However, for the present and in view of the transmissibility of the organism from one host to another, only one is recognized in this book.

Encephalitozoon cuniculi Levaditi, Nicolau, and Schoen, 1923

SYNONYMS. *E. negrii, Nosema cuniculi, N. muris.*

This species occurs worldwide in the domestic rabbit, house mouse, Norway rat, golden hamster, muskrat, guinea pig, cottontail, dog, rat, and probably man. It occurs intracellularly in the brain, kidneys, peritoneal exudate, macrophages, liver, spleen, and other organs. It is common in laboratory rabbits and mice, less so in laboratory rats, apparently not common in guinea pigs or dogs, and rare in man.

In most cases *E. cuniculi* has been found during routine histologic examination of animals being studied for some other purpose or in serologic surveys of presumably healthy animals.

STRUCTURE. The trophozoites, which are $2.0-2.5 \times 0.8-1.2$ μm in tissue sections and up to 4.0×2.5 (mean 2.0×1.2) μm in smears, are straight to slightly curved rods with both ends bluntly rounded but one end a little larger than the other; they are sometimes slightly constricted at or near their midpoint. Round or oval forms occur occasionally. The nucleus is compact, round, oval, or bandlike, about one-fourth to one-third the size of the parasite, and is not central. Pseudocysts or groups of parasites containing 100 or more trophozoites are found within the nerve cells, macrophages, or other tissue cells. Both they and the trophozoites are rarely extracellular.

As seen with the electron microscope, the spores are about 2 μm long and contain a spirally coiled polar filament with 4–5 (usually 4) coils. Proliferative forms (meronts) have a double outer membrane and are often binculeate, dividing to form sporonts. Sporonts divide once to form 2 sporoblasts, which develop into electron-dense spores with a thick, 3-layered wall and a polar filament.

LIFE CYCLE. Multiplication of the trophozoites is apparently by binary fission or perhaps schizogony. The whole life cycle takes about 2 days in rabbit choroid plexus cell cultures (Pakes et al. 1975) or about 5 days in Yoshida rat ascites sarcoma (Petri 1969).

The mode of natural infection is not known with certainty but is undoubtedly via the spores, which have been found in the urine. Congenital infection is said to occur in mice.

E. cuniculi can be transmitted from the mouse, rabbit, rat, or guinea pig to other laboratory animals by intracerebral, intravenous, intraperitoneal, or other parenteral inoculation of infected brain, liver, spleen, or peritoneal exudate (Perrin 1943; Shadduck et al. 1979).

PATHOGENESIS. *E. cuniculi* usually causes an inapparent infection but in some cases has been associated with systemic disease, nephritis, and ascites.The great majority of infections in laboratory animals is inapparent, being discovered on histologic examination carried out for some other reason or as the result of a serologic survey.

It sometimes causes a mild, chronic infection in rabbits, although paralysis and death may occur. The principal lesions in the brain are tiny, focal granulomata made up of epithelioid cells surrounding a tiny area of necrosis. In fatal cases there may be large necrotic areas and perivascular lymphocytic cuffing; the parasites may occur in or around them. Similar granulomatous lesions may be present in the kidneys and other organs. In the kidneys they occur principally in the epithelial cells of the collecting tubules, which they distend and finally rupture before passing out in the urine. The outstanding feature in rabbits is chronic nephritis. There are small, cortical scars on the kidneys that appear histologically as V-shaped, subcapsular depressions that may sometimes extend as far as the medulla. These scars consist of connective tissue with some leukocytes. Vavra et al. (1980) found that *E. cuniculi* caused an 11% difference in weight between infected and uninfected growing rabbits.

In mice and rats the principal lesions are ascites or meningoencephalitis; in experimentally infected animals abdominal enlargement with ascites is common, as is splenomegaly. Nodular, granulomatous lesions, sometimes with central necrosis, occur throughout the brain. There is slight to moderate focal perivascular infiltration by lymphocytes and by a few large mononuclear and plasma cells in the meninges and also in the brain. The parasites may be either within or at the margins of the lesions or even in normal brain tissue at a distance from them. There may be moderate to marked interstitial lymphocytic infiltration in the kidneys, primarily in the cortex. The tubular epithelium may be degenerate or proliferative in the areas of infiltration, and parasites may occur either in the epithelial cells or within the collecting tubules. Similar areas of infiltration may be seen in other organs.

The effects in dogs vary. There may be posterior weakness and incoordination, apathy, rapid tiring, loss of condition, ocular involvement, epileptiform seizures, or signs resembling those of rabies (Plowright 1952; Plowright and Yeoman 1952). Shadduck et al. (1978) observed nonsuppurative nephritis, encephalitis, and segmental vasculitis, with many protozoa in the kidney tubule cells, endothelial cells, and brain.

E. cuniculi has a interesting effect on rat and mouse tumors. It can invade the cells of the transplantable, malignant Yoshida rat ascites sarcoma and kill them, reducing the mortality of affected hosts from more than 95% to 86% (Petri 1966, 1968; Petri and Schidt 1966). It can also invade a variety of tumor cells in mice, retarding the growth of the tumors and significantly prolonging the mouse's life (Arison et al. 1966).

IMMUNITY. There is no cross-immunity between *Toxoplasma* and *E. cuniculi.*

DIAGNOSIS. Encephalitozoonosis is generally diagnosed by finding the causative organisms in tissue sections. Several serologic tests have been found effective for the diagnosis of inapparent encephalitozoonosis, including an indirect immunofluorescence test, an intradermal skin test, an India ink immunoreaction test, and an indirect microagglutination test using a bead substrate.

CULTIVATION. *E. cuniculi* has been cultivated in rabbit choroid plexus tissue culture (Shadduck 1969, 1980; Vavra et al. 1972; Pakes et al. 1975; Shadduck and Polley 1978). It apparently grows best there, although it will grow quite well in HeLa cells; it grows poorly in L cells, pork kidney cells, hamster kidney cells, and chick embryo kidney cells as well as in primary fetal rabbit and mouse kidney cell cultures (Jackson et al. 1974).

TREATMENT. There is no known treatment. Cortisone exacerbates the disease in mice (Vavra et al. 1972).

PREVENTION AND CONTROL. *E. cuniculi* can be eliminated from rabbit colonies by serologic testing of all animals and elimination of the positive ones.

Genus *Nosema* Nägeli, 1857

This genus resembles *Encephalitozoon,* but its nuclei are paired in a diplokaryon arrangement throughout most of the life cycle, including the spore stage.

Nosema connori Sprague, 1974

This species was named for the organism found by Margileth et al. (1973) in an immunologically compromised infant. Sprague (1974) said that it differed from *E. cuniculi* in having binucleate spores measuring 4–4.5 × 2–2.5 μm, compared with the uninucleate spores of *E. cuniculi* that measure 2.5 × 1.5 μm.

Nosema sp. Shadduck, Kelsoe, and Helmke, 1979

This species was found by Shadduck et al. (1979) contaminating primary cultures of baboon placental cells. They said that it was structurally identical to *E. cuniculi* but serologically negative for it, and that it was not *N. connori.*

LITERATURE CITED

For citations before 1965, see the *Index-Catalogue of Medical and Veterinary Zoology, Zoological Record, Biological Abstracts, Protozoological Abstracts,* and *Veterinary Bulletin.*

Arison, R. N., J. A. Cassaro, and M. P. Pruss. 1966. Cancer Res. 26:1915–20.
Jackson, S. J., R. F. Solorzano, and C. C. Middleton. 1974. Proc. 3d Int. Congr. Parasitol. 2:1143–44.
Laird, M. 1979. Proc. 5th Int. Congr. Protozool., 172–75.
Margileth, A. M., A. J. Strano, R. Chandra, R. Neafie, M. Blum, and R. M. McCully. 1973. Arch. Pathol. 95:145–50.
Pakes, S. P., J. A. Shadduck, and A. Cali. 1975. J. Protozool. 22:481–88.
Petri, M. 1966. Acta Pathol. Microbiol. Scand. 66:13–30.
Petri, M. 1968. Acta Pathol. Microbiol. Scand. 73:1–12.
Petri, M. 1969. Prog. Protozool. 3:240–41.
Petri, M., and T. Schidt. 1966. Acta Pathol. Microbiol. Scand. 66:437–46.
Shadduck, J. A. 1969. Science 166:516–17.
Shadduck, J. A. 1980. J. Protozool. 27:202–8.
Shadduck, J. A., R. Bendele, and G. T. Robinson. 1978. Vet. Pathol. 15:449–60.
Shadduck, J. A., G. Kelsoe, and R. J. Helmke. 1979. J. Parasitol. 65:185–88.
Shadduck, J. A., and M. B. Polley. 1978. J. Protozool. 25:491–96.
Shadduck, J. A., W. T. Watson, S. P. Pakes, and A. Cali. 1979. J. Parasitol. 65:123–29.
Sprague, V. 1974. Trans. Am. Microsc. Soc. 93:400–403.
Sprague, V., and S. H. Vernick. 1971. J. Protozool. 18:560–69.
Vavra, J., P. Bedrnik, and J. Cinatl. 1972. Folia Parasitol. 19:349–54.
Vavra, J., J. Chalupsky, J. Oktabec, and P. Bedrnik. 1980. J. Protozool. 27:74A–75A.
WHO (World Health Organization). 1980. WHO/VBC/80.760.

13
Ciliophora

THE ciliates of domestic animals all belong to the phylum Ciliophora. The nuclei of members of this group are unique in the animal kingdom. Every individual (except in a few amicronucleate strains) has a micronucleus that contains a normal set of chromosomes and a macronucleus that contains an indeterminately large number of sets (i.e., is polyploid). The micronucleus is active in reproduction, while the macronucleus has to do with the vegetative functions of the organism.

The ciliates have either simple cilia or compound ciliary cirri or membranelles in at least one stage of their life cycle. (Some protozoa with vesicular nuclei as well as cilia are not considered ciliates by the ciliatologists.) They also have infraciliature in the cortex beneath the pellicle, composed of the ciliary basal granules (kinetosomes) and associated fibrils (kinetodesmata). The infraciliature can be stained with silver, forming the so-called silver-line system. Reproduction is by transverse binary fission, in contrast to the longitudinal fission seen in the flagellates. True sexual reproduction, in which gametes fuse to form a zygote, is absent; but conjugation, in which there is an exchange of micronuclear material between two individuals, may be present.

There are now about 7,200 bona fide named species of Ciliophora, of which about 2,500 are parasitic (Levine et al. 1980). Their classification has recently undergone repeated overhaulings, and the ciliatologists are not done yet (Canella 1977; Lom and Didier 1979; Yankovskii 1980). Corliss (1979) arranged them in 3 classes, 7 subclasses, 23 orders, 36 suborders, and 202 families; he accepted 1,123 generic names (1977) and gave a glossary of terms and concepts useful in ciliate systematics and reviewed the whole group (1979). The classification to suborders was given by Levine et al. (1980); Corliss was responsible for the ciliate part of this classification.

The characteristics of the taxa found in domestic animals have been given in Chapter 1. In the class Kinetofragminophorasida, the oral infraciliature is only slightly distinct from the somatic infraciliature. This class contains 3 orders of veterinary interest. In the order Prostomatidorida, the cytostome is apical or subapical and there is no vestibulum. This order contains about 13 genera and many species of the family Buetschliidae that occur in the rumen and reticulum of ruminants or the large intestine of equids. In the order Trichostomatorida the cytostome is also apical or subapical, but there is a vestibulum. This order contains about 8 genera that occur in the rumen, reticulum, or large intestine (mostly of ruminants or equids), but it also includes the rather ubiquitous genus *Balantidium* that occurs in the large intestine of the pig, man, and other vertebrates. In the order Entodiniomorphidorida, simple somatic ciliature is absent, the body has membranellar tufts or zones of cilia, and the pellicle is firm and sometimes drawn out into spines. This order contains the family Ophryoscolecidae, which contains 13 genera and many species that swarm in the rumen and reticulum of ruminants, and the families Cycloposthiidae, Spirodiniidae, and Ditoxidae, which have many genera and species that swarm in the large intestine of equids. It also contains the genus *Troglodytella* (in the family Troglodytellidae), which occurs in the large intestine of the chimpanzee and other primates (but not man).

A second class is the Oligohymenophorasida, in which the oral ciliature is clearly distinct from the somatic ciliature and whose members have membranelles, peniculi, or polykineties on the left side. This class contains the well-known, free-living vorticellids, *Paramecium, Tetrahymena,* and the fish parasites *Ichthyophthirius* and *Trichodina.*

The third class is the Polyhymenophorasida, whose members have a well-developed adoral zone of membranelles. This class contains the genus *Nyctotherus,* which has species in the large intestine of frogs and various invertebrates and has also been found in the feces of ruminants.

CILIATES OF RUMINANTS

A tremendous number and bewildering variety of ciliates swarm in the rumen and reticulum of ruminants, and a few species occur in the large intestine. Many are holotrichs, with cilia over the whole body, but the most bizarre ones are ophryoscolecids. No attempt will be made here to differentiate all the species, but the genera will be described, the principal species mentioned, and the relations of the different groups to their hosts discussed. None of the species is pathogenic, so far as is known. Further taxonomic and structural information was given by Becker and Talbott (1927), Dogel' (1927), Kofoid and MacLennan (1930, 1932), Polyansky and Strelkov (1938), Lubinsky (1957), and Vasily and Mitchell (1974). Dehority reported on the species in the Alaskan moose, musk-ox, and Dall sheep (1974) and in the water buffalo (1979). He also found that 24 species from cattle were infective for sheep if the latter were fed only alfalfa hay (1978). Grain (1966) did a cytologic and electron microscope study of the holotrich ciliates of ruminants. Coleman (1978) reviewed methods of cultivating rumen entodiniomorphid protozoa. Ogimoto and Imai (1981) wrote an atlas of rumen microbiology.

Order Prostomatorida Schewiakoff, 1896

The cytostome is apical or subapical. A vestibulum is absent. Cilia are generally over the whole body. (See Figs. 13.1 and 13.2.)

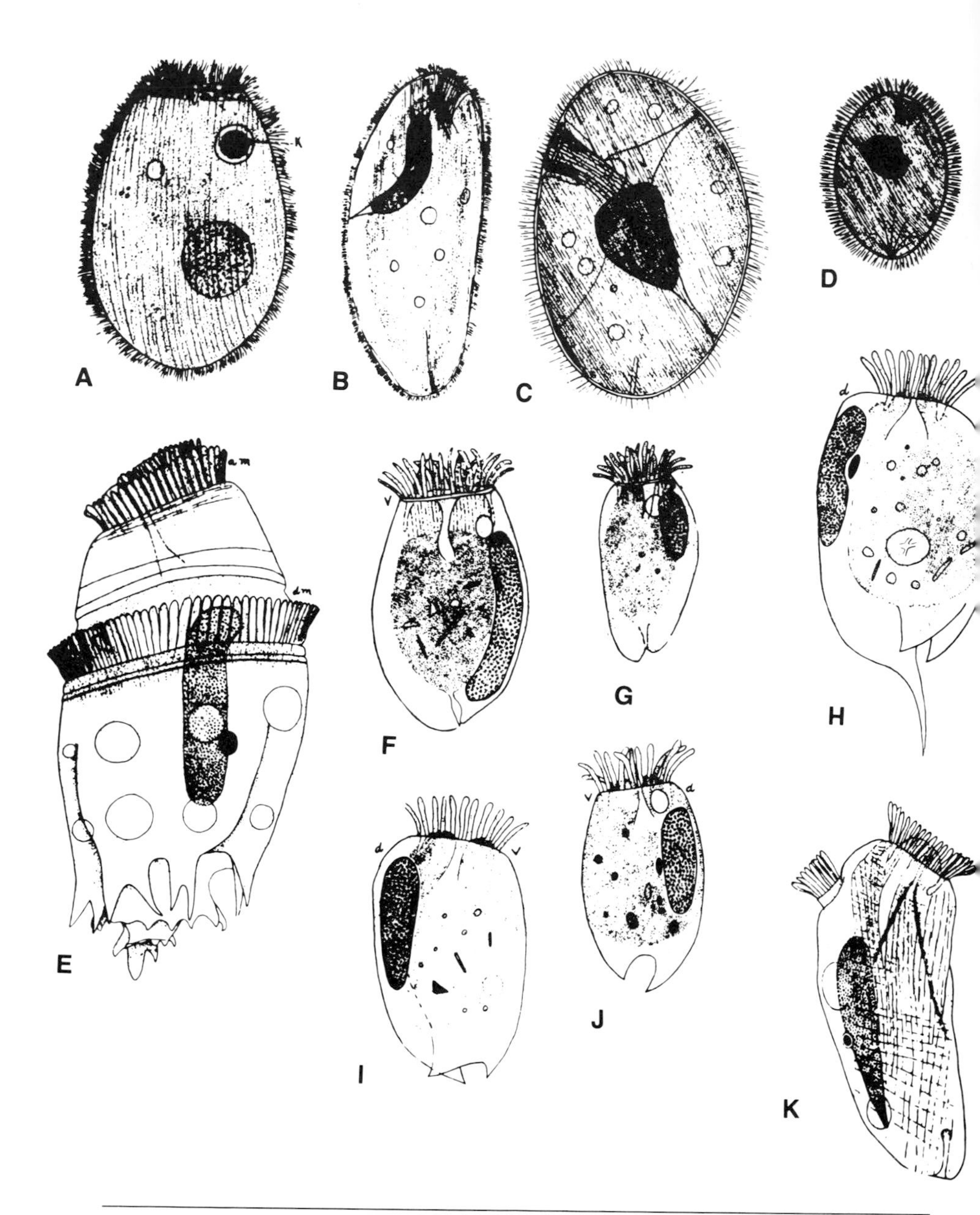

Fig. 13.1. Ciliates of ruminants: A. *Buetschlia parva* (×1,090); B. *Isotricha prostoma* (×320); C. *Isotricha intestinalis* (×640); D. *Dasytricha ruminantium* (×420); E. *Ophryoscolex caudatus* (×425); F. *Entodinium bursa* (×640); G. *Entodinium minimum* (×640); H. *Entodinium caudatum* (×640); I. *Entodinium bicarinatum* (×640); J. *Entodinium furca* (×640); K. *Epidinium ecaudatum* (×425). (From Becker and Talbott 1927, Iowa State College Journal of Science, published by Iowa State University Press)

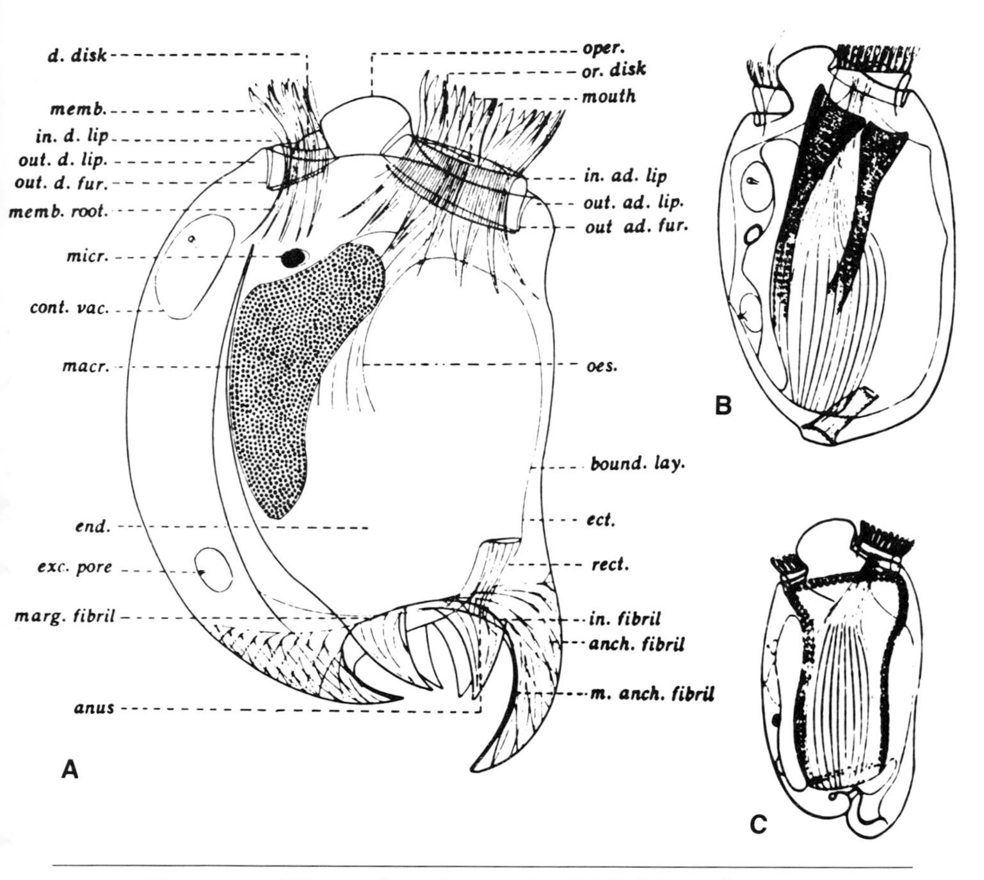

Fig. 13.2. Ciliates of ruminants: A. *Diplodinium dentatum: anch. fibril* = anchoring fibril, *anus* = cytopyge, *bound. lay.* = boundary layer, *cont. vac.* = contractile vacuole, *d. disk* = metoral disk, *ect.* = ectoplasm, *end.* = endoplasm, *exc. pore* = excretory pore, *in. ad. lip* = inner adoral lip, *in. d. lip* = inner metoral lip, *in. fibril* = inner fibril, *macr.* = macronucleus, *m. anch. fibril* = main anchoring fibril, *marg. fibril* = marginal fibril, *memb.* = membranelle, *memb. root* = membranelle root, *micr.* = micronucleus, *mouth* = mouth, *oes.* = esophagus, *oper.* = operculum, *or. disk* = oral disk, *out. ad. fur.* = outer adoral furrow, *out. ad. lip* = outer adoral lip, *out. d. fur.* = outer metoral furrow, *out. d. lip* = outer metoral lip, *rect.* = rectum. B. *Metadinium medium.* C. *Ostracodinium mammosum.* (From Kofoid and MacLennan 1932, Univ. Calif. Publ. Zool. 37:53–152)

Genus *Buetschlia* Schuberg, 1888

The body is ovoid, with a truncate anterior end and a rounded posterior end. There is a circular cytostome at the anterior end but no cytopyge. The body is uniformly ciliated except for long cilia surrounding the cytostome. The ectoplasm at the anterior end is thick, the macronucleus spherical. There is an anterior concretion vacuole (statocyst?).

B. parva Schuberg, 1888, is 30–50 × 20–30 μm.

B. neglecta Schuberg, 1888, resembles *B. parva,* but its posterior end is somewhat pointed and has 4 indentations so that it looks like a cross in cross section; it is 40–60 × 20–30 μm.

B. lanceolata Fiorentini, 1890, is lanceolate, with a collarlike stricture in the anterior fifth of the body; it is 48 × 20 μm. These three species all occur in cattle but are apparently not common, at least in the United States.

B. nana Dogel' and *B. omnivora* Dogel' are found in the rumen of the dromedary.

Order Trichostomatorida Bütschli, 1889

The cytostome is apical or subapical. A vestibulum is present. Somatic ciliature may or may not be reduced.

Genus *Buxtonella* Jameson, 1926

The body is ovoid and uniformly ciliated, with a prominent curved groove bordered by 2 ridges running from end to end. The cytostome is near the anterior end.

B. sulcata Jameson, 1926, is common in the cecum of the ox, zebu, and water buffalo. The trophozoites are 60–138 × 46–100 (mean 100 × 72) μm and have an oval or bean-shaped macronucleus 18–36 × 10–18 (mean 28.5 × 14) μm.

According to Lubinsky (1957), *Balantidium* reported in cattle were actually *B. sulcata.*

Infundibulorium cameli Bozhenko, 1925, which was described from the diarrheic stools of a camel, may be the same species as *B. sulcata.* If so, the name will have to be changed, since *Infundibulorium* has priority.

Genus *Charonina* Strand, 1928

There are 2 caudal and 3 (2?) anterior ciliary zones; an anterior knob is present on the body.

C. ventriculi (Jameson, 1925) Strand, 1928 (syns., *Blepharocorys bovis, B. ventriculi*), occurs in the rumen of the ox in Europe. It is 27–35 × 11–13 μm in silver preparations and has 2 zones of somatic cilia in the anterior part of the body and 2 in the posterior.

C. nuda Hsiung, 1932, occurs in the rumen of the ox.

Genus *Dasytricha* Schuberg, 1888

The body is oval and flattened. The cilia are in spiral, longitudinal rows. There is no karyophore (see *Isotricha*).

D. ruminantium Schuberg, 1888, occurs in the rumen and reticulum of cattle, sheep, and goats. It is 35–75 × 22–40 μm.

Genus *Isotricha* Stein, 1859

The body is oval and flattened, with dense, longitudinal rows of cilia. The cytostome is at or near the anterior end. Several contractile vacuoles are present. The macronucleus is kidney-shaped; it and the micronucleus are connected to each other and suspended by fibrils that constitute the karyophore. Locomotion is toward the rear.

I. prostoma Stein, 1859, the most widely distributed of all ruminant ciliates, occurs in the rumen and reticulum of cattle, sheep, and goats. It is 80–195 × 53–85 μm, and its cytostome is subterminal.

I. intestinalis Stein, 1859, also occurs in the rumen and reticulum of cattle, sheep, and goats. It is 97–130 × 68–88 μm and differs from *I. prostoma* in that its cytostome and nucleus are more posterior.

Order Entodiniomorphidorida Reichenow, 1929

Simple somatic ciliature is absent, replaced by membranellar tufts or zones of cilia. The pellicle is firm, sometimes drawn out into spines. All members of this order in ruminants belong to the family Ophryoscolecidae, in which the ciliary tufts are retractable and are limited principally to the oral or adoral area plus 1 metoral (anterodorsal) group. There may also be skeletal plates. This family contains 18 or more genera, which occur in the rumen of ruminants; 13 of these occur in cattle and sheep. See Lubinsky (1957) for a review.

Genus *Ophryoscolex* Stein, 1858

The body is ovoid, with adoral and metoral zones of membranelles. The metoral zone is some distance from the anterior end and encircles three-fourths of the body circumference at its middle, being broken on the upper right side. There are 3 skeletal plates extending the length of the body on the upper right side and 9–15 contractile vacuoles arranged in an anterior and a posterior circle. The macronucleus is simple and elongate. This genus occurs in the rumen and reticulum of cattle, sheep, goats, and wild sheep.

O. inermis Stein, 1858, occurs principally in the goat but also in the ox. It is 170–190 × 65–100 μm. It differs from the other species of *Orphryoscolex* in having a rounded posterior end, without spines.

O. purkinjei Stein, 1858, occurs in cattle. It is 200 × 80 μm and has 2 or 3 terraces of thornlike appendages or spines encircling the posterior end of the body except for a short gap on the right side; in addition, there is a bifid spine at the posterior end.

O. caudatus Eberlein, 1895, occurs in sheep and cattle. It resembles *O. purkinjei,* but its terminal spine is long and not bifid.

Genus *Entodinium* Stein, 1858

The body is truncate anteriorly, with the adoral zone of membranelles (AZM) at that end. There is no metoral zone of membranelles, and skeletal plates are likewise absent. There are 2–3 rows composed of a few kinetosomes of cilia independent of the AZM. The contractile vacuole is anterior. The macronucleus is cylindrical or sausage-shaped and "dorsal." The micronucleus is anterior to the middle on the upper left side of the macronucleus. This is one of the commonest and most important genera in the rumen and reticulum of cattle, sheep, goats, and other ruminants.

E. bursa Stein, 1858, has a flattened body about 80 × 60 μm. The sausage-shaped macronucleus is four-fifths of the body length, and the micronucleus is pressed closely to it. The body surface has conspicuous longitudinal striations. The contractile vacuole is anterior.

E. minimum Schuberg, 1888, has a flattened body about 40 × 22 μm. The right margin of the body is strongly convex, the left margin almost straight. The body surface has faint longitudinal striations. The macronucleus is about one-third of the body length. The contractile vacuole is anterior.

E. caudatum Stein, 1858, has a flattened body about 30–80 or more μm long. The macronucleus is about half the body length; it is broader anteriorly than posteriorly. The contractile vacuole is near the anterior end of the macronucleus. The upper side of the body is hollowed out to form a groove that broadens posteriorly. There is great variation in the caudal spination.

E. bicarinatum da Cunha, 1914, may be a synonym of *E. caudatum.* It is about 61 × 35 μm, and the upper groove is not as deep as in *E. caudatum.*

E. furca da Cunha, 1914, too, may be a synonym of *E. caudatum.* It has 2 unequal caudal projections, 1 on the left and the other on the right, and is about 52 × 27 μm.

E. dentatum Stein, 1859, is 60–90 × 30–50 μm and has 6 incurved, toothlike posterior projections.

E. rectangulatum Kofoid and MacLennan, 1930, is 23–47 × 23–39 μm. Its body is nearly rectangular as seen from above, except for the caudal spines. The macronucleus is about half the body length and is broader anteriorly than posteriorly. The contractile vacuole is about the middle of the body at the level of the esophagus (i.e., more to the left than that of *E. caudatum*). The upper groove is more marked than that of *E. caudatum,* and its anterior end separates the contractile vacuole from the macronucleus.

E. lobosospinosum Dogel', 1927, is 18–33 × 13–25 μm. Its body is rectangular as seen from above. The macronucleus is about half the body length. The contractile vacuole is on the midline of the upper side of the body on the level of the micronucleus and to the left of the broad upper groove.

E. simulans Lubinsky, 1957, is 27–44 × 21–34 μm. Its body is ovoid as seen from above. The macronucleus is about half the body length. The contractile vacuole is on the midline of the upper side of the body at the level of the micronucleus and to the left of the upper groove. This groove is narrow and long, with a slit-shaped anterior half.

E. longinucleatum Dogel', 1925, is 39–64 × 27–46 μm and has an ellipsoidal, flattened body. The macronucleus usually extends the whole length of the body from the anterior end to the cytopyge. The contractile vacuole is close to the upper side of the macronucleus, slightly anterior to the micronucleus.It also occurs in the white-tailed deer.

E. rostratum Fiorentini, 1889, is 27–51 × 13–23 μm and has a rather long, slim, flattened body with a strongly convex right side and concave left side. The macronucleus is narrow, bandlike, and about half the length of the body. The contractile vacuole is directly anterior to the macronucleus.

E. laterale Kofoid and MacLennan, 1930, is 19–28 × 18–21 μm and has a short, fairly broad, truncated ellipsoidal, flattened body. The macronucleus is broad and wedge-shaped, less than half the body length, and lies anteriorly in the body. The contractile vacuole is in the middle of the upper side.

E. nanellum Dogel', 1922, is 20–32 × 10–18 μm and has an ovoid, flattened body. The macronucleus is thin, wedge-shaped, and longer than half the body length. The contractile vacuole is above the anterior end of the macronucleus.

E. bimastus Dogel', 1927, is 30–46 × 28–40 μm and has a subspherical, flattened body. The macronucleus is flattened, wedge-shaped, and about two-thirds of the body length. The contractile vacuole is above the anterior end of the macronucleus.

E. exiguum Dogel', 1925, is 21–29 × 14–18 μm and has an elongate, oval body. The macronucleus is relatively short and thick, being shorter than half the body length, and lies in the middle of the body.

E. dubardi Buisson, 1923 (syn., *E. simplex* Dogel', 1925, in part), is 30–50 × 20–29 μm and has an oval or elongate oval, flattened body. The macronucleus is more or less band-shaped, with a somewhat broader anterior end. It is about half the body length and lies anteriorly in the body. The contractile vacuole is below the anterior end of the macronucleus. It also occurs in the white-tailed deer.

E. vorax Dogel', 1925, is 60–121 × 40–83 μm and has an oval, plump, thick body. The anterior end is often smaller than the posterior. The macronucleus is sausage-shaped, about half the body length, and lies anteriorly in the body. The contractile vacuole is to the right of the anterior end of the macronucleus.

Quite a few other species of *Entodinium* have been described from various ruminants. Among them are the following, which Kofoid and MacLennan (1930) described from the zebu: *E. ellipsoideum, E. acutonucleatum, E. pisciculum, E. biconcavum, E. bifidum, E. acutum, E. aculeatum, E. brevispinum, E. laterospinum, E. ovoideum, E. rhomboideum, E. gibberosum, E. tricostatum,* and *E. indicum.*

Rothenbacher (1964) described two cases of bovine hepatitis in which rumen protozoa (probably *Ophryoscolex* sp. and *Entodinium* sp.) were present in sections of the liver lesions. The relationship of the protozoa to the hepatitis was not clear; they were probably secondary invaders.

Genus *Epidinium* Crawley, 1924

The body is elongate and twisted around its main axis. The adoral zone of membranelles is at the anterior end, the metoral zone elsewhere. There are several rows of cilia independent of the AZM. There are 3 skeletal plates with secondary plates. The macronucleus is simple and club-shaped, and there are 2 contractile vacuoles. This genus occurs in the rumen and reticulum of cattle, sheep, goats, camels, reindeer, elk, and other ruminants.

E. ecaudatum (Fiorentini, 1889) (syn., *Diplodinium ecaudatum*) is 82–173 × 36–70 μm and has an elongate body with a convex left side and a flat or slightly concave right side. This species has 9 formae that differ principally in their caudal spination: *E. ecaudatum* forma *ecaudatum* has no caudal spines; forma *caudatum* has a single long spine, and the posterior end of its body is narrow; forma *hamatum* has a single long spine, and the posterior end of its body is broad; forma *bulbiferum* has a bulb-shaped appendage instead of a spine; forma *bicaudatum* has 2 spines; forma *tricaudatum* has 3 spines; forma *quadricaudatum* has 4 spines; forma *cattaneoi* has 5 short spines; and forma *fasciculus* has 5 very long spines with greatly swollen bases. All 9 formae occur in the rumen of cattle, the most common being *ecaudatum* and *caudatum*. *Ecaudatum*, *hamatum*, and *cattaneoi* are found in sheep, and the last in goats as well. *Ecaudatum* and *caudatum* also occur in the reindeer, *caudatum* and *hamatum* in camels, and *caudatum* in the elk.

Genus *Eodinium* Kofoid and MacLennan, 1932

The metoral zone of membranelles is at the same level as the adoral zone. Skeletal plates are absent. The macronucleus is a straight, rodlike body near the left edge. Two contractile vacuoles are present.

E. posterovesiculatum (Dogel', 1927) Kofoid and MacLennan, 1932, is 47–60 × 23–30 μm and has a relatively long, flattened body with rounded ends. The macronucleus is very long and straight and has 2 deep depressions on its left side. The micronucleus lies at the posterior one, a contractile vacuole at the anterior one. The second vacuole is posterior to the macronucleus. It occurs in cattle.

E. bilobosum (Dogel', 1927) Kofoid and MacLennan, 1932, is 46–60 × 30–44 μm and has a relatively short, flattened body with 2 caudal spines, one dorsal and the other ventral. The nuclei and vacuoles are similar to those of *E. posterovesiculatum*. It occurs in cattle and sheep.

E. lobatum Kofoid and MacLennan, 1932, is 44–60 × 29–37 μm and has a narrow body. The macronucleus is narrow and rodlike and almost as long as the body. It has 3 large depressions in its left side; the micronucleus lies at the middle one and the contractile vacuoles at the end ones. It occurs in the zebu.

Genus *Diplodinium* Schuberg, 1888

The metoral zone of membranelles is on the same level as the adoral zone. There are rather long rows of large kinetosomes of cilia independent of the AZM. Skeletal plates are absent. The macronucleus is beneath the upper surface of the body; its anterior third is bent to the right at an angle of 30–90°. Two contractile vacuoles are present. In this genus, as in *Entodinium,* there is also considerable variation in the caudal spination.

D. dentatum (Stein, 1858) Schuberg, 1888, is 55–82 × 44–62 μm. Its body has a broad, truncate posterior end with 6 large, relatively heavy, incurved caudal spines. The left side is convex, the right one concave. The macronucleus is 25–50 μm long; it is heavy and rodlike, with the anterior end bent at a 45° angle. The 2 contractile vacuoles lie in the left rib slightly below the midline. It occurs in the ox and zebu.

D. quinquecaudatum Dogel', 1925, is 57–73 × 47–65 μm. It resembles *D. dentatum* but has 5 caudal spines. It occurs in cattle and sheep.

D. anacanthum Dogel', 1927, is 60–124 × 38–72 μm. The posterior end of its body tapers, giving it a somewhat conical appearance. The macronucleus varies a good deal in length. Its anterior third is bent at an angle of about 45°. The 2 contractile vacuoles are on the lower side. This species has 7 formae: *anacanthum, monacanthum, diacanthum, triacanthum, tetracanthum, pentacanthum,* and *anisacanthum,* with 0, 1, 2, 3, 4, 5, and 6 caudal spines, respectively. It occurs in the ox and zebu.

E. psittaceum Dogel', 1927, is 95–155 × 59–105 μm and has a heavy, rounded posteriorly tapering body with a thin ventral spine on the right and a narrow flange on the left of the posterior third of the body. The macronucleus is a stout, rodlike body with its anterior end bent at a 40° angle. The contractile vacuoles lie near the left side. This species occurs in the ox and zebu.

D. bubalidis Dogel', 1925, is 104–195 × 58–98 μm and has an oval body with its largest diameter anterior, a strongly convex left side, and a slightly convex right one. There is a small, longitudinal groove on the posterior part of the upper surface of the body, and a single, thin spine on the right. It occurs in cattle and African antelope.

D. elongatum Dogel', 1927, is 177–205 × 73–100 μm and has an elongate body with weakly convex left and right sides and a narrow groove in the posterior end of the upper surface of the body. It occurs in the ox.

D. laeve Dogel', 1927, is 77–100 × 52–70 μm and has a roughly triangular body with no caudal projections except a small lobe on the right. It occurs in goats.

D. cristagalli Dogel', 1927, is 77–100 × 52–70 μm and has a triangular body with the lower side extended posteriorly to form a prominent fan with 2–7 spines. It occurs in goats.

D. flabellum Kofoid and MacLennan, 1932, is 82–118 × 57–82 μm and has a roughly triangular body with the upper side extended posteriorly to form a prominent fan with 5–7 spines, and with 2 small spines on the posterior left side. It occurs in the zebu.

Genus *Eremoplastron* Kofoid and MacLennan, 1932

The metoral and adoral zones of membranelles are at the same level. There is a single, narrow skeletal plate beneath the upper surface. The macronucleus is triangular or rodlike, often with its anterior end bent to the right. Two contractile vacuoles are present.

E. rostratum (Fiorentini, 1889) Kofoid and MacLennan, 1932 (syn., *Diplodinium helseri*), is 40–63 × 22–47 μm and has a proportionately long, compressed body with a thick flange on the left and a large caudal spine on the right. The macronucleus is rodlike. It occurs in the ox and zebu.

E. neglectum (Dogel', 1925) is 81–124 μm long and has an elongate oval body with the left side strongly convex, the right side slightly convex, a large lobe on the right, and a long, rodlike macronucleus. It occurs in cattle and African antelope.

E. bovis (Dogel', 1927) (syn., *Diplodinium clevelandi*) is 52–100 × 36–57 μm and has an ellipsoidal, compressed body with a somewhat flattened right side, a more strongly convex left side, and a small caudal lobe. The macronucleus is rodlike. It occurs in the ox, zebu, and sheep.

E. monolobum (Dogel', 1927) is 58–83 μm long and has a nearly spherical body with a prominent right lobe and a low, blunt left lobe. The macronucleus is thick and rodlike. It occurs in cattle.

E. dilobum (Dogel', 1927) is 73–101 μm long and has an ellipsoidal, flattened body with one left and one right caudal lobe. The macronucleus is rodlike. It occurs in cattle and sheep.

E. rugosum (Dogel', 1927) is 69–90 μm long and has a short body with a flat or slightly concave right side, a convex left side, and a deep cuticular fold from the cytopyge along the left side of the macronucleus to the region of the metoral zone of membranelles. The right lobe is laterally compressed, with 8–10 shallow indentations in its left border. The macronucleus is long and rodlike. It occurs in cattle.

E. brevispinum Kofoid and MacLennan, 1932, is 72–92 × 42–53 μm and has an ellipsoidal, flattened body with 2 short caudal spines. The macronucleus is rodlike. It occurs in the zebu.

E. magnodentatum Kofoid and MacLennan, 1932, is 58–82 × 30–50 μm and has a rectangular, flattened body with a large, compressed caudal spine on the right and a similar caudal spine on the left. The macronucleus is rodlike. It occurs in the zebu.

Genus *Eudiplodinium* Dogel', 1927

The metoral and adoral zones of membranelles are at the anterior end. There is a single, narrow skeletal plate beneath the upper surface. The macronucleus is rodlike, with its anterior end enlarged to form a hook that opens to the left. The pellicle and ectoplasm are thick. There are 2 contractile vacuoles with heavy membranes and prominent pores.

E. maggii (Fiorentini, 1889) is 104–240 × 63–77 μm and has a roughly triangular body with a smoothly rounded posterior end. It occurs in the rumen and reticulum of the ox and zebu.

Genus *Diploplastron* Kofoid and MacLennan, 1932

The metoral and adoral zones of membranelles are at the anterior end. There are 2 skeletal plates beneath the upper surface. The macronucleus is narrow and rodlike. There are 2 contractile vacuoles below the left surface, separated from the macronucleus.

D. affine (Dogel' and Fedorowa, 1925) is 88–120 × 47–65 μm and is more or less ellipsoidal. It occurs in the rumen of cattle, sheep, and goats.

Genus *Metadinium* Awerinzew and Mutafowa, 1914

The metoral and adoral zones of membranelles are at the anterior end. There are 2 skeletal plates beneath the upper surface, which are sometimes fused posteriorly. The macronucleus has 2–3 left lobes. There are 2 contractile vacuoles. The pellicle and ectoplasm are thick. There are conspicuous esophageal fibrils beneath the left and upper sides.

M. medium Awerinzew and Mutafowa, 1914, is 180–272 × 92–170 μm and has a heavy body with large membranelle zones. The skeletal plates are narrow. It occurs in the rumen of the ox and zebu.

M. tauricum (Dogel' and Fedorowa, 1925) is 188–288 × 70–160 μm and has a heavy body. The skeletal plates are fused posteriorly. The anterior and median lobes of the macronucleus are large, the posterior lobe small. It occurs in the rumen of sheep, goats, and cattle.

M. ypsilon (Dogel', 1925) is 110–152 × 60–72 μm and has an oval, flattened body with a rounded posterior end. The anterior lobe of the macronucleus is small, and there is no posterior lobe. The skeletal plates are fused posteriorly. It occurs in the rumen of cattle.

Genus *Polyplastron* Dogel', 1927

The metoral and adoral zones of membranelles are at the anterior end. There are 2 separate or fused skeletal plates beneath the upper surface and three longitudinal plates with anterior ends connected by crossbars beneath the lower surface. There is a longitudinal row of contractile vacuoles beneath the left surface, and others in other locations.

P. multivesiculatum (Dogel' and Fedorowa, 1925) is 120–190 × 78–140 μm and has an oval body with a smoothly rounded posterior end. There is a row of 4 contractile vacuoles near the macronucleus, plus 2 beneath the left surface, 1 beneath the right surface, and 2 beneath the upper surface. The 2 upper skeletal plates are separate. It occurs in the rumen of cattle and sheep.

P. fenestratum Dogel', 1927, resembles *P. multivesiculatum* except that the upper skeletal plates are partly fused. It occurs in the rumen of cattle.

P. monoscutum Kofoid and MacLennan, 1932, resembles *P. multivesiculatum* except that the upper skeletal plates are completely fused to form a single broad plate. It occurs in the rumen of cattle.

Genus *Elytroplastron* Kofoid and MacLennan, 1932

The metoral and adoral zones of membranelles are at the anterior end. There are 2 skeletal plates beneath the upper surface, a small plate beneath the right surface, and a long plate below the lower surface. The pellicle and ectoplasm are thick. There are conspicuous fibrils beneath the left and upper surfaces.

E. bubali (Dogel', 1928) is 110–160 × 67–97 μm and has an ellipsoidal body with a smoothly rounded posterior end. There are 4 contractile vacuoles along the left midline. It occurs in the rumen of the water buffalo and zebu.

Genus *Ostracodinium* Dogel', 1927

The metoral and adoral zones of membranelles are at the anterior end. There is a broad skeletal plate beneath the upper side of the body, and a row of 2–6 contractile vacuoles beneath the left surface. Heavy pharyngeal fibrils are present that extend to the posterior end.

O. mammosum (Railliet, 1890) is 41–110 × 25–68 μm and has a left caudal lobe and a right lobe that is hollow on the left side. The posterior part of the skeleton extends only two-thirds of the way across the upper side. The macronucleus has a large, shallow depression in the middle of its lower side. There are 3 contractile vacuoles. It occurs in the rumen of the ox and zebu.

O. gracile (Dogel', 1925) is 90–133 × 42–60 μm and has a roughly triangular body with flat right and lower surfaces, convex left and upper surfaces, and a smoothly rounded posterior end. The skeletal plate extends across the upper surface. The macronucleus has 2 lobes. There are 2 contractile vacuoles. It occurs in the rumen of the ox, zebu, sheep, and African antelopes.

O. tenue (Dogel', 1925) is 59–76 × 28–38 μm and has a slender body with a smoothly rounded posterior end. The skeletal plate extends across the upper surface. The macronucleus has an anterior and a median left lobe. There are 2 contractile vacuoles. It occurs in the rumen of cattle and an African antelope.

O. trivesiculatum Kofoid and MacLennan, 1932, is 78–100 × 42–60 μm and has a triangular body with a smoothly rounded posterior end. The skeletal plate extends across the upper side. The macronucleus has a small, shallow depression in the middle of the lower side. There are 3 contractile vacuoles. It occurs in the rumen of the zebu.

O. quadrivesiculatum Kofoid and MacLennan, 1932, is 91–112 × 43–56 μm and has a triangular body with a bluntly rounded posterior end. The skeletal plate extends across the upper side. The macronucleus is elongate and rodlike. There are 4 contractile vacuoles. It occurs in the rumen of the zebu.

O. nanum (Dogel', 1925) is 47–70 × 30–41 μm and has an ellipsoidal body with a slender, right caudal spine. The skeletal plate extends between the macronucleus and the ventral surface. The macronucleus is short and stout. There are 2 small contractile vacuoles. It occurs in the rumen of cattle and African antelopes.

O. gladiator (Dogel', 1925) is 78–112 × 40–55 μm and has a slender body with a long, very narrow, right caudal spine. The skeletal plate extends between the macronucleus and the right side. The macronucleus has a lobe on the left anterior end. There are 2 contractile vacuoles. It occurs in the rumen of cattle and African antelopes.

O. crassum (Dogel', 1925) is 120–142 × 80–100 μm and has a heavy body with a smoothly rounded posterior end. The skeletal plate extends under only half the upper side. The macronucleus is short and stout, with a wide, shallow depression in the anterior half of its left side. There are 2 contractile vacuoles. It occurs in the rumen of cattle and steenbok.

O. obtusum (Dogel' and Fedorowa, 1925) (syn., *Diplodinium hegneri*) is 118–148 × 55–80 μm and has an ellipsoidal, only slightly flattened body, with a smoothly rounded posterior end. The posterior part of the skeleton extends across only two-thirds of the upper side. The macronucleus is elongate and rodlike. There are 6 contractile vacuoles. It occurs in the rumen of cattle and reindeer.

O. venustum Kofoid and MacLennan, 1932, is 76–115 × 41–60 μm and has a triangular body with a small posterior right lobe. The skeletal plate extends beneath the upper surface between the macronucleus and the right side. The macronucleus has 2 left lobes, and there are 2 contractile vacuoles. It occurs in the rumen of the zebu.

O. dogieli Kofoid and MacLennan, 1932, is 92–130 × 48–63 μm and has an ellipsoidal body with a strongly convex left side, a slightly convex right side, and a flattened right lobe lying below the cytopyge. The skeletal plate extends between the macronucleus and the right side. The macronucleus has 2 left lobes (one anterior and one median). There are 2 contractile vacuoles. It occurs in the rumen of the ox.

O. clipeolum Kofoid and MacLennan, 1932, is 92–128 × 50–65 μm and has an ellipsoidal body with a flattened lobe projecting from the right posterior surface below the midline. The skeletal plate extends beneath the upper surface between the macronucleus and the right side. The macronucleus has 2 left lobes, and there are 3 contractile vacuoles. It occurs in the rumen of the zebu.

O. monolobum Dogel', 1927, is 105–150 × 55–77 μm and has a rectangular body with a large right lobe. The skeletal plate extends under only two-thirds of the left side. The macronucleus is elongate and rodlike. There are 5 contractile vacuoles. It occurs in the rumen of the ox.

O. dilobum Dogel', 1927, is 88–140 × 54–78 μm and has an ellipsoidal body with a laterally flattened right lobe and a flattened left lobe. The skeletal plate extends under only two-thirds of the left side. The macronucleus is elongate and rodlike. There are 5 contractile vacuoles. It occurs in the rumen of cattle.

O. rugoloricatum Kofoid and MacLennan, 1932, is 84–125 × 37–58 μm and has a rectangular body with a flattened right lobe. The left side of the exceptionally large skeletal plate turns in and extends toward the middle of the body. The macronucleus is straight and rodlike. There are 3 contractile vacuoles. It occurs in the rumen of the zebu.

Genus *Enoploplastron* Kofoid and MacLennan, 1932

The metoral and adoral zones of membranelles are near the anterior end. There are 3 separate or partially fused skeletal plates beneath the upper and right surfaces of the body. There are 2 contractile vacuoles, and the pharyngeal fibrils are heavy.

E. triloricatum (Dogel', 1925) is 60–112 × 37–70 μm and has an ellipsoidal body with a smoothly rounded posterior end. The skeletal plates are separate. The macronucleus has a shallow depression in the anterior half of its left side. It occurs in the rumen of the ox, reindeer, and an African antelope.

Relations of Rumen Ciliates to Their Hosts

Ciliates swarm in such tremendous numbers in the rumen and reticulum that everyone who has seen them has speculated on their role in their host's nutrition. The problem has been reviewed by Coleman (1975), among others.

The rumen ciliates are obligate anaerobes. The holotrichs *Isotricha* and *Dasytricha* have been cultivated by Sugden and Oxford (1952), Gutierrez (1955), and others. The ophryoscolecids *Diplodinium, Entodinium, Eudiplodinium, Polyplastron,* and *Metadinium* have been cultivated by Hungate (1942, 1943), Sugden (1953), and others, but *Ophryoscolex* has apparently not been cultivated. Coleman (1978) reviewed the cultivation of rumen entodiniomorphid protozoa.

The holotrichs absorb soluble carbohydrates from the medium and convert them into amylopectin, which is stored in ovoid granules 3 × 2 μm resembling small yeast cells. These protozoa are able to utilize glucose, fructose, sucrose, cellobiose, inulin, and levans. In addition, both *Isotricha intestinalis* and *I. prostoma* rapidly ingest small starch granules and are able to metabolize them. *Dasytricha ruminantium* does not ingest starch. Gutierrez and Hungate (1957) found that *D. ruminantium* ingested small cocci and occasionally small rodlike bacteria; they were able to cultivate this species in a medium containing these types of bacteria but not in one without them. Gutierrez (1958) found that *I. prostoma* feeds selectively only on certain rods among the many types of rumen bacteria, but that pure strains do not fulfill all the protozoon's growth requirements.

The holotrichs produce hydrogen, carbon dioxide, and lactic, acetic, and butyric acids, as well as traces of propionic acid.

Many but not all species of *Entodinium* ingest and digest starch. *E. longinucleatum* and *E. acutonucleatum* feed selectively on pollen grains. Certain species of *Entodinium* are the predominant starch-ingesters among the rumen protozoa and are the dominant protozoa in animals on full feed. Among those known to ingest starch are *E. caudatum, E. longinucleatum, E. minimum,* and *E. dubardi.* Little is known about the products of starch fermentation by this genus. Granules of polysaccharide are stored in the outer zone of the endoplasm, but they have never been isolated and identified; it would be difficult to separate them from ingested starch granules.

It has been suggested that carbohydrate metabolism is dependent on intracellular bacteria. Bacteria are ingested by *Entodinium* and appear to be necessary for its nutrition, most likely as a source of nitrogen rather than of prefabricated enzymes. *Epidinium,* like *Entodinium,* ingests starch and also bacteria; its metabolic products are also unknown. *Diplodinium* and related genera (*Eudiplodinium, Polyplastron, Eremoplastron, Metadinium*) ingest and digest cellulose in addition to starch and bacteria. They all contain a cellulase. *M. medium* and *Entodinium* also utilize cellulose.

Diplodinium and related species produce hydrogen, carbon dioxide, and volatile acids. All ophryoscolecid skeletal plates stain brown with iodine and are polysaccharide in nature. The mode of nutrition of *Ophryoscolex* has apparently not been determined, although it is known to ingest starch granules and sometimes cellulose fibers.

Lubinsky (1957) reported that accidental predation on smaller protozoa is a common trait of many of the larger species of Ophryoscolecidae, particularly of *Diplodinium* and related cellulose-feeding genera. Predation is rare in *Ophryoscolex,* however. The prey of these occasional predators consist primarily of spineless smaller species; the spines are thus of value in protecting the smaller ophryoscolecids against ingestion.

The role of the rumen protozoa in their hosts' nutrition is still not clear. Young animals on a milk diet do not have them, but as they grow older and begin to feed on hay and grass, they become infected from protozoa in the saliva of faunated animals, the only way transmission occurs. There are no resistant forms or cysts, and the protozoa are killed when they enter the abomasum.

The relation between the protozoa and their hosts is not symbiotic, since the host does not need the protozoa for survival and indeed gets along perfectly well without them. The fact that defaunation is not harmful does not mean, however, that the protozoa are of no value to their hosts. It means simply that they are not essential.

It has been suggested that the protozoa might harm their hosts (1) by excreting ammonia, which is then not utilized by the rumen bacteria for protein synthesis and is therefore lost to their hosts, (2) by robbing the host of B vitamins, (3) by feeding on and destroying valuable bacteria, or (4) by producing lactic acid and other undesirable intermediate products of carbohydrate metabolism that the rumen bacteria cannot cope with (Oxford 1955). However, this is simply speculation; there is no proof that they are actually harmful.

Rumen protozoa form about 20% of the protein that reaches the abomasum (Hungate 1955). The rumen protozoan and bacterial proteins both have a biological value for rats of 80–81, which is higher than the 72 of brewer's yeast (McNaught et al. 1954). Furthermore, the true digestibility of the protozoan protein is 91%, much higher than that of bacterial (74%) or yeast (84%) proteins. Hence the protozoan protein is nutritionally superior. No amino acid analyses have apparently been carried out on it.

While many of the protozoa store reserve starch (amylopectin), the stored starch is not of much importance for the host's nutrition. About 1% of the carbohydrate required by a mature sheep is supplied from this source.

The protozoa are an important source of volatile fatty acids. Carroll and

Hungate (1954) estimated that about 2.2 kg volatile fatty acids are produced per 100 kg rumen contents in cattle. Gutierrez (1955) calculated that the fermentation acids produced by the rumen holotrichs would constitute a little more than 10% of this amount. If the ophryoscolecids produced an equal amount, then protozoa would provide about 20% of the fermentation products available to their host (Hungate 1955). As Hungate remarked, Gruby and Delafond (who discovered the rumen protozoa in 1843) guessed that they supplied one-fifth the food used by their hosts, and the results of investigation during the next 100 years did not significantly modify this estimate.

Another advantage to the host lies in the fact that the holotrichs take up soluble carbohydrates from the medium and convert them into stored starch, withholding them for a while and then fermenting them for a long time. This smooths out the fermentation process, which would proceed much more irregularly if it depended on bacteria alone (Hungate 1955; Oxford 1955). *Entodinium* and *Epidinium,* too, help smooth out the fermentation process by converting starch into reserve foods. In addition, as Hungate (1955) pointed out, when animals are shifted from hay to grain there is a period of adaptation during which lactic acid is produced explosively by *Streptococcus bovis,* which may be extremely harmful. The adaptation period may reflect the time needed for *Entodinium, Epidinium,* and other bacteria-feeding protozoa to multiply enough to keep the streptococci in check.

Coleman (1975) said that the rumen protozoa and bacteria are mutually beneficial. *Entodinium* spp. digest starch and proteins, making sugars and amino acids available to themselves and the bacteria. The latter keep the redox potential low enough for the protozoa to multiply. The protozoa also ingest the bacteria, killing and digesting them, which increases the turnover rates of carbon and nitrogen. Under appropriate conditions, 1% of the bacteria in the rumen can be killed each minute.

CILIATES OF EQUIDS

Just as great a variety and number of ciliates swarm in the equid cecum and colon as in the ruminant rumen and reticulum. Hsiung (1930) described 51 species of 25 genera in his monograph, while Strelkov (1939) listed 87 species and forms. The fauna of the proximal large intestine (the cecum and ventral colon) differs from that of the distal large intestine (the dorsal and small colons). Strelkov (1939) listed 25 species and forms in the proximal fauna, 43 in the distal fauna, and 7 common to both. Mixing occurs at the pelvic flexure of the colon. All horses do not contain all species. Strelkov (1939) found an average of 7.7 species per horse in the proximal fauna and 16.6 species per horse in the distal fauna. (See also Grain 1966.)

The ciliate population has large daily variations. Adam (1953) obtained counts ranging from 1,000 to 47,000 per ml in the cecum and from 14,000 to 3,072,000 per ml in the ventral colon of a single horse at different times and on different rations.

Almost nothing is known of the relationship of these protozoa to their host; it is most likely that they are simply commensals. No cysts have been reported, and transmission is probably by mouth.

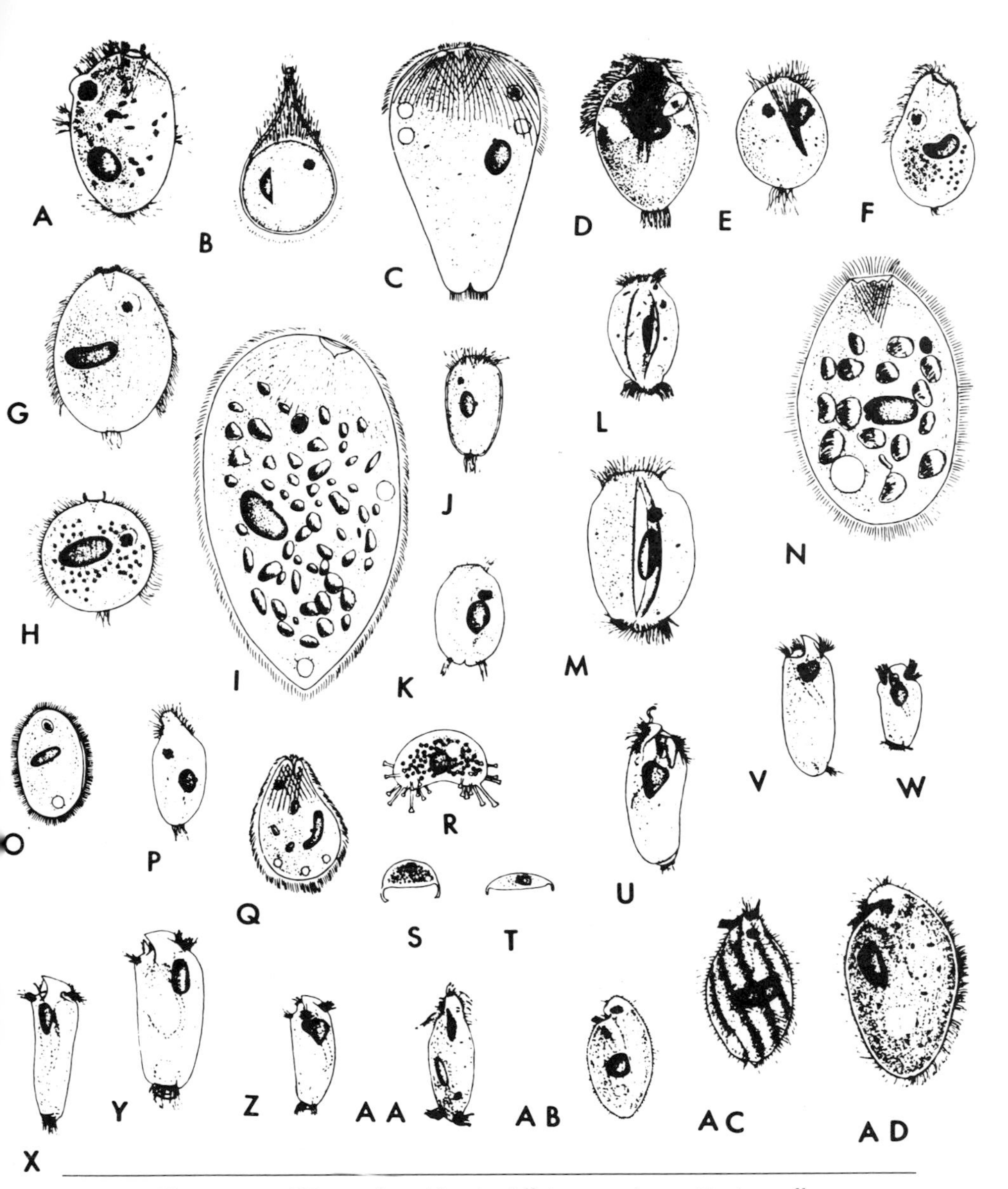

Fig. 13.3. Ciliates of equids: A. *Alloiozona trizona;* B. *Ampullacula ampulla;* C. *Blepharoconus hemiciliatus;* D. *B. cervicalis;* E. *B. benbrooki;* F. *Blepharoprosthium pireum;* G. *Blepharosphaera ellipsoidalis;* H. *B. intestinalis;* I. *Blepharozoum zonatum;* J. *Bundleia postciliata;* K. *Didesmis ovalis;* L. *D. spiralis;* M. *D. quadrata;* N. *Holophryoides ovalis;* O. *Paraisotrichopsis composita;* P. *Polymorphella ampulla;* Q. *Prorodonopsis coli;* R. *Allantosoma intestinalis;* S. *A. dicorniger;* T. *A. brevicorniger;* U. *Blepharocorys uncinata;* V. *B. valvata;* W. *B. jubata;* X. *B. angusta;* Y. *B. curvigula;* Z. *B. cardionucleata;* AA. *Charonina equi;* AB. *Paraisotricha minuta;* AC. *P. beckeri;* AD. *P. colpoidea.* (E, P, and AA ×710; all others ×340) (From Hsiung 1930, in Iowa State College Journal of Science, published by Iowa State University Press)

Order Prostomatidorida Schewiakoff, 1896

Members of this order occur in equids. There are many more genera and species than in ruminants. They have an apical or subapical cytostome, no vestibulum, and generally cilia over the whole body.

Genus *Alloiozona* Hsiung, 1930

The cilia are present in three zones: anterior, equatorial, and posterior.

A. trizona Hsiung, 1930, is ovoid, with both ends rounded, 59–90 × 30–60 μm. The cytostome is at the anterior end, surrounded by a shallow groove provided with short cilia. The cytopharynx is funnel-shaped. The cytopyge is on a knob at the posterior end. The macronucleus is a more or less thick, distinctly granular disk that is not constant in position. The concretion vacuole is large and near the surface in the anterior third of the body. There is usually a small, posterior contractile vacuole. It occurs in the cecum and colon of the horse.

Genus *Ampullacula* Hsiung, 1930

The body is flask-shaped. Its posterior half is covered with fine, short cilia, its neck with longer cilia.

A. ampulla (Fiorentini, 1890) Hsiung, 1930, is about 110 × 40 μm. The cytostome is at the anterior end. It occurs in the cecum of the horse.

Genus *Blepharoconus* Gassovsky, 1919

The body is ovoid. The cytostome is small, and the cytopharynx has rods in its wall. There are cilia on the anterior third to half of the body and at the caudal end. The macronucleus is ovoid. There are 3 contractile vacuoles.

B. hemiciliatus Gassovsky, 1919, has a conical body, 83–135 × 45–65 μm. The macronucleus is nearly spherical. It occurs in the colon of the horse.

B. cervicalis Hsiung, 1930, is ovoid with a blunt anterior and a rounded posterior end, 56–83 × 48–70 μm. There is usually a short neck, which is formed by a slight groove. The macronucleus is more or less disk-shaped, the concretion vacuole small and ellipsoidal. It occurs in the colon of the horse.

B. benbrooki Hsiung, 1930, is ovoid to ellipsoidal with a knoblike anterior end and a rounded posterior one, 21–37 × 17–26 μm. The macronucleus is a thick disk, the concretion vacuole large and ellipsoidal. It occurs in the colon of the horse.

Genus *Blepharoprosthium* Bundle, 1895

The body is piriform, with a contractile anterior half. There are cilia on the anterior half of the body and at the posterior end. The macronucleus is kidney-shaped.

B. pireum Bundle, 1895, is 54–86 × 34–52 μm. The cytostome is anterior, the cytopharynx funnel-shaped. The concretion vacuole contains numerous granules and is found in the anterior half of the body close to the surface.

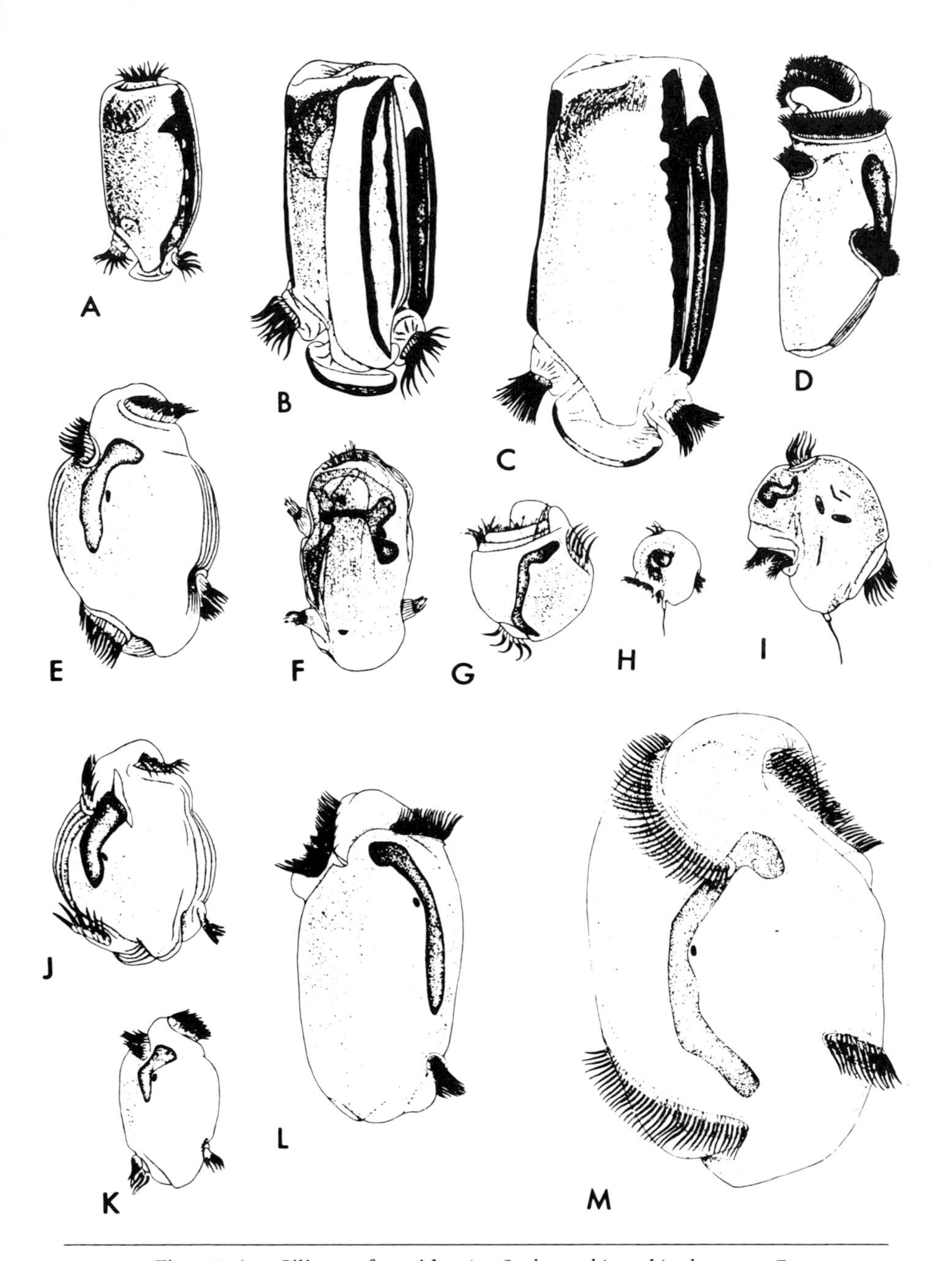

Fig. 13.4. Ciliates of equids: A. *Cycloposthium bipalmatum;* B. *C. scutigerum;* C. *C. edentatum;* D. *Spirodinium equi;* E. *Tetratoxum unifasciculatum;* F. *Tripalmaria dogieli;* G. *Triadinium galea;* H. *T. minimum;* I. *T. caudatum;* J. *Tetratoxum excavatum;* K. *T. parvum;* L. *Ditoxum funinucleum;* M. *Cochliatoxum periachtum* (×340) (From Hsiung 1930, in Iowa State College Journal of Science, published by Iowa State University Press)

There is a contractile vacuole at the posterior end. It occurs in the cecum of the horse.

Genus *Blepharosphaera* Bundle, 1895

The body is spherical or ellipsoidal. Cilia cover the anterior three-fourths of the body, and there is also a caudal tuft of cilia.

B. intestinalis Bundle, 1895, is spherical, 38–74 μm in diameter. Its macronucleus is a thick, ellipsoidal disk. It occurs in the cecum of the colon of the horse.

B. ellipsoidalis Hsiung, 1930, is ellipsoidal, 34–65 × 27–49 μm. Its macronucleus is sausage-shaped. It occurs in the cecum and colon of the horse.

Genus *Blepharozoum* Gassovsky, 1911

The body is ellipsoidal, with an attenuated anterior end, and is uniformly ciliated. The cytostome is near the anterior tip. There are 2–4 contractile vacuoles. The macronucleus is small and kidney-shaped.

B. zonatum Gassovsky, 1919, is 230–245 × 115–122 μm and has an anterior concretion vacuole. It occurs in the cecum of the horse.

Genus *Bundleia* da Cunha and Muniz, 1928

The body is ellipsoidal, with a small cytostome. There are cilia at the anterior and posterior ends, the latter much less numerous than the former.

B. postciliata (Bundle, 1895) da Cunha and Muniz, 1928, has a slightly flattened body with a sharply tapering, truncate anterior end and a truncate posterior end and is 30–56 × 17–32 μm. The cytopharynx is short and funnel-shaped. The macronucleus is ellipsoidal. The concretion vacuole is small and anterior. There is a small contractile vacuole. It occurs in the cecum and colon of the horse.

Genus *Didesmis* Fiorentini, 1890

The anterior end of the body forms a neck behind the large cytostome. There are cilia at the anterior and posterior ends. The macronucleus is ellipsoidal.

D. ovalis Fiorentini, 1890, is oval or rectangular and slightly flattened, with a blunt anterior end and a tapering posterior one. It is 34–55 × 27–40 μm. The cytostome is at the middle of the anterior end, and the cytopharynx is short and funnel-shaped. There is a short neck beyond the cytostome. The concretion vacuole is near the anterior end of the irregularly oval macronucleus. There are 1 or 2 contractile vacuoles. It occurs in the cecum and the colon of the horse.

D. quadrata Fiorentini, 1890, resembles *D. ovalis* but has a deep, wide, highly refractile, longitudinal groove on the dorsal surface. It is 50–90 × 33–68 μm and has a spindle-shaped macronucleus. It occurs in the cecum and colon of the horse.

D. spiralis Hsiung, 1929, resembles *D. quadrata* except that it is spirally shaped. It is 60–94 × 38–54 μm. The dorsal groove runs slightly diagonally to the longitudinal axis. The concretion vacuole contains less than 10 grains. It occurs in the cecum of the horse.

Genus *Holophryoides* Gassovsky, 1919

The body is ovoid and uniformly ciliated, with a comparatively large cytostome at the anterior end. The macronucleus is small and ellipsoidal. The contractile vacuole is subterminal.

H. ovalis (Fiorentini, 1890) Gassovsky, 1919, is 95–140 × 65–90 μm. There is an accumulation of ectoplasm in the anterior part of the body.

Genus *Paraisotrichopsis* Gassovsky, 1919

The body is uniformly ciliated and has a spiral groove from the anterior to the posterior end.

P. composita Gassovsky, 1919, is 43–56 × 31–40 μm and has an elongate macronucleus.

Genus *Polymorphella* Corliss, 1960

The body is flask-shaped, with cilia in the anterior region and a few at the caudal end. The macronucleus is disk-shaped, the contractile vacuole terminal.

P. ampulla (Dogel', 1929) Corliss, 1960, is 22–36 × 13–21 μm. It occurs in the cecum and colon of the horse.

Genus *Prorodonopsis* Gassovsky, 1919

The body is piriform and uniformly ciliated. The macronucleus is sausage-shaped, and there are 3 contractile vacuoles.

P. coli Gassovsky, 1919, is 55–67 × 38–45 μm.

Genus *Sulcoarcus* Hsiung, 1935

The body is ovoid and compressed, with a short spiral groove at the anterior end. The cytostome is at the end of the groove. The cytopyge is terminal. The concretion vacuole is midventral, with the contractile vacuole posterior to it. Cilia are present on the groove, midventral region, and posterior end.

S. pellucidulus Hsiung, 1935, is 33–56 × 30–40 μm.

Order Trichostomatorida Bütschli, 1889

Members of this order occur in both equids and ruminants, but only one genus, *Charonina,* is found in both, and the species in the two types of hosts are all different. Members of this order have an apical or subapical cytostome and also a vestibulum. The somatic ciliature may or may not be reduced.

Genus *Blepharocorys* Bundle, 1895

There are 3 ciliary zones (oral, dorsal, and ventral) at the anterior end and 1 caudal ciliary zone in this genus. There is a deep oral groove near the anterior end.

B. uncinata (Fiorentini, 1890) Bundle, 1895, is elongated and irregular in shape, with a slightly convex dorsal side, a slightly concave ventral side, and more or less rounded ends; it is 55–74 × 22–30 μm. A corkscrewlike anterior process that makes 2 turns projects from the anterior end and also passes through the body dorsal to the cytopharynx, ending just behind it. There is a large, ciliated vestibule at the anterior end that leads to a cytostome opening into a ciliated cytopharynx extending dorsoposteriad and then bends sharply ventral and disappears at the posterior half of the body. The macronucleus is heart-shaped. There is a single posterior contractile vacuole. It occurs in the cecum and colon of the horse.

B. valvata (Fiorentini, 1890) Bundle, 1895, is more or less elliptical and flattened bilaterally, 52–68 × 20–27 μm. The vestibule is small and has a beaklike dorsal plate. The macronucleus is more or less kidney-shaped. It occurs in the cecum and colon of the horse.

B. jubata Bundle, 1895, resembles *B. valvata,* but the dorsal plate guarding the vestibule has 2 teeth. It is 33–60 × 17–23 μm. The cytopharynx extends backward and upward and then again turns backward. The macronucleus is more or less ovoid. It occurs in the cecum and colon of the horse.

B. curvigula Gassovsky, 1919, also resembles *B. valvata,* but its dorsal plate is more or less rhomboid. The long cytopharynx extends backward and upward and finally bends in a smooth 180° curve. The macronucleus is more or less ovoid. It occurs in the colon of the horse.

B. angusta Gassovsky, 1919, resembles *B. valvata,* but is more elongate, 58–78 × 20–25 μm. The dorsal plate is more or less rhomboid, the macronucleus irregular. It occurs in the colon of the horse.

B. cardionucleata Hsiung, 1930, resembles *B. curvigula,* but its macronucleus is heart-shaped with an anterior base and a posterior apex. It is 48–62 × 17–23 μm. It occurs in the colon of the horse.

Genus *Charonina* Strand, 1928

There are 2 caudal and 3 anterior ciliary zones, and an anterior knob is present on the body.

C. equi (Hsiung, 1930) is lanceolate, 30–48 × 10–14 μm. The cytostome occupies nearly the whole ventral side of the anterior knob and leads to a prominent cytopharynx that extends straight down to the middle third of the body. The macronucleus is large and elongate. It occurs in the colon of the horse.

Genus *Ochoterenaia* Chavarria, 1933

There are three ciliary zones at the anterior end and two at the posterior end. One of the latter is borne on a caudal appendage that arises ventral to the cytopyge. There is a beaklike dorsal plate like that of *Blepharocorys.*

O. appendiculata Chavarria, 1933, is more or less elliptical and is flattened bilaterally. It is 58–72 × 24–33 (mean 66 × 28) μm. The vestibule is prominent, the macronucleus more or less kidney-shaped. It occurs in the rectum of the horse.

Genus *Paraisotricha* Fiorentini, 1890

The cilia form more or less spiral longitudinal rows. The contractile vacuole is posterior.

P. colpoidea Fiorentini, 1890, is ovoid, 70–100 × 42–60 μm, and has 34–40 rows of cilia. The macronucleus is a thick, ellipsoidal disk. There is a large concretion vacuole at the anterior end. It occurs in the cecum and colon of the horse.

P. beckeri Hsiung, 1930, resembles *P. colpoidea* but has only 11 rows of cilia. It is 52–98 × 30–52 μm. It occurs in the cecum and colon of the horse.

P. minuta Hsiung, 1930, resembles *P. colpoidea* but has only 20 rows of cilia and is 38–68 × 27–36 μm. It occurs in the cecum and colon of the horse.

Order Suctoriorida Claparède and Lachmann, 1858

Members of this order have suctorial tentacles.

Genus *Allantosoma* Gassovsky, 1919

The body of this suctorian is elongate with 1 or more tentacles at each end but without a lorica or stalk. The macronucleus is ovoid or spherical, the micronucleus compact. There is 1 contractile vacuole. The cytoplasm is often filled with small spheroidal bodies.

A. intestinalis Gassovsky, 1919, has a sausage-shaped body with 3–12 tentacles at each end bearing distinct suckers. It is 33–60 × 18–37 μm. The cytoplasm is filled with small, round bodies. The macronucleus is more or less spherical. It occurs in the cecum and colon of the horse.

A. dicorniger Hsiung, 1928, has a more or less cycloid body with 1 incurved tentacle at each end, 20–33 × 10–20 μm. The end of the tentacle is somewhat boot-shaped. The cytoplasm is filled with granules, the macronucleus subspherical. It occurs in the colon of the horse.

A. brevicorniger Hsiung, 1928, has an elongate, cycloid body with one short, slender, slightly incurved tentacle at each end. It is 23–36 × 7–11 μm. The distal end of the tentacle is rounded, the cytoplasm slightly granular. It occurs in the cecum of the horse.

Order Entodiniomorphidorida Reichenow, 1929

Simple somatic ciliature is absent, replaced by membranellar tufts or zones of cilia. The pellicle is firm and sometimes drawn out into spines. All members of this order in equids belong to the family Cycloposthiidae, in which the adoral ciliature is retractable and the caudal ciliature is nonretractable. There are 2 or more metoral bands of membranelles in addition to the adoral zone. Some genera have skeletal plates.

Genus *Cycloposthium* Bundle, 1895

The body is large and elongate barrel-shaped. The cytostome is in the center of a retractile, conical elevation at the anterior end. The adoral zone of membranelles is conspicuous. On the dorsal and ventral sides there are open ring zones of membranelles near the posterior end. There are 3 rows of kinetosomes, not sharply delimited, of cilia independent of the AZM. The pellicle is ridged, there is a club-shaped skeletal plate, and a row of several contractile vacuoles runs along the band-formed macronucleus.

C. bipalmatum (Fiorentini, 1890) Bundle, 1895, is more or less rectangular, slightly compressed laterally, with a truncate anterior end and a tapering posterior end with a taillike structure. It is 80–127 × 35–57 μm. A longitudinal groove and a light, linear skeletal plate are present on the left side. The macronucleus is hooked anteriorly, the micronucleus located near its middle. There are 4 contractile vacuoles. It occurs in the cecum and colon of the horse.

C. dentiferum Gassovsky, 1919, is 140–220 × 30–110 μm. It resembles *C. bipalmatum* but has a ventral dentiform projection, and the anterior end of the macronucleus is not hooked. The cuticle is not corrugated. A longtitudinal groove is present on the left side, but the linear skeletal plate is quite indistinct. There are 4–6 contractile vacuoles. It occurs in the cecum and colon of the horse.

C. ishikawai Gassovsky, 1919, differs from all other species of the genus in that the posterior arches of the membranelles are nonretractile. It is 230–280 × 110–130 μm.

C. edentatum Strelkov, 1928, resembles *C. bipalmatum* but has 6–7 contractile vacuoles. It is 146–230 × 68–93 μm. It occurs in the cecum and colon of the horse.

C. piscicauda Strelkov, 1928, resembles *C. bipalmatum* but lacks both the longitudinal groove and skeletal plate on the left side. It is 125–190 × 44–80 μm and has 4 or 5 μm contractile vacuoles. Its posterior end forms a tail resembling that of a fish.

C. scutigerum Strelkov, 1928, differs from *C. bipalmatum* in having a shieldlike skeletal plate interrupted by 2 longitudinal grooves on the left side instead of a simple, narrow plate. It is 132–210 × 63–90 μm and has 5 or 6 contractile vacuoles. It occurs in the cecum and colon of the horse.

C. affinae Strelkov, 1928, differs from *C. bipalmatum* in that it has a heavy skeletal plate and that the micronucleus is near the anterior end of the macronucleus. It is 92–141 × 45–58 μm. It occurs in the cecum and colon of the horse.

C. corrugatum Hsiung, 1920, is 135–195 × 70–112 μm. It has a ventral dentiform projection, and its cuticle is corrugated. The anterior end of its macronucleus is not hooked. The linear skeletal plate is quite indistinct, and there are 4 or 5 contractile vacuoles. It occurs in the cecum and colon of the horse.

Genus *Spirodinium* Fiorentini, 1890

The body is elongate and more or less fusiform, with an adoral zone of membranelles at the anterior end. An anterior ciliary zone encircles the body at least once, and a posterior ciliary arch spirals halfway around the body. There is a dorsal cavity of unknown function lined with stiff rods.

S. equi Fiorentini, 1890, is 77–180 × 30–74 μm. Its macronucleus is elongated, with rounded ends, and there is a large contractile vacuole just back of the anterior membranelles. It occurs in the colon of the horse.

Genus *Triadinium* Fiorentini, 1890

The body is more or less helmet-shaped and compressed, with an adoral zone of membranelles at the anterior end. There are ventral and dorsal posterior zones of membranelles; there may or may not be a caudal projection.

T. caudatum Fiorentini, 1890, is 50–105 × 36–85 μm, with a long, slender tail. The macronucleus is bent like a question mark, and there is a single contractile vacuole. It occurs in the colon of the horse.

T. galea Gassovsky, 1919, is 58–88 × 50–70 μm and lacks a tail. It has a long macronucleus running longitudinally along the left surface and 2 contractile vacuoles. It occurs in the colon of the horse.

T. minimum Gassovsky, 1919, is 32–50 × 31–42 μm and has a slender tail. The macronucleus is ellipsoidal. There is a single contractile vacuole. It occurs in the colon of the horse.

Genus *Tetratoxum* Gassovsky, 1919

The body is slightly compressed and has 2 anterior and 2 posterior zones of membranelles.

T. unifasciculatum (Fiorentini, 1890) Gassovsky, 1919, is 104–168 × 62–100 μm. It is irregularly elliptical with both ends rounded and has 7–9 longitudinal, cuticular ridges on both the dorsal and ventral surfaces of the body. Lateral cuticular extensions at the posterior end form 2 caudal sheaths. The macronucleus is elongate with a short hook at the anterior end. There is a large contractile vacuole under its curvature. It occurs in the colon of the horse.

T. excavatum Hsiung, 1930, is 95–135 × 55–90 μm. It differs from *T. unifasciculatum* in having a deep elliptical excavation covered by a flap of cuticle at its anterior end, and its cuticular ridges are more prominent and the adjacent ones farther apart. It occurs in the colon of the horse.

T. parvum Hsiung, 1930, is 67–98 × 39–52 μm. It differs from the other two species in lacking longitudinal cuticular ridges. It occurs in the colon of the horse.

Genus *Tripalmaria* Gassovsky, 1919

There is an adoral zone of membranelles at the anterior end and also 2 dorsal and 1 ventroposterior tuft-formed zones of membranelles. The macronucleus is shaped like an inverted U. A synonym of this genus is *Tricaudalia* Buisson, 1923.

T. dogieli Gassovsky, 1919, is 77–123 × 46–62 μm. Beneath the right side it has skeletal plates forming a horseshoe with its open end directed posteriad. It occurs in the colon of the horse.

Genus *Cochliatoxum* Gassovsky, 1919

There is an adoral zone of membranelles at the anterior end and also 1 anterodorsal, 1 posterodorsal, and 1 posteroventral zone of membranelles. The anterior end of the macronucleus is curved.

C. periachtum Gassovsky, 1919, is more or less cylindrical, with both ends rounded, 210–370 × 130–210 μm. There is a contractile vacuole. It occurs in the colon of the horse.

Genus *Ditoxum* Gassovsky, 1919

There is a large adoral zone of membranelles near the anterior end and also anterodorsal and posterodorsal zones of membranelles. The macronucleus is curved and club-shaped.

D. funinucleum Gassovsky, 1919, is elliptical with both ends rounded, slightly flattened bilaterally, 135–203 × 70–101 μm. It has a single contractile vacuole. It occurs in the colon of the horse.

OTHER CILIATES

Genus *Balantidium* Claparède and Lachmann, 1858

The body is ovoid, ellipsoidal to subcylindrical. The macronucleus is elongated. There is a single micronucleus. The contractile vacuole and cytopyge are terminal.

Many species of *Balantidium* have been named, based on the host in which they occur and on the size and shape of their body and macronucleus. However, many of these are probably not valid.

Balantidium coli (Malmsten, 1857) Stein, 1862

SYNONYMS. *B. aragaoi, B. caviae, B. cunhamunizi, B. philippinensis, B. rhesum, B. simile, B. suis, B. wenrichi.*

This species occurs worldwide in the cecum and colon of the pig, peccary, man, chimpanzee, orangutan, rhesus monkey, cynomolgus monkey, other macaques, guinea pig, and (rarely) dog and rat. It is extremely common in swine, and much less common in man and other primates. The *Balantidium coli* reported from the ox, zebu, water buffalo, and dromedary is more probably a late exconjugant of *Buxtonella sulcata.*

STRUCTURE. The trophozoites are ovoid, 30–150 × 25–120 μm, with a subterminal cytostome at the smaller end. The cytopyge is near the posterior end. The macronucleus is sausage- or kidney-shaped, and the micronucleus lies near the center of one side. There are 2 contractile vacuoles, one terminal and the other near the center of the body. There are many food vacuoles containing starch grains, cell fragments, bacteria, erythrocytes, etc.; starch is the most important food. The body surface is covered by slightly oblique longitudinal rows of cilia. The number of ciliary rows (kineties) varies widely; Krascheninnikow (1962) found 36–106 (mean 71) in 406 individuals from a single pig.

The cysts are spherical to ovoid, 40–60 μm in diameter. They are slightly yellowish or greenish, with hyaline cytoplasm. The cyst wall is composed of 2 membranes. (See Fig. 13.5.)

LIFE CYCLE. *B. coli* reproduces by transverse binary fission (Krascheninnikow and Wenrich 1958). Conjugation also takes place, and resistant cysts are formed.

PATHOGENESIS. In the pig *B. coli* is ordinarily a commensal in the lumen of the large intestine, where it lives on starch, other ingesta, and bacteria. It does not seem able to penetrate the intact intestinal mucosa by itself. Enormous numbers of *Balantidium* may be found in the lumen of the cecum of pigs with normal cecal mucosae. However, once some other organism or condition has initiated a lesion, *Balantidium* may be a secondary invader and may be found

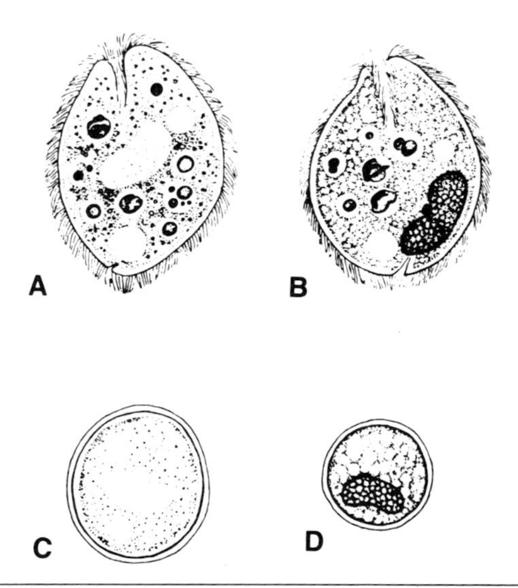

Fig. 13.5. *Balantidium coli:* A. living trophozoite; B. stained trophozoite; C. fresh cyst; D. stained cyst. (×450) (From Kudo, R. R., *Protozoology,* 4th ed., 1954, courtesy of Charles C Thomas, Publisher, Springfield, Ill.)

deep in the ulcer. It produces hyaluronidase, which might help it to enlarge the lesions by attacking the ground substance between the cells (although this enzyme would not help to initiate the lesions).

Balantidium may be pathogenic in man and other primates. It causes diarrhea or dysentery and produces undermining lesions similar to those caused by *Entamoeba histolytica.* The protozoa may be found down to the muscularis mucosae, the ulcers are infiltrated with round cells, and coagulation necrosis and hemorrhagic areas may be present. The protozoa often occur in nests within the tissues or even in the capillaries, lymph ducts, and neighboring lymph nodes. The lesions in the pig and other animals are similar.

B. coli may cause dysentery and intestinal ulceration in the dog (Dikmans 1948; Bailey and Williams 1949).

EPIDEMIOLOGY. *Balantidium* may be transmitted by ingestion of either cysts or trophozoites. The cysts are more resistant to environmental conditions and may remain alive for weeks in pig feces if they do not dry out. The pig is the usual source of infection for man and dogs. Chimpanzees and other primates appear to have their own infection pool.

DIAGNOSIS. *Balantidium* can be easily recognized by microscopic examination of intestinal contents or by histologic examination of intestinal lesions.

CULTIVATION. *B. coli* was first cultivated by Barret and Yarbrough (1922) and has since been cultivated by many workers (Klaas 1974).

TREATMENT. No treatment is necessary in swine. Carbarsone, diodoquin, chlortetracycline, and oxytetracycline have been used in man. Benson, Fremming, and Young (1955) treated chimpanzees with 250 mg carbarsone daily for 10 days, concealing the drug in fruit or fruit juices.

PREVENTION AND CONTROL. Sanitary measures designed to prevent ingestion of cysts or feces should prevent balantidial infections.

Genus *Tetrahymena* Furgason, 1940

The body is piriform and uniformly ciliated with 17 to 42 rows of cilia. The piriform cytostome is near the anterior end; the buccal ciliature consists of 3 membranelles, which lie to the left in the buccal cavity, and a 4th, paroral membrane extending along its right margin. There is a single contractile vacuole.

T. pyriformis (Ehrenberg, 1830) Lwoff, 1947 (syn., *T. geleii*), is 40–60 × 15–30 μm. It is extremely popular in protozoological research, and well over 1,000 papers have been written on it. Although it is normally free-living, it may on rare occasions be a facultative parasite. Knight and McDougle (1944) found it in the digestive tract, infraorbital sinuses, and serous material under the eyelids of chickens in Missouri. It was found only in birds with a vitamin A deficiency. Thompson (1958) infected chicken embryos with *T. pyriformis, T. corlissi,* and *T. vorax.*

Coprophilic Ciliates

A number of ciliates that live in water or soil may contaminate feces and develop coprophilically. They are common in old feces, especially if ever in contact with the ground, but may also appear in feces taken directly from an animal. Cysts ingested by livestock in feeding or drinking may pass through the intestinal tract unharmed, and trophozoites may emerge and develop as the feces stand. Horse and ruminant feces that have been cultured for nematode larvae often contain large numbers of small ciliates, probably including *Colpidium, Chilodonella,* or *Cyclidium.*

Nyctotherus faba Schaudinn, 1899, has been found in human feces on occasion. It belongs to the family Nyctotheridae. Its body is reniform, covered with cilia, and 26–28 μm long. The peristome begins at the anterior end, turns slightly to the right, and ends in the cytostome at the middle of the body. The cytopharynx is a long tube containing an undulating membrane. The macronucleus is about the middle of the body and is spherical. Its chromatin is arranged in 4 or 5 large, solid bodies on the nuclear membrane, while the remainder of the nucleus is chromatin-free.

Noble (1958) found that a *Nyctotherus*-like ciliate about 15–30 μm long appeared in fecal samples from Wyoming sheep and elk after storage at 4 C for about 30 days. A small ciliate about 10–12 μm long also appeared in the elk feces at about the same time. The small ciliates persisted for a few weeks, the *Nyctotherus*-like ones for about twice as long.

Balantiophorus minutus (Schewiakoff, 1893) (syn., *Balantidium minutum*) occurs occasionally in contaminated human feces. It belongs to the family Pleuronemidae. It is ovoid, with the narrow end anterior and bent ventrad, giving the ventral surface a hollowed appearance. It is 12–54 (usually 24–45) × 7–33 μm. The peristome is in the middle of the anterior half of the body. The adoral zone of membranelles on its left, posterior, and right borders forms a saclike structure that is conspicuous when expanded but that can be retracted into the peristome and become invisible. The cytopharynx is funnel-shaped. The body is uniformly covered by 12 rows of setiform cilia, of which only 6 extend anterior to the peristome. The macronucleus is central and ellipsoidal. There is a posterior contractile vacuole.

The taxonomy and bionomics of these and other coprophilic protozoa have been reviewed by Alexeieff (1929) and Watson (1946). The latter listed 51 species of flagellates, 18 of amebae, and 18 of ciliates that have been found in feces; many of these need further study.

Adventitious Ciliates

Free-living ciliates may occasionally be found as commensals or contaminants in various locations in the body. For instance, it is not uncommon to find them in cultures made from washings from the sheath cavity of bulls examined for trichomonads.

LITERATURE CITED

For citations before 1965, see the *Index-Catalogue of Medical and Veterinary Zoology, Zoological Record, Biological Abstracts, Protozoological Abstracts,* and *Veterinary Bulletin*.

Canella, M. F. 1977. Commentaires sur Quelques Analyses pro et contra Personalia I. Editografica-Rastignano. Bologna, Italy.

Coleman, G. S. 1975. In Symbiosis, ed. D. H. Jennings and D. L. Lee. Cambridge, Eng.: Cambridge University Press, 533–58.

______. 1978. In Methods of Cultivating Parasites in Vitro, ed. A. E. R. Taylor and J. R. Baker. New York: Academic, 39–54.

Corliss, J. O. 1977. Trans. Am. Microsc. Soc. 96:104–40.

______. 1979. The Ciliated Protozoa. 2d ed. New York: Pergamon.

David, A., and M. C. John. 1966. Indian Vet. J. 43:269–70.

Dehority, B. A. 1974. J. Protozool. 21:26–32.

______. 1978. J. Protozool. 25:509–13.

______. 1979. J. Protozool. 26:536–44.

Grain, J. 1966. Protistologica 2(1):59–141, 2(2):5–52.

Klaas, J., II. 1974. J. Parasitol. 60:907–10.

Krascheninnikow, S. 1968. Ann. Ukr. Acad. Arts Sci. 11:69–90.

Levine, N. D., J. O. Corliss, F. E. G. Cox, G. Deroux, J. Grain, B. M. Honigberg, G. F. Leedale, A. R. Loeblich III, J. Lom, D. Lynn, E. G. Merinfeld, F. C. Page, G. Poljansky, V. Sprague, J. Vavra, and F. G. Wallace. 1980. J. Protozool. 27:37–58.

Lom, J., and P. Didier. 1979. Proc. 5th Int. Congr. Protozool., 51–58.

Ogimoto, K., and S. Imai. 1981. Atlas of Rumen Microbiology. Tokyo: Japanese Scientific Society Press.

Vasily, D. B., and J. B. Mitchell. 1974. Trans. Am. Microsc. Soc. 93:248–53.

Yankovskii, A. V. 1980. Trudy Zool. Inst. Akad. Nauk SSSR 94:103–21.

14

Laboratory Diagnosis of Protozoan Infections

MANY different techniques have been used for the laboratory diagnosis of protozoan infections and for the study of parasitic protozoa. Only the commonest and those that have been found most useful in my laboratory are given here. Other routine and specialized techniques are given in various textbooks of human parasitology and by Ivens et al. (1978).

Some of these techniques are useful not only for protozoa but also for helminth eggs or larvae. If so, their value for these purposes is mentioned.

MICROSCOPIC EXAMINATION

Direct Microscopic Examination of Wet Fecal Smears

Place a drop of physiologic NaCl solution on a microscope slide. Take up a small amount of feces on the end of a toothpick and mix thoroughly with the salt solution. Do not make too heavy a suspension, or it will be impossible to see objects clearly under the microscope. An emulsion through which newsprint can be read is about right. Place a coverslip on the drop. Examine under the low and high dry powers of the microscope. Some experts even like to use an oil immersion objective to see details such as chromatoid bars.

Flagellates and ciliates can be seen moving about actively. Trichomonads have a characteristic spiral, jerky movement, *Spironucleus* a sort of vibrating, forward one, *Chilomastix* a relatively smooth, progressive one, and most ciliates a characteristic smooth, rapid one. Amebae may move sluggishly or may remain still. Oocysts of coccidia and helminth eggs can be recognized from their shape and size. Many other objects will be seen, some of which may be mistaken for protozoan parasites. These include bacteria, yeasts, fungus spores, the fungus *Blastocystis,* pollen grains, undigested food particles such as starch grains and plant fibers, and ingested pseudoparasites such as grain mites, coccidian oocysts, or nematode eggs in animals that have been eaten by or have defecated in the feed of the animals under examination. In cases of enteritis, red or white blood cells or epithelial cells may be present.

In examining preparations under the microscope, move the slide systematically back and forth or up and down in order to bring every part of the preparation into view.

IODINE STAINING. In order to bring out certain details that are not visible in the living protozoon, wet smears may be stained with iodine by mixing a fecal suspension slightly heavier than that described above with an equal amount of D'Antoni's aqueous iodine solution or of one part Lugol's solution diluted with four parts distilled water.

Direct Microscopic Examination of Intestinal Mucosa

This technique, which can be used only on animals that have been killed and had their intestinal tracts opened, permits a greater amount of material to be examined on a single slide than does the direct examination of diluted feces. It can be used to find the intracellular and extracellular stages of coccidia, other protozoa, small nematodes such as *Strongyloides* and *Capillaria,* small trematodes, cestodes, or cestode scolices, and schistosome eggs.

Make a deep scraping of the suspected intestinal mucosa with a scalpel, toothpick, or similar instrument, or even with the end of a slide. Place the material thus obtained on a microscope slide and cover with a coverslip. Press the coverslip down if necessary to flatten out the preparation and make it thin enough to see through.

To search for *Trichomonas, Giardia, Spironucleus,* and other motile organisms, mix a little physiologic NaCl solution with the scraping before placing the coverslip on it.

Microscopic Diagnosis of *Tritrichomonas foetus* Infections

In heavy infections of female cattle, *T. foetus* can be found by direct microscopic examination of mucus or exudate from the vagina or uterus. In aborted fetuses it can be found in the amniotic or allantoic fluid, fetal membranes, placenta, fetus stomach contents, oral fluid, or other fetal tissues; it occurs most commonly in the stomach contents and the material around the base of the tongue. In bulls, it can be found in the sheath cavity.

Clean the external genitalia thoroughly before taking samples in order to avoid contamination with intestinal or coprophilic protozoa. Take samples from the vagina by introducing about 10 ml physiologic NaCl solution with a bulbed dose syringe and washing it back and forth several times by squeezing the bulb repeatedly. Take samples from the preputial cavity of bulls in the same way, using a long, bulbed pipette or syringe, or introduce a cotton swab into the cavity and rub it around to obtain a sample of exudate; in the latter case, wash off the swab in physiologic NaCl solution.

Allow the washings to stand 1–3 hr, or centrifuge them, before examination. Place a drop of the sediment on a slide, cover with a coverslip, and examine under the microscope.

If trichomonads cannot be found on direct microscopic examination, inoculate some of the washings into CPLM, BGPS, or Diamond's medium and examine after 1, 2, and 4 days' incubation at 37 C.

Sporulation of Coccidian Oocysts

In order to identify coccidia, it is often necessary to allow the oocysts to sporulate (i.e., develop to the infective stage). To permit this, mix feces containing the coccidia with several volumes of 2.5% potassium bichromate solution and place the mixture in a thin layer in a petri dish. The potassium bichromate prevents bacteria from destroying the oocysts. Oxygen is necessary for the oocysts to develop, so the layer of fluid should never be more than a few millimeters thick. In most species, sporocysts and sporozoites form in a few days, but it is well to allow development to proceed for a week (or, for a few species, even longer). If the sporulated oocysts will not be studied immediately, the fecal suspension can be transferred to a bottle and stored in the refrigerator. The oocysts will remain alive for several months or, in some species, as long as a year.

It is best to sporulate coccidian oocysts before they have been subjected to refrigeration, since in some species (apparently a minority), refrigeration of the unsporulated oocysts prevents subsequent sporulation (although it does not harm sporulated oocysts).

MIF (Merthiolate [Thiomersal]-Iodine-Formaldehyde) Stain-Preservation Technique

This technique was introduced by Sapero, Lawless, and Strome (1951) and Sapero and Lawless (1953). It was designed especially to permit identification of human protozoan trophozoites and cysts but can also be used for helminth eggs and for parasites of domestic animals. It is simple and relatively cheap, permits rapid (almost immediate) wet-fixed staining of the smears, and preserves the parasites so that feces can be collected in the field or by untrained persons and shipped to the laboratory for later diagnosis. There is no appreciable loss or deterioration of parasites or cellular exudates for 6 mo or more.

Direct Examination Technique for Fresh Fecal Specimens

1. The MIF stain is composed of tincture of 1:1000 merthiolate No. 99 (Lilly), Lugol's solution (5%), and 40% formaldehyde solution (USP). Since Lugol's solution is unstable, it should be freshly prepared every 3 wk; the amount used should be varied with its age.

   The following amounts (in ml) are recommended:

	1st wk	2nd wk	3rd wk
Lugol's solution	10.0	12.5	15.0
Formaldehyde solution	12.5	12.5	12.5
Tincture of merthiolate (thiomersal)	77.5	75.0	72.5

2. Place 1 ml stain (sufficient for 25 to 30 fecal smears) in a Kahn tube. Place some distilled water in a second tube. Put a small-caliber medicine dropper in each tube.
3. Place 1 drop distilled water at one end of a slide. Add 1 drop MIF stain. Mix.

4. Add a small amount of feces and mix (do not use too much feces, or fixation and staining will be poor). The finished wet smear should be thin enough so that the slide can be tipped on edge without the coverslip sliding.
5. Add a coverslip and examine at once. If it is desired to examine the slide later, ring it with petrolatum to keep the preparation from drying out.

Collection and Preservation of Fecal Specimens in the Field for Subsequent Laboratory Examination

1. Prepare the following stock MIF solution:

Tincture of merthiolate	200 ml
Formaldehyde solution	25 ml
Glycerol	5 ml
Distilled water	250 ml

   Store in a brown bottle.
2. Measure 2.35 ml MIF solution into a standard Kahn tube and stopper with a cork.
3. Measure 0.15 ml 5% Lugol's solution into another Kahn tube and close with a rubber stopper (or keep the Lugol's solution in a bottle and add the proper amount to the MIF solution just before adding the feces in step 4).
4. At the time the fecal sample is collected, pour the MIF solution into the Lugol's solution. *Within a few seconds,* add an amount of feces equal to two medium-sized peas (about 0.25 g) and mix thoroughly with an applicator stick. Do not use too much feces. Stopper the tube and set aside for future examination.
5. To examine, draw off 1 drop of mixed supernatant fluid and feces from the top of the sedimented layer with a medicine dropper and place on a slide. Mix thoroughly, crushing any large particles. Add a coverslip and examine.

PERMANENT FIXING AND STAINING TECHNIQUES

It is often desirable to make permanent preparations of fecal smears or to make hematoxylin-stained slides for detailed study. For this purpose, smears must first be fixed (i.e., the protozoa must be killed by the action of a chemical or a mixture of chemicals that will preserve their structures as nearly as possible in the same form as in life).

Many different techniques are used for fixing and staining tissues, cells, and small organisms. Those given below are especially suitable for protozoa. The standard hematoxylin and eosin stain used routinely for tissue sections is also valuable for protozoa in tissues, but it is so well known that it is not described here. For further information on fixing, sectioning, staining, and mounting techniques, any text or reference book on microscopic techniques may be consulted.

Fixation

Schaudinn's fluid is probably the best all-around fixative for intestinal protozoa and also serves well for other forms. Smears may be made on slides and stained in Coplin jars, or they may be made on coverslips and stained in Columbia jars. The latter method has advantages: smaller amounts of reagents are needed, a neater preparation is obtained (since there is no possibility of a portion of the smear extending beyond the coverslip), and in the completed slide the mounting medium is beneath the smear rather than above it, so the microscope objective can come closer to the smear. The last factor may be of importance when the oil immersion objective is used. Coverslips are fragile, however, and greater care must be exercised in handling them than in handling slides.

Clean the coverslips (or slides) by dipping them in 95% alcohol, and dry them with a clean cloth before use. Coverslips must be handled only by the edges in order to avoid leaving fingerprints.

Place a tiny drop of albumen fixative in the center of the coverslip (or slide), and smear it over the surface with the little finger. (The finger should have previously been cleaned and rid of its oil by dipping it in 95% alcohol and wiping it with a clean cloth). Albumen fixative is used to make the feces adhere to the glass.

Take up a small amount of feces on a toothpick (preferably a round, smooth one) and spread as evenly as possible in a very thin layer over the surface of the coverslip. Do not allow it to dry; drop it immediately into a Columbia jar containing Schaudinn's fluid at room temperature or 37 C. Allow it to remain about 10 min and then transfer to 70% alcohol.

In some cases it may be necessary to mix the feces with a little physiologic NaCl solution in order to make them thin enough to spread well. In other cases the feces are so fluid that if the coverslips are dropped edgewise into the fixative, all the material will come off. To prevent this, put the fixing solution in a small, flat vessel such as a petri dish, and place the coverslip face down on its surface. After a few seconds it can be transferred to a Columbia jar.

After fixation, wash the smear in two changes of 70% alcohol for at least 5 min each. Then transfer to 70% alcohol containing enough iodine to give it a port wine color. Allow to remain at least 10 min (preferably longer). This treatment takes out the excess mercuric chloride, which may otherwise form crystals in the preparation. Then transfer to fresh 70% alcohol. Fixed material may be kept in 70% alcohol indefinitely without injury.

Staining with Heidenhain's Hematoxylin

In order to bring out many of the structures of organisms it is necessary to color them with a dye or dyes. The best and most commonly used dye employed in parasitologic and histologic work is hematoxylin, which is extracted from logwood. Hematoxylin alone has very poor staining properties, and a mordant must be employed to make it effective. Many different formulas have been used for hematoxylin staining solutions. In some, the mordant is mixed with the hematoxylin, while in others it is used separately. Many different compounds are used as mordants, the great majority being salts of heavy metals such as iron, lead, copper, cobalt, tungsten, or molybdenum. One of the best hematoxylin stains is Heidenhain's iron-hematoxylin. A modification of this technique is given below. Starting with the smears in 70% alcohol after passing through iodine, the staining schedule is as follows:

1. 50% alcohol—5 min
2. 30% alcohol—5 min
3. Distilled water—5 min
4. 2% aqueous iron alum (made from violet crystals only)—1 hr
5. Distilled water—1 min
6. 0.5% aqueous hematoxylin—2 hr
7. Distilled water—Rinse
8. Saturated aqueous picric acid—Destain until the structures assume the proper intensity of color. (This process should be controlled by microscopic examination at intervals. For intestinal amebae, 10 min is usually enough; a longer time is needed for large protozoa such as *Balantidium.*)
9. Distilled water—Rinse (2 changes)
10. Tap water—Until all picric acid has come out of the smear. (Change the water at intervals.)
11. 30% alcohol—5 min
12. 50% alcohol—5 min
13. 70% alcohol—5 min

Gradual changes in alcohol concentration are used in all staining and dehydration procedures to avoid distortion of tissues. Hematoxylin-stained smears and sections can be kept in 70% alcohol indefinitely.

In the classical Heidenhain's hematoxylin staining procedure, the stained smears are destained with iron alum. In the above procedure, saturated aqueous picric acid is used instead; this requires a minimum of observation (usually none) during the destaining process, and the resultant stain is dark blue instead of brownish black (as with iron alum).

If desired, longer mordanting and staining times can be used. The smears can be mordanted for 2 hr and stained for 4 hr, or they can be mordanted for 4 hr or overnight and stained overnight or for 24 hr. These longer times give a little more precise staining, but not enough to make them worthwhile for routine purposes.

Counterstaining

If desired, the smears can be counterstained with eosin Y. However, this has a tendency to obscure fine nuclear detail somewhat. To counterstain the smears, transfer them from 70% alcohol to 0.5% eosin Y in 90% alcohol. The pH of this solution should be brought to 5.4–5.6 by adding 4.0 ml of 0.1 N HCl per 100 ml. The acidified solution will not keep more than 10 days to 2 wk. After that its pH will become too high for satisfactory use. Stain for 45 sec to 3 min. Transfer to 95% alcohol to wash out excess dye and then proceed as directed below (eosin is soluble in alcohol).

Mounting

Permanent slides are mounted in a medium that is quite fluid at first but later becomes hard. Most mounting media are immiscible with water, and many with alcohol. Hence, before mounting, all water and alcohol must be removed from the smears. This cannot be done simply by allowing the smears to dry, for such dehydration in air would ruin the preparations by distorting the protozoa. Mounting media that have been employed include natural resins, such as Canada balsam and damar, and synthetic resins, such as euparal, naphrax, permount, and clarite.

Starting with stained coverslips in 70% alcohol, pass them through the following solutions:

1. 95% alcohol—5 min
2. 100% alcohol—5 min
3. 100% alcohol—5 min
4. Toluene—5 min
5. Toluene—5 min

Mount in permount or another resin: (1) place a drop of the resin on a clean slide; (2) place the coverslip slantingly, smear side down, alongside the drop; and (3) gently *lay* it down on the drop, taking care to prevent air bubbles from forming.

Neutral xylene may be used in place of toluene, although it hardens the tissues more. Neutralize the xylene (and balsam, if used) by placing marble chips in the container. If this is not done, the stains will fade after months to years.

Feulgen Stain

The Feulgen nucleal stain, which is used for the detection of deoxyribonucleic acid (DNA), is essentially a modification of the Schiff reaction for aldehydes. When DNA is hydrolyzed by hydrochloric acid, aldehydelike substances are formed that, when treated with colorless fuchsin sulfite, stain purplish red. Whether the reaction is limited to DNA is doubtful, but at any rate, when properly carried out, the Feulgen technique produces a preparation in which only chromatin is stained.

Not all samples of basic fuchsin are satisfactory for the Feulgen stain. Hence, care must be taken to use dye from a batch that has been found satisfactory and certified as such by the Biological Stain Commission.

1. Fix material to be stained by this method for 24 hr in a saturated solution of mercuric chloride containing 2% acetic acid.
2. Wash in running water, and pass through 30%, 50%, and 70% alcohol. *Do not treat with iodine.*
3. Cut and mount sections in the usual manner (if tissue blocks have been

used); run them down to 70% alcohol in the usual manner.

4. Before staining, leave smears and sections in 95% alcohol 48 hr to remove "plasmalogen" substances that may take the stain.
5. To stain, run down through the series of alcohols to distilled water, and place the smears or sections in 1 N HCl at 60 C for 4 min.
6. Wash in in cold 1 N HCl; then rinse with distilled water.
7. Transfer to the decolorized fuchsin solution (see below) and stain 1–3 hr.
8. Wash thoroughly in water containing a little sodium bisulfite plus a few drops HCl.
9. Wash in distilled water.
10. Dehydrate by passing up through the series of alcohols, clear, and mount in permount or another resin.

Bodian Silver Impregnation Technique

This method is superior to ordinary stains for flagella and other diagnostic structures of flagellates. The technique given below is essentially that described by Honigberg (1947). Not all batches of protargol are equally good for this stain; care must be taken to use a sample that has been tested and found satisfactory.

1. Fix in Hollande's or Bouin's solution for 10 min.
2. Wash in in 50% alcohol.
3. Transfer to 30% alcohol and then to distilled water.
4. Bleach in 0.5% aqueous potassium permanganate solution for 5 min.
5. Wash in distilled water.
6. Bleach in 5% aqueous oxalic acid solution for 5 min.
7. Wash several times in distilled water.
8. Place in freshly prepared 1% aqueous protargol solution. (To prepare this solution, place the proper amount of distilled water in a beaker and scatter the protargol powder on its surface; do not stir, heat, or disturb the vessel until the protargol has dissolved.)
9. Keep copper wire or thin copper sheeting in the vessel throughout the staining process. Use 5 g copper per 100 ml protargol solution. Columbia jars contain 10 ml solution. If they are used, it is convenient to place a coil of copper wire weighing 0.5 g in the bottom of each jar before adding the protargol.
10. Stain 1–2 days at room temperature or 37 C in the protargol-copper solution. The staining time and temperature depend on the material being stained and the final intensity desired. If staining is continued for more than 1 day, transfer to fresh protargol solution containing fresh copper for the second day.
11. Wash in distilled water.
12. Place in a solution of 1% hydroquinone in 5% aqueous sodium sulfite for 5–10 min to reduce the silver.
13. Wash several times in distilled water.
14. Place in 1% (or more dilute) aqueous gold chloride solution for 4–5 min.

15. Wash in distilled water.
16. Place in 2% aqueous oxalic acid solution for 2–5 minutes until a purplish color appears.
17. Wash several times in distilled water.
18. Place in 5% sodium thiosulfate solution for 5-10 min.
19. Wash several times in distilled water.
20. Pass up through a graded series of alcohols to dehydrate, clear in toluene or xylene, and mount in permount or balsam.

Giemsa Stain for Tissue Sections

The following technique is the Giemsa-colophonium stain described by Bray and Garnham (1962) for tissue sections.

1. Fix small pieces of tissue in Carnoy's fixative for about 3 hr.
2. Wash in 90% ethyl alcohol.
3. Finish dehydration; infiltrate, embed, section, and mount in the usual manner.
4. Run the sections down through xylene and the series of alcohols into distilled water in the usual manner.
5. Stain for 1 hr in a mixture of 1 volume Giemsa stain, 1 volume absolute methyl alcohol, and 10 volumes distilled water buffered to pH 7.2.
6. Wash in tap water.
7. Differentiate the sections by pouring 15% colophonium resin in acetone onto the wet slides, rocking them back and forth for 3–10 sec, and pouring off the solution. Repeat this process 2 or 3 times until the blue-green dye no longer streams out in the differentiating solution.
8. Wash rapidly with a mixture of 7 parts xylene and 3 parts acetone.
9. Wash several times in xylene.
10. Mount in euparal or another neutral resin. The mounting medium must be neutral; if acid, it will soon decolorize the preparations.

Microscopic Examination of Blood

To search for blood protozoa, thick or thin smears of the blood are first prepared and stained with one or another of the Romanowsky (methylene blue–eosin combination) stains. Thick smears are preferable to thin ones for mammalian blood because their use allows examination of a relatively large amount of blood in a relatively short time. However, they cannot be used for avian blood because of its nucleated erythrocytes. The protozoa may be distorted enough in thick smears so that some practice is needed to differentiate species, especially of the malaria parasites.

Romanowsky stains may be either rapid (e.g., Wright's or Field's stain) or slow (e.g., Giemsa stain). The rapid stains are satisfactory if speed is essential, but they stain unevenly, particularly in thick smears, and are not as precise as the slow stains. Giemsa stain is best for most purposes. Mammalian blood should be stained at pH 7.0–7.2, avian blood at pH 6.76. These pH's can be obtained by using Clark and Lubs phosphate buffers.

Trypanosomes, microfilariae, and most protozoa can be found in fresh, wet, unstained smears, but for critical study they must be stained.

Preparation of Thin Blood Smears

Clean two slides by rinsing them in 95% alcohol and wiping with a clean cloth. (Handle the slides only by the edges to avoid leaving finger marks.) Place a *small* drop of fresh blood at the end of one slide, place the other slide at a 30° angle to the first one, touch the drop of blood with the end of the slanted slide so that the blood flows into the space behind and beneath it, and then draw the slanted slide rather quickly over the length of the first one. The blood should be pulled *behind* the slide and not pushed ahead of it. A thin, even film of blood should result. Wave the slide in the air until it dries (a matter of few seconds if the smear is thin enough). If the smear is to be stained in Giemsa stain, fix it by dipping in absolute methyl alcohol (CP). If the smear is to be stained in Wright's stain, fixation is not necessary since it will take place during the staining process. If the smear is to be stored for more than a day or so before staining, it should be fixed.

While *Leucocytozoon,* microfilariae, and sometimes *Trypanosoma* can be found with the low power of the microscope, the stained smears should be examined with the oil immersion objective for other protozoa. The faster thin smears dry, the less distortion is produced; hence, the most natural-appearing protozoa are at the thin end and around the edges of the smear.

Preparation of Thick Blood Smears

Clean slides as for thin smears. Place a medium-sized drop of blood or several tiny ones on the slide, and mix with a toothpick or the corner of another slide. Allow to dry in air or in an incubator at 37 C; a hair dryer can be used to speed up this process. Thick smears must be laked (i.e., the hemoglobin must be extracted) before being stained. This can be done by placing them in water until their color has disappeared. If Giemsa stain is used and the smears are fresh, laking will take place during the staining process. If the smears are to be stored for more than a day or so before staining, they should be laked and then fixed with absolute methyl alcohol (CP) before storage, since it is often extremely difficult to remove the hemoglobin from smears that have been stored for some time.

Cleaning Immersion Oil from Slides

Stained blood smears are customarily not covered with a coverslip; immersion oil is placed on them for examination. The immersion oil should be removed after the examination has been completed if the prepared slides are not to be discarded. Many people do this by rubbing the slide with lens paper as though polishing silver; this procedure removes not only the oil but also many of the blood cells. The following technique (which I first saw demonstrated by Dr. Joseph A. Long) removes the oil quickly and neatly without disturbing the blood cells, even with slides that have been covered by a coverslip (i.e., removing the oil from a newly mounted slide without removing either the coverslip or the wet mounting medium beneath it).

Fold a small piece (about 5 sq cm) of lens paper twice so that it is four layers thick. Place the lens paper on top of the immersion oil and allow it to take up the oil. Pull it off the slide sideways by a single motion, without rubbing. Fold a second piece of lens paper like the first, and place a drop of xylene on it. Place the wet lens paper on what remains of the oil. Leave it for 1 or 2 sec, and then pull it off the slide sideways in a single motion, without rubbing. When the xylene has evaporated, the slide will be clean and dry. (Sometimes it is necessary to repeat this second step with a fresh piece of lens paper.)

CONCENTRATION OF PROTOZOAN CYSTS FROM FECES

A number of techniques have been developed for the concentration of protozoan cysts and helminth eggs from feces. They are of two general types, flotation and sedimentation. Each has its own advantages.

Flotation Techniques

These techniques make use of solutions of higher specific gravity than protozoan cysts or helminth eggs but of lower specific gravity than most of the fecal debris. When feces are mixed with them, the cysts and eggs float to the top while most of the fecal material remains at the bottom. Flotation techniques are most useful for coccidian oocysts, other protozoan cysts, nematode eggs, and some tapeworm eggs. They are not satisfactory for trematode, acanthocephalan, and other tapeworm eggs.

Many different solutions have been used, and many variations in technique have been proposed. The methods described here all work satisfactorily.

Sugar Flotation

This technique is preferable for general use but is not satisfactory for protozoan cysts other than those of coccidia. Sugar solution is preferable to sodium chloride, sodium nitrate, or other salt solutions except zinc sulfate; it does not crystallize as readily, causes less distortion than salt solutions, and is just as efficient. The following technique is a modification of the DCF (direct centrifugal flotation) technique introduced by Lane (1923).

1. Make a rather heavy suspension of feces in physiologic NaCl solution in a shell vial or other container.
2. Strain through two layers of cheesecloth into a test tube or centrifuge tube, filling the tube almost half full. The lip of the tube must be smooth or an air bubble will form under the coverslip following centrifugation (see step 6).
3. Add an equal volume of Sheather's sugar solution, leaving a small air space at the top. Cover with a plastic coverslip or small piece of card, and invert repeatedly to mix.
4. Add enough additional Sheather's sugar solution to bring the surface of the liquid barely above the top of the tube.
5. Cover with a round coverslip.

6. Centrifuge for 5 min. (If a centrifuge is not available, let stand for 45 min to 1 hr.)
7. Remove the coverslip, place it on a slide, and examine it under the microscope.

(If desired, steps 2 to 4 can be modified by straining the fecal suspension into a second shell vial, mixing it with an equal volume of Sheather's sugar solution, and then filling the centrifuge tube with the mixture.)

Zinc Sulfate Flotation

Zinc sulfate solution has the advantage of concentrating the cysts of protozoa such as *Entamoeba* and *Giardia* without distortion. The following technique is a modification of that introduced by Faust et al. (1938).

1. Make a suspension of feces in physiologic NaCl solution in a shell vial or other container.
2. Strain 4 ml suspension through two layers of cheesecloth into a test tube or centrifuge tube. The lip of the tube must be smooth.
3. Add tap water to within 1 cm of the top of the tube.
4. Mix thoroughly and centrifuge for 5 min.
5. Pour off the supernatant fluid.
6. Add a small amount of zinc sulfate solution and mix with an applicator stick. Add more zinc sulfate solution until the tube is almost full, cover with a plastic coverslip or a small piece of card, and invert repeatedly to mix.
7. Add enough additional zinc sulfate solution to bring the surface of the liquid barely above the top of the tube.
8. Cover with a round coverslip.
9. Centrifuge for 5 min.
10. Remove the coverslip, place it on a slide, and examine it under the microscope.

Hampel (1969) modified this technique for *Giardia* cysts. After the crude fecal particles have been poured off, 0.1–0.2 ml 1% potassium alum solution ($KAl[SO_4]_2 \cdot 12 H_2O$, sp gr 1.010) is layered on the surface of the zinc sulfate–feces mixture and the preparation is again centrifuged for 5 min. Then 1% eosin solution is dropped onto the surface, and the centrifuged material is examined under the microscope. She said this technique permits a 60–70% saving in time and that the cysts remain identifiable for 16–18 hr.

Sedimentation Techniques

Sedimentation techniques can be used for concentration of protozoan cysts and are necessary for the concentration of trematode, acanthocephalan, and some tapeworm eggs, which sink to the bottom of the solutions used in the flotation techniques. A few protozoan cysts, such as the oocysts of *Eimeria leuckarti,* also sink to the bottom.

Since they are essentially washing processes, sedimentation techniques may not concentrate cysts and eggs as much as flotation techniques. Many different sedimentation techniques have been developed; the two described below (FTE and MIFC) appear to be among the best.

FTE (Formalin-Triton-Ether) Sedimentation Technique

This technique was introduced by Ritchie (1948) and modified by Maldonado, Acosta-Matienzo, and Velez-Herrera (1954). The latter considered it the nearest to an all-round diagnostic procedure, since it is highly effective for the detection not only of schistosome, hookworm, whipworm, and ascarid eggs but also of protozoan cysts.

1. Mark off a test tube at the 5-ml and 6-ml levels.
2. Place 5 ml 10% formalin containing a drop of Triton NE in the tube.
3. Add 1 ml feces.
4. Break up the feces thoroughly with a wooden applicator.
5. Strain the suspension through four layers of cheesecloth into a 15-ml conical centrifuge tube. Squeeze the cloth to get out as much liquid as possible.
6. Add 5 ml commercial ether to the suspension in the centrifuge tube. Cover the tube with a plastic coverslip and shake vigorously.
7. Centrifuge (at 2,000 rpm in a horizontal centrifuge with an 8-in. radius from the center to the tip of the tube; if another type of centrifuge is used, change the speed of centrifugation accordingly) for 1 min after the centrifuge has reached its terminal speed.
8. Loosen the plug of detritus at the formalin solution–ether interface with an applicator stick, pour off all the supernatant fluid rapidly, and holding the tube slightly inverted, clean its walls carefully with a piece of clean, dry gauze. (This is done to prevent the liquid and debris on the walls of the tube from sliding down to the bottom and diluting the sediment).
9. Add a drop of physiologic NaCl solution to the sediment to facilitate its removal.
10. Take up the sediment with a pipette (a Stoll pipette works well), place on a slide, add a coverslip, and examine under the microscope.

MIFC (Merthiolate [Thiomersal]-Iodine-Formaldehyde Concentration) Technique

This technique was introduced by Blagg et al. (1955) as a modification of the MIF preservative stain. They found the MIFC technique positive for protozoan trophozoites in 74% of 110 positive human fecal specimens, compared with 55% for the MIF direct smear, and positive for 92% of 226 specimens containing protozoan cysts, compared with 58% positive with the MIF direct smear. Dunn (1968) found that it gave erratic results in concentrating helminth eggs.

1. Prepare an MIF-preserved specimen as described above.
2. When ready to examine, shake the specimen vigorously for 5 sec.

3. Strain through two layers of wet surgical gauze into a 15-ml centrifuge tube.
4. Add 4 ml cold (refrigerated) ether to the centrifuge tube, insert a rubber stopper and shake vigorously. If ether remains on top after shaking, add 1 ml tap water and shake again.
5. Remove the stopper and let stand for 2 min.
6. Centrifuge 1 min at 1,600 rpm. Four layers will appear in the tube: an ether layer on top, a plug of fecal detritus, an MIF layer, and the sediment containing protozoa and helminth eggs on the bottom.
7. Loosen the fecal plug by ringing with an applicator stick.
8. Quickly but carefully pour off all but the bottom layer of sediment.
9. Mix the sediment thoroughly, put a drop on a slide, cover with a coverslip, and examine.

PROTOZOAN CULTURE MEDIA

NNN (Novy, MacNeal, and Nicolle) Medium

This medium, developed for the cultivation of *Leishmania,* can also be used for trypanosomes of the *lewisi* group.

1. Measure or weigh out:

Sodium chloride	6 g
Agar	14 g
Distilled water	900 ml

2. Mix, bring to the boiling point, and place in bacteriologic culture tubes in 5-ml amounts. Sterilize in the autoclave. This is the medium base; it can be stored in the refrigerator.
3. To use, melt the agar in the tubes and cool to 48 C. Add to each tube one-third its volume of sterile, defibrinated rabbit blood. Mix thoroughly by rolling the tube between the palms of the hands.
4. Place the tube on a slant without leaving a butt of medium at the bottom, and allow to solidify. This is best done in the refrigerator or in ice, since more water of condensation is obtained in this way. (The protozoa develop best in the water of condensation at the bottom of the slant.)
5. Seal the tubes to prevent the water of condensation from evaporating and incubate at 37 C for 24 hr to test for sterility before inoculating.
6. Inoculate suspected material into the condensation water and incubate at 22–24 C. Transfer cultures every 1 or 2 wk.

Weinman's Trypanosome Medium

This medium was developed by Weinman (1946) for the cultivation of *Trypanosoma gambiense* and *T. rhodesiense.* It can also be used for other trypanosomes.

1. The base medium is Difco nutrient agar (1.5%):

   Beef extract .... 3 g
   Bacto peptone .... 5 g
   Sodium chloride .... 8 g
   Agar .... 15 g

   Dissolve in 1 L distilled water, bring to pH 7.3, and sterilize by autoclaving.
2. To prepare the culture medium, heat the base medium to melt the agar. Before it has resolidified, add the following aseptically to each 75 ml base medium:

   Citrated human plasma (inactivated at 56 C for 30 min) .... 12.5 ml
   Human red cells .... 12.5 ml

3. Dispense in Kolle flasks or slanted in test tubes. Stopper with rubber corks or seal with Parafilm to retard drying. Store in the refrigerator until used.
4. Inoculate with suspected material and incubate at room temperature. The trypanosomes grow on the surface as small, rounded, colorless, transparent, slightly raised, glistening, moist-appearing colonies 1–2 mm in diameter; they are detectable in 5–10 days or, exceptionally, in 3–4 wk.

RES (Ringer's-Egg-Serum) Medium for Enteric Protozoa

This medium was introduced by Boeck and Drbohlav (1925). Many different useful modifications have been proposed, such as the one described below. The serum may be replaced by egg albumen, for instance, or the Ringer's solution by Locke's solution.

The medium is essentially a coagulated egg slant overlaid with a fluid nutrient solution.

EGG SLANT

1. Mix 12.5 ml Ringer's solution with each egg used. For best results, mix in a Waring blender for 30 sec. If a blender is not used, filter the mixture through cheesecloth.
2. Place 2-ml amounts of the mixture in cotton-stoppered test tubes (or other standard closures for bacteriologic work).
3. Place the tubes upright in a vacuum desiccator. Evacuate the desiccator slowly. As evacuation proceeds, the egg mixture begins to bubble, and within 4 min a dense foam of egg begins to climb in the tubes. Stop the evacuation before the cotton plugs become wet, and allow the tubes to remain in the evacuated desiccator for an hour. The purpose of this treatment is to remove the dissolved air from the medium. If it is allowed to remain, it will bubble out during subsequent sterilization and coagulation, roughening and pitting the slant surface (Levine and Marquardt 1954).
4. Release the vacuum, pack the tubes in baskets, slant them in the autoclave, and inspissate and sterilize them simultaneously at 15 lb pressure for 20 min. Best results are obtained when no butt of medium is left in the tubes.

When this is done, 2 ml of fluid make a slant about 4 cm long in an 18 × 150-mm tube.

FLUID OVERLAY

1. Mix the following aseptically:

Sterile Ringer's solution	500 ml
Sterile 10% glucose solution	10 ml
Sterile serum (e.g., horse, rabbit, ox)	10 ml

2. Add sufficient fluid overlay to each egg slant to cover the whole slant. *Aseptic technique must be used throughout.* Incubate at 37 C for 2 days prior to inoculation to test for sterility.

Balamuth's Ameba Medium

This medium was developed by Balamuth (1946) for enteric amebae but can be used for other enteric protozoa as well.

1. Mix 288 g dehydrated egg yolk with 288 ml distilled water and 1,000 ml physiologic NaCl solution. Mix in a Waring blender or similar instrument until the suspension is smooth.
2. Heat over an open flame in the upper part of a double boiler, stirring constantly, for 5–10 min until coagulation begins.
3. Continue heating over boiling water in the double boiler for about 20 min, until coagulation is complete. Add 160 ml distilled water to replace water lost by evaporation.
4. Filter through a muslin bag. When the bag cools, squeeze it gently to obtain the maximum amount of filtrate.
5. Add enough physiologic NaCl solution to the filtrate to bring its volume to 1,000 ml.
6. Place 500 ml filtrate in each of two Erlenmeyer flasks. Autoclave at 15 lb pressure for 20 min.
7. Chill the flasks by refrigeration overnight or in some other way.
8. Filter while cold through two layers of Whatman qualitative filter paper in a Buchner funnel, using negative pressure. Pour the mixture through the funnel in small amounts, replacing the filter paper frequently.
9. Add an equal volume of Balamuth's buffer solution to the filtrate.
10. Add 5 ml crude liver extract (Lilly No. 408) to each liter of medium.
11. Dispense in 5- to 7-ml amounts in tubes.
12. Autoclave at 15 lb pressure for 20 min.
13. Add a small amount of sterile rice powder aseptically to each tube. Incubate at 37 C for 24 hr to test for sterility. (The medium can be stored in large flasks in the refrigerator after autoclaving; it can be kept for a month or more without deteriorating, but any sediment that forms should be removed by filtration before use.)

CPLM (Cysteine-Peptone-Liver Infusion-Maltose) Medium

This medium was developed by Johnson and Trussell (1943) for *Trichomonas* but can also be used for other enteric protozoa.

LIVER INFUSION

1. Mix the following thoroughly, using a Waring blender if available:

Bacto liver powder	20 g
Distilled water	330 ml

2. Infuse at about 50 C for 1 hr.
3. Heat with stirring at 80 C for 5 min to coagulate the protein.
4. Filter through a Buchner funnel (about 320 ml liver infusion are obtained).

PREPARATION OF FINAL MEDIUM

1. Mix the following, using a Waring blender if available:

Cysteine monohydrochloride	2.4 g
Peptone	32.0 g
Maltose	1.6 g
Agar	1.6 g
Ringer's solution	960 ml

2. Add the liver infusion (from above).
3. Adjust the pH to 7.00 (approximately 20 ml 1.0 N NaOH are needed).
4. Heat to dissolve the agar.
5. Filter through cotton into a 2,000-ml flask.
6. Add 0.7 ml 0.5% methylene blue solution.
7. Place 300-ml amounts in 500-ml Erlenmeyer flasks.
8. Autoclave at 15 lb pressure for 15 min.
9. Add 75 ml sterile inactivated serum to each Erlenmeyer flask.
10. Place 7- to 10-ml amounts aseptically in sterile, plugged test tubes.
11. Incubate at 37 C for 2 days to test for sterility before use.

BGPS (Beef Extract-Glucose-Peptone-Serum) Medium

This medium, introduced by Fitzgerald, Hammond, and Shupe (1954) for use in the diagnosis of *Tritrichomonas foetus* infections, can also be used for other trichomonads.

1. Mix the following in a 3-L flask:

Difco beef extract	3 g
Glucose	10 g
Bacto peptone	10 g
NaCl	1 g

Agar	0.7 g
Distilled water	1,000 ml

2. Dissolve by boiling; after cooling, adjust the pH to 7.4 with 1.0 N NaOH solution.
3. Cover the mouth of the flask with heavy paper and autoclave at 15 lb pressure for 30 min.
4. After cooling, add 20 ml inactivated (at 56 C for 30 min) beef serum aseptically, and mix thoroughly.
5. Dispense in 10-ml amounts into 15-ml culture tubes. Test for sterility by incubating at 37 C for 2 days.
6. Just before inoculation, add 500–1,000 units penicillin and 0.5–1.0 mg streptomycin to each ml of medium and mix thoroughly.
7. Pipette the inoculum onto the top of the medium in such a way as to minimize mixing. The trichomonads migrate to the bottom of the tube, while yeasts and molds tend to remain near the top. Incubate at 39 C for 3–5 days. To examine, remove a sample from the bottom of the tube with a pipette.

Diamond's Trichomonad Medium

This medium was introduced by Diamond (1957) for the axenic cultivation of trichomonads. It can be used successfully for more species than can other media.

1. Mix the following:

Trypticase (BBL)	2.0 g
Yeast extract	1.0 g
Maltose	0.5 g
L-cysteine hydrochloride	0.1 g
Ascorbic acid	0.02 g
K_2HPO_4	0.08 g
KH_2PO_4	0.08 g
Distilled water	90.0 ml

2. Adjust the pH to 6.8–7.0 with 1 N NaOH for all trichomonads except *T. vaginalis;* for this species, adjust the pH to 6.0 with 1 N HCl.
3. Add 0.05 g agar.
4. Autoclave at 15 lb pressure for 10 min.
5. Cool to 48 C, and add the following:

Sheep serum (inactivated at 56 C for 30 min)	10 ml
Potassium penicillin G	100,000 units
Streptomycin sulfate	0.1 g

(The penicillin and streptomycin can be made up in 1 ml sterile distilled water beforehand.)

6. Place 5-ml amounts of the medium aseptically in sterile, stoppered test tubes. Store in the refrigerator up to 14 days or longer.
7. Prior to inoculation, incubate the tubes at 35.5 C for 1 hr.

All the trichomonads that Diamond (1957) cultivated grew well at 35.5 C except *T. gallinarum* and *T. gallinarum*-like species, which grew best at 38.5 C.

I have found that the phosphates are not necessary for the growth of *T. foetus, T. suis, T. gallinae, T. gallinarum,* and several other species.

RSS (Ringer's-Serum-Starch) Medium for *Balantidium*

The following medium is slightly modified from that introduced by Rees (1927) for the cultivation of *Balantidium coli.*

1. Add 25 ml horse, rabbit, or bovine serum aseptically to 500 ml sterile Ringer's solution.
2. Tube in 8-ml amounts, using aseptic technique.
3. To each tube add a 5-mm loop of rice starch that has been sterilized in a large test tube at 15 lb pressure for 30 min.
4. Incubate at 37 C for 48 hr to test for sterility. Store in the refrigerator.
5. Before inoculation, warm the tubes to 37 C by placing them in the incubator. Incubate at 37 C. The protozoa grow in the bottom of the tube.

FORMULAS

Physiologic NaCl Solution

NaCl.. 8.5 g
Distilled water..1,000 ml

D'Antoni's Iodine Solution

Powdered iodine.. 1.5 g
1% aqueous KI solution..100 ml

Allow to stand 4 days before use. This is the stock solution and contains an excess of iodine. Filter small amounts before use. If tightly stoppered, the filtered solution will keep 4 wk before too much iodine volatilizes.

Lugol's Iodine Solution

Potassium iodide.. 10 g
Powdered iodine.. 0.5 g
Distilled water..100 ml

Dissolve the potassium iodide in the water before adding the iodine.

Mayer's Albumen Fixative

Put the whites of several new-laid eggs in a shallow dish. Whip them a little with a fork or wire eggbeater two or three dozen strokes. (Do not beat them until they are white and stiff.) Allow them to stand for about 1 hr; then skim the foam from the top and pour the remaining liquid into a graduated cylinder. Pour in an equal amount of glycerol, and add 1 g sodium salicylate for each 100 ml mixture. Shake thoroughly and filter through paper into a clean bottle. Filtration will require 1 to several weeks. It may be accelerated somewhat by pouring a rather small amount at a time into the filter and replacing the paper every few days. Keep a small quantity in a vial or bottle provided with a glass rod stuck through the cork and into the albumen (useful in placing a drop on a slide).

Carnoy's Fixative

100% ethyl alcohol ....60 ml
Chloroform..30 ml
Acetic acid ....10 ml

Hollande's Fixative

Picric acid .... 4 g
Copper acetate .... 2.5 g
Formalin .... 10.0 ml
Acetic acid.. 1.5 ml
Distilled water..100.0 ml

Schaudinn's Fixative

Saturated aqueous mercuric chloride ....66 ml
95% ethyl alcohol ....33 ml

Add 5% acetic acid immediately before use.

Iron Alum Solution

Ferric ammonium sulfate (violet crystals only) .... 2 g
Distilled water ....100 ml

Filter immediately before use.

Heidenhain's Hematoxylin (Stock Solution)

Hematoxylin.. 10 g
100% ethyl alcohol ....100 ml

Allow to remain 1 mo in a loosely stoppered bottle before use. To make the staining solution, add 0.5 ml of the stock solution to 9.5 ml of distilled water.

Feulgen Stain

1. Dissolve 1 g basic fuchsin (certified as suitable for the Feulgen stain) in 200 ml boiling distilled water.
2. Cool to 50 C.
3. Filter.
4. Add 20 ml 1 N HCl to the filtrate.
5. Cool to 25 C.
6. Add 1 g dried sodium bisulfite ($NaHSO_3$); this liberates sulfurous acid.
7. Allow to stand at room temperature 24 hr until decolorized.
8. Store in the refrigerator in small glass-stoppered bottles filled to the top to exclude air. The solution will keep several weeks. It should be straw-colored; if it is red, discard it.

Sorensen's Phosphate Buffers

To make M/15 Na_2HPO_4 solution, dissolve 9.5 g anhydrous Na_2HPO_4 or 11.9 g $Na_2HPO_4 \cdot 2H_2O$ in 1 L distilled water. To make M/15 KH_2PO_4, dissolve 9.08 g KH_2PO_4 in 1 L distilled water. Store separately in Pyrex, glass-stoppered bottles.

To prepare buffered water for the Giemsa stain, mix the following amounts of the solutions (in ml):

	pH 6.8	pH 7.0	pH 7.2
M/15 Na_2HPO_4	50.0	61.1	72.0
M/15 KH_2PO_4	50.0	38.9	28.0
Distilled water	900.0	900.0	900.0

Balamuth's Buffer Solution

1.0 M K_2HPO_4
(174.180 g K_2HPO_4 in 1,000 ml distilled water)............4.3 parts
1.0 M KH_2PO_4
(136.092 g KH_2PO_4 in 1,000 ml distilled water)............0.7 parts

This is the stock solution. To prepare the final solution used in Balamuth's medium, add 14 parts distilled water to 1 part stock solution.

Ringer's Solution

NaCl	6.5	g
$NaHCO_3$	0.2	g
$CaCl_2 \cdot 2H_2O$	0.16	g
KCl	0.14	g

$NaH_2PO_4 \cdot H_2O$	0.011 g
Distilled water	1,000 ml

Sheather's Sugar Solution

Sucrose (ordinary cane or beet sugar)	500 g
Distilled water	320 g
Phenol (melted in water bath)	6.5 g

Zinc Sulfate Flotation Solution

$ZnSO_4 \cdot 7\ H_2O$	331 g
Distilled water	1,000 ml

The specific gravity of this solution is 1.180.

LITERATURE CITED

For citations before 1965, see the *Index-Catalogue of Medical and Veterinary Zoology, Zoological Record, Biological Abstracts, Protozoological Abstracts,* and *Veterinary Bulletin.*

Dunn, F. L. 1968. Bull. WHO 39:439–49.
Gleason, N. N., and G. R. Healy. 1966. J. Conf. State Prov. Publ. Health Lab. Dir. 24:8–11.
Hampel, M. 1969. Parasitol. Hung. 2:71–74.
Ivens, V., D. L. Mark, and N. D. Levine. 1978. Principal Parasites of Domestic Animals in the United States. Univ. Ill. Coll. Agric. Spec. Publ. 52.

APPENDIX A

Scientific and Common Names of Some Domestic and Wild Animals

Class MAMMALIASIDA

Scientific name	Common name
Order MARSUPIALORIDA	
Didelphis marsupialis	Opossum
Order PRIMATORIDA	
Alouatta villosa	Howler monkey
Ateles geoffroyi	Geoffroy's spider monkey
Cebus capucinus	Capuchin monkey
Cercocebus spp.	Mangabey monkeys
Cercopithecus aethiops	Green guenon; vervet monkey
Cercopithecus mona	Mona monkey
Chimpansee (syn., *Pan*) *troglodytes*	Chimpanzee
Gorilla gorilla	Gorilla
Homo sapiens	Man
Macaca fascicularis	Cynomolgus macaque; kra monkey
Macaca mulatta	Rhesus monkey
Macaca philippinensis	Philippine macaque
Mandrillus sphinx	Mandrill
Papio (syn., *Cynocephalus*) *papio*	Baboon
Pongo pygmaeus (syn., *Simia satyrus*)	Orangutan
Order EDENTATORIDA	
Dasypus novemcinctus	Nine-banded armadillo
Order LAGOMORPHORIDA	
Lepus americanus	Snowshoe hare
Lepus californicus	Black-tailed jackrabbit
Lepus europaeus	European hare
Lepus townsendii	White-tailed jackrabbit
Oryctolagus cuniculus	Domestic rabbit; European wild rabbit
Sylvilagus floridanus	Eastern cottontail
Order RODENTORIDA	
Apodemus sylvaticus	Long-tailed field mouse (European)

Cavia porcellus	Guinea pig
Chinchilla laniger	Chinchilla
Clethrionomys spp.	Red-backed mice
Cricetulus barabensis griseus	Chinese (striped) hamster
Cricetus cricetus	Hamster
Dipodomys spp.	Kangaroo rats
Gerbillus gerbillus	Lesser Egyptian gerbil
Meriones unguiculatus	Mongolian jird; clawed jird
Mesocricetus auratus	Golden hamster
Microtus spp.	Voles
Neotoma spp.	Wood rats
Oryzomys palustris	Swamp rice rat
Peromyscus maniculatus	Deermouse
Praomys (syns., *Mastomys, Rattus*) *coucha*	Multimammate mouse
Rattus norvegicus	Norway rat; laboratory rat
Rattus rattus	Black rat
Rhombomys opimus	Gerbil
Sciurus spp.	Tree squirrels
Sigmodon hispidus	Cotton rat
Spermophilus (syn., *Citellus*) spp.	Ground squirrels; ziesels; susliks
Order CARNIVORIDA	
Alopex lagopus	Arctic fox
Canis dingo	Dingo
Canis familiaris	Dog
Canis latrans	Coyote
Canis lupus	Gray wolf
Euarctos americanus	Black bear
Felis catus	Domestic cat
Felis concolor	Mountain lion; puma
Leo (syn., *Felis*) *leo*	Lion
Leo (syn., *Panthera*) *tigris*	Tiger
Lynx canadensis	Lynx
Lynx rufus	Bobcat
Martes americana	Marten
Mephitis mephitis	Striped skunk
Mustela erminea	Ermine
Mustela frenata	Long-tailed weasel
Mustela putorius furo	Ferret
Mustela vison	Mink
Procyon lotor	Raccoon
Spilogale spp.	Spotted skunks
Urocyon cinereoargenteus	Gray fox
Ursus arctos (syn., *U. horribilis*)	Grizzly bear; Alaskan brown bear
Vulpes vulpes (syn., *V. fulva*)	Red fox (North American and European)
Order PERISSODACTYLORIDA	
Asinus asinus	Ass; donkey
Equus caballus	Horse

Order ARTIODACTYLORIDA	
Alces alces	Moose
Antilocapra americana	Pronghorn
Bison bison	Bison
Bos taurus indicus	Zebu
Bos taurus taurus	Ox
Bubalus bubalis	Water buffalo; carabao
Camelus bactrianus	Bactrian camel; two-humped camel
Camelus dromedarius	Dromedary; one-humped camel
Capra hircus	Domestic goat
Capreolus capreolus	Roe deer
Cervus elaphus (syn., *C. canadensis*)	Red deer (European); wapiti; elk
Dama dama	Fallow deer
Lama glama	Llama
Mazama americana	Red brocket
Odocoileus hemionus	Black-tailed deer; mule deer
Odocoileus virginianus	White-tailed deer
Oreamnos americanus	Mountain goat
Ovibos moschatus	Musk ox
Ovis ammon	Argali
Ovis aries	Domestic sheep
Ovis canadensis	Mountain sheep; Rocky Mountain bighorn sheep
Ovis musimon	Mouflon
Ovis vignei	Urial
Rangifer tarandus	Caribou; reindeer
Rupicapra rupicapra	Chamois
Sus scrofa	Domestic pig; wild boar
Syncerus (syn., *Bubalus*) *caffer*	African buffalo
Order PROBOSCIDORIDA	
Elaphus indicus	Indian elephant
Loxodonta africana	African elephant
Class AVEASIDA	
Order ANSERORIDA	
Anas platyrhynchos	Domestic duck; wild mallard
Anser albifrons	White-fronted goose
Anser anser (syn., *A. cinereus*)	Domestic goose; graylag goose
Branta canadensis	Canada goose
Cairina moschata	Muscovy duck
Cygnus olor	Swan
Order GALLORIDA	
Alectoris graeca	Chukar partridge
Bonasa umbellus	Ruffed grouse
Colinus virginianus	Bobwhite
Gallus gallus	Chicken
Meleagris gallopavo	Turkey

Numida meleagris	Guinea fowl
Pavo cristatus	Peafowl
Perdix perdix	Gray partridge
Phasianus colchicus	Ring-necked pheasant
Order COLUMBORIDA	
Columba fasciata	Band-tailed pigeon
Columba livia	Domestic pigeon
Streptopelia chinensis	Spotted dove
Streptopelia turtur	Turtledove (European)
Zenaidura macroura	Mourning dove
Order PASSERORIDA	
Passer domesticus	House sparrow; English sparrow
Serinus canarius	Canary
Order STRUTHIONORIDA	
Struthio camelus	Ostrich

APPENDIX B

Some Protozoan Parasites Reported from Various Hosts

CAT (*Felis catus*)
  Flagellates
    Giardia cati
    Pentatrichomonas hominis
    Tetratrichomonas felistomae
    Trypanosoma rangeli
  Amebae
    Entamoeba gingivalis
    E. histolytica
  Apicomplexa
    Babesia cati
    B. felis
    Besnoitia besnoiti
    B. darlingi
    B. wallacei
    Cryptosporidium muris
    Hepatozoon felis
    Isospora felis
    I. rivolta
    Sarcocystis cuniculi
    S. cymruensis
    S. fusiformis
    S. gigantea
    S. hirsuta
    S. horvathi
    S. leporum
    S. muris
    S. porcifelis
    Theileria felis
    Toxoplasma gondii
    T. hammondi
  Microspora
    Encephalitozoon cuniculi

CHICKEN (*Gallus gallus*)
  Flagellates
    Chilomastix gallinarum
    Histomonas meleagridis
    Pentatrichomonas sp.
    Tetratrichomonas gallinarum
    Trichomonas gallinae
    Tritrichomonas eberthi
    Trypanosoma calmettei
    T. gallinarum
  Amebae
    Endolimax gregariniformis
    Entamoeba gallinarum
  Apicomplexa
    Arthrocystis galli
    Babesia moshkovskii
    Cryptosporidium meleagridis
    Eimeria acervulina
    E. brunetti
    E. hagani
    E. maxima
    E. mitis
    E. mivati
    E. necatrix
    E. praecox
    E. tenella
    Isospora gallinae
    Leucocytozoon andrewsi
    L. caulleryi
    L. sabrazesi
    Plasmodium gallinaceum
    P. juxtanucleare
    Sarcocystis horvathi
    Toxoplasma gondii
    Wenyonella gallinae

Dog (*Canis familiaris*)
  Flagellates
    Giardia canis
    Pentatrichomonas hominis

Leishmania braziliensis panamensis
L. donovani
L. infantum
L. peruviana
L. tropica
Tetratrichomonas canistomae
Tetratrichomonas sp.
Tritrichomonas sp.
Trypanosoma b. brucei
T. brucei evansi
T. c. congolense
T. c. cruzi
T. rangeli
Amebae
Acanthamoeba sp.
Entamoeba coli
E. gingivalis
E. hartmanni
E. histolytica
Hartmannella sp.
Apicomplexa
Babesia canis
B. gibsoni
B. vogeli
Eimeria canis
E. rayii
Hepatozoon canis
Isospora burrowsi
I. canis
I. neorivolta
I. ohioensis
Isospora sp.
Sarcocystis bertrami
S. capracanis
S. capreoli
S. cruzi
S. equicanis
S. fayeri
S. hemionilatrantis
S. horvathi
S. levinei
S. miescheriana
S. tenella
Toxoplasma bahiensis
T. gondii
Microspora
Encephalitozoon cuniculi
Ciliate
Balantidium coli

DUCK (*Anas platyrhynchos*)
Flagellates
Cochlosoma anatis
Protrichomonas anatis
Spironucleus sp.
Tetratrichomonas anatis
Amebae
Endolimax gregariniformis
Entamoeba anatis
Apicomplexa
Eimeria anatis
E. battakhi
E. danailova
E. saitamae
E. schachdagica
Haemoproteus nettionis
Isospora sp.
Leucocytozoon simondi
Sarcocystis rileyi
Tyzzeria perniciosa
Wenyonella anatis

GOAT (*Capra hircus*)
Flagellates
Callimastix frontalis
Chilomastix caprae
Giardia caprae
Monocercomonoides caprae
Oikomonas communis
Sphaeromonas communis
Trypanosoma b. brucei
T. brucei evansi
T. c. congolense
T. theodori
T. vivax uniforme
T. v. vivax
Amebae
Entamoeba caprae
E. dilimani
E. ovis
E. wenyoni
Apicomplexa
Babesia motasi
B. ovis
B. taylori
Besnoitia besnoiti
Eimeria alijevi
E. apsheronica
E. arloingi
E. caprina
E. caprovina
E. christenseni
E. gilruthi
E. hirci

E. jolchijevi
E. kocharii
E. ninakohlyakimovae
E. pallida
E. punctata
Sarcocystis capracanis
S. moulei
Theileria lestoquardi
T. ovis
Toxoplasma gondii
Ciliates
Dasytricha ruminantium
Isotricha intestinalis
I. prostoma
Ophryoscolex inermis

GOOSE (*Anser anser*)
Flagellate
Tetratrichomonas anseris
Ameba
Endolimax gregariniformis
Apicomplexa
Eimeria anseris
E. kotlani
E. nocens
E. stigmosa
E. truncata
Haemoproteus nettionis
Isospora sp.
Leucocytozoon simondi
Tyzzeria parvula

GUINEA FOWL (*Numida meleagris*)
Flagellates
Histomonas meleagridis
Pentatrichomonas sp.
Tetratrichomonas gallinarum
Trypanosoma numidae
Ameba
Endolimax gregariniformis
Apicomplexa
Eimeria gorakhpuri
E. grenieri
E. numidae
Leucocytozoon caulleryi
Plasmodium fallax

GUINEA PIG (*Cavia porcellus*)
Flagellates
Caviomonas mobilis
Chilomastix intestinalis
C. megamorpha
C. wenrichi
Chilomitus caviae
C. conexus
Enteromonas caviae
Giardia caviae
Hexamastix caviae
H. robustus
Leishmania enriettii
Monocercomonoides caviae
M. exilis
M. quadrifunilis
M. wenrichi
Proteromonas brevifilia
Sphaeromonas communis
Retortamonas caviae
Tritrichomonas caviae
Tritrichomonas sp.
Amebae
Endolimax caviae
Entamoeba caviae
Apicomplexa
Eimeria caviae
Klossiella cobayae
Sarcocystis caviae
Toxoplasma gondii
T. hammondi
Microspora
Encephalitozoon cuniculi
Ciliates
Balantidium coli
Cyathodinium conicum
C. piriforme

HORSE (*Equus caballus*)
Flagellates
Callimastix equi
Cercomonas equi
Chilomastix equi
Giardia equi
Oikomonas equi
Trichomonas equibuccalis
Tritrichomonas equi
Trypanosoma b. brucei
T. brucei equiperdum
T. brucei evansi
T. c. congolense
T. vivax viennei
T. v. vivax
Amebae
Entamoeba equi
E. equibuccalis
E. gedoelsti

Apicomplexa
Babesia caballi
B. equi
Besnoitia bennetti
Eimeria leuckarti
E. solipedum
E. uniungulati
Klossiella equi
Sarcocystis bertrami
S. equicanis
S. fayeri
Toxoplasma gondii
Ciliates
Allantosoma brevicorniger
A. dicorniger
A. intestinalis
Alloiozona trizona
Ampullacula ampulla
Blepharoconus benbrooki
B. cervicalis
B. hemiciliatus
Blepharocorys angusta
B. cardionucleata
B. curvigula
B. jubata
B. uncinata
B. valvata
Blepharoprosthium pireum
Blepharosphaera ellipsoidalis
B. intestinalis
Blepharozoum zonatum
Bundleia postciliata
Charonina equi
Cochliatoxum periachtum
Cycloposthium affinae
C. bipalmatum
C. corrugatum
C. dentiferum
C. edentatum
C. ishikawai
C. piscicauda
C. scutigerum
Didesmis ovalis
D. quadrata
D. spiralis
Ditoxum funinucleum
Holophryoides ovalis
Ochoterenaia appendiculata
Paraisotricha beckeri
P. colpoidea
P. minuta
Paraisotrichopsis composita
Polymorphella ampulla
Prorodonopsis coli
Spirodinium equi
Sulcoarcus pellucidulus
Tetratoxum excavatum
T. parvum
T. unifasciculatum
Triadinium caudatum
T. galea
T. minimum
Tripalmaria dogieli

MAN (*Homo sapiens*)
Flagellates
Chilomastix mesnili
Dientamoeba fragilis
Enteromonas hominis
Giardia lamblia
Leishmania aethiopica
L. b. braziliensis
L. b. guyanensis
L. b. panamensis
L. chagasi
L. donovani
L. infantum
L. major
L. mexicana amazonensis
L. m. mexicana
L. mexicana pifanoi
L. peruviana
L. tropica
Pentatrichomonas hominis
Retortamonas intestinalis
Trichomitus fecalis
Trichomonas tenax
T. vaginalis
Trypanosoma brucei gambiense
T. b. rhodesiense
T. c. cruzi
T. minasense
T. rangeli
Amebae
Acanthamoeba astronyxis
A. culbertsoni
Endolimax nana
Entamoeba chattoni
E. coli
E. gingivalis
E. hartmanni
E. histolytica
Hartmannella sp.
Iodamoeba buetschlii

Naegleria fowleri
Trimastigamoeba philippinensis
Apicomplexa
Babesia divergens
B. microti
Cryptosporidium muris
Isospora belli
I. natalensis
Plasmodium cynomolgi
P. falciparum
P. knowlesi
P. malariae
P. ovale
P. simium
P. vivax
Sarcocystis hominis
S. suihominis
Toxoplasma gondii
Microspora
Encephalitozoon cuniculi
Nosema connori
Ciliate
Balantidium coli

MOUSE, house or laboratory (*Mus musculus*)
Flagellates
Chilomastix bettencourti
Giardia muris
G. simoni
Octomitus pulcher
Pentatrichomonas hominis
Spironucleus muris
Tetratrichomonas microti
Trichomitus wenyoni
Tritrichomonas minuta
T. muris
Trypanosoma musculi
Ameba
Entamoeba muris
Apicomplexa
Cryptosporidium muris
Eimeria falciformis
E. ferrisi
E. hansonorum
E. hindlei
E. keilini
E. krijgsmanni
E. musculi
E. papillata
E. schueffneri
E. vermiformis
Hepatozoon musculi
Klossiella muris
Sarcocystis dispersa
S. muris
S. scotti
S. sebeki
Toxoplasma gondii
T. hammondi
Microspora
Encephalitozoon cuniculi

OX (*Bos t. taurus*)
Flagellates
Callimastix frontalis
Giardia bovis
Monocercomonas ruminantium
Monocercomonoides bovis
Oikomonas communis
O. minima
Pentatrichomonas hominis
Protrichomonas ruminantium
Sphaeromonas communis
Tetratrichomonas buttreyi
T. pavlovi
Tritrichomonas enteris
T. foetus
Tritrichomonas sp.
Trypanosoma b. brucei
T. b. evansi
T. c. congolense
T. theileri
T. vivax uniforme
T. vivax viennei
T. v. vivax
Amebae
Acanthamoeba polyphaga
Entamoeba bovis
E. histolytica
Hartmannella (?) sp.
Vahlkampfia lobospinosa
Apicomplexa
Babesia bigemina
B. bovis
B. divergens
B. major
Besnoitia besnoiti
Cryptosporidium muris
Eimeria alabamensis
E. auburnensis
E. bovis
E. brasiliensis
E. bukidnonensis

E. canadensis
E. cylindrica
E. ellipsoidalis
E. illinoisensis
E. pellita
E. subspherica
E. thianethi
E. wyomingensis
E. zuernii
Isospora sp.
Sarcocystis cruzi
S. hirsuta
S. hominis
Theileria annulata
T. mutans
T. parva
T. velifera
Toxoplasma gondii

Ciliates
Buetschlia lanceolata
B. nana
B. neglecta
B. parva
Buxtonella sulcata
Charonina nuda
C. ventriculi
Dasytricha ruminantium
Diplodinium anacanthum
D. bubalidis
D. dentatum
D. elongatum
D. psittaceum
D. quinquecaudatum
Diploplastron affine
Endoploplastron triloricatum
Entodinium bicarinatum
E. bimastus
E. bursa
E. caudatum
E. dentatum
E. dubardi
E. exiguum
E. furca
E. laterale
E. lobospinosum
E. longinucleatum
E. minimum
E. nanellum
E. rectangulatum
E. rostratum
E. simulans
E. vorax
Eodinium bilobosum
E. lobatum
E. posterovesiculatum
Epidinium ecaudatum
Eremoplastron bovis
E. dilobum
E. monolobum
E. neglectum
E. rostratum
E. rugosum
Eudiplodinium maggii
Isotricha intestinalis
I. prostoma
Metadinium medium
M. ypsilon
Ophryoscolex caudatus
O. purkinjei
Ostracodinium crassum
O. dilobum
O. dogieli
O. gladiator
O. gracile
O. mammosum
O. nanum
O. obtusum
Polyplastron fenestratum
P. monoscutum
P. multivesiculatum

PIG (*Sus scrofa*)

Flagellates
Chilomastix mesnili
Enteromonas suis
Tetratrichomonas buttreyi
Trichomitus rotunda
Tritrichomonas suis
Trypanosoma b. brucei
T. b. evansi
T. c. congolense
T. c. simiae
T. suis

Amebae
Endolimax nana
Entamoeba coli
E. histolytica
E. suigingivalis
E. suis
Iodamoeba buetschlii
Vahlkampfia avara
V. inornata

Apicomplexa
Adelina sp.

Babesia perroncitoi
B. trautmanni
Cryptosporidium muris
Eimeria betica
E. debliecki
E. guevarai
E. neodebliecki
E. perminuta
E. polita
E. porci
E. residualis
E. scabra
E. spinosa
E. suis
Isospora almataensis
I. neyrai
I. suis
Isospora sp.
Sarcocystis miescheriana
S. porcifelis
S. suihominis
Toxoplasma gondii
Ciliate
Balantidium coli

PIGEON (*Columba livia*)
Flagellates
Spironucleus columbae
Trichomonas gallinae
Trypanosoma hannai
Ameba
Acanthamoeba polyphaga
Apicomplexa
Eimeria columbae
E. labbeana
E. tropicalis
Haemoproteus columbae
H. sacharovi
Leucocytozoon marchouxi
Plasmodium relictum
Toxoplasma gondii
Wenyonella columbae

RABBIT, domestic or laboratory (*Oryctolagus cuniculus*)
Flagellates
Chilomastix cuniculi
Giardia duodenalis
Monocercomonas cuniculi
Retortamonas cuniculi
Trypanosoma nabiasi
Amebae
Acanthamoeba polyphaga
Entamoeba cuniculi
Apicomplexa
Cryptosporidium muris
Eimeria coecicola
E. elongata
E. exigua
E. intestinalis
E. irresidua
E. magna
E. matsubayashii
E. media
E. nagpurensis
E. neoleporis
E. perforans
E. piriformis
E. stiedai
Hepatozoon cuniculi
Sarcocystis cuniculi
Toxoplasma gondii
Microspora
Encephalitozoon cuniculi

RAT, Norway or laboratory (*Rattus norvegicus*)
Flagellates
Chilomastix bettencourti
Enteromonas hominis
Giardia muris
G. simoni
Monocercomonoides sp.
Octomitus pulcher
Pentatrichomonas hominis
Spironucleus muris
Tetratrichomonas microti
Trichomitus wenyoni
Tritrichomonas minuta
T. muris
Trypanosoma conorhini
Amebae
Endolimax ratti
Entamoeba histolytica
E. muris
Apicomplexa
Besnoitia wallacei
Eimeria miyairii
E. nieschulzi
E. separata
Hepatozoon epsteini
H. muris
Sarcocystis cymruensis
S. murinotechis

S. singaporensis
S. villivillosi
S. zamani
Toxoplasma gondii
T. hammondi
Microspora
Encephalitozoon cuniculi
Ciliate
Balantidium coli

RHESUS MONKEY (*Macaca mulatta*)
Flagellates
Chilomastix mesnili
Dientamoeba fragilis
Enteromonas hominis
Pentatrichomonas hominis
Retortamonas intestinalis
Spironucleus pitheci
Tetratrichomonas macacovaginae
Trichomitus wenyoni
Trichomonas tenax
Amebae
Endolimax nana
Entamoeba chattoni
E. coli
E. gingivalis
E. hartmanni
E. histolytica
Iodamoeba buetschlii
Plasmodium cynomolgi
P. inui
Sarcocystis hominis
S. kortei
S. nesbitti
S. suihominis
Toxoplasma gondii
Ciliate
Balantidium coli

SHEEP (*Ovis aries*)
Flagellates
Callimastix frontalis
Giardia capri
Protrichomonas ruminantium
Retortamonas ovis
Tetratrichomonas ovis
Trypanosoma b. brucei
T. c. congolense
T. melophagium
T. vivax uniforme
T. v. vivax
Ameba
Entamoeba ovis
Apicomplexa
Babesia crassa
B. foliata
B. motasi
B. ovis
Cryptosporidium muris
Eimeria ahsata
E. crandallis
E. faurei
E. gilruthi
E. gonzalezi
E. granulosa
E. intricata
E. marsica
E. ovina
E. ovinoidalis
E. pallida
E. parva
E. punctata
E. weybridgensis
Isospora sp.
Sarcocystis gigantea
S. tenella
Theileria lestoquardi
T. ovis
Toxoplasma gondii
Ciliates
Dasytricha ruminantium
Isotricha intestinalis
I. prostoma
Ophryoscolex caudatus

TURKEY (*Meleagris gallopavo*)
Flagellates
Chilomastix gallinarum
Cochlosoma sp.
Histomonas meleagridis
Parahistomonas wenrichi
Pentatrichomonas sp.
Spironucleus meleagridis
Tetratrichomonas gallinarum
Trichomonas gallinae
Tritrichomonas eberthi
Amebae
Acanthamoeba polyphaga
Entamoeba gallinarum
Hartmannella sp.
Vahlkampfia enterica
Apicomplexa

Babesia moshkovskii
Cryptosporidium meleagridis
Eimeria adenoeides
E. dispersa
E. gallopavonis
E. innocua
E. meleagridis
E. meleagrimitis
E. subrotunda
Haemoproteus meleagridis
Isospora heissini
Leucocytozoon smithi
Plasmodium durae
P. griffithsi
P. hermani

INDEX

Page numbers for definitions are in italics